ADVANCES IN VIBRATION ANALYSIS RESEARCH

Edited by **Farzad Ebrahimi**

Advances in Vibration Analysis Research
Edited by Farzad Ebrahimi

Published by InTech
Janeza Trdine 9, 51000 Rijeka, Croatia

Publishing Process Manager Ivana Lorkovic

Technical Editor Teodora Smiljanic

Cover Designer InTech Design Team

Additional hard copies can be obtained from orders@intechopen.com

Advances in Vibration Analysis Research, Edited by Farzad Ebrahimi
p. cm.
ISBN 978-953-307-209-8

For EU product safety concerns:
IN TECH d.o.o., Prolaz Marije Krucifikse Kozulić 3, 51000 Rijeka, Croatia,
info@intechopen.com or visit our website at intechopen.com.

Meet the editor

Dr. Farzad Ebrahimi was born in Qazvin, Iran, in 1979. He graduated in Mechanical engineering, from the University of Tehran, Iran, in 2002. He received his Msc and PhD in Mechanical engineering, with a specialization in applied design from the University of Tehran, Iran, in 2009. Since 2002, he has been working at the "Smart Materials and Structures Lab" Research Center of the Faculty of Mechanical Engineering at the University of Tehran, where he is a researcher of smart functionally graded materials and structures. He is involved in several international journals as editor and reviewer. He serves on the editorial board for Far East Journal of mechanical engineering and physics. He is the author of the books "Smart functionally graded plates" and "Progress in analysis of functionally graded structures", Nova Science Publishers, NY. His research interests focus on areas of smart materials and structures, plate and shell theory, vibration analysis of continuous systems, composite materials and structures, functionally graded materials and structures, finite element analysis and fracture mechanics and he has published several researches in these fields. His research in these areas has been presented at international conferences and has appeared in academic journals such as Journal of Mechanical Science and Technology; Smart Materials and Structures; European Journal of Mechanics; Archive of Applied Mechanics and Thin Walled Structures.

Contents

Preface

Vibrations are extremely important in all areas of human activities, for all sciences, technologies and industrial applications. Sometimes these vibrations are harmless, often they can be noticed as noise or cause wear. Vibrations, if they are not desired, can be dangerous. But sensibly organized and controlled vibrations may be pleasant (think of all kinds of music) or vitally important (heartbeat). In any case, understanding and analysis of vibrations are crucial.

This book reports on the state of the art research and development findings on this very broad matter through 22 original and innovative research studies exhibiting various investigation directions.

In particular, it introduces recent research results on many important issues at the vibration analysis field such as vibration analysis of structural members like beams and plates especially made of composite or functionally graded materials using analytical and finite element method and shows some results on applications in vibration analysis of framed structures, masonry structures and building vibration problems due to human rhythmic activities.

It also presents related themes in the field of vibration analysis of internal combustion engines, electrical machines, shafts, rotors and gear units and some other interesting topics like vibration analysis of carbon nanotube mass sensors, sensitivity analysis of spatial multibody systems, analysis of microparts dynamics, defect detection of tubes and vocal fold vibrations and introduces harmonic balance; topology-based transformation and independent coordinate coupling methods.

In summary, this book covers a wide range of interesting topics of vibration analysis.

The advantage of the book **vibration analysis** is its open access fully searchable by anyone anywhere, and in this way it provides the forum for dissemination and exchange of the latest scientific information on theoretical as well as applied areas of knowledge in the field of vibration analysis.

The present book is a result of contributions of experts from international scientific community working in different aspects of vibration analysis. The introductions, data, and references in this book will help the readers know more about this topic and help them explore this exciting and fast-evolving field.

The text is addressed not only to researchers, but also to professional engineers, students and other experts in a variety of disciplines, both academic and industrial seeking to gain a better understanding of what has been done in the field recently, and what kind of open problems are in this area.

I hope that readers will find the book useful and inspiring by examining the recent developments in vibration analysis.

Tehran, February 2011

Farzad Ebrahimi
Mechanical Engineering Department
University of Tehran

1

Transverse Vibration Analysis of Euler-Bernoulli Beams Using Analytical Approximate Techniques

Safa Bozkurt Coşkun[1], Mehmet Tarik Atay[2] and Baki Öztürk[3]
[1]Kocaeli University, Faculty of Engineering, Department of Civil Engineering 41380 Kocaeli,
[2]Niğde University, Faculty of Arts and Science, Department of Mathematics 51200 Niğde,
[3]Niğde University, Faculty of Engineering, Department of Civil Engineering 51200 Niğde, Turkey

1. Introduction

The vibration problems of uniform and nonuniform Euler-Bernoulli beams have been solved analytically or approximately [1-5] for various end conditions. In order to calculate fundamental natural frequencies and related mode shapes, well known variational techniques such as Rayleigh_Ritz and Galerkin methods have been applied in the past. Besides these techniques, some discretized numerical methods were also applied to beam vibration analysis successfully.
Recently, by the emergence of new and innovative semi analytical approximation methods, research on this subject has gained momentum. Among these studies, Liu and Gurram [6] used He's Variational Iteration Method to analyze the free vibration of an Euler-Bernoulli beam under various supporting conditions. Similarly, Lai et al [7] used Adomian Decomposition Method (ADM) as an innovative eigenvalue solver for free vibration of Euler-Bernoulli beam again under various supporting conditions. By doing some mathematical elaborations on the method, the authors obtained i^{th} natural frequencies and modes shapes one at a time. Hsu et al. [8] again used Modified Adomian Decomposition Method to solve free vibration of non-uniform Euler-Bernoulli beams with general elastically end conditions. Ozgumus and Kaya [9] used a new analytical approximation method namely Differential Transforms Method to analyze flapwise bending vibration analysis of double tapered rotating Euler-Bernoulli beam. Hsu et al. [10] also used Modified Adomian Decomposition Method, a new analytical approximation method, to solve eigenvalue problem for free vibration of uniform Timoshenko beams. Ho and Chen [11] studied the problem of free transverse vibration of an axially loaded non-uniform spinning twisted Timoshenko beam using Differential Transform Method. Another researcher, Register [12] found a general expression for the modal frequencies of a beam with symmetric spring boundary conditions. In addition, Wang [13] studied the dynamic analysis of generally supported beam. Yieh [14] determined the natural frequencies and natural

modes of the Euler_Bernoulli beam using the singular value decomposition method. Also, Kim [15] studied the vibration of uniform beams with generally restrained boundary conditions. Naguleswaran [16] derived an approximate solution to the transverse vibration of the uniform Euler-Bernoulli beam under linearly varying axial force. Chen and Ho [17] studied the problem of transverse vibration of rotating twisted Timoshenko beams under axial loading using differential transform method to obtain natural frequencies and mode shapes.

In this study, transverse vibration analysis of uniform and nonuniform Euler-Bernoulli beams will be briefly explained and demonstrated with some examples by using some of these novel approaches. To this aim, the theory and analytical techniques about lateral vibration of Euler-Bernoulli beams will be explained first, and then the methods used in the analysis will be described. Finally, some case studies will be presented by using the proposed techniques and the advantages of those methods will be discussed.

2. Transverse vibration of the beams

2.1 Formulation of the problem

Lateral vibration of beams is governed by well-known Bernoulli-Euler equation. To develop the governing equation, consider the free body diagram of a beam element in bending shown in Fig.1. In this figure, *M(x,t)* is the bending moment, *Q(x,t)* is the shear force, and *f(x,t)* is the external force per unit length acting on the beam.

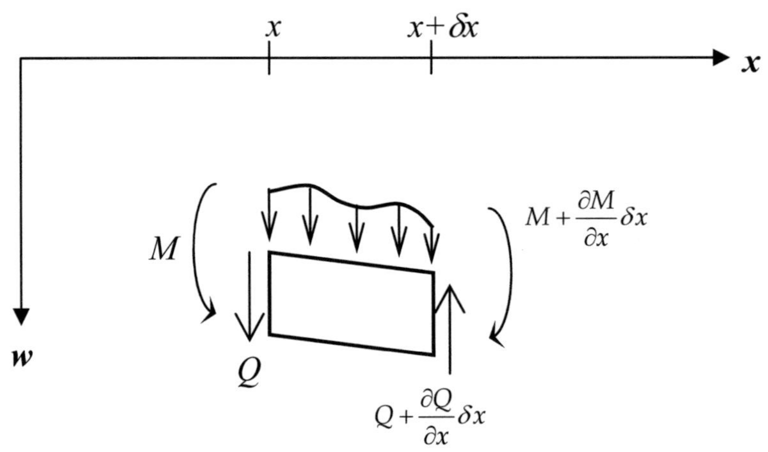

Fig. 1. Free-body diagram of a beam element in bending

Equilibrium condition of moments leads to the following equation:

$$M + Q\delta x - \left(M + \frac{\partial M}{\partial x} \right) = 0 \tag{1}$$

or

$$Q = \frac{\partial M}{\partial x} = \frac{\partial}{\partial x}\left(EI \frac{\partial^2 w}{\partial x^2} \right) \tag{2}$$

Since a uniform beam is not assumed in the formulation, *I(x)* will be variable along beam length.
The equation of motion in the tranverse direction for the beam element is:

$$(\rho A\delta x)\frac{\partial^2 w}{\partial t^2} = f(x,t)\delta x + Q - \left(Q + \frac{\partial Q}{\partial x}\right) \tag{3}$$

In Eq.(3), ρ is mass density of the material of the beam. After simplifications, Eq.(3) can be rewritten as follows:

$$\rho A\frac{\partial^2 w}{\partial t^2} + \frac{\partial Q}{\partial x} = f(x,t) \tag{4}$$

In view of Eq.(2), governing equation for forced transverse vibration is obtained as below which is the well known Euler-Bernoulli equation.

$$\frac{\partial^2}{\partial x^2}\left(EI\frac{\partial^2 w}{\partial x^2}\right) + \rho A\frac{\partial^2 w}{\partial t^2} = f(x,t) \tag{5}$$

For a uniform beam Eq.(5) reduces to

$$EI\frac{\partial^4 w}{\partial x^4} + \rho A\frac{\partial^2 w}{\partial t^2} = f(x,t) \tag{6}$$

For the free vibration case, *i.e. f(x,t)=0*, the equation of motion becomes

$$\frac{\partial^2}{\partial x^2}\left(EI\frac{\partial^2 w}{\partial x^2}\right) + \rho A\frac{\partial^2 w}{\partial t^2} = 0 \tag{7}$$

If the beam is uniform, i.e. EI is constant, the equation of motion in Eq.(7) reduces to

$$c^2\frac{\partial^4 w}{\partial x^4} + \frac{\partial^2 w}{\partial t^2} = 0 \tag{8}$$

where

$$c = \sqrt{\frac{EI}{\rho A}}\,. \tag{9}$$

Transverse vibration of beams is an initial-boundary value problem. Hence, both initial and boundary conditions are required to obtain a unique solution *w(x,t)*. Since the equation involves a second order derivative with respect to time and a fourth order derivative with respect to a space coordinate, two initial conditions and four boundary conditions are needed.

2.2 Modal analysis

The solution to problem given by Eq.(5) can be produced by, first obtaining the natural frequencies and mode shapes and then expressing the general solution as a summation of modal responses. In each mode, the system will vibrate in a fixed shape ratio which leads to providing a separable displacement function into two separate time and space functions. This approach is the same for both free and forced vibration problems. Hence, the displacement function *w(x,t)* can be defined by the following form.

$$w(x,t) = Y(x)T(t) \tag{10}$$

Consider the free vibration problem for a uniform beam, *i.e. EI* is constant. The governing equation for this specific case previously was given in Eq.(8). The free vibration solution will be obtained by inserting Eq.(10) into Eq.(8) and rearranging it as

$$\frac{c^2}{Y(x)}\frac{\partial^4 Y(x)}{\partial x^4} = -\frac{1}{T(t)}\frac{\partial^2 T(t)}{\partial t^2} = \omega^2 \tag{11}$$

where c is defined in Eq.(9) and ω^2 is defined as constant. Eq.(11) can be rearranged as two ordinary differential equations as

$$\frac{d^4 Y(x)}{dx^4} - \lambda^4 Y(x) = 0 \tag{12}$$

$$\frac{d^2 T(t)}{dt^2} + \omega^2 T(t) = 0 \tag{13}$$

where

$$\lambda^4 = \frac{\omega^2}{c^2} \tag{14}$$

General solution of Eq.(12) is a mode shape and given by

$$Y(x) = C_1 \cosh \lambda x + C_2 \sinh \lambda x + C_3 \cos \lambda x + C_4 \sin \lambda x \tag{15}$$

The constants C_1, C_2, C_3 and C_4 can be found from the end conditions of the beam. Then, the natural frequencies of the beam are obtained from Eq.(14) as

$$\omega = \lambda^2 c \tag{16}$$

Inserting Eq.(9) into Eq.(16) with rearranging leads to

$$\omega = (\lambda L)^2 \sqrt{\frac{EI}{\rho A L^4}} \tag{17}$$

2.3 Boundary conditions

The common boundary conditions related to beam's ends are as follows:

2.3.1 Simply supported (pinned) end

Deflection = 0 $\qquad\qquad Y = 0$

Bending Moment = 0 $$EI\frac{\partial^2 Y}{\partial x^2}=0$$

2.3.2 Fixed (clamped) end

Deflection = 0 $$Y=0$$

Slope = 0 $$\frac{\partial Y}{\partial x}=0$$

2.3.3 Free end

Bending Moment = 0 $$EI\frac{\partial^2 Y}{\partial x^2}=0$$

Shear Force = 0 $$\frac{\partial}{\partial x}\left(EI\frac{\partial^2 Y}{\partial x^2}\right)=0$$

2.3.4 Sliding end

Slope = 0 $$\frac{\partial Y}{\partial x}=0$$

Shear Force = 0 $$\frac{\partial}{\partial x}\left(EI\frac{\partial^2 Y}{\partial x^2}\right)=0$$

The exact frequencies for lateral vibration of the beams with different end conditions will not be computed due to the procedure explained here. Since, the motivation of this chapter is the demonstration of the use of analytical approximate techniques in the analysis of bending vibration of beams, available exact results related to the selected case studies will be directly taken from [5,18]. The reader can refer to these references for further details in analytical derivations of the exact results.

2.4 The methods used in the analysis of transverse vibration of beams

Analytical approximate solution techniques are used widely to solve nonlinear ordinary or partial differential equations, integro-differential equations, delay equations, etc. Main advantage of employing such techniques is that the problems are considered in a more realistic manner and the solution obtained is a continuous function which is not the case for the solutions obtained by discretized solution techniques. Hence these methods are computationally much more efficient in the solution of those equations.

The methods that will be used throughout the study are, Adomian Decomposition Method (ADM), Variational Iteration Method (VIM) and Homotopy Perturbation Method (HPM). Below, each technique will be explained and then all will be applied to several problems related to the topic of the article.

2.4.1 Adomian Decomposition Method (ADM)

In the ADM a differential equation of the following form is considered

$$Lu + Ru + Nu = g(x) \tag{18}$$

where L is the linear operator which is highest order derivative, R is the remainder of linear operator including derivatives of less order than L, Nu represents the nonlinear terms and g is the source term. Eq.(18) can be rearranged as

$$Lu = g(x) - Ru - Nu \tag{19}$$

Applying the inverse operator L^{-1} to both sides of Eq.(19) employing given conditions we obtain

$$u = L^{-1}\{g(x)\} - L^{-1}(Ru) - L^{-1}(Nu) \tag{20}$$

After integrating source term and combining it with the terms arising from given conditions of the problem, a function $f(x)$ is defined in the equation as

$$u = f(x) - L^{-1}(Ru) - L^{-1}(Nu) \tag{21}$$

The nonlinear operator $Nu = F(u)$ is represented by an infinite series of specially generated (Adomian) polynomials for the specific nonlinearity. Assuming Nu is analytic we write

$$F(u) = \sum_{k=0}^{\infty} A_k \tag{22}$$

The polynomials A_k's are generated for all kinds of nonlinearity so that they depend only on u_o to u_k components and can be produced by the following algorithm.

$$A_0 = F(u_0) \tag{23}$$

$$A_1 = u_1 F'(u_0) \tag{24}$$

$$A_2 = u_2 F'(u_0) + \frac{1}{2!} u_1^2 F''(u_0) \tag{25}$$

$$A_3 = u_3 F'(u_0) + u_1 u_2 F''(u_0) + \frac{1}{3!} u_1^3 F'''(u_0) \tag{26}$$

$$\vdots$$

The reader can refer to [19,20] for the algorithms used in formulating Adomian polynomials. The solution $u(x)$ is defined by the following series

$$u = \sum_{k=0}^{\infty} u_k \tag{27}$$

where the components of the series are determined recursively as follows:

$$u_0 = f(x) \tag{28}$$

$$u_{k+1} = -L^{-1}(Ru_k) - L^{-1}(A_k), \quad k \geq 0 \tag{29}$$

2.4.2 Variational Iteration Method (VIM)

According to VIM, the following differential equation may be considered:

$$Lu + Nu = g(x) \tag{30}$$

where L is a linear operator, and N is a nonlinear operator, and $g(x)$ is an inhomogeneous source term. Based on VIM, a correct functional can be constructed as follows:

$$u_{n+1} = u_n + \int_0^x \lambda(\xi)\{Lu_n(\xi) + N\tilde{u}_n(\xi) - g(\xi)\}\, d\xi \tag{31}$$

where λ is a general Lagrangian multiplier, which can be identified optimally via the variational theory, the subscript n denotes the nth-order approximation, $\tilde{u}$ is considered as a restricted variation *i.e.* $\delta\tilde{u} = 0$. By solving the differential equation for λ obtained from Eq.(31) in view of $\delta\tilde{u} = 0$ with respect to its boundary conditions, Lagrangian multiplier $\lambda(\xi)$ can be obtained. For further details of the method the reader can refer to [21].

2.4.3 Homotopy Perturbation Method (HPM)

HPM provides an analytical approximate solution for problems at hand as other explained techniques. Brief theoretical steps for the equation of following type can be given as

$$L(u) + N(u) = f(r) \ , \ r \in \Omega \tag{32}$$

with boundary conditions $B(u, \partial u/\partial n) = 0$. In Eq.(8) L is a linear operator, N is nonlinear operator, B is a boundary operator, and $f(r)$ is a known analytic function. HPM defines homotopy as

$$v(r,p) = \Omega \times [0,1] \rightarrow R \tag{33}$$

which satisfies following inequalities:

$$H(v,p) = (1-p)[L(v) - L(u_0)] + p[L(v) + N(v) - f(r)] = 0 \tag{34}$$

or

$$H(v,p) = L(v) - L(u_0) + pL(u_0) + p[N(v) - f(r)] = 0 \tag{35}$$

where $r \in \Omega$ and $p \in [0,1]$ is an imbedding parameter, u_0 is an initial approximation which satisfies the boundary conditions. Obviously, from Eq.(34) and Eq.(35) , we have :

$$H(v,0) = L(v) - L(u_0) = 0 \tag{36}$$

$$H(v,1) = L(v) + N(v) - f(r) = 0 \tag{37}$$

As p changing from zero to unity is that of $v(r,p)$ from u_0 to $u(r)$. In topology, this deformation $L(v) - L(u_0)$ and $L(v) + N(v) - f(r)$ are called homotopic. The basic assumption is that the solutions of Eq.(34) and Eq.(35) can be expressed as a power series in p such that:

$$v = v_0 + pv_1 + p^2 v_2 + p^3 v_3 + ... \tag{38}$$

The approximate solution of $L(u) + N(u) = f(r)$, $r \in \Omega$ can be obtained as:

$$u = \lim_{p \to 1} v = v_0 + v_1 + v_2 + v_3 + ... \tag{39}$$

The convergence of the series in Eq.(39) has been proved in [22]. The method is described in detail in references [22-25].

2.5 Case studies

2.5.1 Free vibration of a uniform beam

The governing equation for this case was previously given in Eq.(12). ADM, VIM and HPM will be applied to this equation in order to compute the natural frequencies for the free vibration of a beam with constant flexural stiffness, *i.e.* constant *EI*, and its corresponding mode shapes. To this aim, five different beam configurations are defined with its end conditions. These are *PP*, the beam with both ends pinned, *CC*, the beam with both ends clamped, *CP*, the beam with one end clamped and one end pinned, *CF*, the beam with one end clamped and one end free, *CS*, the beam with one end clamped and one and sliding. The boundary conditions associated with these configurations was given previously in text. Below, the formulations by using ADM, VIM and HPM are given and then applied to the governing equation of the problem.

2.5.1.1 Formulation of the algorithms

2.5.1.1.1 ADM

The linear operator and its inverse operator for Eq.(12) is

$$L(\cdot) = \frac{d^4}{dx^4}(\cdot) \tag{40}$$

$$L^{-1}(\cdot) = \int_0^x \int_0^x \int_0^x \int_0^x (\cdot)\, dx\, dx\, dx\, dx \tag{41}$$

To keep the formulation a general one for all configurations to be considered, the boundary conditions are chosen as $Y(0) = A$, $Y'(0) = B$, $Y''(0) = C$ and $Y'''(0) = D$. Suitable values should be replaced in the formulation with these constants. For example, $A = 0$ and $C = 0$ should be inserted for the PP beam. Hence, the equation to be solved and the recursive algorithm can be given as

$$LY = \lambda^4 Y \tag{42}$$

$$Y = A + Bx + C\frac{x^2}{2!} + D\frac{x^3}{3!} + L^{-1}(\lambda^4 Y) \tag{43}$$

$$Y_{n+1} = L^{-1}(\lambda^4 Y_n), \quad n \geq 0 \tag{44}$$

Finally, the solution is defined by

$$Y = Y_0 + Y_1 + Y_2 + Y_3 + \ldots \tag{45}$$

2.5.1.1.2 VIM

Based on the formulation given previously, Lagrange multiplier λ would be obtained for the governing equation, *i.e.* Eq.(12), as

$$\lambda(\xi) = \frac{(\xi - x)^3}{3!} \tag{46}$$

An iterative algorithm can be constructed inserting Lagrange multiplier and governing equation into the formulation given in Eq.(31) as

$$Y_{n+1} = Y_n + \int_0^x \lambda(\xi)\left\{Y_n^{iv}(\xi) - \lambda^4\tilde{Y}_n(\xi)\right\} d\xi \tag{47}$$

Initial approximation for the algorithm is chosen as the solution of $LY = 0$ which is a cubic polynomial with four unknowns which will be determined by the end conditions of the beam.

2.5.1.1.3 HPM

Based on the formulation, Eq.(12) can be divided into two parts as

$$LY = Y^{iv} \tag{48}$$

$$NY = -\lambda^4 Y \tag{49}$$

The solution can be expressed as a power series in p such that

$$Y = Y_0 + pY_1 + p^2Y_2 + p^3Y_3 + \ldots \tag{50}$$

Inserting Eq.(50) into Eq.(35) provides a solution algorithm as

$$Y_0^{iv} - y_0^{iv} = 0 \tag{51}$$

$$Y_1^{iv} + y_0^{iv} - \lambda^4 Y_0 = 0 \tag{52}$$

$$Y_n - \lambda^4 Y_{n-1} = 0, \quad n \geq 2 \tag{53}$$

Hence, an approximate solution would be obtained as

$$Y = Y_0 + Y_1 + Y_2 + Y_3 + \ldots \tag{54}$$

Initial guess is very important for the convergence of solution in HPM. A cubic polynomial with four unknown coefficients can be chosen as an initial guess which was shown previously to be an effective one in problems related to Euler beams and columns [26-31].

2.5.1.2 Computation of natural frequencies

By the use of described algorithms, an iterative procedure is conducted and a polynomial including the unknown coefficients coming from the initial guess is produced as a solution to the governing equation. Besides four unknowns from initial guess, an additional unknown λ also exists in the solution. Applying each boundary condition to the solution produces a linear algebraic system of equations which can be defined in matrix form as

$$[M(\lambda)]\{\alpha\} = \{0\} \tag{55}$$

where $\{\alpha\} = \langle A,B,C,D\rangle^T$. For a nontrivial solution, determinant of coefficient matrix must be zero. Determinant of matrix $[M(\lambda)]$ yields a characteristic equation in terms of λ. Positive real roots of this equation are the natural free vibration frequencies for the beam with specified end conditions.

2.5.1.3 Determination of vibration mode shapes

Vibration mode shapes for the beams can also be obtained from the polynomial approximations by the methods considered in this study. Introducing, the natural frequencies into the solution, normalized polynomial eigenfunctions for the mode shapes are obtained from

$$\bar{Y}_j = \frac{Y_N\left(x,\lambda_j\right)}{\left[\int_0^1 \left|Y_N\left(x,\lambda_j\right)\right|^2 dx\right]^{1/2}} \quad , \; j=1,2,3,... \tag{56}$$

The same approach can be employed to predict mode shapes for the cases including variable flexural stiffness.

2.5.1.4 Orthogonality of mode shapes

Normalized mode shapes obtained from Eq.(56) should be orthogonal. These modes can be shown to satisfy the following condition.

$$\int Y_i Y_j \, dx = \begin{cases} 0, & i \neq j \\ 1, & i = j \end{cases} \tag{57}$$

2.5.1.5 Results of the analysis

After applying the procedures explained in the text, the following results are obtained for the natural frequencies and mode shapes. Comparison with the exact solutions is also provided that one can observe an excellent agreement between the exact results and computed results.

Ten iterations are conducted for each method and computed λL values are compared with the corresponding exact values for the first three modes of vibration in the following table.

From the table it can be seen that computed values are highly accurate which show that the techniques used in the analysis are very effective. Natural frequencies can be easily obtained by inserting the values in Table 1 into Eq.(17).

The free vibration mode shapes of uniform beam for the first three mode are also depicted in the following figures. Since the obtained mode shapes coincide with the exact ones, to prevent a possible confusion to the reader, the exact mode shapes and the computed ones are not shown separately in these figures. The mode shapes for the free vibration of a uniform beam for five different configurations are given between Figs.2-6.

Beam	Mode	Exact	ADM	VIM	HPM
P-P	1	3.14159265 (π)	3.14159265	3.14159265	3.14159265
	2	6.28318531 (2π)	6.28318531	6.28318531	6.28318531
	3	9.42477796 (3π)	9.42477796	9.4247796	9.4247796
C-C	1	4.730041	4.73004074	4.73004074	4.73004074
	2	7.853205	7.85320462	7.85320462	7.85320462
	3	10.995608	10.99560784	10.99560784	10.99560784
C-P	1	3.926602	3.92660231	3.92660231	3.92660231
	2	7.068583	7.06858275	7.06858275	7.06858275
	3	10.210176	10.21017612	10.21017612	10.21017612
C-F	1	1.875104	1.87510407	1.87510407	1.87510407
	2	4.694091	4.69409113	4.69409113	4.69409113
	3	7.854757	7.85475744	7.85475744	7.85475744
C-S	1	2.365020	2.36502037	2.36502037	2.36502037
	2	5.497806	5.49780392	5.49780392	5.49780392
	3	8.639380	8.63937983	8.63937983	8.63937983

Table 1. Comparison of λL values for the uniform beam

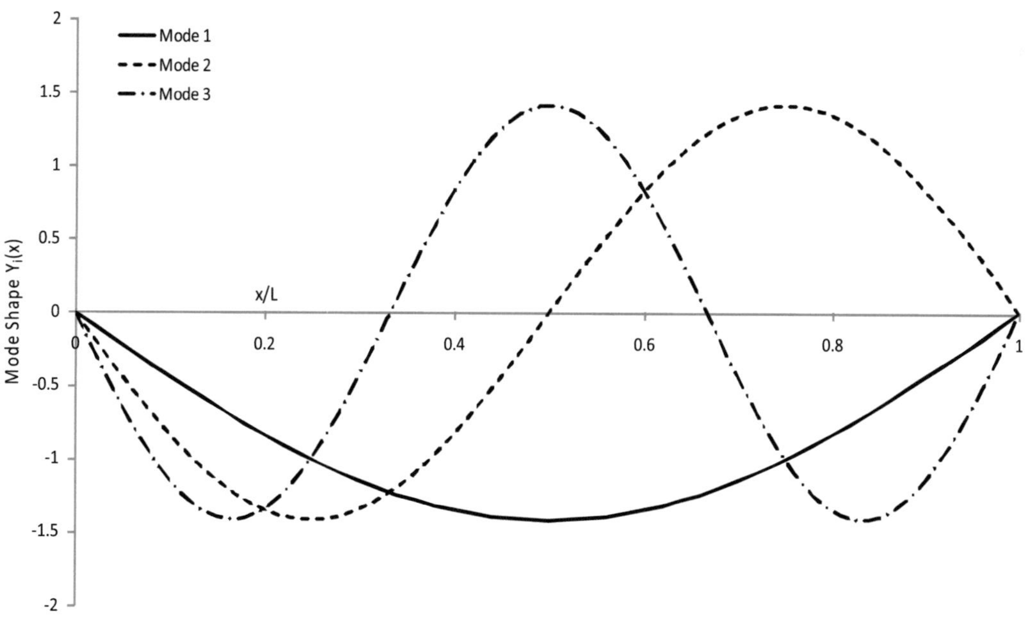

Fig. 2. Free vibration modes of PP beam.

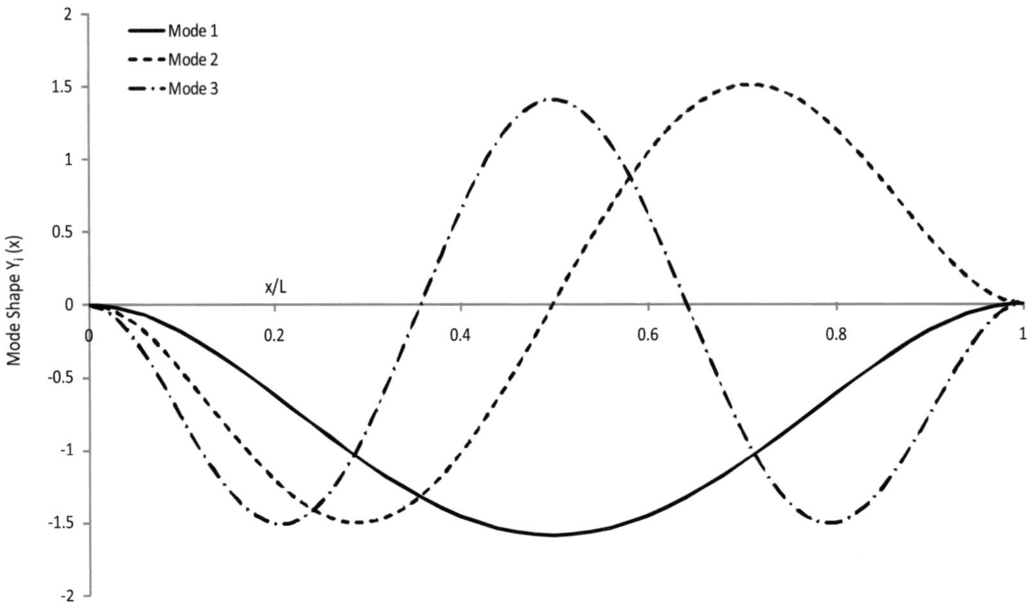

Fig. 3. Free vibration modes of CC beam.

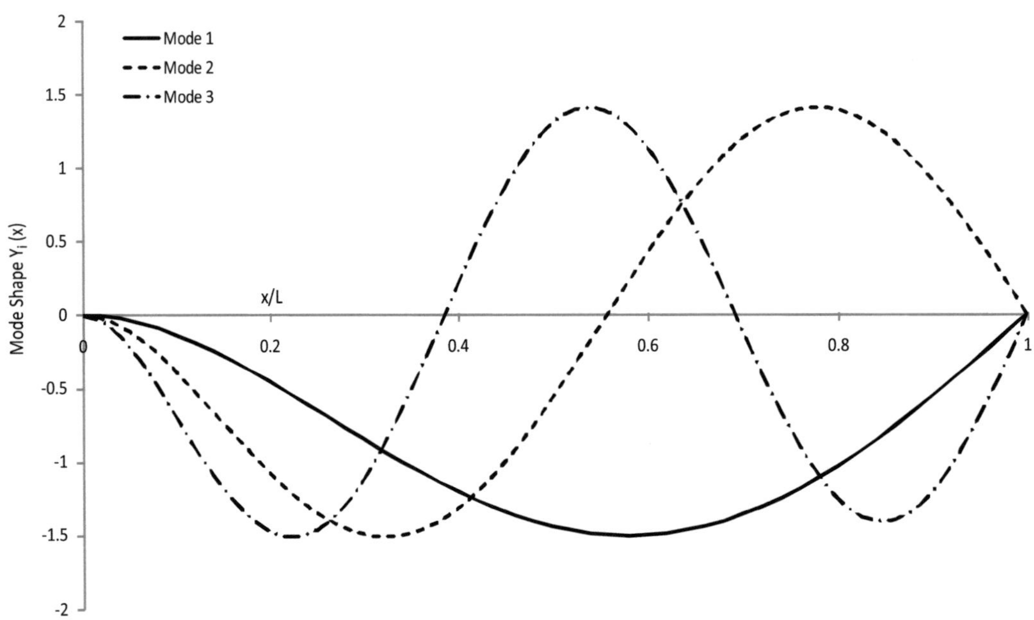

Fig. 4. Free vibration modes of CP beam.

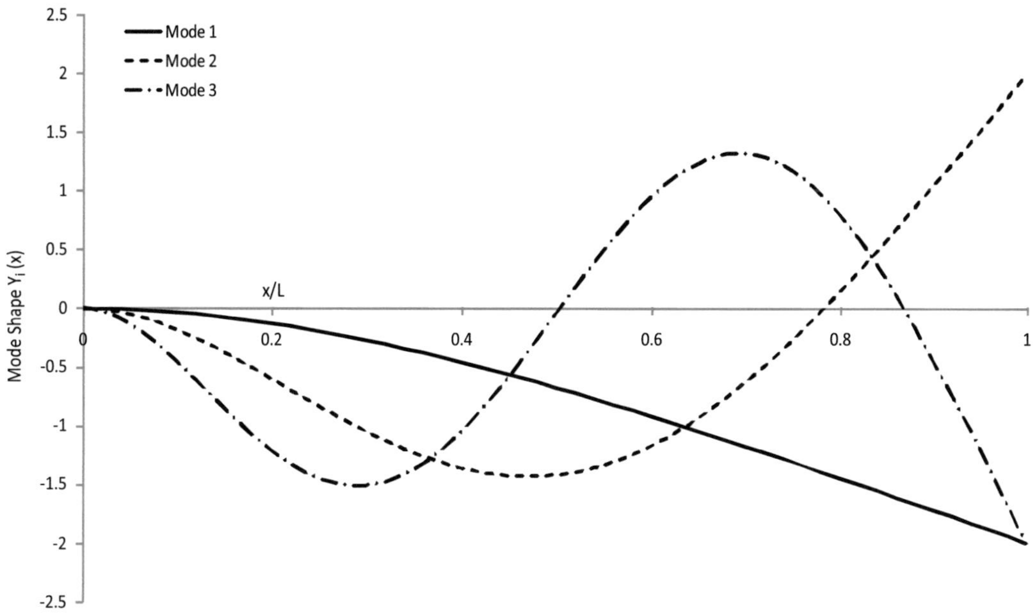

Fig. 5. Free vibration modes of CF beam.

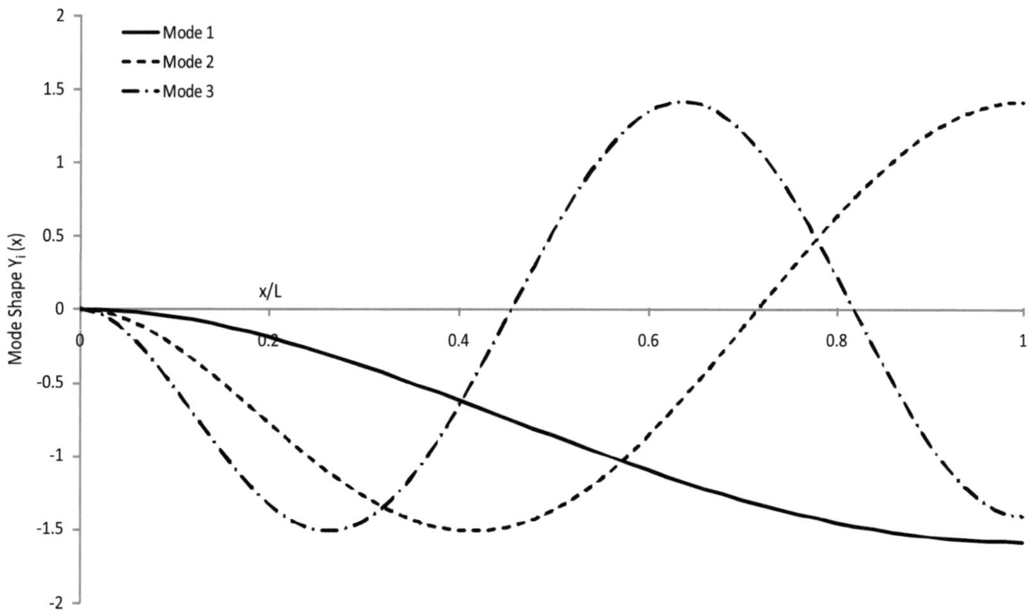

Fig. 6. Free vibration modes of CF beam.

Orthogonality condition given in Eq.(57) for each mode will also be shown to be satisfied. To this aim, the resulting polynomials representing normalized eigenfunctions are integrated according to the orthogonality condition and following results are obtained.
The PP Beam:

$$\int Y_i Y_j \, dx = \begin{bmatrix} 1.0000000000000018 & 3.133937506642793*10^{-14} & 1.1716394903869283*10^{-12} \\ & 1.0000000000011495 & -1.2402960384615706*10^{-11} \\ & & 1.0000000002542724 \end{bmatrix}$$

The CC Beam:

$$\int Y_i Y_j \, dx = \begin{bmatrix} 1.0000000000000218 & -3.2594265231428034*10^{-13} & 3.0586251883350275*10^{-11} \\ & 0.9999999999825311 & -4.152039340197406*10^{-10} \\ & & 0.9999999986384138 \end{bmatrix}$$

The CP Beam:

$$\int Y_i Y_j \, dx = \begin{bmatrix} 1.0000000000000027 & -1.1266760906960104*10^{-13} & 3.757083743946838*10^{-12} \\ & 0.9999999999991402 & -5.469593759847241*10^{-11} \\ & & 1.000000001594055 \end{bmatrix}$$

The CF Beam:

$$\int Y_i Y_j \, dx = \begin{bmatrix} 1.0000000000000000 & 1.134001985461197*10^{-15} & 5.844267022420876*10^{-14} \\ & 1.0000000000000178 & 4.1094000558822104*10^{-13} \\ & & 0.9999999999969831 \end{bmatrix}$$

The CS Beam:

$$\int Y_i Y_j \, dx = \begin{bmatrix} 1.0000000000000009 & -1.067231239470151*10^{-15} & -2.57978811982526*10^{-13} \\ & 1.0000000000002232 & -2.422143056441983*10^{-13} \\ & & 1.0000000000643874 \end{bmatrix}$$

From these results it can be clearly observed that the orthogonality condition is perfectly satisfied for each configuration of the beam.
The analysis for the lateral free vibration of the uniform beam is completed. Now, these techniques will be applied to a circular rod having variable cross-section along its length.

2.5.2 Free vibration of a rod with variable cross-section

A circular rod having a radius changing linearly is considered in this case. Such a rod is shown below in Fig.7. The function representing the radius would be as

$$R(x) = R_0(1 - bx) \tag{58}$$

where R_o is the radius at the left end, L is the length of the rod and $bL \leq 1$.

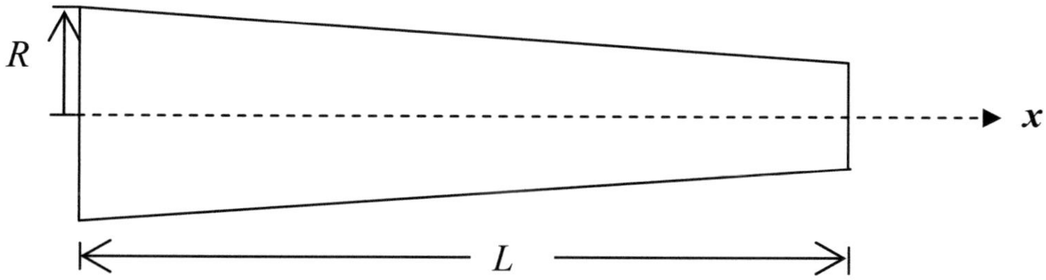

Fig. 7. Circular rod with variable cross-section

Employing Eq.(58), cross-sectional area and moment of inertia for a section at an arbitrary point x becomes:

$$A(x) = A_0(1-bx)^2 \tag{59}$$

$$I(x) = I_0(1-bx)^4 \tag{60}$$

where

$$A_0 = \pi R_0^2 \tag{61}$$

$$I_0 = \frac{\pi R_0^4}{4} \tag{62}$$

Free vibration equation of the rod was previously given in Eq.(7) as

$$\frac{\partial^2}{\partial x^2}\left(EI\frac{\partial^2 w}{\partial x^2}\right) + \rho A \frac{\partial^2 w}{\partial t^2} = 0$$

After the application of separation of variables technique by defining the displacement function as $w(x,t) = Y(x)T(t)$, the equation to obtain natural frequencies and mode shapes becomes

$$\frac{d^2}{dx^2}\left(EI(x)\frac{d^2Y}{dx^2}\right) - \omega^2 \rho A(x) Y = 0 \tag{63}$$

2.5.2.1 Formulation of the algorithms

2.5.2.1.1 ADM

Application of ADM to Eq.(63) leads to the following

$$Y^{iv} - 8b\psi_1(x)Y''' + 12b^2\psi_2(x)Y'' - \lambda_0^4\psi_2(x)Y = 0 \tag{64}$$

where

$$\psi_1(x) = \frac{1}{1-bx} \tag{65}$$

$$\psi_2(x) = \frac{1}{(1-bx)^2} \tag{66}$$

$$\lambda_0^4 = \frac{\omega^2}{c_0^2} \tag{67}$$

$$c_0 = \sqrt{\frac{EI_0}{\rho A_0}} \tag{68}$$

Once λ_o is provided by ADM, natural vibration frequencies for the rod can be easily found from the equation below.

$$\omega = (\lambda_0 L)^2 \sqrt{\frac{EI_0}{\rho A_0 L^4}} \tag{69}$$

ADM gives the following formulation with the previously defined fourth order linear operator.

$$Y = A + Bx + C\frac{x^2}{2!} + D\frac{x^3}{3!} + L^{-1}\left(8b\psi_1(x)Y''' - 12b^2\psi_2(x)Y'' + \lambda_0^4\psi_2(x)Y\right) \tag{70}$$

2.5.2.1.2 VIM

Lagrange multiplier is the same as used in the uniform beam case due to the fourth order derivative in Eq.(64). Hence an algorithm by using VIM can be constructed as

$$Y_{n+1} = Y_n + \int_0^x \lambda(\xi)\left\{Y_n^{iv} - 8b\psi_1(x)\tilde{Y}_n''' + 12b^2\psi_2(x)\tilde{Y}_n'' - \lambda_0^4\psi_2(x)\tilde{Y}_n\right\} d\xi \tag{71}$$

2.5.2.1.3 HPM

Application of HPM to Eq.(64) produce following set of recursive equations as the solution algorithm.

$$Y_0^{iv} - y_0^{iv} = 0 \tag{72}$$

$$Y_1^{iv} + y_0^{iv} - 8b\psi_1(x)Y_0''' + 12b^2\psi_2(x)Y_0'' - \lambda_0^4\psi_2(x)Y_0 = 0 \tag{73}$$

$$Y_n - 8b\psi_1(x)Y_{n-1}''' + 12b^2\psi_2(x)Y_{n-1}'' - \lambda_0^4\psi_2(x)Y_{n-1} = 0, \quad n \geq 2 \tag{74}$$

2.5.2.2 Results of the analysis

After applying the proposed formulations, the following results are obtained for the natural frequencies and mode shapes. Ten iterations are conducted for each method and computed $\lambda_o L$ values are given for the first three modes of vibration in the following table.
The free vibration mode shapes of the rod for the first three modes are also depicted in the following figures. The mode shapes for predefined five different configurations are given

between Figs. 8-12. To demonstrate the effect of variable cross-section in the results, a comparison is made with normalized mode shapes for a uniform rod which are given between Figs.2-6.

Beam	Mode	ADM	VIM	HPM
P-P	1	2.97061902	2.97061902	2.97061902
	2	5.95530352	5.95530352	5.95530352
	3	8.93099026	8.93099026	8.93099026
C-C	1	4.48292606	4.48292606	4.48292606
	2	7.44076320	7.44076320	7.44076320
	3	10.41682600	10.41682600	10.41682600
C-P	1	3.80402043	3.80402043	3.80402043
	2	6.74289447	6.74289447	6.74289447
	3	9.70480586	9.70480586	9.70480586
C-F	1	1.96344512	1.96344512	1.96344512
	2	4.58876313	4.58876313	4.58876313
	3	7.52531208	7.52531208	7.52531208
C-S	1	2.35500726	2.35500726	2.35500726
	2	5.26125511	5.26125511	5.26125511
	3	8.21783948	8.21783948	8.21783948

Table 2. Comparison of $\lambda_o L$ values for the variable cross-section rod

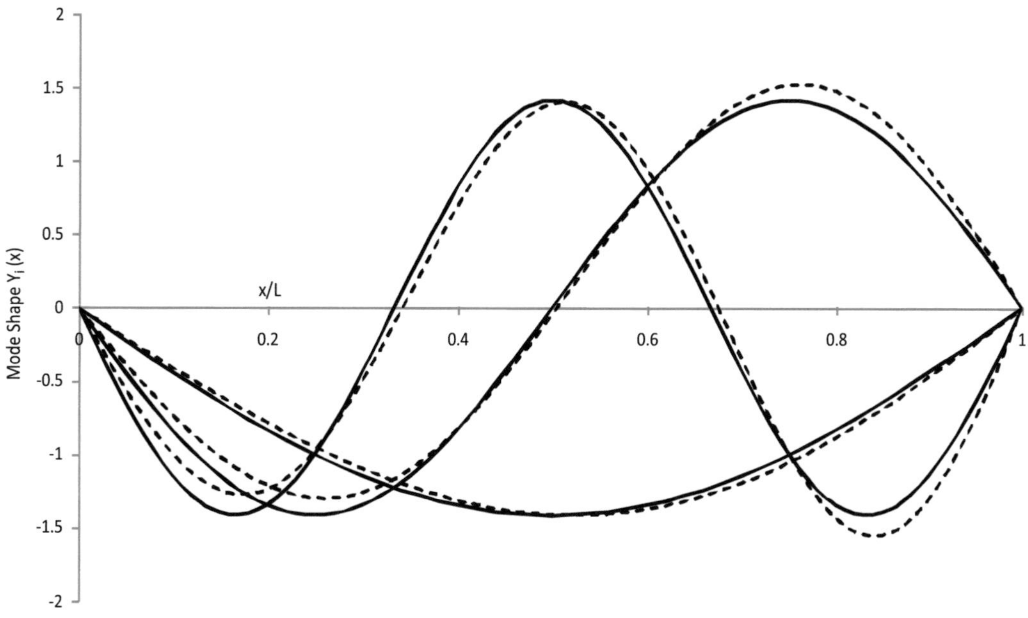

Fig. 8. Free vibration modes of PP rod (- - - -variable cross section —— uniform rod).

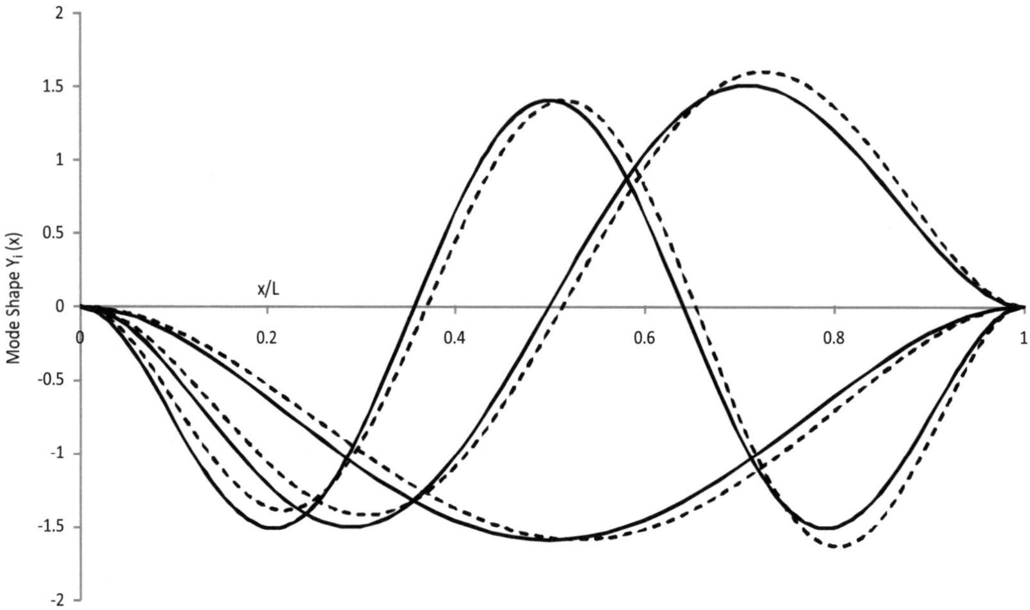

Fig. 9. Free vibration modes of CC rod (_ _ _ _variable cross section ____uniform rod).

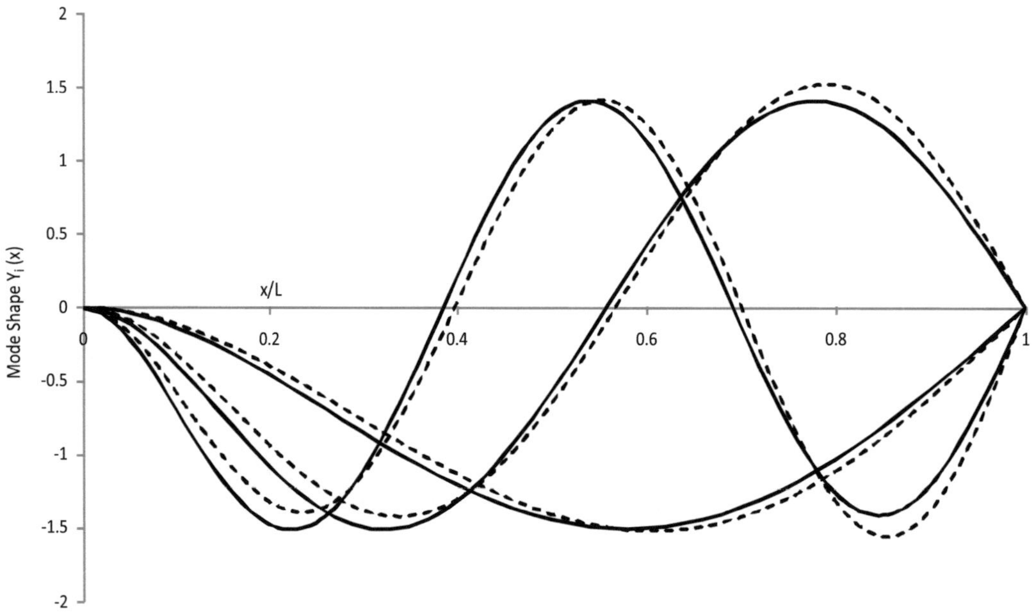

Fig. 10. Free vibration modes of CP rod (_ _ _ _variable cross section ____uniform rod).

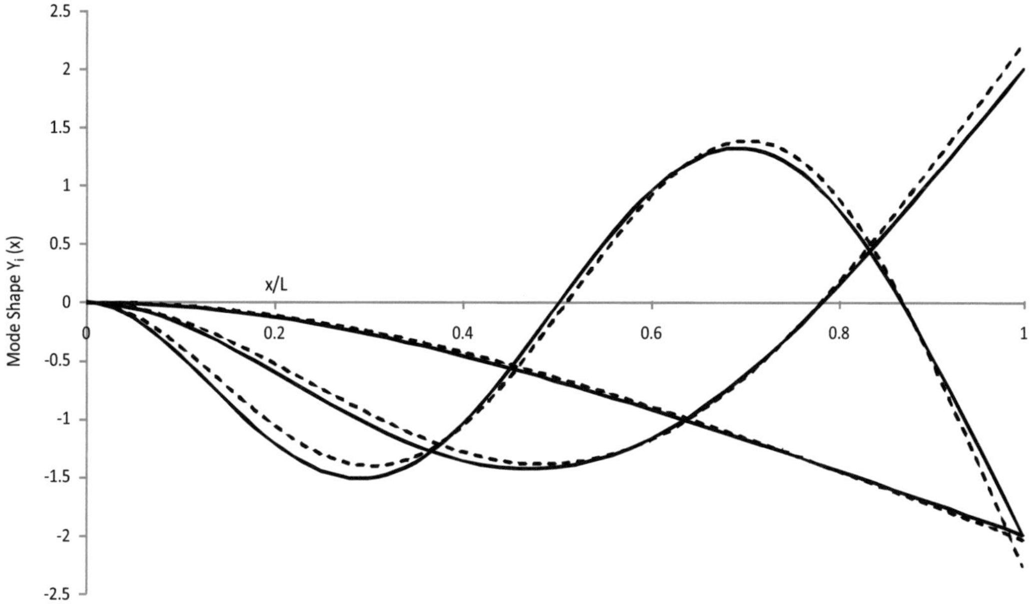

Fig. 11. Free vibration modes of CF rod (_ _ _ _ variable cross section ____ uniform rod).

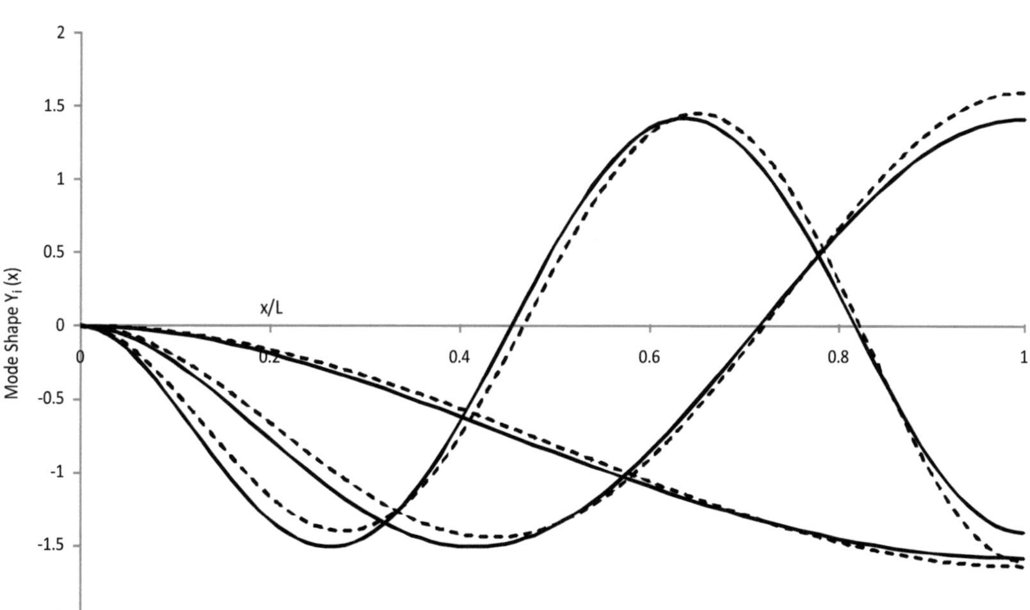

Fig. 12. Free vibration modes of CS rod (- - - -variable cross section —— uniform rod).

3. Conclusion

In this article, some analytical approximation techniques were employed in the transverse vibration analysis of beams. In a variety of such techniques, the most used ones, namely ADM, VIM and HPM were chosen for use in the computations. First, a brief theoretical knowledge was given in the text and then all of the methods were applied to selected cases. Since the exact values for the free vibration of a uniform beam was available, the analyses were started for that case. Results showed an excellent agreement with the exact ones that all three methods were highly effective in the computation of natural frequencies and vibration mode shapes. Orthogonality of the mode shapes was also proven. Finally, ADM, VIM and HPM were applied to the free vibration analysis of a rod having variable cross section. To this aim, a rod with linearly changing radius was chosen and natural frequencies with their corresponding mode shapes were obtained easily.

The study has shown that ADM, VIM and HPM can be used effectively in the analysis of vibration problems. It is possible to construct easy-to-use algorithms which are highly accurate and computationally efficient.

4. References

[1] L. Meirovitch, Fundamentals of Vibrations, International Edition, McGraw-Hill, 2001.

[2] A. Dimarogonas, Vibration for Engineers, 2nd ed., Prentice-Hall, Inc., 1996.

[3] W. Weaver, S.P. Timoshenko, D.H. Young, Vibration Problems in Engineering, 5th ed., John Wiley & Sons, Inc., 1990.

[4] W. T. Thomson, Theory of Vibration with Applications, 2nd ed., 1981.

[5] S. S. Rao , Mechanical Vibrations, 3rd ed. Addison-Wesley Publishing Company 1995.

[6] Y. Liu, C. S. Gurram, The use of He's variational iterationmethod for obtaining the free vibration of an Euler-Beam beam, Matematical and Computer Modelling, 50(2009) 1545-1552.

[7] H-L. Lai, J-C., Hsu, C-K. Chen, An innovative eigenvalue problem solver for free vibration of Euler-Bernoulli beam by using the Adomian Decomposition Method., Computers and Mathematics with Applications 56 (2008) 3204-3220.

[8] J-C. Hsu, H-Y Lai,C.K. Chen, Free Vibration of non-uniform Euler-Bernoulli beams with general elastically end constraints using Adomian modifiad decomposition method, Journal of Sound and Vibration, 318 (2008) 965-981.

[9] O. O. Ozgumus, M. O. Kaya, Flapwise bending vibration analysis of double tapered rotating Euler-Bernoulli beam by using the Diferential Transform Method, Mechanica 41 (2006) 661-670.

[10] J-C. Hsu, H-Y. Lai, C-K. Chen, An innovative eigenvalue problem solver for free vibration of Timoshenko beams by using the Adomian Decomposition Method, Journal of Sound and Vibration 325 (2009) 451-470.

[11] S. H. Ho, C. K. Chen, Free transverse vibration of an axially loaded non-uniform spinning twisted Timoshenko beam using Differential Transform, International Journal of Mechanical Sciences 48 (2006) 1323-1331.

[12] A.H. Register, A note on the vibrations of generally restrained, end-loaded beams, Journal of Sound and Vibration 172 (4) (1994) 561_571.

[13] J.T.S. Wang, C.C. Lin, Dynamic analysis of generally supported beams using Fourier series, Journal of Sound and Vibration 196 (3) (1996) 285_293.
[14] W. Yeih, J.T. Chen, C.M. Chang, Applications of dual MRM for determining the natural frequencies and natural modes of an Euler_Bernoulli beam using the singular value decomposition method, Engineering Analysis with Boundary Elements 23 (1999) 339_360.
[15] H.K. Kim, M.S. Kim, Vibration of beams with generally restrained boundary conditions using Fourier series, Journal of Sound and Vibration 245 (5) (2001) 771_784.
[16] S. Naguleswaran, Transverse vibration of an uniform Euler_Bernoulli beam under linearly varying axial force, Journal of Sound and Vibration 275 (2004) 47_57.
[17] C.K. Chen, S. H. Ho, Transverse vibration of rotating twisted Timoshenko beams under axial loading using differential transform, International Journal of Mechanical Sciences 41 (1999) 1339- 1356.
[18] C.W. deSilva, Vibration: Fundamentals and Practice, CRC Press, Boca Raton, Florida, 2000.
[19] G. Adomian, Solving Frontier Problems of Physics: The Decomposition Method, Kluwer, Boston, MA, 1994.
[20] G. Adomian, A review of the decomposition method and some recent results for nonlinear equation, Math. Comput. Modell., 13(7), 1992, 17-43.
[21] J.H. He, Variational iteration method: a kind of nonlinear analytical technique, Int. J. Nonlin. Mech., 34, 1999, 699-708.
[22] J.H. He, A coupling method of a homotopy technique and a perturbation technique for non-linear problems, Int. J. Nonlin. Mech., 35, 2000, 37-43.
[23] J.H. He, An elemantary introduction to the homotopy perturbation method, Computers and Mathematics with Applications, 57(2009), 410-412.
[24] J.H. He, New interpretation of homotopy perturbation method, International Journal of Modern Physics B, 20(2006), 2561-2568.
[25] J.H. He, The homotopy perturbation method for solving boundary problems, Phys. Lett. A, 350 (2006), 87-88.
[26] S.B. Coskun, "Determination of critical buckling loads for Euler columns of variable flexural stiffness with a continuous elastic restraint using Homotopy Perturbation Method", Int. Journal Nonlinear Sci. and Numer. Simulation, 10(2) (2009), 191-197.
[27] S.B Coskun, "Analysis of tilt-buckling of Euler columns with varying flexural stiffness using homotopy perturbation method", Mathematical Modelling and Analysis, 15(3), (2010), 275-286.
[28] M.T. Atay, "Determination of critical buckling loads for variable stiffness Euler Columns using Homotopy Perturbation Method", Int. Journal Nonlinear Sci. and Numer. Simulation, 10(2) (2009), 199-206.
[29] B. Öztürk, S.B. Coskun, "The homotopy perturbation method for free vibration analysis of beam on elastic foundation", Structural Engineering and Mechanics, An Int. Journal, 37(4), (2010).
[30] B. Öztürk, "Free vibration analysis of beam on elastic foundation by variational iteration method", International journal of Nonlinear Science and Numerical Simulation, 10 (10) 2009, 1255-1262.

[31] B. Öztürk, S.B. Coskun, M.Z. Koc, M.T. Atay, Homotopy perturbation method for free vibration analysis of beams on elastic foundations, IOP Conf. Ser.: Mater. Sci. Engr., Volume: 10, Number:1, *9th World Congress on Computational Mechanics and 4th Asian Pasific Congress on Computational Mechanics*, Sydney, Australia (2010).

2

Vibration Analysis of Beams with and without Cracks Using the Composite Element Model

Z.R. Lu, M. Huang and J.K. Liu
Sun Yat-sen University
P.R. China

1. Introduction

Beams are fundamental models for the structural elements of many engineering applications and have been studied extensively. There are many examples of structures that may be modeled with beam-like elements, for instance, long span bridges, tall buildings, and robot arms.

The vibration of Euler–Bernoulli beams with one step change in cross-section has been well studied. Jang and Bert (1989) derived the frequency equations for combinations of classical end supports as fourth order determinants equated to zero. Balasubramanian and Subramanian (1985) investigated the performance of a four-degree-of-freedom per node element in the vibration analysis of a stepped cantilever. De Rosa (1994) studied the vibration of a stepped beam with elastic end supports. Recently, Koplow et al. (2006) presented closed form solutions for the dynamic response of Euler–Bernoulli beams with step changes in cross section.

There are also some works on the vibration of beams with more than one step change in cross-section. Bapat and Bapat (1987) proposed the transfer matrix approach for beams with n-steps but provided no numerical results. Lee and Bergman (1994) used the dynamic flexibility method to derive the frequency equation of a beam with n-step changes in cross-section. Jaworski and Dowell (2008) carried out a study for the free vibration of a cantilevered beam with multiple steps and compared the results of several theoretical methods with experiment.

A new method is presented to analyze the free and forced vibrations of beams with either a single step change or multiple step changes using the composite element method (CEM) (Zeng, 1998; Lu & Law, 2009). The correctness and accuracy of the proposed method are verified by some examples in the existing literatures. The presence of cracks in the structural components, for instance, beams can have a significant influence on the dynamic responses of the whole structure; it can lead to the catastrophic failure of the structure. To predict the failure, vibration monitoring can be used to detect changes in the dynamic responses and/or dynamic characteristics of the structure. Knowledge of the effects of cracks on the vibration of the structure is of importance. Efficient techniques for the forward analysis of cracked beams are required. To this end, the composite element method is then extended for free and forced vibration analysis of cracked beams.

The principal advantage of the proposed method is that it does not need to partition the stepped beam into uniform beam segments between any two successive discontinuity points and the whole beam can be treated as a uniform beam. Moreover, the presented work can easily be extended to cracked beams with an arbitrary number of non-uniform segments.

2. Theory

2.1 Introduction to Composite Element Method (CEM)

The composite element is a relatively new tool for finite element modeling. This method is basically a combination of the conventional finite element method (FEM) and the highly precise classical theory (CT). In the composite element method, the displacement field is expressed as the sum of the finite element displacement and the shape functions from the classical theory. The displacement field of the CEM can be written as

$$u_{CEM}(x,t) = u_{FEM}(x,t) + u_{CT}(x,t) \tag{1}$$

where $u_{FEM}(x,t)$ and $u_{CT}(x,t)$ are the individual displacement fields from the FEM and CT, respectively.
Taking a planar beam element as an example, the first term of the CEM displacement field can be expressed as the product of the shape function vector of the conventional finite element method $N(x)$ and the nodal displacement vector q

$$u_{FEM}(x,t) = N(x)q(t) \tag{2}$$

where $q(t) = [v_1(t), \theta_1(t), v_2(t), \theta_2(t)]^T$ and 'v' and 'θ' represent the transverse and rotational displacements, respectively.
The second term $u_{CT}(x,t)$ is obtained by the multiplication of the analytical mode shapes with a vector of N coefficients c (also called the c degrees-of-freedom or c-coordinates).

$$u_{CT}(x,t) = \sum_{i=1}^{N} \varphi_i(x) c_i(t) \tag{3}$$

where φ_i (i=1,2,...N) is the analytical shape function of the beam. Different analytical shape functions are used according to the boundary conditions of the beam.
Like the FEM, the CEM can be refined using the h-refinement technique by increasing the number of finite elements. Moreover, it can also be refined through the c-refinement method, by increasing the number of shape functions. Here, we apply the c-refinement from the CEM, where the beam needs only to be discretized into one element. This will reduce the total number of degrees-of-freedom in the FEM.
The displacement field of the CEM for the Euler-Bernoulli beam element can be written from Equations (1) to (3) as

$$u_{CEM}(x,t) = S(x)Q(t) \tag{4}$$

where $S(x) = [N_1(x), N_2(x), N_3(x), N_4(x), \phi_1(x), \phi_2(x), ..., \phi_N(x)]$ is the generalized shape function of the CEM, $Q(t) = [v_1(t), \theta_1(t), v_2(t), \theta_2(t), c_1(t), c_2(t), ..., c_N(t)]^T$ is the vector of generalized displacements, and N is the number of shape functions used from the classical theory.

2.2 Vibration analysis for stepped beams without crack

Figure 1 shows the sketch of a beam with n steps, the height of the beam $d(x)$ with n step changes in cross section is expressed as

$$d(x) = \begin{cases} d_1 & 0 \le x < L_1 \\ d_2 & L_1 \le x \le L_2 \\ & \vdots \\ d_n & L_{n-1} \le x \le L_n \end{cases} \tag{5}$$

It is assumed that the beam has aligned neutral axis, the flexibility of the beam $EI(x)$ can be expressed as

$$EI(x) = \begin{cases} \dfrac{wd_1^3}{12} & 0 \le x < L_1 \\ \dfrac{wd_2^3}{12} & L_1 \le x \le L_2 \\ & \vdots \\ \dfrac{wd_n^3}{12} & L_{n-1} \le x \le L_n \end{cases} \tag{6}$$

where w is the width of the beam. For the stepped beam with misaligned neutral axes, the expression of $EI(x)$ can not expressed simply as shown in Equation (6).
The beam mass per unit length is

$$m(x) = \begin{cases} \rho w d_1 & 0 \le x < L_1 \\ \rho w d_2 & L_1 \le x \le L_2 \\ & \vdots \\ \rho w d_n & L_{n-1} \le x \le L_n \end{cases} \tag{7}$$

where ρ is the mass density of the beam.
The elemental stiffness matrix of the stepped beam can be obtained from the following equation

$$\mathbf{K_e} = \int_0^L \frac{d^2 S^T}{dx^2} EI(x) \frac{d^2 S}{dx^2} dx = \begin{bmatrix} [k_{qq}] & [k_{qc}] \\ [k_{cq}] & [k_{cc}] \end{bmatrix} \tag{8}$$

where the submatrix $[k_{qq}]$ corresponds to the element stiffness matrix from the FEM for the stepped beam; the submatrix $[k_{qc}]$ corresponds to the coupling terms of the q-dofs and the c-dofs; submatrix $[k_{cq}]$ is a transpose matrix of $[k_{qc}]$, and the submatrix $[k_{cc}]$ corresponds to the c-dofs and is a diagonal matrix.
The consistent elemental mass matrix can be expressed as

$$\mathbf{M_e} = \int_0^L S(x)^T m(x) S(x) dx = \begin{bmatrix} [m_{qq}] & [m_{qc}] \\ [m_{cq}] & [m_{cc}] \end{bmatrix} \tag{9}$$

where the submatrix $[m_{qq}]$ corresponds to the elemental mass matrix from the FEM for the stepped beam; the submatrix $[m_{qc}]$ corresponds to the coupling terms of the q-dofs and the c-dofs; submatrix $[m_{cq}]$ is a transpose matrix of $[m_{qc}]$, and the submatrix $[m_{cc}]$ corresponds to the c-dofs and is a diagonal matrix.

After introducing the boundary conditions, this can be performed by setting the associated degrees-of-freedom in the systematic stiffness matrix $\mathbf{K}$ to be a large number, say, 10^{12}, the governing equation for free vibration of the beam can be expressed as

$$(\mathbf{K}-\omega^2\mathbf{M})V=0 \tag{10}$$

where $\mathbf{K}$ and $\mathbf{M}$ are system stiffness and mass matrices, respectively, ω is the circular frequency, from which and the natural frequencies are identified. The ith normalized mode shapes of the stepped beam can be expressed as

$$\Psi_i=\sum_{i=1}^{4}N_iV_i+\sum_{i=1}^{N}\varphi_iV_{i+4} \tag{11}$$

The equation of motion of the forced vibration of the beam with n steps when expressed in terms of the composite element method is

$$\mathbf{M}\ddot{Q}+\mathbf{C}\dot{Q}+\mathbf{K}Q=f(t) \tag{12}$$

where $\mathbf{M}$ and $\mathbf{K}$ are the system mass and stiffness matrices, which are the same as those shown in Equation (10), $\mathbf{C}$ is the damping matrix which represents a Rayleigh damping model,say, $\mathbf{C}=a_1\mathbf{M}+a_2\mathbf{K}$, a_1 and a_2 are constants to be determined from two modal damping ratios. For an external force *F(t)* acting at the location x_F from the left support, the generalized force vector $f(t)$ can be expressed as

$$f(t)=\begin{bmatrix}N_1(x_F) & N_2(x_F) & N_3(x_F) & N_4(x_F) & \phi_1(x_F) & \cdots & \phi_n(x_F)\end{bmatrix}^T F(t) \tag{13}$$

The generalized acceleration $\ddot{Q}$, velocity $\dot{Q}$ and displacement Q of the stepped beam can be obtained from Equation (12) by direct integration. The physical acceleration $\ddot{u}(x,t)$ is obtained from

$$\ddot{u}(x,t)=[S(x)]^T\ddot{Q} \tag{14}$$

The physical velocity and displacement can be obtained in a similar way, i.e.

$$\dot{u}(x,t)=[S(x)]^T\dot{Q}\,, \tag{15a}$$

$$u(x,t)=[S(x)]^T Q \tag{15b}$$

2.3 The crack model

Numerous crack models for a cracked beam can be found in the literature. The simplest one is a reduced stiffness (or increased flexibility) in a finite element to simulate a small crack in the element (Pandey et al., 1991; Pandy & Biswas, 1994). Another simple approach is to divide the cracked beam into two beam segments joined by a rotational spring that

represents the cracked section (Rizos et al., 1990; Chaudhari & Maiti, 2000). Christides and Barr (1984) developed the one-dimensional vibration theory for the lateral vibration of a cracked Euler-Bernoulli beam with one or more pairs of symmetric cracks.
According to Christides and Barr(1984), the variation of bending stiffness $EI_d(x)$ along the cracked beam length takes up the form of

$$EI_d(x) = \frac{EI_0}{1+(c-1)\exp(-2\alpha|x-x_c|/d)} \tag{16}$$

where E is the Young's modulus of the beam, $I_0 = wd^3/12$ is the second moment of area of the intact beam, $c = 1/(1-C_r)^3$, $C_r = d_c/d$ is the crack depth ratio and d_c and d are the depth of crack and the beam, respectively, x_c is the location of the crack. α is a constant which governs the rate of decay and it is estimated by Christides and Barr from experiments to be 0.667. According to Lu and Law (2009), this parameter needs to be adjusted to be 1.426.

2.4 Vibration analysis for beams with crack(s)

The elemental stiffness matrix of the cracked beam can be obtained from the following equation

$$\mathbf{K_e} = \int_0^L \frac{d^2S^T}{dx^2} EI_d(x) \frac{d^2S}{dx^2} dx = \begin{bmatrix} [k_{qq}] & [k_{qc}] \\ [k_{cq}] & [k_{cc}] \end{bmatrix} \tag{17}$$

It is assumed that the existence of crack does not affect the elemental mass matrix, the elemental mass matrix can be expressed in the similar way with the intact beam

$$\mathbf{M_e} = \int_0^L S(x)^T m(x) S(x) dx = \begin{bmatrix} [m_{qq}] & [m_{qc}] \\ [m_{cq}] & [m_{cc}] \end{bmatrix} \tag{18}$$

The equation of motion of the forced vibration of a cracked beam with n cracks when expressed in terms of the composite element method is

$$\mathbf{M}\ddot{Q} + \mathbf{C}\dot{Q} + \mathbf{K}(x_{L_1}, d_{c_1}, \ldots, x_{L_i}, d_{c_i}, \ldots x_{L_n}, d_{c_n})Q = f(t) \tag{19}$$

3. Applications Information

3.1 Free and forced vibration analysis for beam without crack

3.1.1 Free vibration analysis for a free-free beam with a single step

The free vibration of the free-free beam studied in Koplow et al. (2006) is restudied using the CEM and the results are compared with those in Koplow et al. Figure 2 shows the geometry of the beam under study. The material has a mass density of $\rho = 2830 kg/m^3$, and a Young's modulus of $E = 71.7GPa$. In the CEM when 350 numbers of c-dofs are used, the first three natural frequencies are converged. The first three natural frequencies of the beam are 291.9Hz, 1176.2Hz and 1795.7Hz, respectively. The calculated natural frequencies from the CEM are very close to the experimental values in Koplow et al. when the test is measured at location A in Figure 2, which are 291Hz, 1165Hz and 1771Hz, respectively. The relative errors between the CEM and the experimental values of the three natural frequencies are 0.31%, 0.96% and 1.39%, respectively. This shows the proposed method is accuracte.

3.1.2 Free vibration analysis for a cantilever beam with a several steps

The cantilever beam studied in Jaworski and Dowell (2008) is restudied to further check the accuracy and effectiveness of the proposed method. Figure 3 shows the dimensions of the beam under study. The parameters of the beam under study are: $E = 60.6GPa$ and $\rho = 2664kg/m^3$. In the CEM model of the beam, the beam is discretized into one element and 350 terms of c-dofs are used in the calculation. The first and second flapwise (out-of-plane) bending mode frequencies are calculated to be 10.758 Hz and 67.553 Hz, and the first chordwise (in-plane) bending mode frequency is 54.699 Hz. The results from the CEM agree well with the theoretical results in Jaworski and Dowell using Euler-Bernoulli theory, as shown in Table 1.

3.1.3 Forced vibration analysis for a cantilever beam with two steps

In this section, the forced vibration analysis for the stepped beam is investigated. The dynamic responses of the beam under external force are obtained from the CEM and the results are compared with those from the FEM. Figure 4 shows the cantilever beam under study. The parameters of the beam under study are $E = 69.6GPa$ and $\rho = 2700kg/m^3$. A sinusoidal external force is assumed to act at free end of the beam with a magnitude of 1 N and at a frequency of 10 Hz. The time step is 0.005 second in calculating the dynamic response. The Rayleigh damping model is adopted in the calculation with 0.01 and 0.02 as the first two modal damping ratios. In the CEM model, the beam is discretized into one element and 350 c-dofs are used in the calculation of the dynamic responses. Figure 5 shows the displacement response, velocity response and acceleration response at the free end of the beam. In order to check the accuracy of the responses from the CEM, a forced vibration analysis for the beam is conducted using the FEM. The beam is discretized into 90 Euler-Bernoulli beam elements with a total of 182 dofs. The corresponding responses from the FEM and the CEM are compared in Figure 5. This indicates the accuracy of the proposed method for forced vibration of multiple stepped beam. Figure 6 gives a close view between the responses from two methods. From this figure, one can see that the two time histories in every subplot are virtually coincident indicating the excellent agreement between the time histories.

3.2 Free and forced vibration analysis for beam with crack

3.2.1 Free vibration analysis for a uniform cantilever beam with a single crack

An experimental work in Sinha et al. (2002) is re-examined. The geometric parameters of the beam are: length 996mm, width 50mm, depth 25mm, material properties of the beam are: Young's modulus $E = 69.79GPa$, mass density $\rho = 2600kg/m^3$. The beam is discretized into one element and ten shape functions are used in the calculation with the total degrees-of-freedom in the CEM equals 14, while the total degrees-of-freedom in the finite element model is 34 for the beam in Sinha et al. The crack depth in the beams varies in three stages of 4mm, 8mm and 12mm. The comparison of predicted natural frequencies of the beam from the proposed model and those in Sinha et al. and the experimental results are shown in Table 2. The proposed model, in general, gives better results than the model in Sinha et al.

since the latter crack model is a linear approximation of the theoretical crack model of Christides and Barr.

3.2.2 Free vibration analysis for a cantilever beam with multiple cracks

The last beam above is studied again with a new crack introduced. The first crack is at 595mm from the left end with a fixed crack depth of 12mm while the second crack is at 800mm from the left end with the crack depth varying from 4mm to 12mm in step of 4mm. Table 3 gives the first five natural frequencies of the beam by CEM method and compares with those from Sinha et al. and the experimental measurement. The results from CEM are found closer to the experimental prediction than those in Sinha et al. The above comparisons show that the CEM approach of modeling a beam with crack(s) is accurate for the vibration analysis. A significant advantage of the model is the much lesser number of DOFs in the resulting finite element model of the structure.

3.2.3 Forced vibration analysis for a cracked simply supported beam

The forced vibration analysis for a simply supported cracked beam is conducted in this section. The effects of the presence of crack on the dynamic response of the beam is investigated. The parameters of the beam under study are taken as: Young's modulus $E = 28\,\text{GPa}$, width w=200mm, depth d=200mm, length L=8.0m, mass density $\rho = 2500 kg / m^3$. Two cases are investigated in the following.

Effect of crack depth on the dynamic response

An impulsive force is assumed to act at mid-span of the beam with a magnitude of 100*N*, the force starts to act on the beam from the beginning and lasts for 0.1 second. The time step is 0.002 *s* in calculating the dynamic response. Rayleigh damping model is adopted in the calculation with 0.01 and 0.02 as the first two modal damping ratios.

Figure 7 shows comparison on the acceleration response at the 1/4 span of the beam for different crack depth. The crack is assumed to be at the mid-span of the beam. From this figure, one can see that the crack depth has significant effect on the dynamic response of the beam.

Effect of crack location on the dynamic response

Figure 8 shows comparison on the acceleration response at the 1/4 span of the beam for different crack locations with a fixed crack depth $d_c / d = 0.3$. The crack is assumed to be at the 0.1 L, 0.2L, 0.3L, 0.4L, and 0.5L of the beam. From this figure, one can see that the response changes with the crack location.

These studies show that the effect of the crack on the dynamic response is significant, so it is feasible to identify crack from measured structural dynamic responses.

4. Conclusion

The composite element method is proposed for both free and forced vibration analyses of beams with multiple steps. As the composite beam element is of a one-element-one-member configuration, modeling with this type of element would not need to take into account the discontinuity between different parts of the beam. The accuracy of this new composite element has been compared satisfactorily with existing results. One advantage of the method proposed is that it can be extended easily to deal with beams consisting of an arbitrary number of non-uniform segments. Regarding the free and forced vibration analysis for cracked beam using composite element, modelling with this type of element would allow the automatic inclusion of interaction effect between adjacent local damages in the finite element model. The accuracy of the present method has been compared satisfactory with existing model and experimental results.

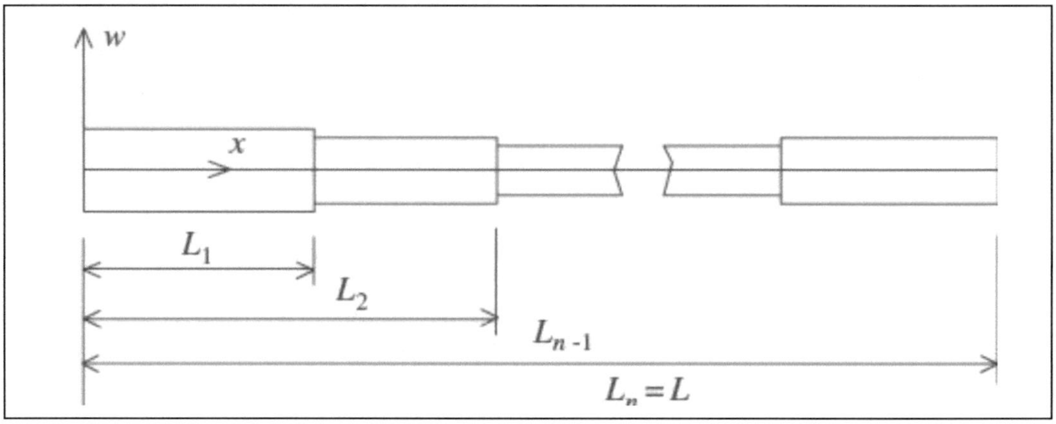

Fig. 1. Sketch of the stepped free-free beam with n segments

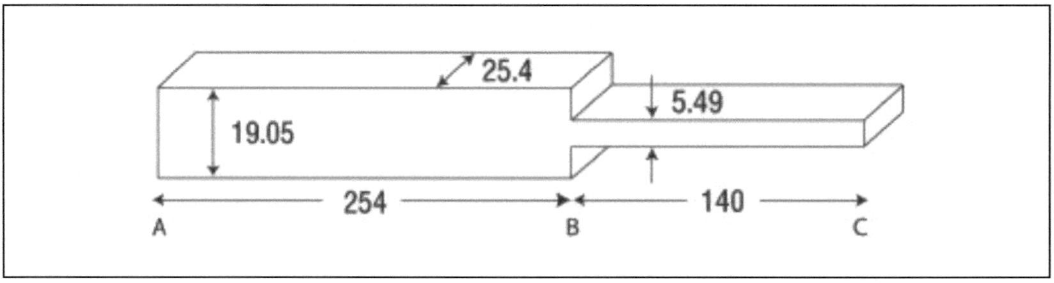

Fig. 2. Sketch of the stepped free-free beam in Koplow et al. (2006). Dimension in millimetre

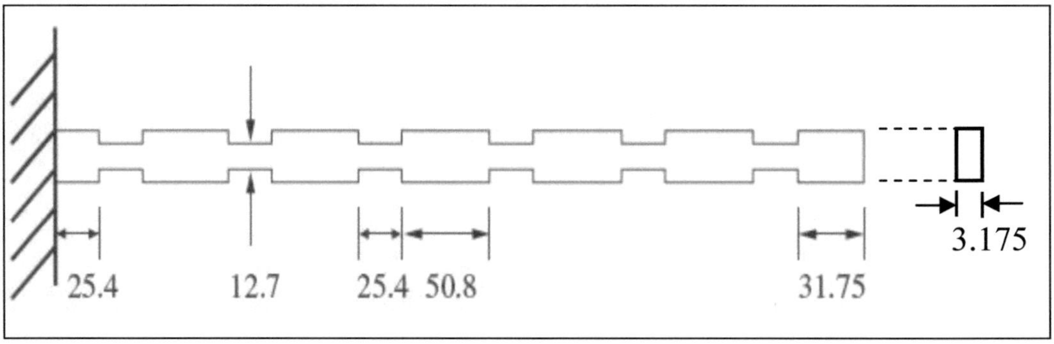

Fig. 3. Cantilever beam in Jaworski and Dowell (2008) with up and down steps. Dimension in millimeter

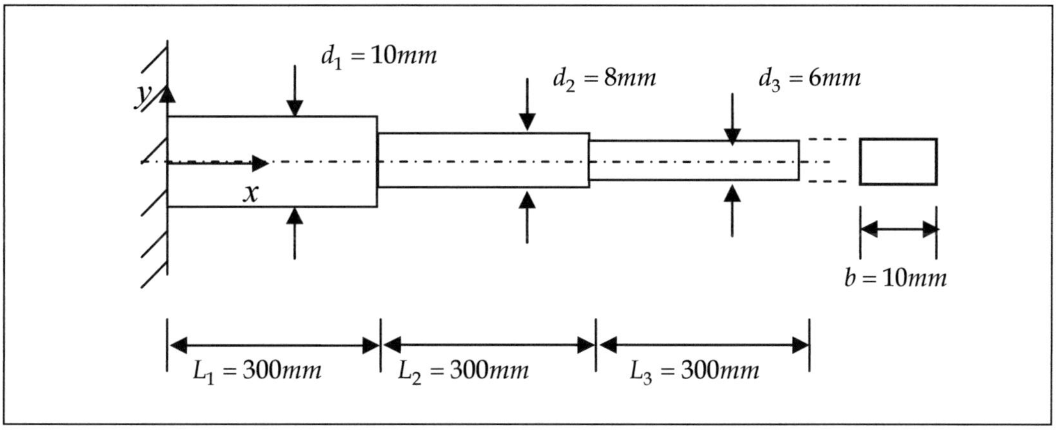

Fig. 4. Sketch of the stepped cantilever beam (dimensions are not scaled)

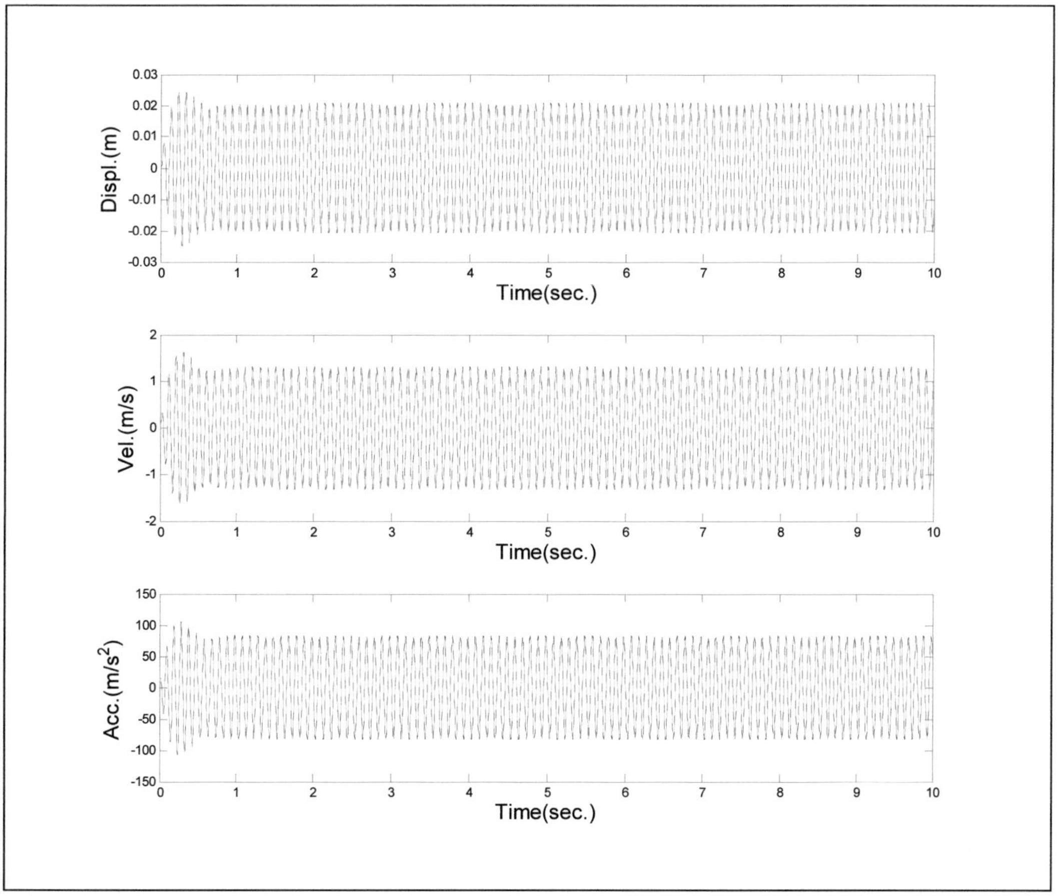

Fig. 5. Forced vibration dynamic response comparison between the CEM and FEM(- Solid: CEM; -- Dashed: FEM)

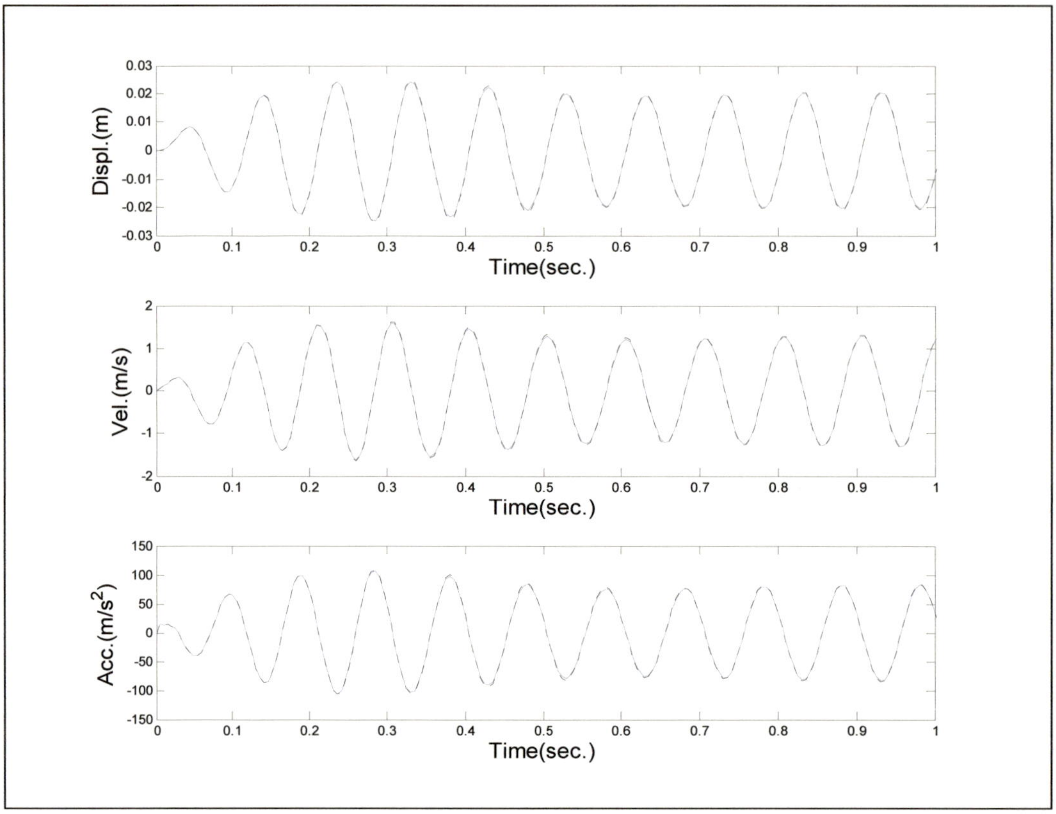

Fig. 6. Forced vibration dynamic response comparison between the CEM and FEM (a close view; - Solid: CEM, -- Dashed: FEM)

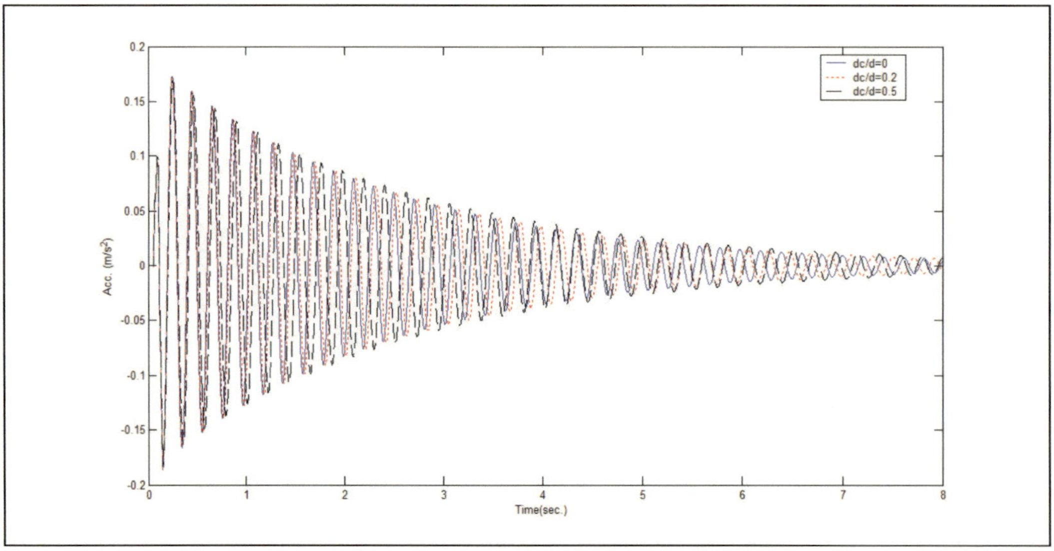

Fig. 7. Comparison on the dynamic responses for different crack depth

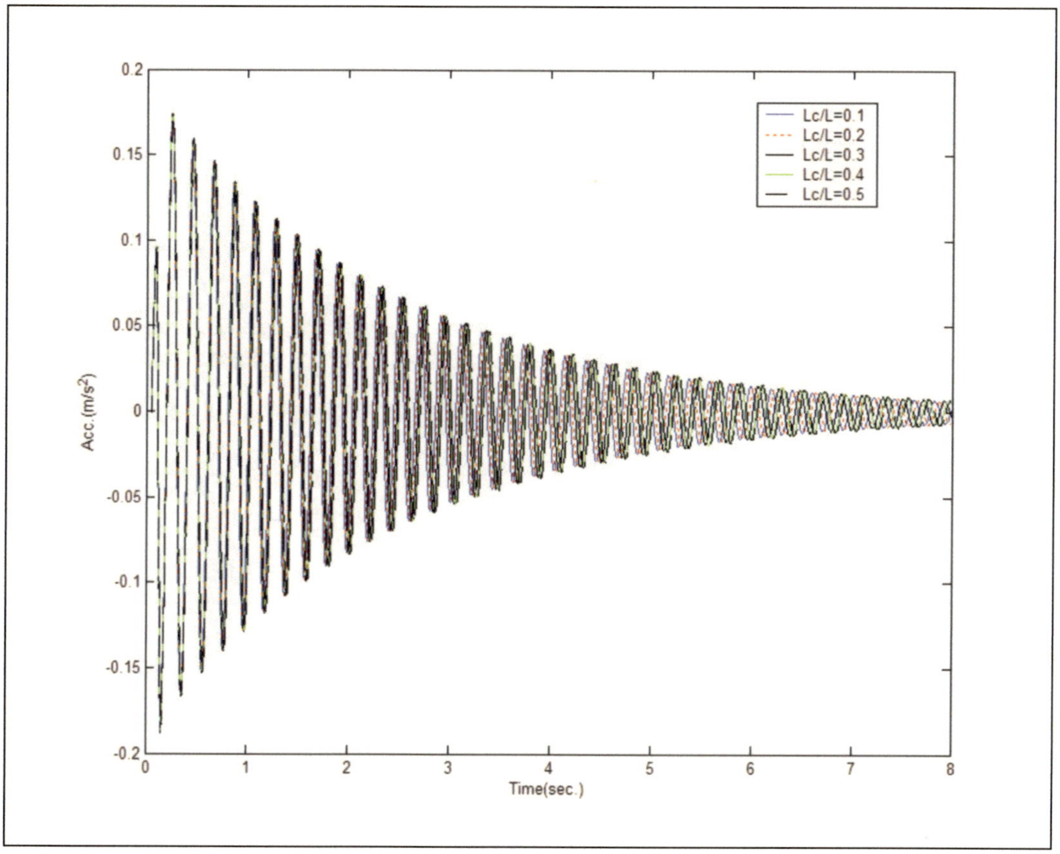

Fig. 8. Comparison on the dynamic responses for different crack location

Mode	Present	Jaworski and Dowell (2008)						
	CEM	Rayleigh-Ritz	CMA	ANSYS				Experiment
	Euler	Euler	Euler	Euler	Timoshenko	2D Shell	3D Solid	
ω_{1B}	10.758	10.752	10.816	10.775	10.745	10.44	10.46	10.63
ω_{2B}	67.553	67.429	67.463	67.469	67.456	65.54	65.70	66.75
ω_{1C}	54.699	54.795	54.985	54.469	54.429	49.62	49.83	49.38

Table 1. Natural frequencies [Hz] comparison for the stepped beam in Jaworski and Dowell (2008)

Note: ω_{1B} , ω_{2B} are the first and second out-of-plane bending mode frequencies, respectively.

ω_{1c} denotes the first in-plane bending mode frequency.

CMA represents component modal analysis.

Mode	No crack		$d_{c1} = 4$ mm at $x_1 = 595$ mm			$d_{c1} = 8$ mm at $x_1 = 595$ mm			$d_{c1} = 12$ mm at $x_1 = 595$ mm		
	Exp.	Proposed	Exp.	Sinha et al.(2002)	Proposed	Exp.	Sinha et al.(2002)	Proposed	Exp.	Sinha et al.(2002)	Proposed
1	40.000	39.770	39.688	39.379	39.490	39.375	39.094	39.242	39.063	38.857	38.869
2	109.688	109.340	109.063	108.206	108.633	108.125	107.132	107.670	105.938	106.278	106.293
3	215.000	214.795	215.000	214.087	214.230	214.688	213.825	213.986	214.375	213.622	213.631
4	355.000	354.853	354.688	353.107	353.683	353.438	351.872	352.524	350.625	350.881	350.921
5	528.750	529.601	527.188	524.696	526.540	522.812	520.452	522.448	513.125	517.219	517.003

Table 2. Comparison of natural frequencies (Hz) of the aluminium free-free beam with one crack in Sinha et al.(2002)

Mode	No crack		$d_{c1} = 12$ mm at $x_1 = 595$ mm $d_{c2} = 4$ mm at $x_2 = 800$ mm			$d_{c1} = 12$ mm at $x_1 = 595$ mm $d_{c2} = 8$ mm at $x_2 = 800$ mm			$d_{c1} = 12$ mm at $x_1 = 595$ mm $d_{c2} = 12$ mm at $x_2 = 800$ mm		
	Exp.	Proposed	Exp.	Sinha et al.(2002)	Proposed	Exp.	Sinha et al.(2002)	Proposed	Exp.	Sinha et al.(2002)	Proposed
1	40.000	39.770	38.750	38.352	38.607	38.437	37.897	38.246	37.500	37.513	37.703
2	109.688	109.340	105.938	105.890	106.196	105.938	105.510	106.062	105.625	105.559	105.858
3	215.000	214.795	213.750	212.207	212.786	212.813	210.897	211.643	210.000	209.815	209.975
4	355.000	354.853	350.000	348.920	349.843	349.063	347.235	348.410	345.625	345.876	346.374
5	528.750	529.601	512.500	514.575	514.735	511.250	512.903	513.044	507.500	510.560	510.633

Table 3. Comparison of natural frequencies (Hz) of the aluminium free-free beam with two cracks in Sinha et al.(2002)

5. References

Balasubramanian T.S. & Subramanian G. (1985). On the performance of a four-degree-of-freedom per node element for stepped beam analysis and higher frequency estimation. *Journal of Sound and Vibration,* Vol.99, No.4, 563–567, ISSN: 0022-460X.

Bapat C.N & Bapat C. (1987). Natural frequencies of a beam with non-classical boundary conditions and concentrated masses. *Journal of Sound and Vibration,* Vol.112, No.1, 177–182, ISSN: 0022-460X.

Chaudhari T.D & Maiti S.K. (2000). A study of vibration geometrically segmented beams with and without crack. *International Journal of Solids and Structures,* Vol. 37, 761-779, ISSN: 0020-7683.

Christides A. & Barr A.D.S. (1984). One dimensional theory of cracked Bernoulli-Euler beams. *International Journal of Mechanical Science,* Vol. 26, 639-648, ISSN: 0020-7403.

De Rosa M.A. (1994). Free vibration of stepped beams with elastic ends. *Journal of Sound and Vibration,* Vol.173, No.4, 557–563, ISSN: 0022-460X.

Jang S.K. & Bert C.W. (1989). Free vibrations of stepped beams: exact and numerical solutions. *Journal of Sound and Vibration,* Vol. 130, No.2, 342–346, ISSN: 0022-460X.

Jaworski J.W. & Dowell E.H. (2008). Free vibration of a cantilevered beam with multiple steps: Comparison of several theoretical methods with experiment. *Journal of Sound and Vibration,* Vol.312, No. 4-5, 713-725, ISSN: 0022-460X.

Lee J. & Bergman L.A. (1994). Vibration of stepped beams and rectangular plates by an elemental dynamic flexibility method. *Journal of Sound and Vibration,* Vol. 171, No.5, 617–640, ISSN: 0022-460X.

Lu Z.R. & Law S.S. (2009). Dynamic condition assessment of a cracked beam with the composite element model. *Mechanical Systems and Signal Processing,* Vol. 23, No. 3, 415-431, ISSN: 0888-3270.

Michael A. Koplow, Abhijit Bhattacharyya and Brian P. Mann. (2006). Closed form solutions for the dynamic response of Euler–Bernoulli beams with step changes in cross section. *Journal of Sound and Vibration,* Vol.295, No.1-2, 214-225, ISSN: 0022-460X.

Pandey A.K., Biswas M. & Samman M.M. (1991). Damage detection from change in curvature mode shapes. *Journal of Sound and Vibration,* Vol.145, 321-332, ISSN: 0022-460X.

Pandey A.K & Biswas M. (1994). Damage detection in structures using change in flexibility. *Journal of Sound and Vibration,* Vol.169, 3-17, ISSN: 0022-460X.

Rizos P.F; Aspragathos N. & Dimarogonas A.D. (1990). Identification of crack location and magnitude in a cantilever beam. *Journal of Sound and Vibration,* Vol.138, 381-388, ISSN: 0022-460X.

Sinha J.K; Friswell M.I & Edwards S. (2002). Simplified models for the location of cracks in beam structures using measured vibration data. *Journal of Sound and Vibration,* Vol. 251, No.1, 13-38, ISSN: 0022-460X.

Zeng P. (1998). Composite element method for vibration analysis of structure, Part II: C^1 Element (Beam). *Journal of Sound and Vibration*, Vol. 218, No.4, 658-696, ISSN: 0022-460X.

3

Free Vibration Analysis of Curved Sandwich Beams: A Dynamic Finite Element

Seyed M. Hashemi and Ernest J. Adique
Ryerson University
Canada

1. Introduction

Applications of sandwich construction and composites continue to expand. They are used in a number of industries such as the aerospace, automotive, marine and even sports equipment. Sandwich construction offers designers high strength to weight ratios, as well as good buckling resistance, formability to complex shapes and easy reparability, which are of extremely high importance in aerospace applications. Due to their many advantages over traditional aerospace materials, the analysis of sandwich beams has been investigated by a large number of authors for more than four decades. Sandwich construction can also offer energy and vibration damping when a visco-elastic core layer is used. However, such non-conservative systems are not the focus of the present study.

The most common sandwich structure is composed of two thin face sheets with a thicker lightweight, low-stiffness core. Common materials used for the face layers are metals and composite while the core is often made of foam or a honeycomb structure made of metal. It is very important that the core, although weaker than the face layers, be strong enough to resist crushing. The current trend in the aerospace industry of using composites and sandwich material, to lighten aircraft in an attempt to make them more fuel efficient, has led to further recent researches on development of reliable methods to predict the vibration behaviour of sandwich structures.

In the late 1960s, pioneering works in the field of vibration analysis of viscously damped sandwich beams (Di Taranto, 1965, and Mead and Marcus, 1968) used classical methods to solve the governing differential equations of motion, leading to the natural frequencies and mode shapes of the system. Ahmed (1971) applied the finite element method (FEM) to a curved sandwich beam with an elastic core and performed a comparative study of several different formulations in order to compare their performances in determining the natural frequencies and mode shapes for various different beam configurations. Interest in the vibration behaviour of sandwich beams has seen resurgence in the past decade with the availability of more powerful computing systems. This has allowed for more complex finite element models to be developed. Sainsbury and Zhang (1999), Baber et al. (1998), and Fasana and Marchesiello (2001) are just some among many researchers who investigated FEM application in the analysis of visco-elastically damped sandwich beams. The Dynamic Stiffness Method (DSM), which employs symbolic computation to combine all the governing differential equations of motion into a single ordinary differential equation, has also been well established. Banerjee and his co-workers (1995-2007) and Howson and Zare (2005) have published numerous papers on DSM illustrating its successful application to numerous homogeneous and sandwich/composite beam configurations, with a number of papers focusing on elastic-core sandwich beams. It is worth noting that in all the above-mentioned sandwich element models, the beam motion is assumed to exhibit coupled bending-axial motion only, with no torsional or out-of-plane motion. Also, the layers are assumed to be perfectly and rigidly joined together and the interaction of the different materials at the interfaces is ignored. Although it is known that bonding such very much different materials will cause stress at the interfaces, the study of their interactions and behaviour at the bonding site is another research topic altogether and is beyond the scope of the present Chapter.

Another important factor that largely affects the results of the sandwich beam analysis is the assumed vibration behaviour of the layers. The simplest sandwich beam model utilizes Euler-Bernoulli theory for the face layers and only allows the core to deform only in shear. This assumption has been widely used in several DSM and FEM studies such as those by Banerjee (2003), Ahmed (1971,1972), Mead and Markus (1968), Fasana and Marchesiello (2001), Baber et al. (1998), and in earlier papers by the authors; see e.g., Adique & Hashemi (2007), and Hashemi & Adique (2009). In more recent publications, Banerjee derived two new DSM models which exploit more complex displacement fields. In the first and simpler of the two (Banerjee & Sobey, 2005), the core bending is governed by Timoshenko beam theory, whereas the face plates are modeled as Rayleigh beams. To the authors' best knowledge, the most comprehensive sandwich beam theory was developed and used by Banerjee et al. (2007), where all three layers are modeled as Timoshenko beams. However, increasing the complexity of the model also significantly increases the amount of numerical and symbolic computation in order to achieve the complete formulation.

Classical FEM method has a proven track record and is the most commonly used method for structural analysis. It is a systematic approach, leading to element stiffness and mass matrices, easily adaptable to a wide range of problems. The polynomial shape functions are used to approximate the displacement fields, resulting in a linear eigenproblem, whose solutions are the natural frequencies of the system. Most commercial FEM-based structural analysis software also offer multi-layered elements that can be used to model layered composite materials and sandwich construction (e.g., ANSYS® and MSC NASTRAN/PATRAN®). As a numerical formulation, however, the versatility of the FEM theory comes with a drawback; the accuracy of its results depends on the number of elements used in the model. This is the most evident when FEM is used to evaluate system behaviour at higher frequencies, where a large number of elements are needed to achieve accurate results.

Dynamic Stiffness Matrix (DSM) method, on the other hand, provides an analytical solution to the free vibration problem, achieved by combining the coupled governing differential equations of motion of the system into a single higher order ordinary differential equation. Enforcing the boundary conditions then leads to the system's DSM and the most general closed form solution is then sought. The DSM formulation results in a non-linear eigenvalue problem and the bi-section method, combined with the root counting algorithm developed by Wittrick & Williams (1971), is then used as a solution technique. DSM provides exact results (i.e., closed form solution) for any of the natural frequencies of the beam, or beam-structure, with the use of a single continuous element characterized by an infinite number of degrees of freedom. However, the DSM methods is limited to special cases, for which the closed form solution of the governing differential equation is known; e.g., systems with constant geometric and material properties and only a certain number of boundary conditions.

The Dynamic Finite Element (DFE) method is a hybrid formulation that blends the well-established classical FEM with the DSM theory in order to achieve a model that possesses all the best traits of both methods, while trying to minimize the effects of their limitations; i.e., to fuse the adaptability of classical FEM with the accuracy of DSM. Therefore, the approximation space is defined using frequency dependent trigonometric basis functions to obtain the appropriate interpolation functions with constant parameters over the length of the element. DFE theory was first developed by Hashemi (1998), and its application has ever since been extended by him and his coworkers to the vibration analysis of intact (Hashemi et al.,1999, and Hashemi & Richard, 2000a,b) and defective homogeneous (Hashemi et al., 2008), sandwich (Adique & Hashemi, 2007-2009, and Hashemi & Adique, 2009, 2010) and laminated composite beam configurations (Hashemi & Borneman, 2005, 2004, and Hashemi & Roach, 2008a,b) exhibiting diverse geometric and material couplings. DFE follows a very similar procedure as FEM by first applying the weighted residual method to the differential equations of motion. Next, the element stiffness matrices are derived by discretizing the integral form of the equations of motion. For FEM, the polynomial interpolation functions are used to express the field variables, which in turn are introduced into the integral form of the equations of motion and the integrations are carried out and evaluated in order to obtain the element matrices. At this point, DFE applies an additional set of integration by parts to the element equations, introduces the Dynamic Trigonometric Shape Functions (DTSFs), and then carries out the integrations to form the element matrices. In the case of a three-layered sandwich beam, the closed form solutions to the uncoupled parts of the equations of motion are used as the basis functions of the approximation space to develop the DTSFs. The assembly of the global stiffness matrix from the element matrices follows the same procedure for FEM, DSM and DFE methods. Like DSM, the DFE results in a non-linear eigenvalue problem, however, unlike DSM, it is not limited to uniform/stepped geometry and can be readily extended to beam configurations with variable material and geometric parameters; see e.g., Hashemi (1998).

In the this Chapter, we derive a DFE formulation for the free vibration analysis of curved sandwich beams and test it against FEM and DSM to show that DFE is another viable tool for structural vibration analysis. The face layers are assumed to behave according to Euler-Bernoulli theory and the core deforms in shear only, as was also studied by Ahmed (1971,1972). The authors have previously developed DFE models for two straight, 3-layered, sandwich beam configurations; a symmetric sandwich beam, where the face layers are assumed to follow Euler-Bernoulli theory and core is allowed to deform in shear only (Adique & Hashemi, 2007, and Hashemi & Adique, 2009), and a more general non-symmetric model, where the core layer of the beam behaves according to Timoshenko theory while the faces adhere to Rayleigh beam theory (Adique & Hashemi, 2008, 2009). The latter model not only can analyze sandwich beams, where all three layers possess widely different material and geometric properties, but also it has shown to be a quasi-exact formulation (Hashemi & Adique, 2010) when the core is made of a soft material.

2. Mathematical model

Figure 1 below shows the notation and corresponding coordinate system used for a symmetrical curved three-layered sandwich beam with a length of S and radius R at the mid-plane of the beam. The thicknesses of the inner and outer face layers are t while the thickness of the core is represented by t_c. In the coordinate system shown, the z-axis is the normal co-ordinate measured from the centre of each layer and the y-axis is the circumferential coordinate and coincides with the centreline of the beam. The beam only deflects in the y-z plane. The top and bottom faces, in this case, are modelled as Euler-Bernoulli beams, while the core is assumed to have only shear rigidity (e.g., the stresses in the core in the longitudinal direction are zero). The centreline displacements of layers 1 and 3 are v_1 and v_2, respectively. The main focus of the model is flexural vibration, w, and is common among all three layers, which leads to the assumption $v_1 = -v_2 = -v$.

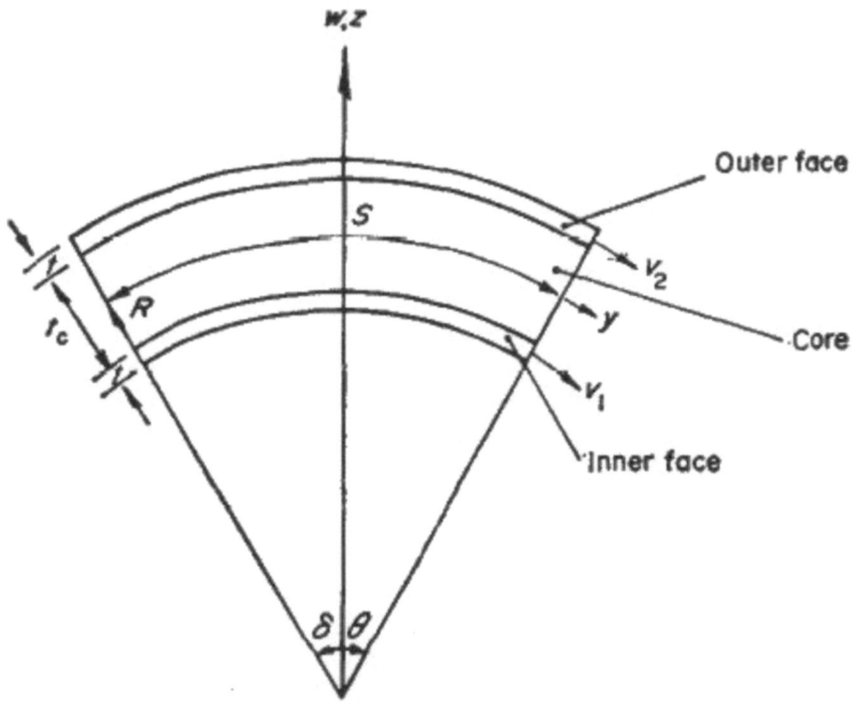

Fig. 1. Coordinate system and notation for curved symmetric three-layered sandwich beams

For the beam model developed, the following assumptions made (Ahmed, 1971):

- All displacements and strains are so small that the theory of linear elasticity still applies.
- The face materials are homogeneous and elastic, while the core material is assumed to be homogeneous, orthotropic and rigid in the z-direction.
- The transverse displacement w does not vary throughout the thickness of the beam.
- The shear within the faces is negligible.
- The bending strain within the core is negligible.
- There is no slippage or delamination between the layers during deformation.

Using the model and assumptions described above, Ahmed (1971) used the principle of minimum potential energy to obtain the differential equations of motion and corresponding boundary conditions. For free vibration analysis, the assumption of simple harmonic motion is used, leading to the following form of the differential equations of motion for a curved symmetrical sandwich beam (Ahmed, 1971):

$$\frac{\partial^2 v}{\partial y^2}+\frac{1}{\alpha^2}(\omega^2 Q_1-4\beta^2)v-\frac{2h\beta^2}{\alpha^2}\frac{\partial w}{\partial y}=0, \tag{1}$$

$$\frac{\partial^4 w}{\partial y^4}-\frac{h^2\beta^2}{\gamma^2}\frac{\partial^2 w}{\partial y^2}+\frac{1}{\gamma^2}(\frac{\alpha^2}{R^2}-\omega^2 Q_2)w-\frac{2h\beta^2}{\gamma^2}\frac{\partial v}{\partial y}=0, \tag{2}$$

where

$$\begin{aligned}&\alpha^2=2Et, \quad \beta^2=(1/t_c+t_c/4R^2)G_c, \quad \gamma^2=Et^3/6,\\ &h=t+t_c, \quad Q_1=2t\rho_f+t_c\rho_c/3, \quad Q_2=2t\rho_f+t_c\rho_c.\end{aligned} \tag{3}$$

In the equations above, $v(y)$ and $w(y)$ are the amplitudes of the sinusoidally varying circumferential and radial displacements, respectively. E is the Young's modulus of the face layers, G_c is the shear modulus of the core layer, and ρ and ρ_c are the mass densities of the face and core materials, respectively. The appropriate boundary conditions are imposed at $y=0$ and $y=S$. For example, for

- clamped at $y = 0$ and $y = S$; $v = w = \partial w/\partial y = 0$.
- simply supported at $y = 0$ and $y = S$; $\partial v/\partial y = w = \partial^2 w/\partial y^2 = 0$.
- cantilever configuration; at $y = 0$: $v=w=\partial w/\partial y=0$; and at $y=S$: $\partial v/\partial y=\partial^2 w/\partial y^2=0$ and a resultant force term of $[2\gamma^2\partial^3 w/\partial y^3+2\beta^2 h(2v+h\partial w/\partial y)]=0$, …

For harmonic oscillation, the weak form of the governing equations (1) and (2) are obtained by applying a Galerkin-type integral formulation, based on the weighted-residual method. The method involves the use of integration by parts on different elements of the governing differential equations and then the discretization of the beam length into a number of two-node beam elements (Figure 2).

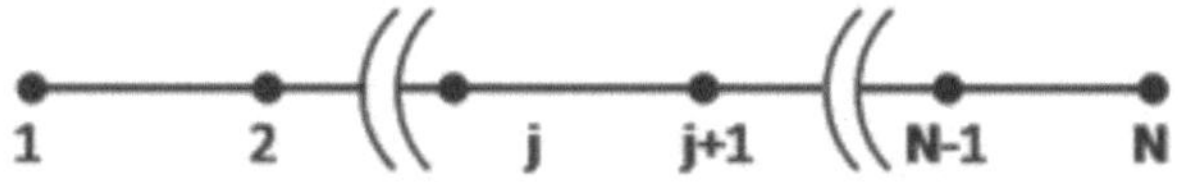

Fig. 2. Domain discretized by N number of 2-noded elements

Applying the appropriate number of integration by parts to the governing equations and discretization lead to the following form (in the equations below, primes denote integration with respect to y):

$$W_v^k=\int_0^l \delta v'\alpha^2 v' dy-\int_0^l \delta v(\omega^2 Q_1-4\beta^2)v dy+\int_0^l \delta v 2h\beta^2 w' dy \tag{4}$$

$$W_w^k=\int_0^l \delta w''\gamma^2 w'' dy+\int_0^l \delta w' h^2\beta^2 w' dy+\int_0^l \delta w(\alpha^2/R^2-\omega^2 Q_2)w dy+\int_0^l \delta w' 2h\beta^2 v dy \tag{5}$$

All of the resulting global boundary terms produced by integration by parts before discretization in the equations above are equal to zero. The above equations are known as the *element* Galerkin-type weak form associated to the discretized equations (4) and (5) and also satisfy the principle of virtual work:

$$W = W_{INT} - W_{EXT} = (W_v + W_w) - W_{EXT} = 0 \tag{6}$$

For the free vibration analysis, $W_{EXT} = 0$, and

$$W_{INT} = \sum_{k=1}^{\text{Number of Elements}} W^k; \text{ where } W^k = W_v^k + W_w^k \tag{7}$$

In the equations above, δv and δw are the test- or weighting -functions, both defined in the same approximation spaces as v and w, respectively. Each element is defined by nodes j and $j+1$ with the corresponding co-ordinates ($l=x_{j+1}-x_j$). The admissibility condition for finite element approximation is controlled by the undiscretized forms of equations (4) and (5).

3. Finite elements method (FEM) derivations

Two different FEM models were derived for the curved beam model. The first one has three degrees of freedom (DOFs) per node and uses a linear approximation for the axial displacement and a Hermite type polynomial approximation for the bending displacement.

$$v(y) = < N(y)_v > \{v_j \;\; v_{j+1}\} = N_{1v}(y)v_j + N_{2v}(y)v_{j+1} \tag{8}$$

$$w(y) = < N(y)_w > \{w_j \;\; w'_j \;\; w_{j+1} \;\; w'_{j+1}\} = N_{1w}(y)w_j + N_{2w}(y)w'_j + N_{3w}(y)w_{j+1} + N_{4w}(y)w'_{j+1} \tag{9}$$

In the equations above, v_j, v_{j+1}, w_j and w_{j+1} are the nodal values at j and $j+1$ corresponding to the circumferential and radial displacements, respectively (these can be likened to the axial and flexural displacements for a straight beam). w_j' and w'_{j+1} represent the nodal values of the rate of change of the radial displacements with respect to x (which can be likened to the bending slope for a straight beam). The same approximations were also used for δv and δw, respectively. The first FEM formulation is achieved when the nodal approximations expressed by equations (8) and (9) are applied to simplify equations (4) and (5). Similar approximations are also used for the corresponding test functions, δv and δw, and the integrations are performed to arrive at the classical linear (in ω^2) eigenvalue problem as functions of constant mass and stiffness matrices, which can be solved using programs such as Matlab®.

In the second FEM model the number of DOFs per node is increased to four and Hermite-type polynomial approximations are used for both the axial and bending displacements.

$$v(y) = < N(y)_v > \{v_j \;\; v'_j \;\; v_{j+1} \;\; v'_{j+1}\} = N_{1v}(y)v_j + N_{2v}(y)v'_{j+1} + N_{3v}(y)v_{j+1} + N_{4v}(y)v'_{j+1} \tag{10}$$

$$w(y) = < N(y)_w > \{w_j \;\; w'_j \;\; w_{j+1} \;\; w'_{j+1}\} = N_{1w}(y)w_j + N_{2w}(y)w'_j + N_{3w}(y)w_{j+1} + N_{4w}(y)w'_{j+1} \tag{11}$$

In the equations above, v_j, v_{j+1}, w_j and w_{j+1} are the nodal values at j and $j+1$ corresponding to the circumferential and radial displacements, respectively. v_j', v'_{j+1}, w_j' and w'_{j+1} are the nodal values at j and $j+1$ for the rate of change with respect to y for the circumferential and radial displacements, respectively. The same approximations are also used for δv and δw. The second FEM formulation applies equations (10) and (11) to simplify equations (4) and (5) to produce the linear (in ω^2) eigenvalue problem as a function of constant mass and stiffness matrices, which can again be solved using programs such as Matlab®. For the current research, both FEM models were programmed using Matlab®.

4. Dynamic finite element (DFE) formulation

In order to obtain the DFE formulation, an additional set of integration by parts are applied to the element equations (4) and (5) leading to:

$$W_V^k = \underbrace{-\int_0^l (\delta v''\alpha^2 + \delta v\omega^2 Q_1) v dy}_{(*)} + \underbrace{\int_0^l \delta v(4\beta^2) v dy + [\delta v'\alpha^2 v]_0^l}_{[k]_V^k \text{ Uncoupled}} + \underbrace{\int_0^l (\delta v 2h\beta^2) w' dy}_{[k_{VW}]_{2\times4} \text{ Coupling}} \tag{12}$$

$$\begin{aligned} W_W^k = & \underbrace{\int_0^l (\delta w''''\gamma^2 - \delta w'' h^2\beta^2 + \delta w(\alpha^2 / R^2 - \omega^2 Q_1)) w dy}_{(**)} + \\ & \underbrace{[\delta w' h^2\beta^2 w]_0^l + [\delta w''\gamma^2 w']_0^l - [\delta w'''\gamma^2 w]_0^l}_{[k]_W^k \text{ Uncoupled}} + \underbrace{\int_0^l \delta w'(2h\beta^2) v dy}_{[k_{WV}]_{4\times2} \text{ Coupling}} \end{aligned} \tag{13}$$

Equation (12) and (13) are simply a different, yet equivalent, way of evaluating equations (4) and (5) at the element level. The follwing non-nodal approximations are defined

$$\delta v =< P(y) >_V \{\delta a\}; \quad v =< P(y) >_V \{a\}; \tag{14}$$

$$\delta w =< P(y) >_W \{\delta b\}; \quad w =< P(y) >_V \{b\}, \tag{15}$$

where $\{a\}$ and {b} are the generalized co-ordinates for v and w, respectively, with the basis functions of approximation space expressed as:

$$< P(y) >_V = \langle \cos(\varepsilon y) \quad \sin(\varepsilon y)/\varepsilon \rangle; \tag{16}$$

$$< P(y) >_W = \left\langle \cos(\sigma y) \quad \frac{\sin(\sigma y)}{\sigma} \quad \frac{\cosh(\tau y) - \cos(\sigma y)}{\sigma^2 + \tau^2} \quad \frac{\sinh(\tau y) - \sin(\sigma y)}{\sigma^3 + \tau^3} \right\rangle, \tag{17}$$

where ε, σ and τ (shown below) are calculated based on the characteristic equations (*) and (**) in expressions (12) and (13) being reduced to zero.

$$\varepsilon = \sqrt{\omega^2 Q_1 / \alpha^2}; \quad \sigma, \tau = \sqrt{\frac{h^2\beta^2 \pm \sqrt{(h^2\beta^2)^2 - 4\gamma^2(\alpha^2 / R^2 - \omega^2 Q_2)}}{2\gamma^2}} \tag{18}$$

The non-nodal approximations (14) and (15) are made for δv, v, δw and w so that the integral terms (*) and (**) in expressions (12) and (13) become zero. The former term has a 2nd-order characteristic equation of the form $A_1 D^2 + B_1 \omega^2 = 0$, whereas the latter one has a 4th-order characteristic equation of the form $A_2 D^4 - B_2 D^2 + C_2 \omega^2 = 0$. Solving (*) and (**) yields the solution to the uncoupled parts of (12) and (13), which are subsequently used as the dynamic basis functions of approximation space to derive the DTSFs. The nodal approximations for *element* variables, $v(y)$ and $w(y)$, are then written as:

$$v =< P(y) >_V [\mathrm{P_n}]_V^{-1} \{\mathrm{u_n}\}_V = < N(y) >_V \{v_1 \quad v_2\}; \tag{19}$$

$$w =< P(y) >_V [\mathrm{P_n}]_W^{-1} \{\mathrm{u_n}\}_W = < N(y) >_W \{w_1 \quad w'_1 \quad w_2 \quad w'_2\}; \tag{20}$$

where $<N(y)_v>$ and $<N(y)_w>$ are the dynamic (frequency-dependent), trigonometric, shape functions, DTSFs, of the approximation space. Similar expressions are also written for the weighting functions, $\delta v(y)$ and $\delta w(y)$. Substituting the above nodal approximations into (12) and (13) and carrying out the integrations and term evaluations leads to the following matrix form:

$$\mathrm{W}^{\mathrm{k}} = ([k]_V^k{}_{\text{Uncoupled}} + [k]_W^k{}_{\text{Uncoupled}} + [k]_V^k{}_{\text{Coupling}})\{u_n\} = [k(\omega)]^k\{u_n\} \tag{21}$$

where $[k(\omega)]^k$ represents the frequency-dependent *element* dynamic stiffness matrix for coupled bending-axial vibrations of a curved symmetric sandwich beam element k. The appendix provides a more in-depth description of the process used to obtain the element matrices. The standard assembly method is used to obtain the global equation:

$$\mathrm{W}= \sum_{k=1}^{\text{Number of Elements}} W^k =< \delta U > [k(\omega)]\{U\} = 0 \tag{22}$$

where $[k(\omega)]$ is the global, overall, dynamic Stiffness Matrix (DSM), and $\{U\}$ stands for the vector of global DOFs of the system.

Matlab® program was used in the calculation of the integral terms for the element dynamic stiffness matrix. It is worth noting that Matlab® performs the calculations using complex arithmetics and as a result some of the elements in the matrix $[K]^k{}_{Coupling}$ are complex. However, the resulting dynamic stiffness matrix $[k(\omega)]$ is real and symmetric, with the imaginary parts of each element being zero.

It should also be pointed out that in equation (12) an integral term containing " $\delta v(4\beta^2)vdy$ ", was purposely left out of (*). This term represents the effect of the shear from the core on the face layers (S_{CF}), and its inclusion in (*) would change the trigonometric basis functions to purely hyperbolic functions. This, in turn, makes it impossible to find the solution to the free vibration problem. However, above a given frequency, the excluded integral term can be included in the (*) term (using, e.g., an 'if' statement) without any convergence problems. For the test cases being studied here, the critical frequency is much higher than the range being studied. Therefore, the S_{CF} term is simply evaluated separately and using the originally proposed basis functions (16) and (17).

5. Numerical tests and results

The DFE is used to compute the natural frequencies and modes of curved symmetrical sandwich beams. The solution to the problem lies in finding the system eigenvalues (natural frequencies, ω), and eigenvectors (natural modes). A simple determinant search method is utilized to compute the natural frequencies of the system. The beam considered has a span of S = 0.7112 m, a radius of curvature of R = 4.225 m, with the top and bottom faces having thicknesses of t = 0.4572 mm, and a core thickness of t_c = 12.7 mm. The material properties of the face layers are: E = 68.9 GPa and ρ_f = 2680 kg/m³, while the core has properties of G_c = 82.68 MPa and ρ_c = 32.8 kg/m³.

5.1 Cantilever end conditions

The first test case investigates the natural frequencies of the beam described above, with cantilever end conditions. The DFE and FEM results (Table 1) are presented and compared with those reported by Ahmed (1971), obtained from a 10-element FEM model of 2-noded 8-DOFs beam elements. The model developed by Ahmed employs polynomial cubic Hermite shape functions for the approximation space of the field variables v, v', w and w'.

ω_n rad/s	FEM, 10-Elem. Ahmed, 1971	DFE 20-Elem.	DFE 30- Elem.	DFE 40-Elem.	FEM; 3-DOF/node		FEM; 4-DOF/node	
					20-Elem.	40-Elem.	20-Elem.	40-Elem.
ω_1	1124.69	1124.69	1121.93	1121.8	1121.67	1121.61	1121.61	1121.61
ω_2	1671.33	1678.87	1671.89	1668.37	1668.25	1665.67	1665.48	1664.98
ω_3	3430.62	3451.98	3420.38	3408.88	3420.32	3402.97	3402.41	3398.51
ω_4	5868.50	5901.80	5838.65	5817.10	5860.33	5811.82	5811.07	5799.69
ω_5	8664.51	8695.93	8600.30	8567.37	8659.42	8566.24	8564.74	8524.02

Table 1. Natural frequencies (rad/s) of a clamped- free curved symmetric sandwich beam

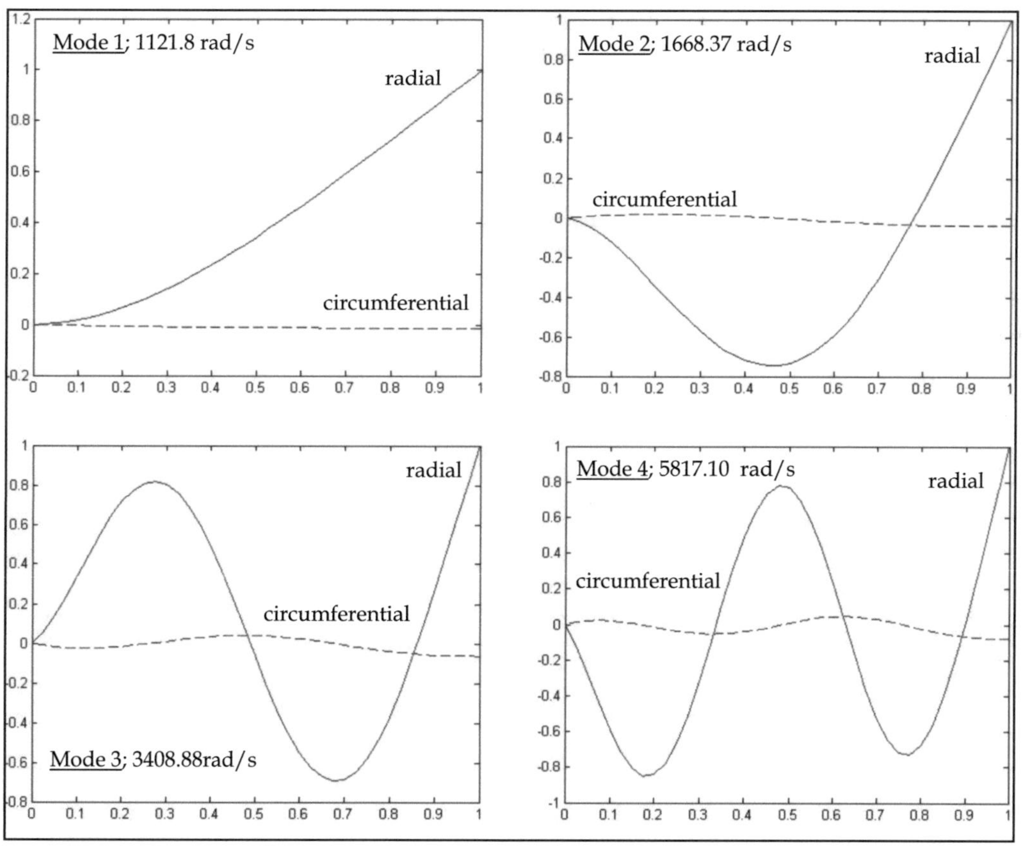

Fig. 3. First four ormalized modes for cantilever curved symmetric sandwich beam

The frequency results for the FEM and DFE models agree very well with one another with the maximum difference of 1.53% for the fifth natural frequency for 20-element models

when comparing the DFE and the 4-DOF FEM. For the 40-element models, the largest difference is 0.51% again for the fifth mode when comparing the DFE and 4-DOF FEM. Also, the first four normalized modes were computed using DFE model for the cantilevered curved sandwich beam and are shown in the Figure 3 below, generated using a 40-element DFE model. The curved beam has a large radius of curvature compared to its span, so the mode shapes of a straight beam can be used as a rough guideline to gauge the acceptability of the current modes. The frequency values used in the calculations of the mode shapes of the beam are 99.99% of the natural frequencies because the displacements cannot be evaluated as the true value of the natural frequency is approached.
As can be seen in Figures 3, all the mode shapes are dominated by radial displacements. This was expected as the bending stiffness of the beam is much smaller than its axial stiffness and the primary concern of the equations derived by Ahmed was to study the flexural behaviour of the beam (The undeformed shape of the beam was not included in the figures above because the beam's short length (0.7112) with respect to its large radius of curvature (4.225 m) would make the beam appear nearly straight).

5.2 Clamped-Clamped (C-C) end conditions

The next test case uses the same beam properties as the previous example, with clamped-clamped end conditions. The results of the DFE, and 3- and 4-DOF/node FEM formulations along with those reported by Ahmed (1971,1972) are listed in Table 2 below. For the first set of results from Ahmed (1971), shown in the second column of Table 2 below, each node has 4-DOFs. The 10-element FEM model developed employs similar polynomial Hermite shape functions such as those found in equations (10) and (11) for the approximation space of the field variables v, v', w and w', respectively. The results from Ahmed (1972), shown in the third column of Table 2, are from a 10-element FEM model where each node has 6-DOFs. The DOFs, in this case, are associated with circumferential displacement (v and v'), radial displacement (w and w') and transverse shear in the x-y plane (φ and φ', which none of the derived models takes into account). For each of the displacements, a Hermite polynomial shape function similar to expressions (10) and (11) was used to define the approximation space for both the field variables and weighting - or test - functions.

ω_n	FEM						DFE		
	10 Elements Ahmed, 1971, 1972		3-DOF		4-DOF		20 Elem.	30 Elem.	40 Elem.
	4-DOF	6-DOF	20-Elem	40-Elem	20-Elem	40-Elem			
ω_1	1658.76	1507.96	1653.73	1649.96	1649.84	1648.96	1665.67	1655.62	1652.23
ω_2	3279.82	2978.23	3272.97	3249.60	3250.92	3244.20	3295.53	3263.30	3252.30
ω_3	5585.75	5296.73	5563.57	5502.19	5508.34	5488.74	5580.10	5520.47	5499.99
ω_4	8243.54	7872.83	8208.29	8093.94	8107.70	8069.37	8203.96	8112.91	8081.62
ω_5	11102.4	10662.6	11054.8	10878.2	10900.1	10839.1	11020.1	10896.0	10853.0

Table 2. Natural frequencies (rad/s) of a clamped- clamped curved symmetric sandwich beam

Table 2 above, shows that for the first two natural frequencies, the DFE results are slightly larger than those obtained from both FEM formulations, but for the 3rd-5th frequencies, the DFE values are smaller than those found by the 3-DOF FEM formulation but larger than the

4-DOF FEM formulation. For 20-element FEM models, the largest difference is 1.4% seen between the 3-DOF and 4-DOF FEM formulations (in the 5th natural frequencies), but when the number of elements is increased to 40, the difference reduces to 0.36%, which is still the largest when comparing all three models.

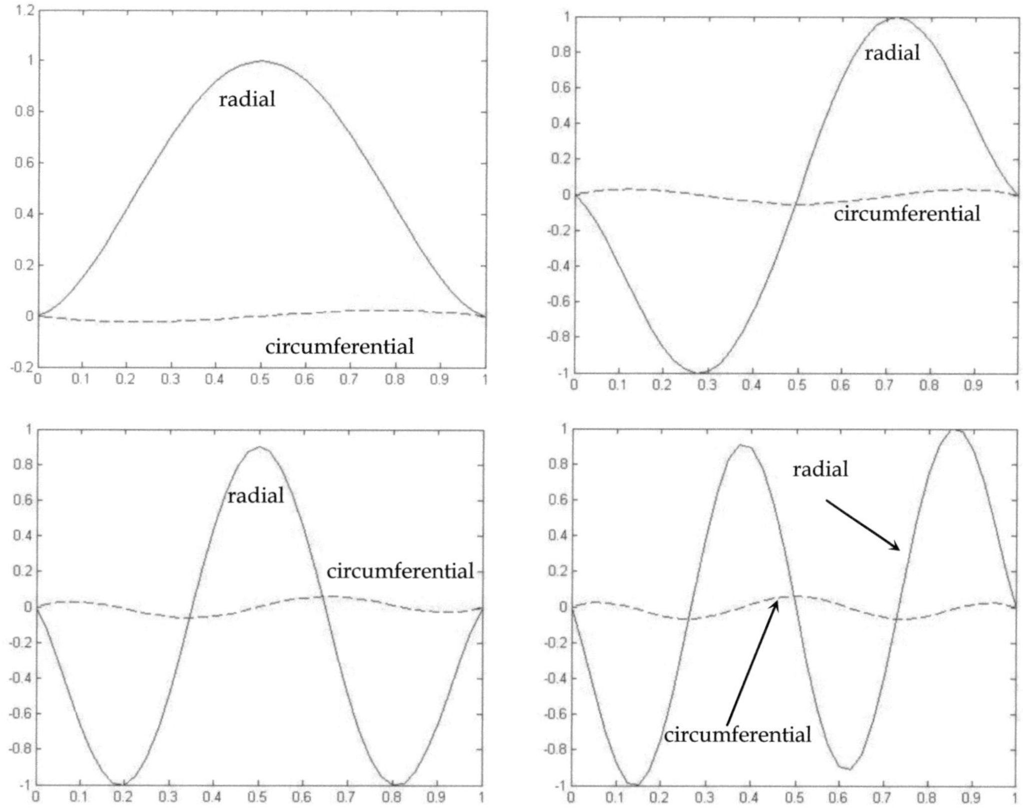

Fig. 4. First four ormalized modes for clamped-clamped curved symmetric sandwich beam

The largest difference when comparing the 40-element DFE and 3-DOF FEM models is 0.23% for the 5th natural frequency with the rest of the error being smaller. When comparing the 40-element DFE and 4-DOF FEM models, the largest error is 0.25% for the 2nd mode. The dramatic decrease in the discrepancies of the three models indicates that they are all converging to nearly the same values for the natural frequencies. When comparing the results to those of Ahmed, it can be seen that they agree very well with the 4-DOF model, although, they are smaller in value. The main reason for this is that Ahmed only used 10 elements and an increase in the number of elements used would give lower values. From Ahmed's results for the 6-DOF model, it can be seen that they are considerably lower than all the calculated values. When comparing the DFE to Ahmed's 6-DOF formulation, the largest differences can be seen for the first two natural frequencies with a difference of 9.56% and 9.20%, respectively. For the 3rd, 4th and 5th frequencies, the difference between the DFE and Ahmed 6-DOF formulation is 3.84%, 2.65% and 1.79%, respectively. Ahmed (1971) states that the difference in values is most likely due to the differences in formulations

between the two models. The equations of motion upon which the DFE is based on ignores the shear of the face layers and the bending and axial stiffness of the core while the 6-DOF formulation takes all of these factors into account.
The normalized natural modes of the curved sandwich beam, generated using a 40-element DFE model, are shown in Figures 4. As expected, the mode shapes for the curved symmetrical sandwich beam with clamped-clamped end conditions exhibit mainly radial displacement. Some circumferential displacement is also observed but is small when compared the magnitude of the radial displacement. This can be explained by the fact that the beam's axial stiffness is much higher than its bending stiffness. Also, the mode shapes conform to the clamped-clamped boundary conditions applied to the beam; the radial and circumferential displacements are zero at the end points, as is also the slope.

5.3 Simply supported-Simply supported (S-S) end conditions

The third numerical case uses the beam described earlier in the chapter with both ends simply supported. The DFE, 3- and 4-DOF FEM formulations are used to calculate the beam's natural frequencies and mode shapes. The results of these models are listed along with those reported by Ahmed (1971), obtained using a 10-element FEM model with 4-DOFs per node (see Table 3). The FEM model developed by Ahmed uses polynomial Hermite shape functions similar to equations (10) and (11) for the approximation space of the field variables v, v', w and w', respectively.
As can be seen from the 2nd row in Table 3, there is a good agreement between all the 20-element models, with the biggest discrepancy being between the DFE and the 4-DOF FEM formulations; the FEM 1st natural frequency is only 0.41% smaller than that obtained from the DFE. However, when the remaining frequencies are examined, the growing difference can be observed for the higher modes. When comparing the 20-element DFE and the 20-element 3-DOF FEM formulations, the largest difference is for the 2nd natural frequency, with the FEM value being 1.21% smaller than the DFE result. The difference between the DFE and 3-DOF FEM results decreases with increasing mode number.

ω_n	FEM					DFE		
	4DOF; 10-Elem. Ahmed, 1971	3DOF		4DOF				
		20-Elem.	40-Elem.	20-Elem.	40-Elem.	20-Elem.	30-Elem.	40-Elem.
ω_1	1253.5	1248.60	1248.34	1248.34	1248.34	1253.50	1250.35	1249.47
ω_2	2475.58	2471.74	2466.65	2464.89	2464.89	2501.96	2480.60	2472.87
ω_3	4687.26	4690.84	4669.22	4662.06	4662.06	4746.95	4697.94	4680.97
ω_4	7382.74	7405.49	7354.72	7337.82	7337.82	7478.88	7397.82	7370.11
ω_5	10298.1	10351.3	10261.4	10231.4	10231.4	10433.9	10318.9	10279.0

Table 3. Natural frequencies (rad/s) of a simply-supported curved symmetric sandwich beam

Increasing the number of elements from 20 to 40, reduces the difference between the two models for the 2nd frequency to 0.25% remaining the maximum and the difference for the other frequencies decreasing with the increase in mode number.
Comparing the 20-element DFE and the 4-DOF FEM models, the trend is reversed; the two values are closest for the 1st natural frequency and increase with the higher modes with the

largest difference being for the 5th frequency, where the FEM value is 1.94% smaller than that of the DFE. When the number of elements used in the model is increased to 40, the agreement between the two formulations becomes much better with the maximum relative error being 0.46% for the 5th frequency. Increasing the number of elements from 20 to 40 considerably reduces the relative error between all the models; i.e., convergence. For the 1st natural frequency, there is a perfect match between Ahmed's results and the 20-element DFE model. But with the increase in the mode number, the difference between the DFE and Ahmed's results grow to a maximum of 1.32% for the 5th natural frequency.

As seen in Table 3 above, increasing the number of elements in the DFE to 40 reduces the values of all the DFE frequencies lower than those reported by Ahmed; the maximum difference is now in the 1st mode, with the DFE frequency 0.32% smaller than the value reported by Ahmed. Although increasing the number of elements seems to have gone in the opposite direction of what it was intended, it should be noted that Ahmed (1971) only used 10 elements in the reported FEM results and based on the trend observed, increasing the number of elements will lower the values of the frequencies, better matching the DFE results.

Using the 40-element DFE model, the mode shapes are calculated and illustrated in Figures 5 below. The mode shapes were found using values 99.99% of the actual natural frequencies

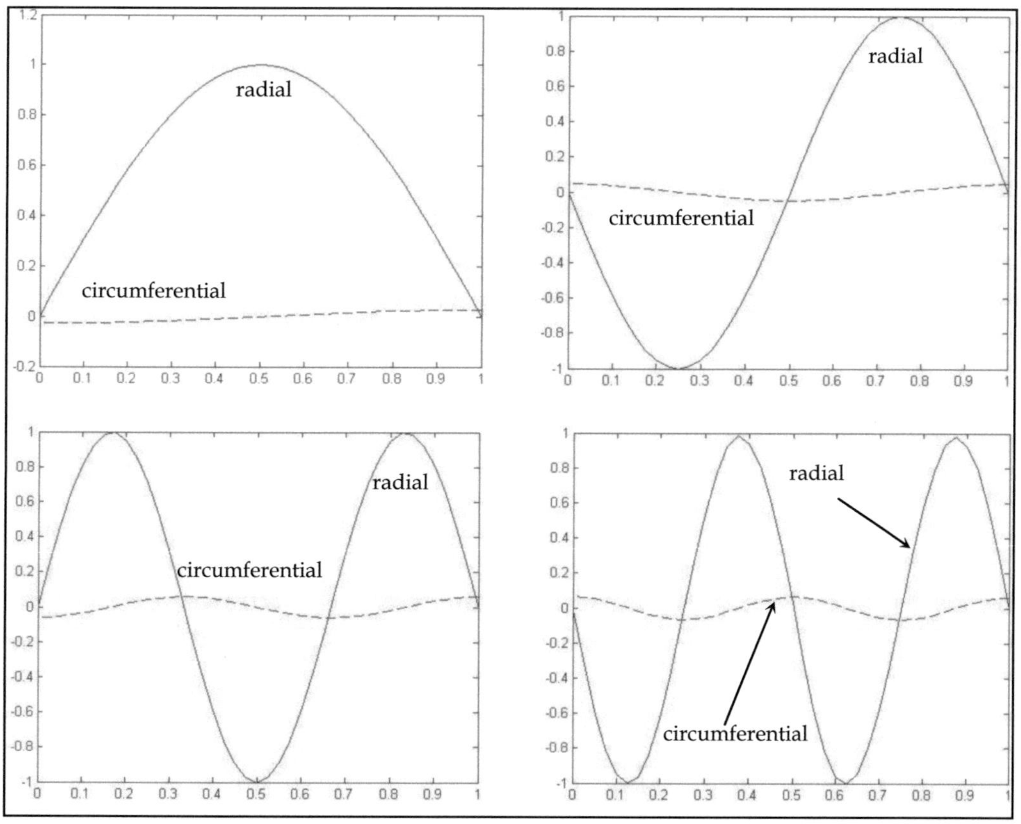

Fig. 5. First four normalized modes for clamped-clamped curved symmetric sandwich beam

of the system because displacements of the system become impossible to evaluate at the values near the natural frequencies. As can be seen from Figures 5, the mode shapes for the curved symmetric sandwich beam with simply supported end conditions are dominated by radial displacement which is the expected result due to the beam's high axial stiffness in comparison to its bending stiffness. It is worth noting that at the end points some axial displacement is observed. This is in accordance with the fact that for the simply supported end condition, the circumferential displacement is not forced to zero, giving the possibility of a non-zero value for displacement at the end points.

5.4 Simply-Supported (S-S) straight symmetric sandwich beam

In the final numerical test, the curved symmetrical sandwich beam formulation is applied to a straight beam case. The beam has a length of S = 0.9144 m, radius $R = \infty$, with face thickness t = 0.4572 mm and core thickness t_c = 12.7 mm. The mechanical properties of the face layers are: E = 68.9 GPa and ρ_f = 2680 kg/m^3, while the core has properties of G_c = 82.68 MPa and ρ_c = 32.8 kg/m^3. The natural frequencies of the beam are calculated using the DFE method as well as the 3-DOF and 4-DOF FEM formulations and compared to the data published by Ahmed (1971) (see Table 4). In the case of a straight beam, the radial displacement and circumferential displacements directly translate into the flexural and axial displacements, respectively.

ω_n	FEM					DFE		
	Ahmed,1971 4DOF 10-Elem	3DOF		4DOF		20-Elem.	30-Elem.	40-Elem.
		20-Elem.	40-Elem.	20-Elem	40-Elem.			
ω_1	361.35	359.27	359.02	358.90	358.90	370.02	363.55	361.41
ω_3	2938.6	2940.5	2924.3	2918.9	2918.9	3012.4	2958.6	2952.72
ω_5	6980.6	7044.7	6966.0	6939.9	6939.8	7169.2	6993.5	6987.1
ω_7	11574.	11740.	11559.	11498.	11498.	11885.	11667	11591.
ω_9	16299.	16582.	16284.	16184.	16182.	16729.	16423.	16316.

Table 4. Natural frequencies (rad/s) of a simply-supported straight symmetric sandwich beam

6. Conclusion

Based on the theory developed by Ahmed (1971,1972) and the weak integral form of the differential equations of motion, a dynamic finite element (DFE) formulation for the free vibration analysis of symmetric curved sandwich beams has been developed. The DFE formulation models the face layer as Euler-Bernoulli beams and allows the core to deform in shear only. The DFE formulation is used to calculate the natural frequencies and mode shapes for four separate test cases. In the first three cases the same curved beam, with different end conditions, are used: cantilever, both ends clamped and lastly, both ends simply supported. The final test case used the DFE formulation to determine the natural frequencies of a simply supported straight sandwich beam.

All the numerical tests show satisfactory agreement between the results for the developed DFE, FEM and those published in literature. For all test studies, when a similar number of elements are used, the DFE matched more closely with the 3-DOF FEM formulation than with Ahmed's 4-DOF FEM results. The reason for this is that the DFE is derived from the 3-

DOF FEM formulation and such a trend is expected. Ahmed (1971) goes on to explain that the addition of an extra degree of freedom for each node has a tendency to lower the overall stiffness of a sandwich beam element causing an overall reduction in values of the natural frequencies. The mode shapes determined by the DFE formulation match the expectations based on previous knowledge on the behaviour of straight sandwich beams. The results of the DFE theory and methodology applied to the analysis of a curved symmetric sandwich beam demonstrate that DFE can be successfully extended from a straight beam case to produce a more general formulation. The proposed DFE is equally applicable to the piecewise uniform (i.e., stepped) configurations and beam-structures. It is also possible to further extend the DFE formulation to more complex configurations and to model geometric non-uniformity and material changes over the length of the beam.

7. Acknowledgement

The support provided by Natural Science and Engineering Research Council of Canada (NSERC), Ontario Graduate Scholarship (OGS) Program, and High Performance Computing Virtual Laboratory (HPCVL)/Sun Microsystems is also gratefully acknowledged.

8. Appendix: development of DFE Stiffness matrices for curved symmetric Euler-Bernoulli/Shear sandwich beam

The Dynamic Finite Element stiffness matrix for a symmetric curved sandwich beam is developed from equations (12) and (13) found in Section 4. Applying the approximations for the element variables, $v(y)$ and $w(y)$, and the test functions, $\delta v(y)$ and $\delta w(y)$, as shown in expressions (19) and (20) to *element* integral equations (12) and (13) yield the *element* DFE stiffness matrix defined in equation (21).

First, let us consider the *element* virtual work corresponding to the circumferential displacement, $v(y)$. Based on the governing differential equation (1), the critical value, or *changeover frequency*, is then determined from

$$\omega^2 Q_1 - 4\beta^2 = 0 \tag{A1}$$

For the frequencies below the *changeover frequency*, the element integral equation (12) can be expressed as:

$$W_V^k = \underbrace{-\int_0^l (\delta v''\alpha^2 + \delta v\omega^2 Q_1)v dy}_{(*)} + \underbrace{\int_0^l \delta v(4\beta^2)v dy + [\delta v'\alpha^2 v]_0^l}_{[k]_V^k \text{ Uncoupled}} + \underbrace{\int_0^l (\delta v 2h\beta^2)w' dy}_{[k_{VW}]_{2\times4} \text{ Coupling}} \qquad \text{(12 repeated)}$$

where the first integral term, (*) vanishes due to the choice of the trigonometric basis function for $v(y)$, as stated in:

$$< P(y) >_V = \langle \cos(\varepsilon y) \;\; \sin(\varepsilon y)/\varepsilon \rangle; \qquad \text{(16 repeated)}$$

The next two terms, produce a symmetric 2x2 matrix $[\ k\ ]_V^k$ that contains all the uncoupled stiffness matrix elements associated with the displacement $v(y)$. The inclusion of S_{CF} term in

(*) would make the solution to the corresponding characteristic equation (also used as basis functions of approximation space) change form trigonometric to purely hyperbolic functions. This, in turn, would lead to solution divergence of the DFE formulation, where natural frequencies of the system cannot be reached using the determinant search method.

For the test cases examined here, the *changeover frequency* for the faces is well above the range of frequencies being studied; therefore, the S_{CF} term, representing the shear effect from the core on the face layers, is kept out of the integral term (*) and evaluated as a part of the second term, $[\,k\,]_V{}^k$.

For the frequencies above the *changeover frequency*, the element integral equation can be re-written as:

$$W_V^k = \underbrace{-\int_0^l (\delta v''\alpha^2 + \delta v(\omega^2 Q_1 - 4\beta^2)v dy}_{(*)} + \underbrace{[\delta v'\alpha^2 v]_0^l}_{[k]_V^k \text{ Uncoupled}} + \underbrace{\int_0^l (\delta v 2h\beta^2)w' dy}_{[k_{VW}]_{2\times 4} \text{ Coupling}} \tag{A2}$$

where the S_{CF} term is included in the integral term (*), which vanishes due to the choice of purely trigonometric basis functions for *v(y)*, similar to (16). The next term, then produces a symmetric 2x2 matrix $[\,k\,]_V{}^k$ that contains all the uncoupled stiffness matrix elements associated with the displacement *v(y)* and the final term, produces a 2x4 matrix $[k_{VW}]$ that contain all the terms that couple the displacement *v(y)* with *w(y)*.

$$[k]_V^k = \begin{bmatrix} k_V(1,1) & k_V(1,2) \\ sym. & k_V(2,2) \end{bmatrix} \tag{A3}$$

$$[k_{VW}]^k = \begin{bmatrix} k_{VW}(1,1) & k_{VW}(1,2) & k_{VW}(1,3) & k_{VW}(1,4) \\ k_{VW}(2,1) & k_{VW}(2,2) & k_{VW}(2,3) & k_{VW}(2,4) \end{bmatrix} \tag{A4}$$

Now considering equations (13):

$$W_W^k = \underbrace{\int_0^l (\delta w''''\gamma^2 - \delta w'' h^2\beta^2 + \delta w(\alpha^2/R^2 - \omega^2 Q_1))w dy}_{(**)} +$$

$$\underbrace{[\delta w' h^2\beta^2 w]_0^l + [\delta w''\gamma^2 w']_0^l - [\delta w'''\gamma^2 w]_0^l}_{[k]_W^k \text{ Uncoupled}} + \underbrace{\int_0^l \delta w'(2h\beta^2)v dy}_{[k_{WV}]_{4\times 2} \text{ Coupling}} \tag{13 repeated}$$

The first integral term, (**), in equation (13), vanishes due to the choice of mixed trigonometric-hyperbolic basis functions for *w(y)*, similar to (17):

$$< P(y) >_W = \left\langle \cos(\sigma y) \quad \frac{\sin(\sigma y)}{\sigma} \quad \frac{\cosh(\tau y) - \cos(\sigma y)}{\sigma^2 + \tau^2} \quad \frac{\sinh(\tau y) - \sin(\sigma y)}{\sigma^3 + \tau^3} \right\rangle, \tag{17 repeated}$$

The next three terms, produce a symmetric 4x4 matrix $[k]_W{}^k$ that contain all the uncoupled stiffness matrix elements associated with the displacement *w(y)*. The final term, produces a 4x2 matrix $[k_{WV}]$ that contain all the terms that couple the displacement *w(y)* with *v(y)*. It is important to note that $[k_{WV}] = [k_{VW}]^T$.

$$[k]_W^k = \begin{bmatrix} k_W(1,1) & k_W(1,2) & k_W(1,3) & k_W(1,4) \\ & k_W(2,2) & k_W(2,3) & k_W(2,4) \\ & & k_W(3,3) & k_W(3,4) \\ sym. & & & k_W(4,4) \end{bmatrix} \tag{A5}$$

$$[k_{WV}]^k = \begin{bmatrix} k_{WV}(1,1) & k_{WV}(1,2) \\ k_{WV}(2,1) & k_{WV}(2,2) \\ k_{WV}(3,1) & k_{WV}(3,2) \\ k_{WV}(4,1) & k_{WV}(4,2) \end{bmatrix} \tag{A6}$$

Matrices (A3), (A4), (A5) and (A6) are added according to equation (21) in order to obtain the 6x6 element stiffness matrix for a symmetric straight sandwich beam.

$$[k]^k = \begin{bmatrix} k_V(1,1) & k_{VW}(1,1) & k_{VW}(1,2) & k_V(1,2) & k_{VW}(1,3) & k_{VW}(1,4) \\ & k_W(1,1) & k_W(1,2) & k_{WV}(1,2) & k_W(1,3) & k_W(1,4) \\ & & k_W(2,2) & k_{WV}(2,2) & k_W(2,3) & k_W(2,4) \\ & & & k_V(2,2) & k_W(2,3) & k_W(2,4) \\ & & & & k_W(3,3) & k_W(3,4) \\ sym. & & & & & k_W(4,4) \end{bmatrix} \tag{A7}$$

9. References

Adique E. & Hashemi S.M. (2007). Free Vibration of Sandwich Beams using the Dynamic Finite Element Method", in B.H.V. Topping, (Editor), "Proceedings of the Eleventh International Conference on Civil, Structural and Environmental Engineering Computing", Civil-Comp Press, Stirlingshire, UK, Paper 118, 2007. doi:10.4203/ccp.86.118, St. Julians, Malta. 18-21 Sept 2007.

Adique E. & Hashemi S.M. (2008). Dynamic Finite Element Formulation and Free Vibration Analysis of a Three-layered Sandwich Beam," Proceedings of The 7th Joint Canada-Japan Workshop on Composite Materials, July 28-31, 2008, Fujisawa, Kanagawa, Japan, pp. 93-100.

Adique E. & Hashemi S.M. (2009). A Super-Convergent Formulation for Dynamic Analysis of Soft-Core Sandwich Beams", in B.H.V. Topping, L.F. Costa Neves, R.C. Barros, (Editors), "Proceedings of the 12th International Conference on Civil, Structural and Environmental Engineering Computing", Civil-Comp Press, Stirlingshire, UK, Paper 98, 2009. doi:10.4203/ccp.91.98, Funchal, Madeira Island, 1-4 Sept. 2009.

Ahmed, K. M. (1971). Free vibration of curved sandwich beams by the method of finite elements. *Journal of Sound and Vibration,* Vol. 18, No. 1, (September 1971) 61-74, ISSN: 0022-460X.

Ahmed, K. M. (1972). Dynamic analysis of sandwich beams. *Journal of Sound and Vibration*, Vol. 21, No. 3, (April 1972) 263-276, ISSN: 0022-460X.

Baber, T.T.; Maddox, R.A. & Orozco, C.E. (1998). A finite element model for harmonically excited viscoelastic sandwich beams. *Computers &Structures*, Vol. 66, No. 1, (January 1998) 105-113, ISSN: 0045-7949.

Banerjee, J. R. (1999). Explicit frequency equation and mode shapes of a cantilever beam coupled in bending and torsion. *Journal of Sound and Vibration*, Vol. 224, No. 2, (July 1999) 267-281, ISSN: 0022-460X.

Banerjee, J. R. (2001). Explicit analytical expressions for frequency equation and mode shapes of composite beams. *International Journal of Solids and Structures*, Vol. 38, No. 14 (April 2001) 2415-2426, ISSN: 0045-7949.

Banerjee, J. R. (2001). Frequency equation and mode shape formulae for composite Timoshenko beams. *Composite Structures*, Vol. 51, No. 4, (May 2001) 381-388, ISSN: 0045-7949.

Banerjee, J. R. (2001). Dynamic stiffness formulation and free vibration analysis of centrifugally stiffened Timoshenko beams. *Journal of Sound and Vibration* , Vol. 247, No. 1, (October 2001) 97-115, ISSN: 0022-460X.

Banerjee, J. R. (2003). Free vibration of sandwich using the dynamic stiffness method. *Computers &Structures*, Vol. 81, No. 18-19 (August 2003) 1915-1922, ISSN: 0045-7949.

Banerjee, J. R.; Cheung, C. W.; Morishima, R.; Perera, M. & Njuguna, J. (2007). Free vibration of a three-layered sandwich beam using the dynamic stiffness method and experiment. *International Journal of Solids and Structures*, Vol. 44, No. 22-23 (November 2007) 7543-7563, ISSN: 0045-7949.

Banerjee, J. R. and Sobey, A. J. (2005). Dynamic stiffness formulation and free vibration analysis of a three-layered sandwich beam. *International Journal of Solids and Structures*, Vol. 42, No. 8, (2005) 2181-2197, ISSN: 0045-7949.

Banerjee, J. R. And Su, H. (2004). Development of a dynamic stiffness matrix for free vibration analysis of spinning beams. *Computers &Structures*, Vol. 82, No. 23-24 (September - October 2004) 2189-2197, ISSN: 0045-7949.

Banerjee, J. R. and Su, H. (2006). Dynamic stiffness formulation and free vibration analysis of a spinning composite beam. *Computers &Structures*, Vol. 84, No. 19-20, (July 2006) 1208-1214, ISSN: 0045-7949.

Banerjee, J. R. & Williams, F.W. (1996). Exact dynamic stiffness matrix for composite Timoshenko beams with applications. *Journal of Sound and Vibration*, Vol. 194, No. 4, (July 1996), 573-585, ISSN: 0022-460X.

Banerjee, J. R. & Williams, F.W. (1995). Free vibration of composite beams – an exact method using symbolic computation. *Journal of Aircraft* , Vol. 32, No. 3, (1995) 636-642, ISSN: 0021-8669.

Di Taranto, R. A. (1965). Theory of vibratory bending for elastic and viscoelastic layered finite length beams. *Journal of Applied Mechanics*, Vol. 87, (1965) 881-886, ISSN: 0021-8936 (Print), eISSN: 1528-9036.

Fasana, A. & Marchesiello, S. (2001). Rayleigh-Ritz analysis of sandwich beams. *Journal of Sound and Vibration*, Vol. 241, No. 4, 643-652, ISSN: 0022-460X.

Hashemi, S. M. (1998). *Free Vibration Analysis Of Rotating Beam-Like Structures: A Dynamic Finite Element Approach*. Ph.D. Dissertation, Department of Mechanical Engineering, Laval University, Québec, Canada.

Hashemi, S. M. (2002). The use of frequency dependent trigonometric shape functions in vibration analysis of beam structures - bridging the gap between FEM and exact DSM formulations. *Asian Journal of Civil Engineering*, Vol. 3, No. 3&4, (2002) 33-56, ISSN: 15630854.

Hashemi, S. M. & Adique, E.J. (2009). Free Vibration analysis of Sandwich Beams: A Dynamic Finite Element, *International Journal of Vehicle Structures & Systems (IJVSS),* Vol. 1, No 4, (November 2009) 59-65, ISSN: 0975-3060 (Print), 0975-3540 (Online).

Hashemi, S. M. & Adique, E.J (2010). A Quasi-Exact Dynamic Finite Element for Free Vibration Analysis of Sandwich Beams, *Applied Composite Materials,* Vol. 17, No. 2, (April 2010) 259-269, ISSN: 0929-189X (print version, 1573-4897 (electronic version), doi:10.1007/s10443-009-9109-3.

Hashemi, S. M. & Borneman, S. R. (2004). Vibration analysis of composite wings undergoing material and geometric couplings: a dynamic finite element formulation. *CD Proceedings of the 2004 ASME International Mechanical Engineering Congress (IMECE 2004,) Aerospace Division,* pp 1-7, November 2004, Anaheim, CA, USA.

Hashemi, S. M. and Borneman, S. R. (2005). A dynamic finite element formulation for the vibration analysis of laminated tapered composite beams. *CD Proceedings of the Sixth Canadian-International Composites Conference (CanCom),* pp. 1-13, August 2005, Vancouver, BC, Canada.

Hashemi, S. M.; Borneman, S. R. & Alighanbari, H. (2008). Vibration analysis of cracked composite beams: a dynamic finite element. *International Review of Aerospace Engineering (I.RE.AS.E.),* Vol. 1, No. 1, (February 2008) 110-121, ISSN: 1973-7459.

Hashemi, S. M.; Richard, M. J. & Dhatt, G. (1999). A new dynamic finite elements (DFE) formulation for lateral free vibrations of Euler-Bernoulli spinning beams using trigonometric shape functions. *Journal of Sound and Vibration,* Vol. 220, No. 4, (March 1999) 601-624, ISSN: 0022-460X.

Hashemi, S. M. & Richard, M. J. (2000a). Free vibration analysis of axially loaded bending-torsion coupled beams - a dynamic finite element (DFE). *Computers and Structures* , Vol. 77, No. 6, (August 2000) 711-724, ISSN: 0045-7949.

Hashemi, S. M. & Richard, M. J. (2000b). A dynamic finite element (DFE) for free vibrations of bending-torsion coupled beams. *Aerospace Science and Technology* , Vol. 4, No. 1, (January 2000) 41-55, ISSN: 1270-9638.

Hashemi, S. M. & Roach, A. (2008a). A dynamic finite element for coupled extensional-torsional vibrations of uniform composite thin-walled beams. *International Review of Aerospace Engineering (I.RE.AS.E.)* , Vol. 1, No. 2, (April 2008) 234-245, ISSN: 1973-7459.

Hashemi, S.M. & Roach, A. (2008b). Free vibration of helical springs using a dynamic finite element mesh reduction technique. *International Review of Mechanical Engineering,* Vol. 2, No. 3, (May 2008) 435-449 , ISSN: 1970 - 8734.

Howson, W. P. & Zare, A. (2005). Exact dynamic stiffness matrix for flexural vibration of three-layered sandwich beams. *Journal of Sound and Vibration*, Vol. 282, No. 3-5, (April 2005) 753-767, ISSN: 0022-460X.

Mead, D. J. and Markus, S. (1968). The forced vibration of a three-layer, damped sandwich beam with arbitrary boundary conditions. *Journal of Sound and Vibration*, Vol. 10, No. 2, (September 1968) 163-175, ISSN: 0022-460X.

Sainsbury, M. G. & Zhang, Q. J. (1999). The Galerkin element method applied to the vibration of damped sandwich beams. *Computers and Structures* , Vol. 71, No. 3, (May 1999) 239-256, ISSN: 0045-7949.

Wittrick, W. H. & Williams, F. W. (1971). A general algorithm for computing natural frequencies of elastic structures. *Quarterly Journal of Mechanics and Applied Mathematics*, Vol. 24, No. 3, (August 1971) 263-284, Online ISSN 1464-3855 - Print ISSN 0033-5614.

4

Some Complicating Effects in the Vibration of Composite Beams

Metin Aydogdu[1], Vedat Taskin[1], Tolga Aksencer[1], Pınar Aydan Demirhan[1] and Seckin Filiz[2]
[1]*Trakya University Department of Mechanical Engineering, Edirne,*
[2]*Trakya University Natural Sciences Institute, Edirne,*
Turkey

1. Introduction

In the last 50-60 years, use of composite structures in engineering applications has increased. Due to this fact many studies have been conducted related with composite structures (such as: shells, plates and beams).

Bending, buckling and free vibration analysis of composite structures has taken considerable attention. Beams are one of these structures that are used in mechanical, civil and aeronautical engineering applications (such: robot arms, helicopter rotors and mechanisms). Considering these applications free vibration problem of the composite beams are studied in the previous studies. Kapania & Raciti, 1989 investigated the nonlinear vibrations of un-symmetrically laminated composite beams. Chandashekhara et al., 1990 studied the free vibration of symmetric composite beams. Chandrashekhara & Bangera, 1993 investigated the free vibration of angle-ply composite beams by a higher-order shear deformation theory. They used the shear flexible finite element method. Krishnaswamy et al., 1992 solved the generally layered composite beam vibration problems. Chen et al., 2004 used the state-space based differential quadrature method to study the free vibration of generally laminated composite beams. Solution methods for composite beam vibration problems depend on the boundary conditions, some analytical (Chandrashekhara et al., 1990, Abramovich, 1992, Krishnaswamy et al., 1992, Abramovic & Livshits, 1994, Khdeir & Reddy, 1994, Eisenberger et al., 1995, Marur & Kant, 1996, Kant et al., 1998, Shi & Lam, 1999, Yıldırım et al., 1999, Yıldırım, 2000, Matsunaga, 2001, Kameswara et al., 2001, Banerjee, 2001, Chandrashekhara & Bangera, 1992, Ramtekkar et al., 2002, Murthy et al., 2005, Arya, 2003, Karama et al., 1998, Aydogdu, 2005, 2006) solution procedures have been used.

Many factors can affect the vibrations of beams, in particular the attached springs and masses, axial loads and dampers. This type of complicating effects is considered in the vibration problem of isotropic beams. Gürgöze and his collogues studied vibration of isotropic beam with attached mass, spring and dumpers (Gürgöze, 1986, Gürgöze, 1996, Gürgöze & Erol, 2004). Vibration of Euler-Bernoulli beam carrying two particles and several particles investigated by Naguleswaran, 2001, 2002. Nonlinear vibrations of beam-mass system with different boundary conditions are investigated by Ozkaya & Pakdemirli, 1999, Ozkaya et. al., 1997. They used multiscale perturbation technique in their solutions.

It is interesting to note that, although mass or spring attached composite beams are used or can be used in some engineering applications, their vibration problem is not generally considered in the previous studies. Vibration of symmetrically laminated clamped-free beam with a mass at the free end is studied by Chandrashekhara & Bangera, 1993.
The aim of present study is to fill this gap. Therefore in this study vibration of composite beams with attached mass or springs is studied. After driving equations of motion different boundary conditions, lamination angles, attached mass or spring are considered in detail.

2. Equation of motion

In this study, equations of motion of composite beams will be derived from Classical Laminated Plate Theory (CLPT). For CLPT following displacement field is generally assumed:

$$\begin{aligned} U(x,z;t) &= u(x,t) - zw_{,x} \\ V(x,z;t) &= v(x,t) - zw_{,y} \\ W(x,z;t) &= w(x,t) \end{aligned} \tag{1}$$

where U,V and W are displacement components of a point of the plate in the x, y and z directions respectively and u, v and w are the displacement components of a point of the beam in the midplane again in the x, y and z directions respectively. The comma after a letter denotes partial derivative with respect to x and y. The Hooke's law can be written in the following form using CLPT:

$$\begin{bmatrix} \sigma_x \\ \sigma_y \\ \tau_{xy} \end{bmatrix} = \begin{bmatrix} Q_{11} & Q_{12} & Q_{16} \\ Q_{21} & Q_{22} & Q_{26} \\ Q_{61} & Q_{62} & Q_{66} \end{bmatrix} \begin{bmatrix} \varepsilon_x \\ \varepsilon_y \\ \gamma_{xy} \end{bmatrix} \tag{2}$$

where σ_x and σ_y are the in-plane normal stress components in the x and y directions respectively, τ_{xy} is the shear stress in the x-y plane, ε_x, ε_y and γ_{xy} are normal strains and shear strain respectively and Q_{ij} are the reduced transformed rigidities (Jones, 1975). These strains are defined in the following form:

$$\varepsilon_x = \frac{\partial U}{\partial x}, \quad \varepsilon_y = \frac{\partial V}{\partial y} \qquad \gamma_{xy} = \frac{\partial U}{\partial y} + \frac{\partial V}{\partial x} \tag{3}$$

Applying Hamilton principle leads to the following equations of motion for laminated composite plate.

$$\begin{aligned} N_{x,x} + N_{xy,y} &= \rho u_{,tt} \\ N_{xy,x} + N_{y,y} &= \rho v_{,tt} \\ M_{x,xx} + 2M_{xy,xy} + M_{y,yy} &= \rho w_{,tt} \end{aligned} \tag{4}$$

where the force and moment resultants are defined in the following form.

$$(N_x, N_y, N_{xy}) = \int_{-h/2}^{h/2} (\sigma_x, \sigma_y, \tau_{xy}) dz \quad (5)$$

$$(M_x, M_y, M_{xy}) = \int_{-h/2}^{h/2} (\sigma_x, \sigma_y, \tau_{xy}) z dz \quad (6)$$

These force and moment results can also be written in the following form:

$$\begin{bmatrix} N_x \\ N_y \\ N_{xy} \\ M_x \\ M_y \\ M_{xy} \end{bmatrix} = \begin{bmatrix} A_{11} & A_{12} & A_{16} & B_{11} & B_{12} & B_{16} \\ A_{12} & A_{22} & A_{26} & B_{12} & B_{22} & B_{26} \\ A_{16} & A_{26} & A_{66} & B_{16} & B_{26} & B_{66} \\ B_{11} & B_{12} & B_{16} & D_{11} & D_{12} & D_{16} \\ B_{12} & B_{22} & B_{26} & D_{12} & D_{22} & D_{26} \\ B_{16} & B_{26} & B_{66} & D_{12} & D_{26} & D_{66} \end{bmatrix} \begin{bmatrix} u_{,x} \\ v_{,y} \\ u_{,x} + v_{,y} \\ -w_{,xx} \\ -w_{,yy} \\ -2w_{,xy} \end{bmatrix} \quad (7)$$

where extensional, coupling and bending rigidities are defined as follows:

$$\begin{aligned} A_{ij} &= \int_{-h/2}^{h/2} Q_{ij}^{(k)} dz \\ Bij &= \int_{-h/2}^{h/2} Q_{ij}^{(k)} z dz \\ Dij &= \int_{-h/2}^{h/2} Q_{ij}^{(k)} z^2 dz \end{aligned} \quad (8)$$

Now, consider a laminated composite beam with length L, width b and thickness h. Equations of motion of laminated composite beams can be derived from Eq.(4) assuming $N_y=N_{xy}=M_y=M_{xy}=0$.

$$\begin{aligned} N_{x,x} &= \rho u_{,tt} \\ M_{x,xx} &= \rho w_{,tt} \end{aligned} \quad (9)$$

Eq.(7) can be inverted in the following form:

$$\begin{bmatrix} u_{,x} \\ \varepsilon_y \\ \gamma_{xy} \\ \kappa_x \\ \kappa_y \\ \kappa_{xy} \end{bmatrix} = \begin{bmatrix} A^*_{11} & A^*_{12} & A^*_{16} & B^*_{11} & B^*_{12} & B^*_{16} \\ A^*_{12} & A^*_{22} & A^*_{26} & B^*_{12} & B^*_{22} & B^*_{26} \\ A^*_{16} & A^*_{26} & A^*_{66} & B^*_{16} & B^*_{26} & B^*_{66} \\ B^*_{11} & B^*_{12} & B^*_{16} & D^*_{11} & D^*_{12} & D^*_{16} \\ B^*_{12} & B^*_{22} & B^*_{26} & D^*_{12} & D^*_{22} & D^*_{26} \\ B^*_{16} & B^*_{26} & B^*_{66} & D^*_{16} & D^*_{26} & D^*_{66} \end{bmatrix} \begin{bmatrix} N_x \\ 0 \\ 0 \\ M_x \\ 0 \\ 0 \end{bmatrix} \quad (10)$$

where A_{ij}^* ,B_{ij}^* ,D_{ij}^* are the members of inverse of rigidity matrix given in Eq.(7). Eq.(10) can be written in the following form.

$$u_{,x} = A_{11}^* N_x + B_{11}^* M_x$$
$$w_{,xx} = B_{11}^* N_x + D_{11}^* M_x \tag{11}$$

Eq.(11) can be solved in term of N_x and M_x.

$$M_x = \frac{B_{11}^*}{[(B_{11}^*)^2 - (A_{11}^* D_{11}^*)]} u_{,x} + \frac{A_{11}^*}{[(B_{11}^*)^2 - (A_{11}^* D_{11}^*)]} w_{,xx} \tag{12}$$

$$N_x = \frac{B_{11}^* u_{,x} - B_{11}^{*\,2} [\frac{B_{11}^*}{[(B_{11}^*)^2 - (A_{11}^* D_{11}^*)]} u_{,x} + \frac{A_{11}^*}{[(B_{11}^*)^2 - (A_{11}^* D_{11}^*)]} w_{,xx}]}{A_{11}^* B_{11}^*} \tag{13}$$

Inserting equations (12)-(13) in equation (9) yields to:

$$\bar{A}\frac{\partial^2 u}{\partial x^2} + \bar{B}\frac{\partial^3 w}{\partial x^3} = \rho \frac{\partial^2 u}{\partial t^2}$$
$$\bar{D}\frac{\partial^4 w}{\partial x^4} + \bar{B}\frac{\partial^3 u}{\partial x^3} = -\rho \frac{\partial^2 w}{\partial t^2} \tag{14}$$

Where $\bar{A}$, $\bar{B}$ and $\bar{D}$ are defined in the following form.

$$\bar{A} = \frac{1}{A_{11}^*} - \frac{(B_{11}^*)^2}{[(B_{11}^*)^2 - (A_{11}^* D_{11}^*)]} \tag{15}$$

$$\bar{B} = -\frac{B_{11}^*}{[(B_{11}^*)^2 - (A_{11}^* D_{11}^*)]} \tag{16}$$

$$\bar{D} = \frac{A_{11}^*}{[(B_{11}^*)^2 - (A_{11}^* D_{11}^*)]} \tag{17}$$

Eqs. (14) are the equations of motion of generally laminated composite beam for the assumptions $N_y=N_{xy}=M_y=M_{xy}=0$. Boundary conditions of the generally laminated composite beams can be written in the following form:

$$\mathrm{S}: \mathrm{w} = \mathrm{M_x} = \mathrm{N_x} = 0$$
$$\mathrm{C}: \mathrm{w} = \mathrm{w_{,x}} = \mathrm{u} = 0 \tag{18}$$
$$\mathrm{F}: \mathrm{M_x} = \mathrm{Q_x} = \mathrm{N_x} = 0$$

2.1 Symmetrically laminated composite beams

For symmetrically laminated composite beams coupling terms B_{ij}'s are zero. Then Eq. (14) takes the following form.

$$\bar{D}\frac{\partial^4 w}{\partial x^4} = -\rho\frac{\partial^2 w}{\partial t^2} \tag{19}$$

General solution of Eq.(19) can be written in the following form:

$$w(x) = A\sin(\Omega x) + B\cos(\Omega x) + C\sinh(\Omega x) + D\cosh(\Omega x) \tag{20}$$

Where A,B,C and D are undetermined coefficients, $\Omega^4 = \rho\omega^2 L^4 / E_2 h^3$ is non-dimensional frequency parameter. Using boundary conditions given in Eq.(18) following Eigenvalue determinants are obtained for different boundary conditions:

H-H boundary condition:

Following condition exists between undetermined coefficients given in Eq.(20): B=D=0

$$\begin{vmatrix} \sin(\Omega) & \sinh(\Omega) \\ -\Omega^2\sin(\Omega) & \Omega^2\sinh(\Omega) \end{vmatrix} = 0 \tag{21}$$

C-H boundary condition:

Following condition exists between undetermined coefficients given in Eq.(20): D=-B, C=-A:

$$\begin{vmatrix} \sin(\Omega)-\sinh(\Omega) & \cos(\Omega)-\cosh(\Omega) \\ -\Omega^2\sin(\Omega)-\Omega^2\sinh(\Omega) & -\Omega^2\cos(\Omega)-\Omega^2\cosh(\Omega) \end{vmatrix} = 0 \tag{22}$$

C-C boundary condition:

Following condition exists between undetermined coefficients given in Eq.(20): D=-B, C=-A:

$$\begin{vmatrix} \sin(\Omega)-\sinh(\Omega) & \cos(\Omega)-\cosh(\Omega) \\ \Omega\cos(\Omega)-\Omega\cosh(\Omega) & -\Omega\sin(\Omega)-\Omega\sinh(\Omega) \end{vmatrix} = 0 \tag{23}$$

C-F boundary condition:

Following condition exists between undetermined coefficients given in Eq.(20): D=-B, C=-A

$$\begin{vmatrix} -\Omega^2\sin(\Omega)-\Omega^2\sinh(\Omega) & -\Omega^2\cos(\Omega)-\Omega^2\cosh(\Omega) \\ -\Omega^3\cos(\Omega)-\Omega^3\cosh(\Omega) & -\Omega^3\sin(\Omega)-\Omega^3\sinh(\Omega) \end{vmatrix} = 0 \tag{24}$$

F-F boundary condition:

Following condition exists between undetermined coefficients given in Eq.(20): D=B, C=A:

$$\begin{vmatrix} -\Omega^2\sin(\Omega) & \Omega^2\sinh(\Omega) \\ -\Omega^2\cos(\Omega) & \Omega^2\cosh(\Omega) \end{vmatrix} = 0 \tag{25}$$

H-F boundary condition:

Following condition exists between undetermined coefficients given in Eq.(20): B=D=0:

$$\begin{vmatrix} -\Omega^2\sin(\Omega)+\Omega^2\sinh(\Omega & -\Omega^2\cos(\Omega)+\Omega^2\cosh(\Omega) \\ -\Omega^3\cos(\Omega)+\Omega^3\cosh(\Omega) & \Omega^3\sin(\Omega)+\Omega^3\sinh(\Omega) \end{vmatrix} = 0 \tag{26}$$

Solution of each determinant equation given in Eq.(21)-Eq.(26) gives frequency parameter of symmetrically laminated composite beams.

2.2 Symmetrically laminated beams with attached mass or spring

Now consider a symmetrically laminated composite beam with attached mass or spring (figure 1). Where η is length of first part of the beam. In order to investigate vibration of two portion composite beam Eq.(20) is written for each portion in the following form:

$$\begin{aligned} w_1(x) &= A_1\sin(\Omega x)+B_1\cos(\Omega x)+C_1\sinh(\Omega x)+D_1\cosh(\Omega x) \\ w_2(x) &= A_2\sin(\Omega x)+B_2\cos(\Omega x)+C_2\sinh(\Omega x)+D_2\cosh(\Omega x) \end{aligned} \quad (27)$$

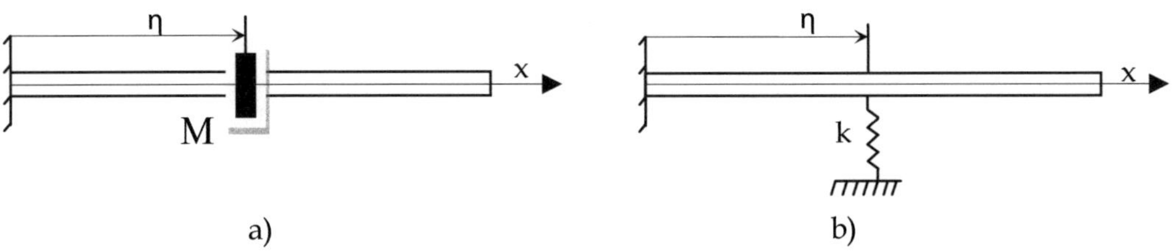

Fig. 1. Composite beam with attached mass (a) and spring (b).

Continuity conditions of the beam at x=η can be written in the following form:

$$\begin{aligned} & w_1(\eta,t)=w_2(\eta,t), \\ & w_1^{'}(\eta,t)=w_2^{'}(\eta,t) \\ & w_1^{''}(\eta,t)=w_2^{''}(\eta,t), \\ & w_1^{'''}(\eta,t)-w_2^{'''}(\eta,t)+\alpha_m\Omega^4 w_1(\eta,t)=0 \qquad \text{(mass)} \\ & \Omega^3 w_1^{'''}(\eta,t)-\Omega^3 w_2^{'''}(\eta,t)-\alpha_s w_1(\eta,t)=0 \qquad \text{(spring)} \end{aligned} \quad (28)$$

Where dimensionless mass and spring parameter are defined in the following form:

$$\alpha_m=\frac{M}{\rho_0 L}, \quad \alpha_s=\frac{kL}{AE}$$

Using boundary conditions Eq.(18) and continuity conditions Eq.(28) following equations are obtained for different boundary conditions and composite beams with attached mass and spring at different position.

H-H boundary condition:

Following condition exists between undetermined coefficients given in Eq.(27): $B_1=D_1=0$:

$$\begin{vmatrix} 0 & 0 & S(\Omega) & C(\Omega) & Sh(\Omega) & Ch(\Omega) \\ 0 & 0 & -S(\Omega) & -C(\Omega) & Sh(\Omega) & Ch(\Omega) \\ S(\eta\Omega) & Sh(\eta\Omega) & -S(\eta\Omega) & -C(\eta\Omega) & -Sh(\eta\Omega) & -Ch(\eta\Omega) \\ C(\eta\Omega) & Ch(\eta\Omega) & -C(\eta\Omega) & -S(\eta\Omega) & -Ch(\eta\Omega) & -S(\eta\Omega) \\ -S(\eta\Omega) & Sh(\eta\Omega) & S(\eta\Omega) & C(\eta\Omega) & -Sh(\eta\Omega) & -Ch(\eta\Omega) \\ A61 & A62 & A63 & A64 & A65 & A66 \end{vmatrix}=0 \quad (29)$$

With spring

$A61 = -\Omega^3 C(\eta\Omega) - \alpha_S S(\eta\Omega)$

$A62 = \Omega^3 Ch(\eta\Omega) - \alpha_S Sh(\eta\Omega)$

$A63 = \Omega^3 C(\eta\Omega)$

$A64 = -\Omega^3 S(\eta\Omega)$

$A65 = -\Omega^3 Ch(\eta\Omega)$

$A66 = -\Omega^3 Sh(\eta\Omega)$

With mass

$A61 = -C(\eta\Omega) + \alpha_m \Omega S(\eta\Omega)$

$A62 = Ch(\eta\Omega) + \alpha_m \Omega Sh(\eta\Omega)$

$A63 = Ch(\eta\Omega)$

$A64 = -S(\eta\Omega)$

$A65 = -Ch(\eta\Omega)$

$A66 = -Sh(\eta\Omega)$

H-C boundary condition:
Following condition exists between undetermined coefficients given in Eq.(27): $B_1=D_1=0$:

$$\begin{vmatrix} 0 & 0 & S(\Omega) & C(\Omega) & Sh(\Omega) & Ch(\Omega) \\ 0 & 0 & C(\Omega) & -S(\Omega) & Ch(\Omega) & Sh(\Omega) \\ S(\eta\Omega) & Sh(\eta\Omega) & -S(\eta\Omega) & -C(\eta\Omega) & -Sh(\eta\Omega) & -Ch(\eta\Omega) \\ C(\eta\Omega) & Ch(\eta\Omega) & -C(\eta\Omega) & -S(\eta\Omega) & -Ch(\eta\Omega) & -S(\eta\Omega) \\ -S(\eta\Omega) & Sh(\eta\Omega) & S(\eta\Omega) & C(\eta\Omega) & -Sh(\eta\Omega) & -Ch(\eta\Omega) \\ A61 & A62 & A63 & A64 & A65 & A66 \end{vmatrix} = 0 \quad (30)$$

With mass

$A61 = -C(\eta\Omega) + \alpha_m \Omega S(\eta\Omega)$

$A62 = Ch(\eta\Omega) + \alpha_m \Omega Sh(\eta\Omega)$

$A63 = C(\eta\Omega)$ $A64 = -S(\eta\Omega)$

$A65 = -Ch(\eta\Omega)$ $A66 = -Sh(\eta\Omega)$

With spring

$A61 = -\Omega^3 C(\eta\Omega) - \alpha_S S(\eta\Omega)$

$A62 = \Omega^3 Ch(\eta\Omega) - \alpha_S Sh(\eta\Omega)$

$A63 = \Omega^3 C(\eta\Omega)$ $A64 = -\Omega^3 S(\eta\Omega)$

$A65 = -\Omega^3 Ch(\eta\Omega)$ $A66 = -\Omega^3 Sh(\eta\Omega)$

C-C boundary condition:
Following condition exists between undetermined coefficients given in Eq.(27): $D_1=-B_1$, $C_1=-A_1$:

$$\begin{vmatrix} 0 & 0 & S(\Omega) & C(\Omega) & Sh(\Omega) & Ch(\Omega) \\ 0 & 0 & C(\Omega) & -S(\Omega) & Ch(\Omega) & Sh(\Omega) \\ S(\eta\Omega)-Sh(\eta\Omega) & C(\eta\Omega)-Ch(\eta\Omega) & -S(\eta\Omega) & -C(\eta\Omega) & -Sh(\eta\Omega) & -Ch(\eta\Omega) \\ C(\eta\Omega)-Ch(\eta\Omega) & -S(\eta\Omega)-Sh(\eta\Omega) & -C(\eta\Omega) & -S(\eta\Omega) & -Ch(\eta\Omega) & -S(\eta\Omega) \\ -S(\eta\Omega)-Sh(\eta\Omega) & -C(\eta\Omega)-Ch(\eta\Omega) & S(\eta\Omega) & C(\eta\Omega) & -Sh(\eta\Omega) & -Ch(\eta\Omega) \\ A61 & A62 & A63 & A64 & A65 & A66 \end{vmatrix} = 0 \quad (31)$$

With mass

$$A61 = -C(\eta\Omega) - Ch(\eta\Omega) + \alpha_m \Omega S(\eta\Omega) - \alpha_m \Omega Sh(\eta\Omega)$$

$$A62 = S(\eta\Omega) - Sh(\eta\Omega) + \alpha_m \Omega C(\eta\Omega) - \alpha_m \Omega Ch(\eta\Omega)$$

$$A63 = C(\eta\Omega) \qquad A64 = -S(\eta\Omega)$$

$$A65 = -Ch(\eta\Omega) \qquad A66 = -Sh(\eta\Omega)$$

With spring

$$A61 = -\Omega^3 C(\eta\Omega) - \Omega^3 Ch(\eta\Omega) - \alpha_s S(\eta\Omega) + \alpha_s Sh(\eta\Omega)$$

$$A62 = \Omega^3 S(\eta\Omega) - \Omega^3 Sh(\eta\Omega) - \alpha_s C(\eta\Omega) + \alpha_s Ch(\eta\Omega)$$

$$A63 = \Omega^3 C(\eta\Omega) \qquad A64 = -\Omega^3 S(\eta\Omega)$$

$$A65 = -\Omega^3 Ch(\eta\Omega) \qquad A66 = -\Omega^3 Sh(\eta\Omega)$$

C-F boundary condition:

Following condition exists between undetermined coefficients given in Eq.(27): $D_1=-B_1$, $C_1=-A_1$:

$$\begin{vmatrix} 0 & 0 & -S(\Omega) & -C(\Omega) & Sh(\Omega)) & Ch(\Omega) \\ 0 & 0 & -C(\Omega) & S(\Omega) & Ch(\Omega) & Sh(\Omega) \\ S(\eta\Omega)-Sh(\eta\Omega) & C(\eta\Omega)-Ch(\eta\Omega) & -S(\eta\Omega) & -C(\eta\Omega) & -Sh(\eta\Omega) & -Ch(\eta\Omega) \\ C(\eta\Omega)-Ch(\eta\Omega) & -S(\eta\Omega)-Sh(\eta\Omega) & -C(\eta\Omega) & -S(\eta\Omega) & -Ch(\eta\Omega) & -S(\eta\Omega) \\ -S(\eta\Omega)-Sh(\eta\Omega) & -C(\eta\Omega)-Ch(\eta\Omega) & S(\eta\Omega) & C(\eta\Omega) & -Sh(\eta\Omega) & -Ch(\eta\Omega) \\ A61 & A62 & A63 & A64 & A65 & A66 \end{vmatrix} = 0 \quad (32)$$

With mass

$$A61 = -C(\eta\Omega) - Ch(\eta\Omega) + \alpha_m \Omega S(\eta\Omega) - \alpha_m \Omega Sh(\eta\Omega)$$

$$A62 = S(\eta\Omega) - Sh(\eta\Omega) + \alpha_m \Omega C(\eta\Omega) - \alpha_m \Omega Ch(\eta\Omega)$$

$$A63 = C(\eta\Omega)$$

$$A64 = -S(\eta\Omega)$$

$$A65 = -Ch(\eta\Omega)$$

$$A66 = -Sh(\eta\Omega)$$

With spring

$$A61 = -\Omega^3 C(\eta\Omega) - \Omega^3 Ch(\eta\Omega) - \alpha_s S(\eta\Omega) + \alpha_s Sh(\eta\Omega)$$

$$A62 = \Omega^3 S(\eta\Omega) - \Omega^3 Sh(\eta\Omega) - \alpha_s C(\eta\Omega) + \alpha_s Ch(\eta\Omega)$$

$$A63 = \Omega^3 C(\eta\Omega)$$

$$A64 = -\Omega^3 S(\eta\Omega)$$

$$A65 = -\Omega^3 Ch(\eta\Omega)$$

$$A66 = -\Omega^3 Sh(\eta\Omega)$$

F-F boundary condition

Following condition exists between undetermined coefficients given in Eq.(27): $D_1=B_1$, $C_1=A_1$:

$$\begin{vmatrix} 0 & 0 & -S(\Omega) & -C(\Omega) & Sh(\Omega)) & Ch(\Omega) \\ 0 & 0 & -C(\Omega) & S(\Omega) & Ch(\Omega) & Sh(\Omega) \\ S(\eta\Omega)+Sh(\eta\Omega) & C(\eta\Omega)+Ch(\eta\Omega) & -S(\eta\Omega) & -C(\eta\Omega) & -Sh(\eta\Omega) & -Ch(\eta\Omega) \\ C(\eta\Omega)+Ch(\eta\Omega) & -S(\eta\Omega)+Sh(\eta\Omega) & -C(\eta\Omega) & S(\eta\Omega) & -Ch(\eta\Omega) & -Sh(\eta\Omega) \\ -S(\eta\Omega)+Sh(\eta\Omega) & -C(\eta\Omega)+Ch(\eta\Omega) & S(\eta\Omega) & C(\eta\Omega) & -Sh(\eta\Omega) & -Ch(\eta\Omega) \\ A61 & A62 & A63 & A64 & A65 & A66 \end{vmatrix} = 0 \quad (33)$$

With mass

$$A61 = -C(\eta\Omega) - Ch(\eta\Omega) + \alpha_m \Omega S(\eta\Omega) + \alpha_m \Omega Sh(\eta\Omega)$$
$$A62 = S(\eta\Omega) - Sh(\eta\Omega) + \alpha_m \Omega C(\eta\Omega) + \alpha_m \Omega Ch(\eta\Omega)$$
$$A63 = C(\eta\Omega) \qquad A64 = -S(\eta\Omega)$$
$$A65 = -Ch(\eta\Omega) \qquad A66 = -Sh(\eta\Omega)$$

With spring

$$A61 = -\Omega^3 C(\eta\Omega) + \Omega^3 Ch(\eta\Omega) - \alpha_s S(\eta\Omega) - \alpha_s Sh(\eta\Omega)$$
$$A62 = \Omega^3 S(\eta\Omega) + \Omega^3 Sh(\eta\Omega) - \alpha_s C(\eta\Omega) + \alpha_s Ch(\eta\Omega)$$
$$A63 = \Omega^3 C(\eta\Omega) \qquad A64 = -\Omega^3 S(\eta\Omega)$$
$$A65 = -\Omega^3 Ch(\eta\Omega) \qquad A66 = -\Omega^3 Sh(\eta\Omega)$$

H-F boundary condition:
Following condition exists between undetermined coefficients given in Eq.(27): $B_1=D_1=0$:

$$\begin{vmatrix} 0 & 0 & -S(\Omega) & -C(\Omega) & Sh(\Omega)) & Ch(\Omega) \\ 0 & 0 & -C(\Omega) & S(\Omega) & Ch(\Omega) & Sh(\Omega) \\ S(\eta\Omega) & Sh(\eta\Omega) & -S(\eta\Omega) & -C(\eta\Omega) & -Sh(\eta\Omega) & -Ch(\eta\Omega) \\ C(\eta\Omega) & Ch(\eta\Omega) & -C(\eta\Omega) & S(\eta\Omega) & -Ch(\eta\Omega) & -Sh(\eta\Omega) \\ -S(\eta\Omega) & Sh(\eta\Omega) & S(\eta\Omega) & C(\eta\Omega) & -Sh(\eta\Omega) & -Ch(\eta\Omega) \\ A61 & A62 & A63 & A64 & A65 & A66 \end{vmatrix} = 0 \qquad (34)$$

With mass

$$A61 = -C(\eta\Omega) + \alpha_m \Omega S(\eta\Omega)$$
$$A62 = Ch(\eta\Omega) + \alpha_m \Omega Sh(\eta\Omega)$$
$$A63 = C(\eta\Omega)$$
$$A64 = -S(\eta\Omega)$$
$$A65 = -Ch(\eta\Omega)$$
$$A66 = -Sh(\eta\Omega)$$

With spring

$$A61 = -\Omega^3 C(\eta\Omega) - \alpha_s S(\eta\Omega)$$
$$A62 = \Omega^3 Ch(\eta\Omega) - \alpha_s Sh(\eta\Omega)$$
$$A63 = \Omega^3 C(\eta\Omega)$$
$$A64 = -\Omega^3 S(\eta\Omega)$$
$$A65 = -\Omega^3 Ch(\eta\Omega)$$
$$A66 = -\Omega^3 Sh(\eta\Omega)$$

Solution of each determinant equation given in Eq.(29)-Eq.(34) gives frequency parameter of symmetrically laminated beams with attached point mass or spring at the different location of the beam.

3. Numerical results

In this section, firstly, numerical results are given for vibration of composite beams with or without attached mass or springs. In order to check validity of present results first five flexural vibration frequencies of laminated composite beams are compared with previous results (Reddy, 1997) and good agreement is observed between two results. After checking validity of present formulation, vibration of composite beams with attached mass or spring is investigated for different boundary conditions. Material properties are chosen as: $E_1=25E_2$, $G_{12}=0.5E_2$ and $\nu_{12}=0.3$. Obtained parametrical results are given in figures. In order to completeness of present study, first five frequency of symmetric three layer $(\theta/-\theta/\theta)$ composite beams are given in Fig.2. According to Fig. 2, dimensionless frequency parameters decrease with increasing lamination angle θ. This is due to decrease in rigidities D_{ij} with increasing θ. The frequency gap is narrowing for higher θ, so this type of beams should be carefully designed. Highest frequencies are obtained for C-C and F-F boundary conditions where as lowest one is obtained for C-F boundary condition.

Variation of frequency ratio of composite beams with attached mass to composite beam without mass (Ω_m/Ω_0) is depicted in Fig.3 for different boundary conditions. According to this figure, ratio of frequencies is insensitive to lamination angle θ. The lowest frequencies generally are most affected by attached mass. Influence of attached mass is decreasing with increasing mode number. This fact can be explained by considering mode shapes of vibrating composite beams. For H-H, C-C, H-C and F-F beams $\eta=0.25$ is a nodal point for fourth frequency, therefore this frequency is not affected by attached mass as expected. Highest %40 and lowest %20 changes are observed for frequencies for different boundary conditions.

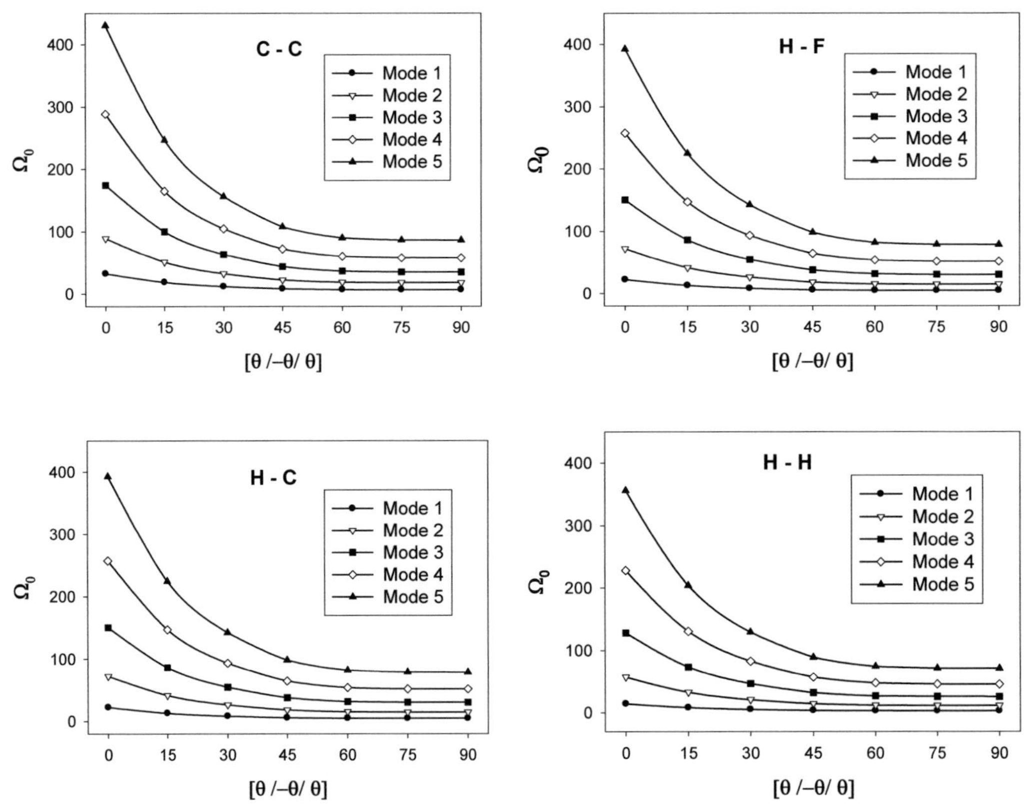

Fig. 2. Variation of frequency parameter of composite beam with lamination angle θ.

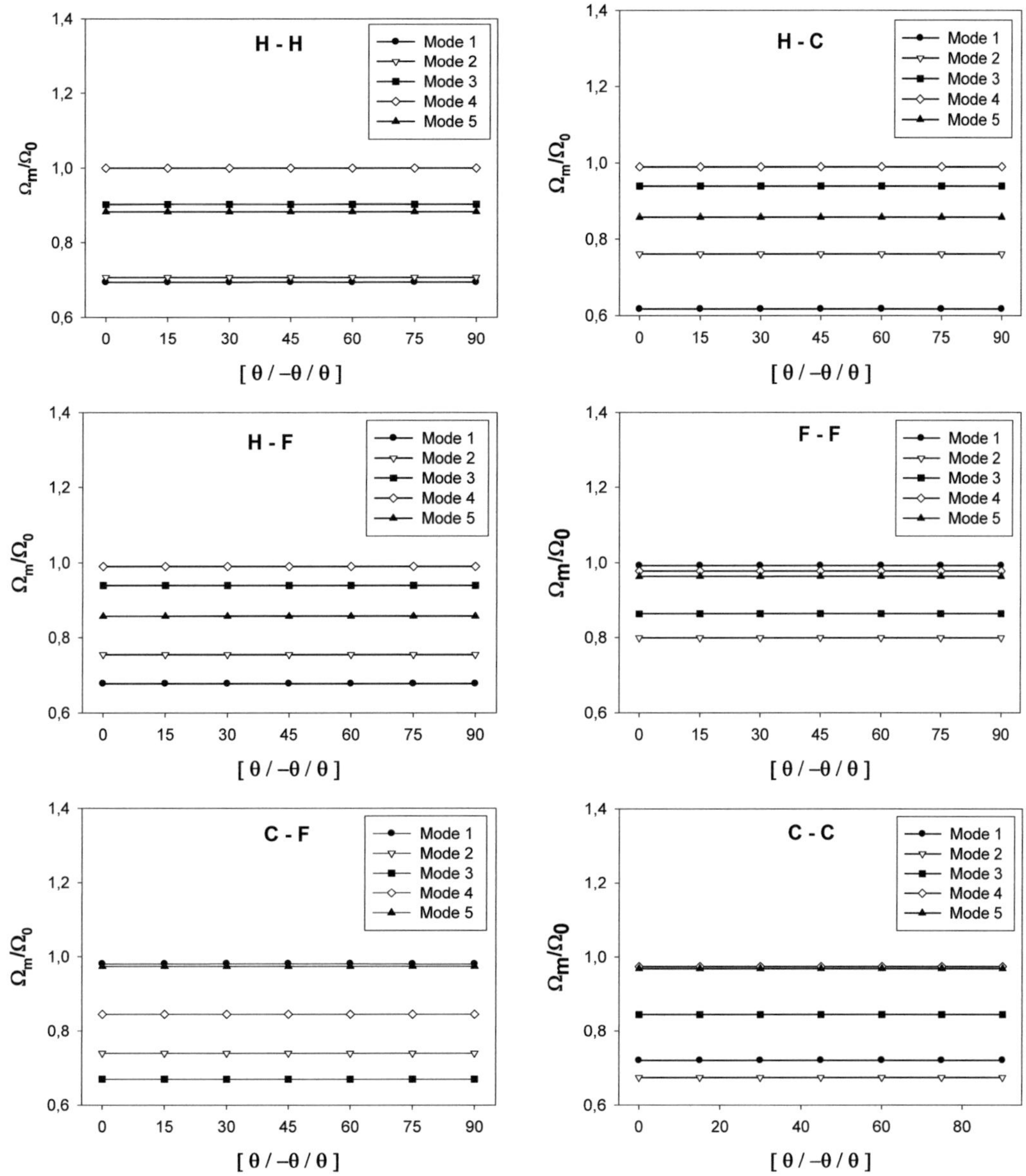

Fig. 3. Variation of frequency ratio of composite beam with lamination angle for α_m=1 and η=0,25 .

Variation of frequency ratio of composite beams with attached spring to composite beam without spring (Ω_s/Ω_0) is given in Fig.4 for different boundary conditions. According to this figure, ratio of frequencies is insensitive to lamination angle θ. Effect of attached spring on the frequency ratio is negligible for composite beams with at least one clamped edge. The beams with F-F and H-F boundary conditions are most affected by attached mass. For these boundary conditions spring behaves like a hinged boundary condition and decreases frequency of composite beam.

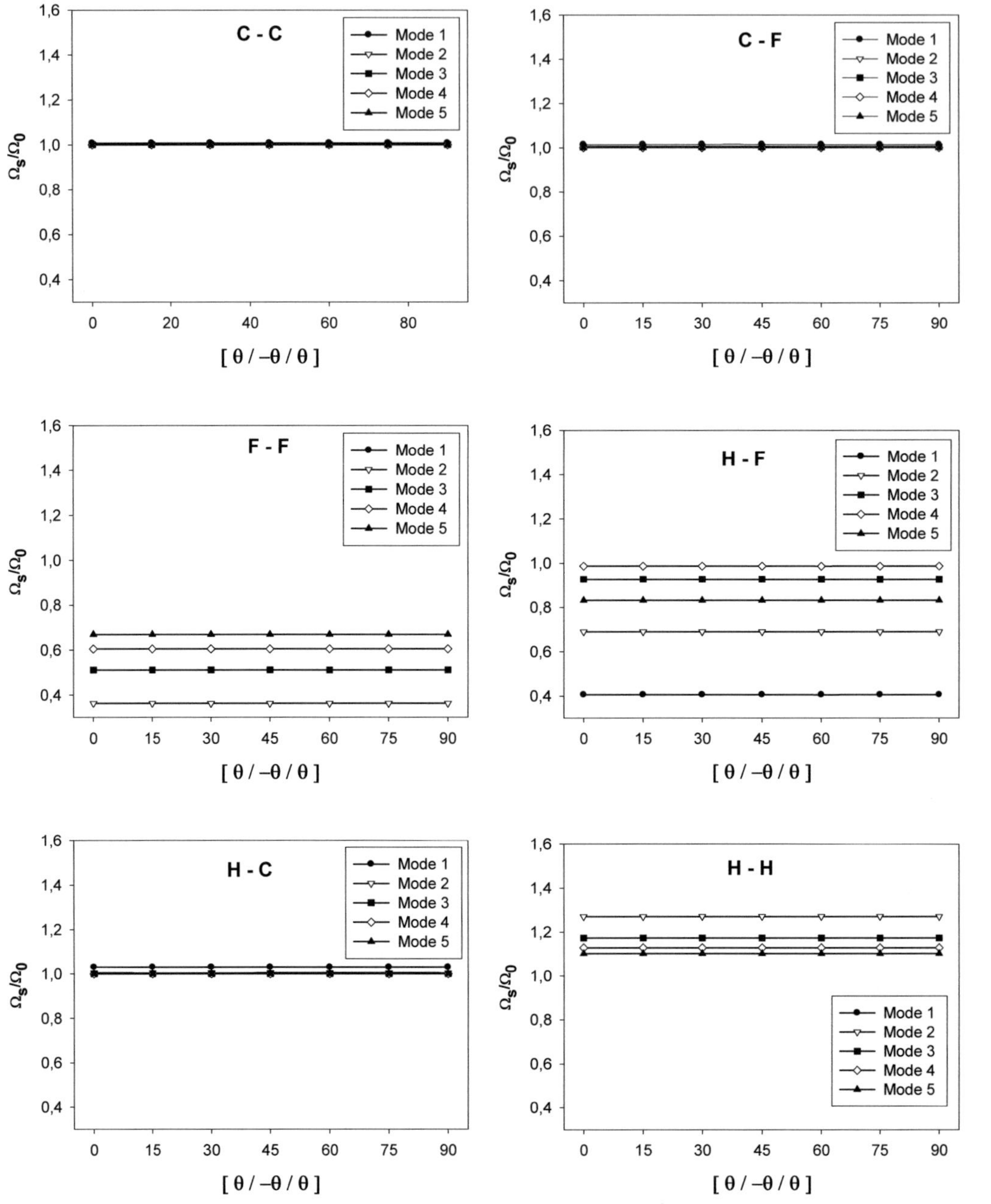

Fig. 4. Variation of frequency ratio of symmetric angle-ply composite beam with lamination angle for α_s=10 and η=0,25.

In Fig. 5, variation of frequency ratio with α_m is given for three layer symmetric angle-ply ($30^0/-30^0/30^0$) composite beams. Increasing α_m decreases frequency of the composite beam. Different decreases are observed for different boundary conditions.

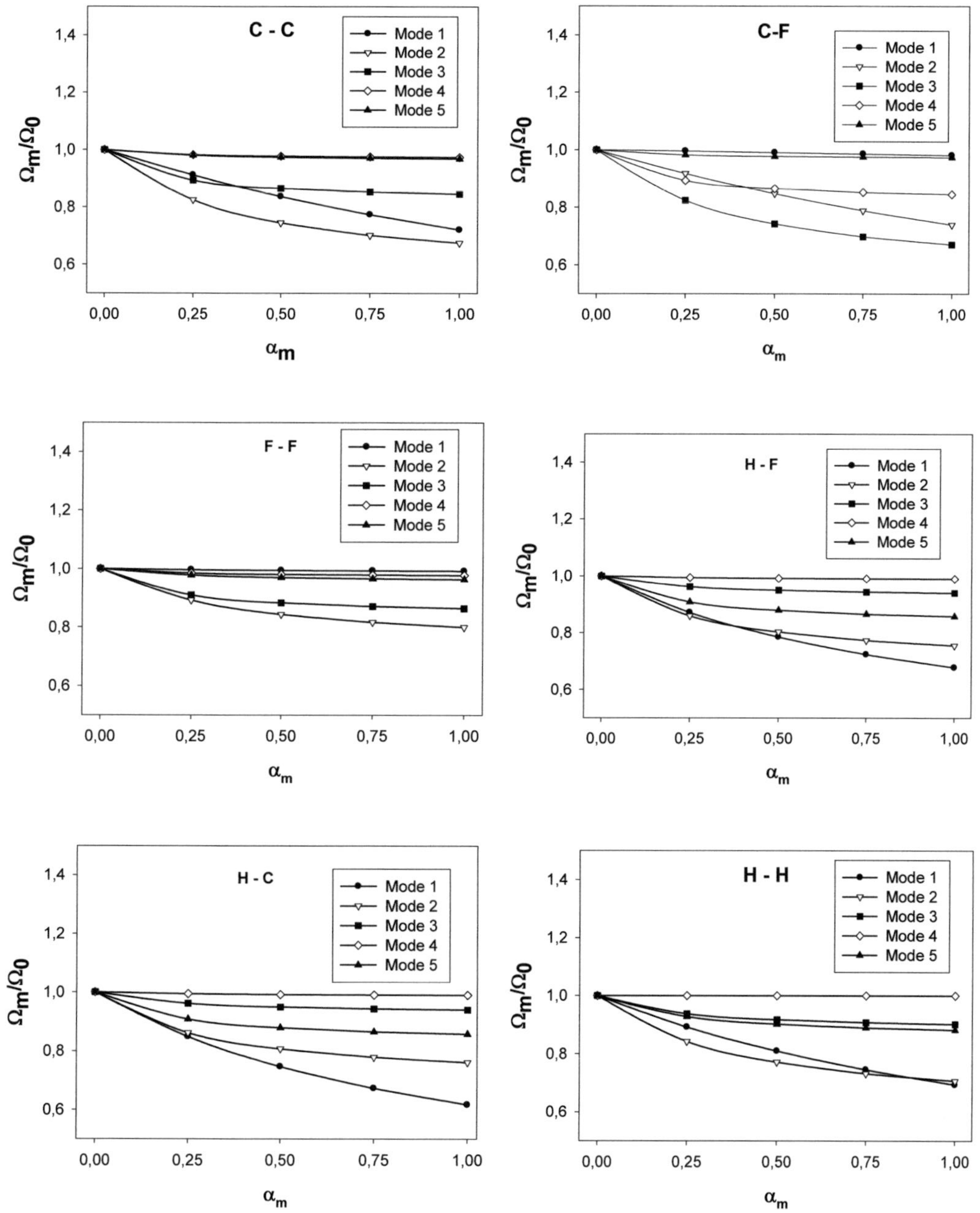

Fig. 5. Variation of frequency ratio of symmetric angle-ply composite beam ($30^0/-30^0/30^0$) with α_m for η=0,25.

Variation of frequency ratio with α_s is given in Fig. 6 for three layer symmetric angle-ply ($30^0/-30^0/30^0$) composite beams. Increasing α_s decreases frequency of the composite beam for F-F and H-F boundary conditions. For these two boundary conditions zero frequencies exist for rigid body motions. Attaching a spring prevents from rigid body motion and these zero frequencies turn two none zero frequencies. Other boundary conditions are insensitive to increase of α_s for given range.

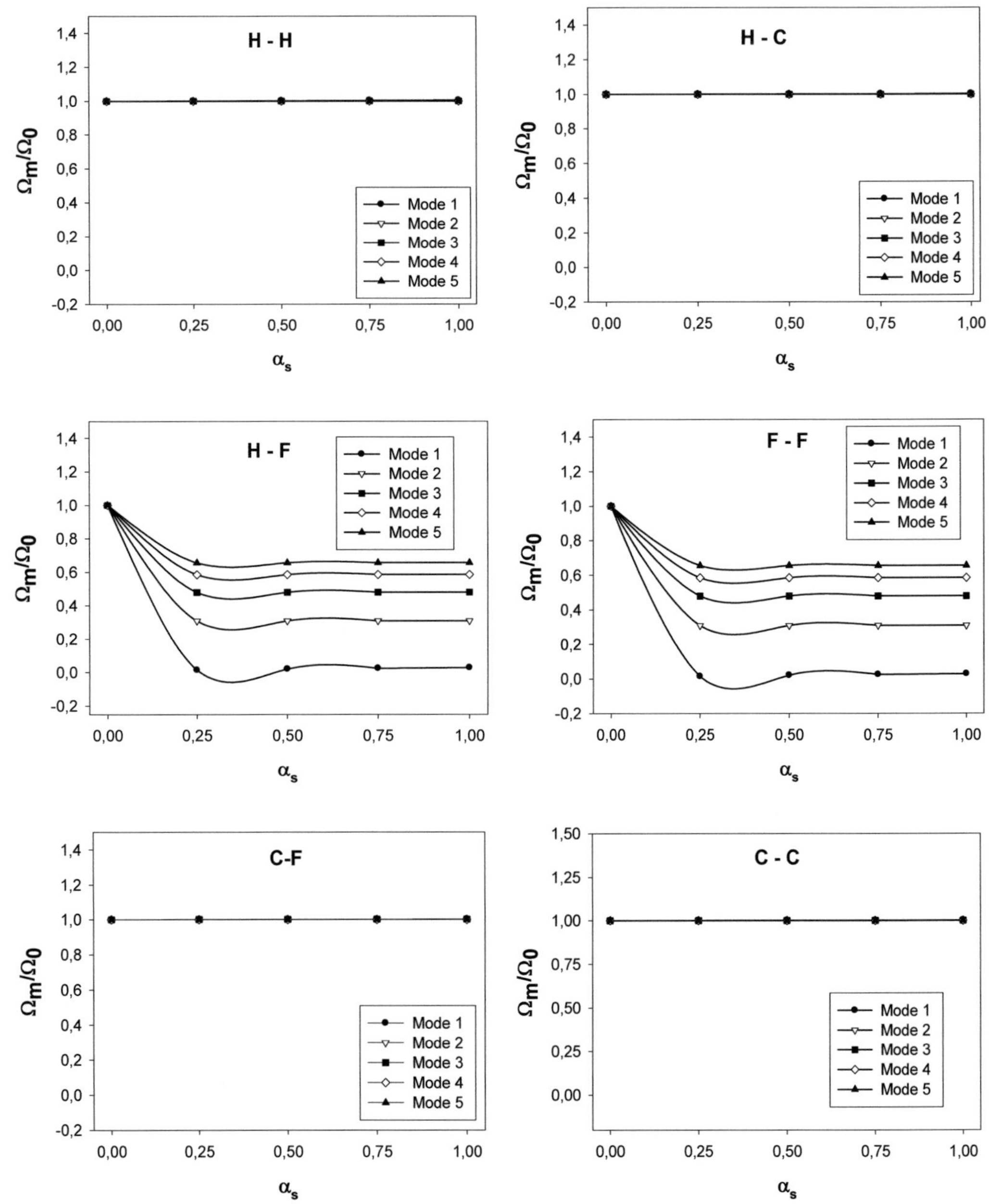

Fig. 6. Variation of frequency ratio of symmetric angle-ply composite beam (30^0/-30^0/30^0) with α_s for η=0,25.

In Fig. 7, variation of frequency ratio of composite beam with η for α_m=1 and η=0.25 are given for three layer symmetric angle-ply (30^0/-30^0/30^0) composite beams. Generally, lower frequencies are most affected by position of attached mass. Forth frequency is not affected by position of attached mass for boundary conditions other than F-F and F-H. This is due to nodal points coincides with position of attached masses.

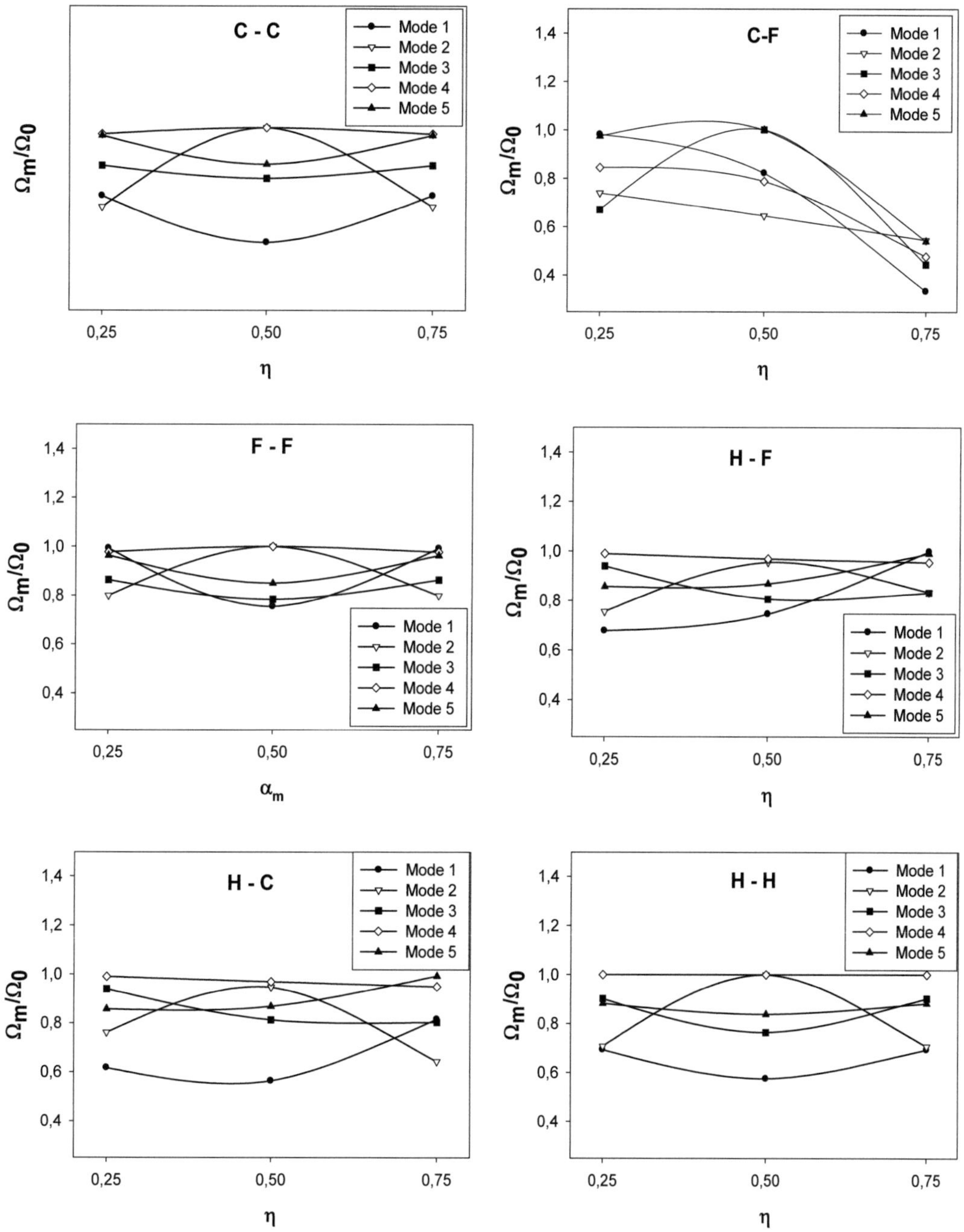

Fig. 7. Variation of frequency ratio of symmetric angle-ply composite beam ($30^0/-30^0/30^0$) with η for α_m=1and η=0,25.

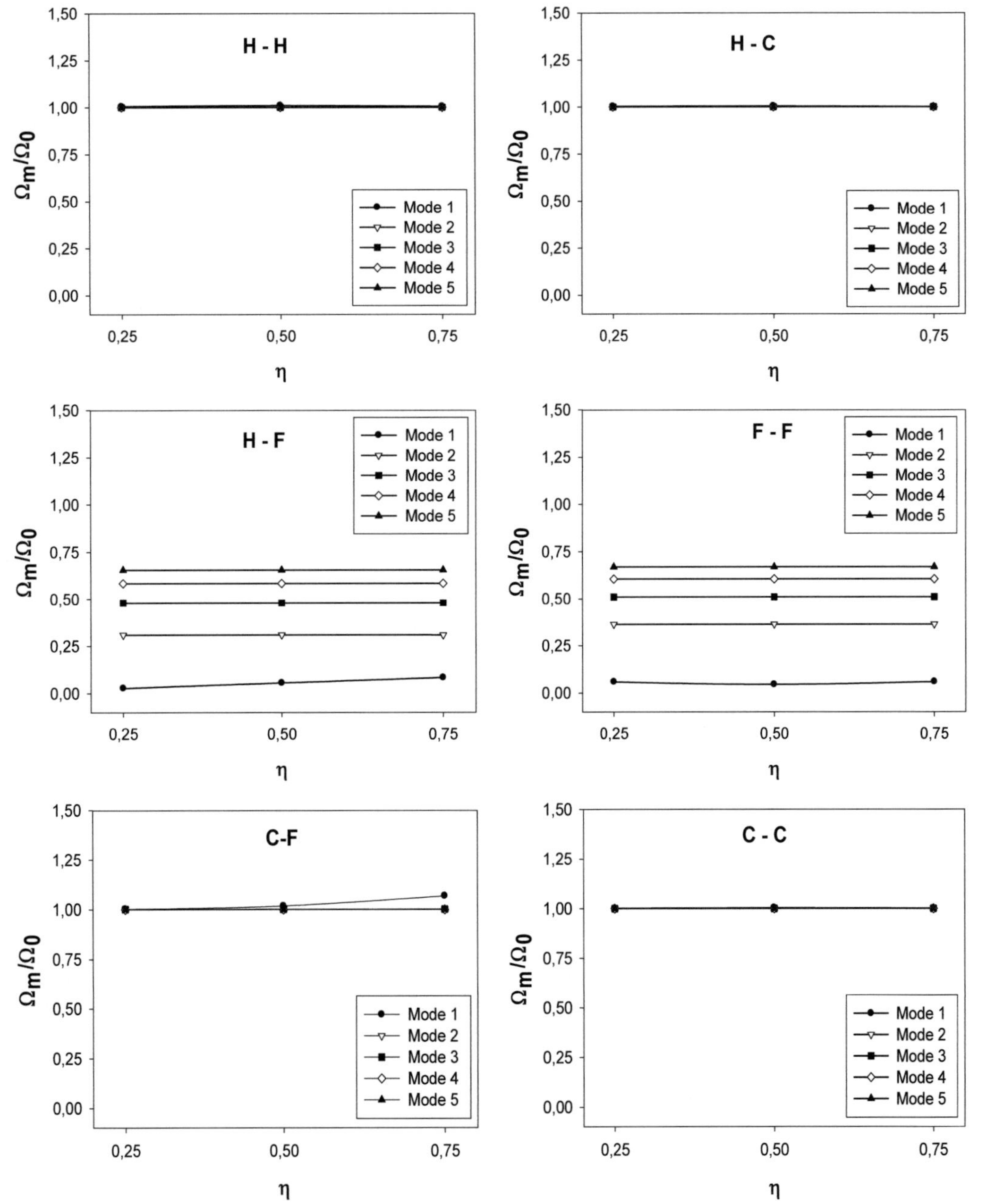

Fig. 8. Variation of frequency ratio of symmetric angle-ply composite beam ($30^0/-30^0/30^0$) with η for α_s=1 and η=0,25.

Variation of frequency ratio of composite beam with η for α_s=1 and η=0.25 are given in Fig. 8 for three layer symmetric angle-ply ($30^0/-30^0/30^0$) composite beams. Similar to Fig. 7 generally lower frequencies are most affected by position of attached spring. Forth frequency is not affected by position of attached spring for boundary conditions other than F-F and F-H. This is due to nodal points coincides with position of attached spring.

In Fig. 9-10, variation of frequency parameter of composite beam with lamination angle for α_s=1, α_m=1 and η=0.25 for different number of layers (single, three and four layer) are given respectively. First frequencies are insensitive to number of layers but for the fourth frequencies higher frequencies are obtained with increasing number of layers.

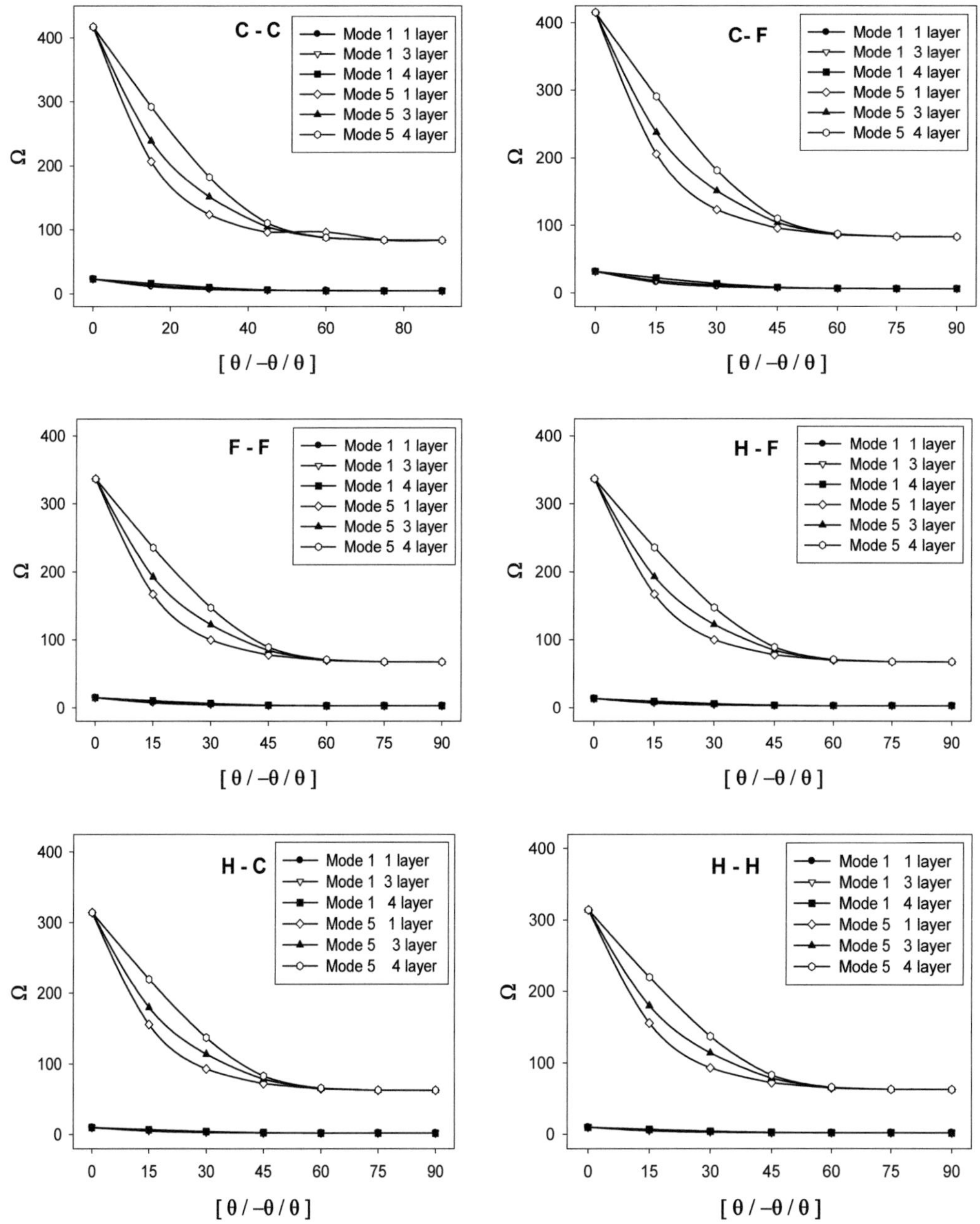

Fig. 9. Variation of frequency parameter of symmetric angle-ply composite beam with lamination angle for α_s=1 and η=0,25 for different number of layers.

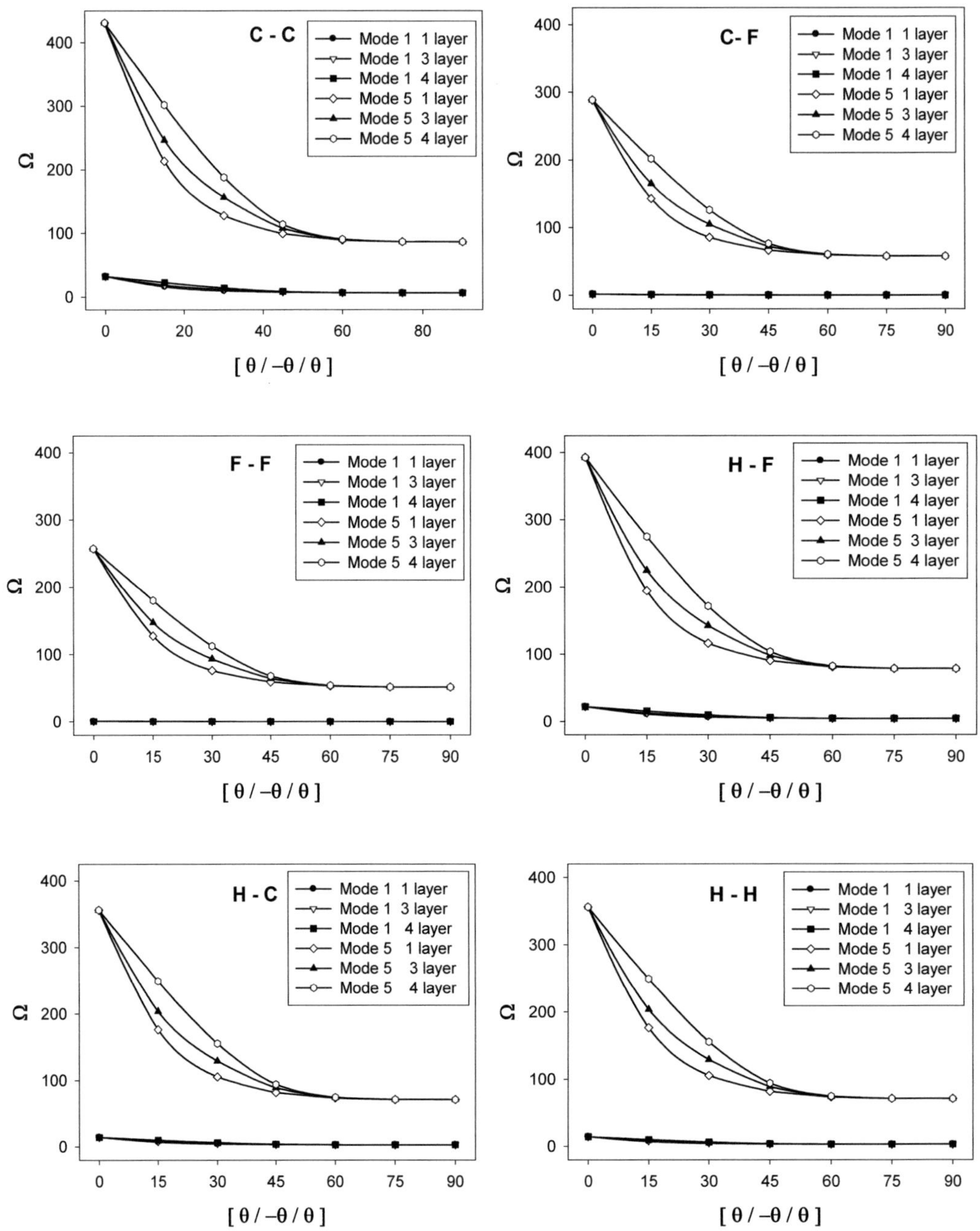

Fig. 10. Variation of frequency parameter of symmetric angle-ply composite beam with lamination angle for α_m=1 and η=0,25 for different number of layers.

In Fig 11-12 variation of frequency ratio of three layer cross-ply composite beams with η is given for α_m=1 and α_s=1 respectively. Generally similar behavior is observed with symmetric angle-ply and cross-ply composite beams.

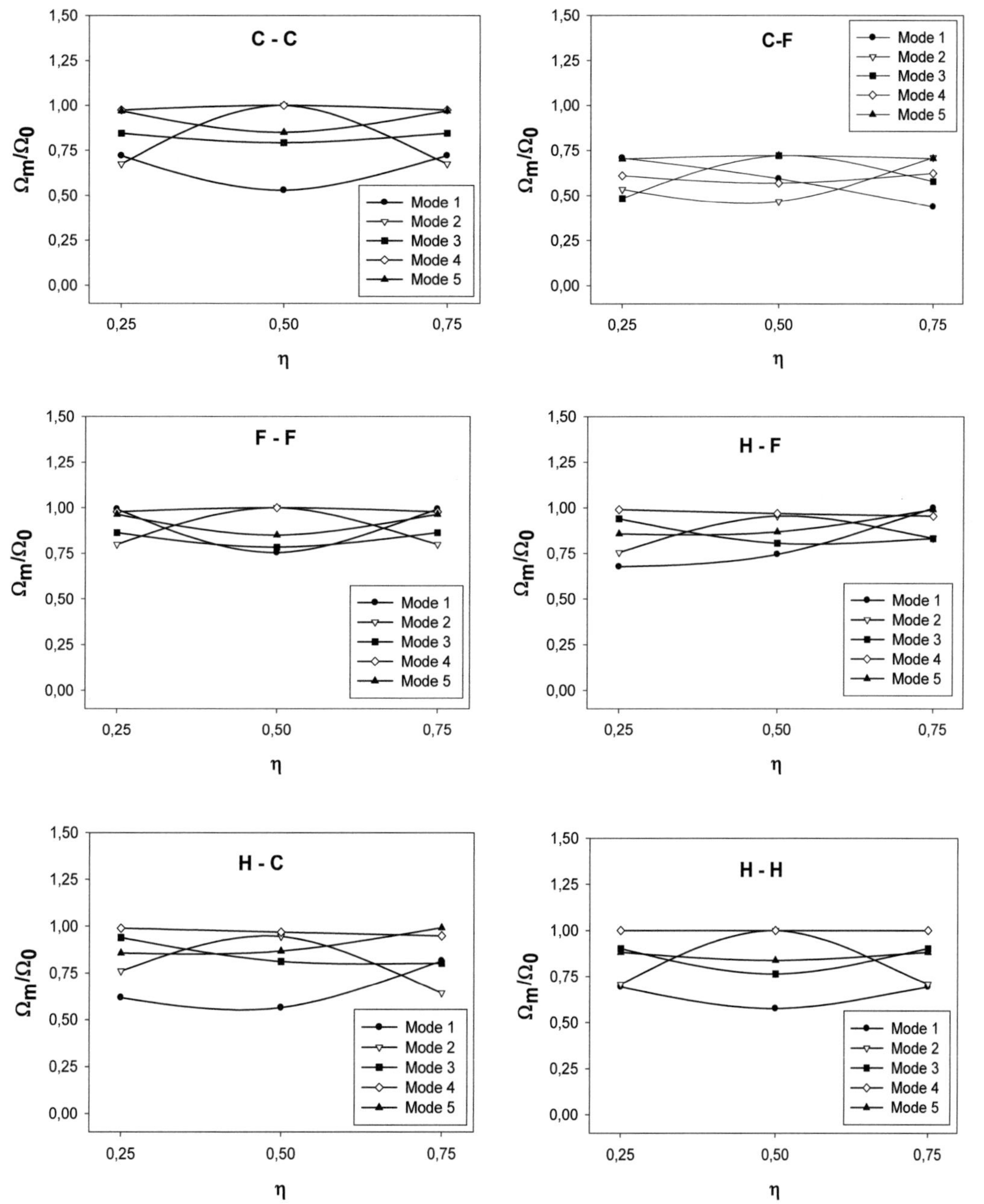

Fig. 11. Variation of frequency ratio of symmetric cross-ply composite beam ($0^0/90^0/0^0$) with η for α_m=1and η=0,25 .

4. Conclusion

In this study, vibration of laminated composite beams with attached mass or spring is studied using classical lamination theory. First five flexural frequencies of composite beams

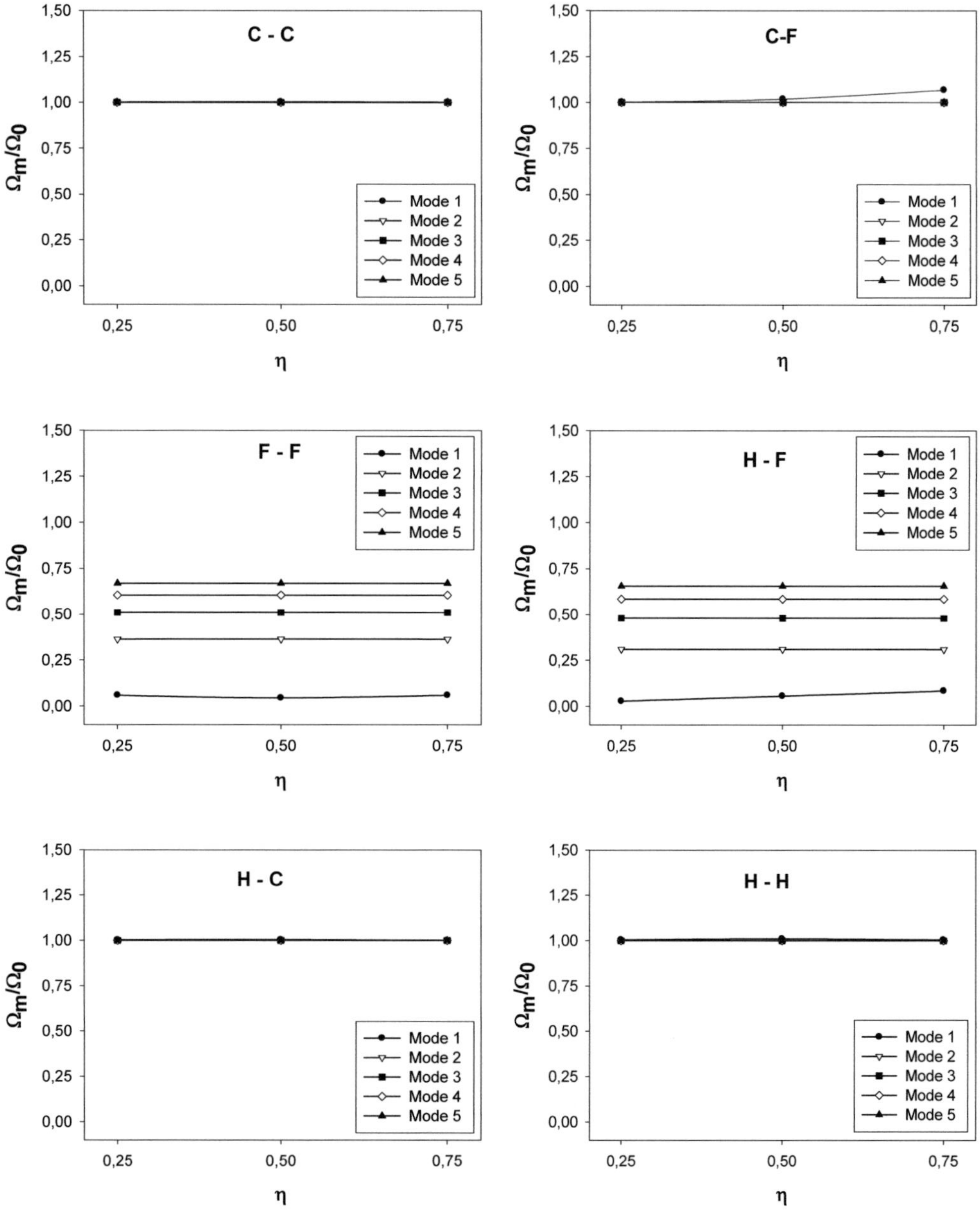

Fig. 12. Variation of frequency ratio of symmetric cross-ply composite beam ($0^0/90^0/0^0$) with η for α_s=1and η=0,25.

are obtained for different boundary conditions, attached mass, spring and their different positions. It is obtained that attaching mass reduces frequency of composite beams whereas attaching spring increases frequency of composite beams. Some modes do not change depending on position of attached spring or mass. This study can be extended to anti-symmetric composite beams and shear deformation effects can be added in the future studies.

5. References

Abramovich H. (1992) Shear deformation and rotary inertia effects of vibrating composite beams. *Composite Structures*, 20, 165-73.

Abramovich H. & Livshits A. (1994) Free vibrations of non-symmetric cross-ply laminated composite beams, *Journal of Sound and Vibration*, 176, 5, 597-612.

Arya H. (2003) A new zig-zag model for laminated composite beams:free vibration analysis. *Journal of Sound and Vibration*, 264, 2, 485-490.

Aydogdu M. (2005) Vibration analysis of cross-ply laminated beams with general boundary conditions by Ritz method, *International Journal of Mechanical Sciences*, 47, 1740-1755.

Aydogdu M. (2006) Free vibration analysis of angle-ply laminated beams with general boundary conditions, *Journal of Reinforced Plastics and Composites*, 25, 15, 1571-1583.

Banerjee JR. (2001) Frequency equation and mode shape formulae for composite Timoshenko beams. *Composite Structures* , 51, 4, 381-388.

Chandrashekhara, K., Krishnamurty, K. & Roy, S. (1990). Free vibration of composite beams including rotary inertia and shear deformation, *Composite Structures*, 14, 4, 269-279.

Chandrashekhara, K. & Bangera, K.M. (1992). Free vibration of composite beams Using a refined shear flexible beam element, *Computers and Structures*, 43, 4, 719-727.

Chandrashekhara K. & Bangera K. M. (1993) Vibration of symmetrically laminated clamped-free beam with a mass at the free end, *Journal of Sound and Vibration* 160, 1, 93-101.

Chen, W.Q., Lv, C. F. & Bian, Z. G. (2004). Free Vibration Analysis of Generally Laminated Beams via State-space-based Differential Quadrature, *Composite Structures*, 63, 1, 417-425.

Eisenberger M., Abramovich H. & Shulepov O. (1995), Dynamic stiffness analysis of laminated beams using a first order shear deformation theory, *Composite Structures*, 31, 4, 265-271.

Gürgöze M. (1986) On the approximate determination of the fundamental frequency of a restrained cantilever beam carrying a tip heavy body Original Research Article, *Journal of Sound and Vibration*, 105, 3, 443-449.

Gürgöze M. (1996) On the eigenfrequencies of a cantilever beam with attached tip mass and a spring-mass system, Original Research Article *Journal of Sound and Vibration*, 190, 2, 149-162.

Gürgöze M. & Erol H. (2004) On laterally vibrating beams carrying tip masses, coupled by several double spring–mass systems., Journal of Sound and Vibration, 269, 1-2, 431-438.

Jones R.M. (1975), Mechanics of composite materials. New York: *Hemisphere Publishing Corporation*,.

Kapani, R.K. & Raciti, S. (1989). Recent advances in analysis of laminated beams and plates, Part II: Vibrations and Wave Propagation, *AIAA Journal*, 27, 935-946.

Karama M, Abou Harb B, Mistou S & Caperaa S. (1998) Bending, buckling and free vibration of laminated composite with a transverse stress continuity model. *Composite Part B-Eng*, 29, 3, 223-234.

Kameswara Rao M., Desai Y.M. & Chitnis M.R. (2001) Free vibrations of laminated beams using mixed theory. *Composite Structures*, 52, 2, 149-160.

Kant T, Marur S.R. & Rao GS. (1998) Analytical solution to the dynamic analysis of laminated beams using higher order refined theory. *Composite Structures*, 40, 1, 1-9.

Khdeir A.A.& Reddy J.N. (1994) Free vibration of cross-ply laminated beams with arbitrary boundary conditions. *International Journal of Engineering Science,* 32, 12, 1971-1980.

Krishnaswamy, K., Chandrashekhara, K. & Wu.W.Z.B. (1992). Analytical solutions to vibration of generally layered composite beams, *Composite Structures,* 14, 4, 269-279.

Marur S.R., Kant T. (1996) Free vibration analysis of fiber reinforced composite beams using higher order theories and finite element modelling., *Journal of Sound Vibration,* 194, 3, 337-351.

Matsunaga H. (2001) Vibration and buckling of multilayered composite beams according to higher order deformation theories., *Journal of Sound and Vibration,* 246, 1, 47-62.

Murthy MVVS, Roy Mahapatra D, Badarinarayana K & Gopalakrishnan S. (2005) A refined higher order finite element for asymmetric composite beams. *Composite Structures,* 67, 1, 27-35.

Nagulaeswaran S., (2001) Transverse vibrations of Euler-Bernoulli uniform beam carrying two particles in-span Original Research Article *International Journal of Mechanical Sciences,* 43, 12, 2737-2752.

Naguleswaran S. (2002) Transverse vibrations of an Euler–Bernoulli uniform beam carrying several particles Original Research Article International Journal of Mechanical Science, 44, 12, 2463-2478.

Ramtekkar GS, Desai YM & Shah A.H. (2002) Natural vibrations of laminated composite beams by using mixed finite element modelling., *Journal of Sound Vibration.* 257, 4, 635-651.

Reddy, J.N., (1997). Mechanics of Laminated Composite Plates and Shells Theory and Analysis, Junuthula Narasimha (2nd Ed.), 182-186, CRC Press, ISBN: 0-8493-1592, USA.

Özkaya E, Pakdemirli M & Öz H.R. (1997) Non-linear vibrations of a beam-mass system under different boundary conditions, *Journal of Sound and Vibration* 199, 4, 679-696.

Özkaya E & Pakdemirli M.(1999) Non-linear vibrations of a beam-mass system with both

Shi G. & Lam K.Y. (1999) Finite element vibration analysis of composite beams based on higher order beam theory. *Journal of Sound and Vibration,* 219, 4, 707-721.

Yıldırım V., Sancaktar E. & Kıral E. (1999) Comparison of the in-plane natural frequencies of symmetric cross-ply laminated beams based on the Bernoulli-Euler and Timoshenko Beam theories. *Journal of Applied Mechanics-T ASME,* 66, 2, 410-417.

Yıldırım V. (2000) Effect of the longitudinal to transverse moduli ratio on the in-plane natural frequencies of symmetric cross-ply laminated beams by the stiffness method. *Composite Structures,* 50, 3, 319-326.

5

Independent Coordinate Coupling Method for Free Vibration Analysis of a Plate With Holes

Moon Kyu Kwak and Seok Heo
Dongguk University
Republic of Korea

1. Introduction

A rectangular plate with a rectangular or a circular hole has been widely used as a substructure for ship, airplane, and plant. Uniform circular and annular plates have been also widely used as structural components for various industrial applications and their dynamic behaviors can be described by exact solutions. However, the vibration characteristics of a circular plate with an eccentric circular hole cannot be analyzed easily. The vibration characteristics of a rectangular plate with a hole can be solved by either the Rayleigh-Ritz method or the finite element method. The Rayleigh-Ritz method is an effective method when the rectangular plate has a rectangular hole. However, it cannot be easily applied to the case of a rectangular plate with a circular hole since the admissible functions for the rectangular hole domain do not permit closed-form integrals. The finite element method is a versatile tool for structural vibration analysis and therefore, can be applied to any of the cases mentioned above. But it does not permit qualitative analysis and requires enormous computational time.

Tremendous amount of research has been carried out on the free vibration of various problems involving various shape and method. Monahan et al.(1970) applied the finite element method to a clamped rectangular plate with a rectangular hole and verified the numerical results by experiments. Paramasivam(1973) used the finite difference method for a simply-supported and clamped rectangular plate with a rectangular hole. There are many research works concerning plate with a single hole but a few work on plate with multiple holes. Aksu and Ali(1976) also used the finite difference method to analyze a rectangular plate with more than two holes. Rajamani and Prabhakaran(1977) assumed that the effect of a hole is equivalent to an externally applied loading and carried out a numerical analysis based on this assumption for a composite plate. Rajamani and Prabhakaran(1977) investigated the effect of a hole on the natural vibration characteristics of isotropic and orthotropic plates with simply-supported and clamped boundary conditions. Ali and Atwal(1980) applied the Rayleigh-Ritz method to a simply-supported rectangular plate with a rectangular hole, using the static deflection curves for a uniform loading as admissible functions. Lam et al.(1989) divided the rectangular plate with a hole into several sub areas and applied the modified Rayleigh-Ritz method. Lam and Hung(1990) applied the same method to a stiffened plate. The admissible functions used in (Lam et al. 1989, Lam and Hung 1990) are the orthogonal polynomial functions proposed by Bhat(1985, 1990). Laura et al.(1997) calculated the natural vibration characteristics of a simply-supported rectangular plate with a rectangular hole by the classical Rayleigh-Ritz method. Sakiyama et al.(2003) analyzed the natural vibration characteristics of an orthotropic plate with a square hole by means of the Green function assuming the hole as an extremely thin plate.

The vibration analysis of a rectangular plate with a circular hole does not lend an easy approach since the geometry of the hole is not the same as the geometry of the rectangular plate. Takahashi(1958) used the classical Rayleigh-Ritz method after deriving the total energy by subtracting the energy of the hole from the energy of the whole plate. He employed the eigenfunctions of a uniform beam as admissible functions. Joga-Rao and Pickett(1961) proposed the use of algebraic polynomial functions and biharmonic singular functions. Kumai(1952), Hegarty(1975), Eastep and Hemmig(1978), and Nagaya(1951) used the point-matching method for the analysis of a rectangular plate with a circular hole. The point-matching method employed the polar coordinate system based on the circular hole and the boundary conditions were satisfied along the points located on the sides of the rectangular plate. Lee and Kim(1984) carried out vibration experiments on the rectangular plates with a hole in air and water. Kim et al.(1987) performed the theoretical analysis on a stiffened rectangular plate with a hole. Avalos and Laura(2003) calculated the natural frequency of a simply-supported rectangular plate with two rectangular holes using the classical Rayleigh-Ritz method. Lee et al.(1994) analyzed a square plate with two collinear circular holes using the classical Rayleigh-Ritz method.

A circular plate with en eccentric circular hole has been treated by various methods. Nagaya(1980) developed an analytical method which utilizes a coordinate system whose origin is at the center of the eccentric hole and an infinite series to represent the outer boundary curve. Khurasia and Rawtani(1978) studied the effect of the eccentricity of the hole on the vibration characteristics of the circular plate by using the triangular finite element method. Lin(1982) used an analytical method based on the transformation of Bessel functions to calculate the free transverse vibrations of uniform circular plates and membranes with eccentric holes. Laura et al.(2006) applied the Rayleigh-Ritz method to circular plates restrained against rotation with an eccentric circular perforation with a free edge. Cheng et al.(2003) used the finite element analysis code, Nastran, to analyze the effects of the hole eccentricity, hole size and boundary condition on the vibration modes of annular-like plates. Lee et al.(2007) used an indirect formulation in conjunction with degenerate kernels and Fourier series to solve for the natural frequencies and modes of circular plates with multiple circular holes and verified the finite element solution by using ABAQUS. Zhong and Yu(2007) formulated a weak-form quadrature element method to study the flexural vibrations of an eccentric annular Mindlin plate.

Recently, Kwak et al.(2005, 2006, 2007), and Heo and Kwak(2008) presented a new method called the Independent Coordinate Coupling Method(ICCM) for the free vibration analysis of a rectangular plate with a rectangular or a circular hole. This method utilizes independent coordinates for the global and local domains and the transformation matrix between the local and global coordinates which is obtained by imposing a kinematical relation on the displacement matching condition inside the hole domain. In the Rayleigh-Ritz method, the effect of the hole can be considered by the subtraction of the energy for the hole domain in deriving the total energy. In doing so, the previous researches considered only the global coordinate system for the integration. The ICCM is advantageous because it does not need to use a complex integration process to determine the total energy of the plate with a hole. The ICCM can be also applied to a circular plate with an eccentric hole. The numerical results obtained by the ICCM were compared to the numerical results of the classical approach, the finite element method, and the experimental results. The numerical results show the efficacy of the proposed method.

2. Rayleigh-Ritz method for free vibration analysis of rectangular plate

Let us consider a rectangular plate with side lengths a in the X direction and b in the Y direction. The kinetic and potential energies of the rectangular plate can be expressed as

$$T_R = \frac{1}{2}\rho h \int_0^a \int_0^b \dot{w}_r^2 dxdy \tag{1}$$

$$V_R = \frac{1}{2}D\int_0^a \int_0^b \left[\left(\frac{\partial^2 w_r}{\partial x^2}\right)^2 + \left(\frac{\partial^2 w_r}{\partial y^2}\right)^2 + 2\nu\left(\frac{\partial^2 w_r}{\partial x^2}\frac{\partial^2 w_r}{\partial y^2}\right) + 2(1-\nu)\left(\frac{\partial^2 w_r}{\partial x \partial y}\right)^2 \right] dxdy \tag{2}$$

where $w_r = w_r(x,y,t)$ represents the deflection of the plate, ρ the mass density, h the thickness, $D = Eh^3/12(1-v^2)$, E the Young's modulus, and ν the Poisson's ratio.
By using the non-dimensional variables, $\xi = x/a$, $\eta = y/b$ and the assumed mode method, the deflection of the plate can be expressed as

$$w_r(\xi,\eta,t) = \Phi_r(\xi,\eta)q_r(t) \tag{3}$$

where $\Phi_r(\xi,\eta) = [\Phi_{r1}\,\Phi_{r2}\cdots\Phi_{rm}]$ is a $1\times m$ matrix consisting of the admissible functions and $q_r(t) = [q_{r1}\,q_{r2}\cdots q_{rm}]^T$ is a $m\times 1$ vector consisting of generalized coordinates, in which m is the number of admissible functions used for the approximation of the deflection. Inserting Eq. (3) into Eqs. (1) and (2) results in Eq. (4).

$$T_R = \frac{1}{2}\dot{q}_r^T M_r \dot{q}_r\,,\quad V_R = \frac{1}{2}q_r^T K_r q_r \tag{4a,b}$$

where

$$M_r = \rho hab\,\bar{M}_r\,,\quad K_r = \frac{Db}{a^3}\bar{K}_r \tag{5a,b}$$

In which

$$\bar{M}_r = \int_0^1 \int_0^1 \Phi_r^T \Phi_r \, d\xi d\eta \tag{6a}$$

$$\begin{aligned}\bar{K}_r = \int_0^1 \int_0^1 \Bigg[& \frac{\partial^2 \Phi_r^T}{\partial \xi^2}\frac{\partial^2 \Phi_r}{\partial \xi^2} + \alpha^4 \frac{\partial^2 \Phi_r^T}{\partial \eta^2}\frac{\partial^2 \Phi_r}{\partial \eta^2} + \nu\alpha^2\left(\frac{\partial^2 \Phi_r^T}{\partial \xi^2}\frac{\partial^2 \Phi_r}{\partial \eta^2} + \frac{\partial^2 \Phi_r^T}{\partial \eta^2}\frac{\partial^2 \Phi_r}{\partial \xi^2}\right) \\ & + 2(1-\nu)\alpha^2 \frac{\partial^2 \Phi_r^T}{\partial \xi \partial \eta}\frac{\partial^2 \Phi_r}{\partial \xi \partial \eta}\Bigg] d\xi d\eta \end{aligned} \tag{6b}$$

$\bar{M}_r$, $\bar{K}_r$ represent the non-dimensionalized mass and stiffness matrices, respectively, and $\alpha = a/b$ represents the aspect ratio of the plate. The equation of motion can be derived by inserting Eq. (4) into the Lagrange's equation and the eigenvalue problem can be expressed as

$$\left[K_r - \omega^2 M_r\right]A = 0 \tag{7}$$

If we use the non-dimensionalized mass and stiffness matrices introduced in Eq. (5), the eigenvalue problem given by Eq. (7) can be also non-dimensionalized.

$$\left[\bar{K}_r - \bar{\omega}^2 \bar{M}_r\right]A = 0 \tag{8}$$

where $\bar{\omega}$ is the non-dimensionalized natural frequency, which has the relationship with the natural frequency as follows:

$$\bar{\omega} = \omega\sqrt{\frac{\rho h a^4}{D}} \tag{9}$$

To calculate the mass and stiffness matrices given by Eq. (6) easily, the admissible function matrix given by Eq. (3) needs to be expressed in terms of admissible function matrices in each direction.

$$\Phi_{ri}(\xi,\eta) = \phi_i(\xi)\psi_i(\eta),\ i = 1,2,\ldots,m \tag{10}$$

Then, the non-dimensionalized mass and stiffness matrices given by Eq. (6) can be expressed as [Kwak and Han(2007)]

$$\left(\bar{M}_r\right)_{ij} = X_{ij}Y_{ij} \tag{11a}$$

$$\left(\bar{K}_r\right)_{ij} = \hat{X}_{ij}Y_{ij} + \alpha^4 X_{ij}\hat{Y}_{ij} + \alpha^2\nu\left(\tilde{X}_{ji}\tilde{Y}_{ij} + \tilde{X}_{ij}\tilde{Y}_{ji}\right) + \alpha^2(1-\nu)\bar{X}_{ij}\bar{Y}_{ij},\quad i,j = 1,2,\ldots,m \tag{11b}$$

where

$$X_{ij} = \int_0^1 \phi_i\phi_j d\xi\,,\ \bar{X}_{ij} = \int_0^1 \phi_i'\phi_j' d\xi\,,\ \hat{X}_{ij} = \int_0^1 \phi_i''\phi_j'' d\xi\,,\ \tilde{X}_{ij} = \int_0^1 \phi_i\phi_j'' d\xi \tag{12a-d}$$

$$Y_{ij} = \int_0^1 \psi_i\psi_j d\eta\,,\ \bar{Y}_{ij} = \int_0^1 \psi_i'\psi_j' d\eta\,,\ \hat{Y}_{ij} = \int_0^1 \psi_i''\psi_j'' d\eta\,,\ \tilde{Y}_{ij} = \int_0^1 \psi_i\psi_j'' d\eta,\ i,j = 1,2,\ldots,m \tag{12e-h}$$

If n admissible functions are used in the X and Y directions and the combination of admissible functions are used, a total of n^2 admissible functions can be obtained, which yields $m = n^2$. If each type of admissible functions are considered as $\chi_i\,(i = 1,2,\ldots,n)$ and $\gamma_i\,(i = 1,2,\ldots,n)$, then the relationship of between the sequence of the admissible function introduced in Eq. (10) and those of separated admissible functions can be expressed as

$$\phi_k = \begin{cases} \chi_1 & 1 \le k \le n \\ \chi_2 & n+1 \le k \le 2n \\ \chi_3 & 2n+1 \le k \le 3n \\ \vdots & \\ \chi_n & (n-1)n+1 \le k \le n^2 \end{cases}\ ,\ \psi_k = \begin{cases} \gamma_k & 1 \le k \le n \\ \gamma_{k-n} & n+1 \le k \le 2n \\ \gamma_{k-2n} & 2n+1 \le k \le 3n \\ \vdots & \\ \gamma_{k-(n-1)n} & (n-1)n+1 \le k \le n^2 \end{cases} \tag{13a,b}$$

Therefore, instead of integrating $m^2 = n^4$ elements in Eq. (12), n^2 integrations and matrix rearrangement will suffice. First, let us calculate the following.

$$\Sigma_{ij} = \int_0^1 \chi_i \chi_j d\xi\,,\ \bar{\Sigma}_{ij} = \int_0^1 \chi_i' \chi_j' d\xi\,,\ \hat{\Sigma}_{ij} = \int_0^1 \chi_i'' \chi_j'' d\xi\,,\ \ \tilde{\Sigma}_{ij} = \int_0^1 \chi_i \chi_j'' d\xi \qquad \text{(14a-d)}$$

$$\Gamma_{ij} = \int_0^1 \gamma_i \gamma_j d\eta\,,\ \bar{\Gamma}_{ij} = \int_0^1 \gamma_i' \gamma_j' d\eta\ \ \hat{\Gamma}_{ij} = \int_0^1 \gamma_i'' \gamma_j'' d\eta\,,\ \tilde{\Gamma}_{ij} = \int_0^1 \gamma_i \gamma_j'' d\eta,\ i,j = 1,2,\ldots,n \qquad \text{(14e-h)}$$

And then the matrices given by Eq. (12) can be derived as follows:

$$X = \begin{bmatrix} \Sigma_{11}\bar{I} & \Sigma_{12}\bar{I} & \cdots & \Sigma_{1n}\bar{I} \\ \Sigma_{21}\bar{I} & \Sigma_{22}\bar{I} & \cdots & \Sigma_{2n}\bar{I} \\ \vdots & \vdots & \ddots & \vdots \\ \Sigma_{n1}\bar{I} & \Sigma_{n2}\bar{I} & \cdots & \Sigma_{nn}\bar{I} \end{bmatrix},\ \bar{X} = \begin{bmatrix} \bar{\Sigma}_{11}\bar{I} & \bar{\Sigma}_{12}\bar{I} & \cdots & \bar{\Sigma}_{1n}\bar{I} \\ \bar{\Sigma}_{21}\bar{I} & \bar{\Sigma}_{22}\bar{I} & \cdots & \bar{\Sigma}_{2n}\bar{I} \\ \vdots & \vdots & \ddots & \vdots \\ \bar{\Sigma}_{n1}\bar{I} & \bar{\Sigma}_{n2}\bar{I} & \cdots & \bar{\Sigma}_{nn}\bar{I} \end{bmatrix} \qquad \text{(15a,b)}$$

$$\hat{X} = \begin{bmatrix} \hat{\Sigma}_{11}\bar{I} & \hat{\Sigma}_{12}\bar{I} & \cdots & \hat{\Sigma}_{1n}\bar{I} \\ \hat{\Sigma}_{21}\bar{I} & \hat{\Sigma}_{22}\bar{I} & \cdots & \hat{\Sigma}_{2n}\bar{I} \\ \vdots & \vdots & \ddots & \vdots \\ \hat{\Sigma}_{n1}\bar{I} & \hat{\Sigma}_{n2}\bar{I} & \cdots & \hat{\Sigma}_{nn}\bar{I} \end{bmatrix},\ \tilde{X} = \begin{bmatrix} \tilde{\Sigma}_{11}\bar{I} & \tilde{\Sigma}_{12}\bar{I} & \cdots & \tilde{\Sigma}_{1n}\bar{I} \\ \tilde{\Sigma}_{21}\bar{I} & \tilde{\Sigma}_{22}\bar{I} & \cdots & \tilde{\Sigma}_{2n}\bar{I} \\ \vdots & \vdots & \ddots & \vdots \\ \tilde{\Sigma}_{n1}\bar{I} & \tilde{\Sigma}_{n2}\bar{I} & \cdots & \tilde{\Sigma}_{nn}\bar{I} \end{bmatrix} \qquad \text{(15c,d)}$$

$$Y = \begin{bmatrix} \Gamma & \Gamma & \cdots & \Gamma \\ \Gamma & \Gamma & \cdots & \Gamma \\ \vdots & \vdots & \ddots & \vdots \\ \Gamma & \Gamma & \cdots & \Gamma \end{bmatrix},\ \bar{Y} = \begin{bmatrix} \bar{\Gamma} & \bar{\Gamma} & \cdots & \bar{\Gamma} \\ \bar{\Gamma} & \bar{\Gamma} & \cdots & \bar{\Gamma} \\ \vdots & \vdots & \ddots & \vdots \\ \bar{\Gamma} & \bar{\Gamma} & \cdots & \bar{\Gamma} \end{bmatrix}, \qquad \text{(15e,f)}$$

$$\hat{Y} = \begin{bmatrix} \hat{\Gamma} & \hat{\Gamma} & \cdots & \hat{\Gamma} \\ \hat{\Gamma} & \hat{\Gamma} & \cdots & \hat{\Gamma} \\ \vdots & \vdots & \ddots & \vdots \\ \hat{\Gamma} & \hat{\Gamma} & \cdots & \hat{\Gamma} \end{bmatrix},\ \hat{Y} = \begin{bmatrix} \hat{\Gamma} & \hat{\Gamma} & \cdots & \hat{\Gamma} \\ \hat{\Gamma} & \hat{\Gamma} & \cdots & \hat{\Gamma} \\ \vdots & \vdots & \ddots & \vdots \\ \hat{\Gamma} & \hat{\Gamma} & \cdots & \hat{\Gamma} \end{bmatrix} \qquad \text{(15g,h)}$$

where $\bar{I}$ is an $n \times n$ matrix full of ones.

Let us consider the simply-supported case in the X direction. In this case, the eigenfunction of the uniform beam can be used as an admissible function.

$$\chi_i = \sqrt{2}\sin i\pi\xi,\ i = 1,2,\ldots n \qquad \text{(16)}$$

In the case of the clamped condition in the X direction, the eigenfunction of a clamped-clamped uniform beam can be used.

$$\chi_i = \cosh\lambda_i\xi - \cos\lambda_i\xi - \sigma_i(\sinh\lambda_i\xi - \sin\lambda_i\xi),\ \ i = 1,2,\ldots,n \qquad \text{(17)}$$

where λ_i =4.730, 7.853, 10.996, 14.137,... and $\sigma_i = (\cosh\lambda_i - \cos\lambda_i)/(\sinh\lambda_i - \sin\lambda_i)$. In the case of a free-edge condition in the X direction, we can use the eigenfunction of a free-free uniform beam.

$$\chi_1 = 1 \,,\ \chi_2 = \sqrt{12}\left(\xi - \frac{1}{2}\right) \tag{18a,b}$$

$$\chi_{i+2} = \cosh \lambda_i \xi + \cos \lambda_i \xi - \sigma_i (\sinh \lambda_i \xi + \sin \lambda_i \xi) \,,\ i = 1,2,\ldots n-2 \tag{18c}$$

where λ_i and σ_i are the same as the ones for the clamped-clamped beam, and the first and the second modes represent the rigid-body modes. Σ_{ij} , $\bar{\Sigma}_{ij}$, $\hat{\Sigma}_{ij}$, $\tilde{\Sigma}_{ij}$ for each case are given in the work of Kwak and Han(2007).
For the admissible functions in the y direction, γ_i, the same method can be applied. The combination of different admissible functions can yield various boundary conditions.

3. Rayleigh-Ritz method for free vibration analysis of circular plate

Let us consider a uniform circular plate with radius, R, and thickness, h. The kinetic and potential energies can be expressed as follows:

$$T_C = \frac{1}{2}\rho h \int_0^{2\pi} \int_0^R \dot{w}_c^2 \, r dr d\theta \tag{19a}$$

$$V_C = \frac{1}{2} D \int_0^{2\pi} \int_0^R \left\{ \left(\frac{\partial^2 w_c}{\partial r^2} + \frac{1}{r}\frac{\partial w_c}{\partial r} + \frac{1}{r^2}\frac{\partial^2 w_c}{\partial \theta^2} \right)^2 - 2(1-\nu)\left[\left(\frac{\partial^2 w_c}{\partial r^2} \right)\left(\frac{1}{r}\frac{\partial w_c}{\partial r} + \frac{1}{r^2}\frac{\partial^2 w_c}{\partial \theta^2} \right) - \left(\frac{1}{r}\frac{\partial^2 w_c}{\partial r \partial \theta} - \frac{1}{r^2}\frac{\partial^2 w_c}{\partial \theta^2} \right)^2 \right] \right\} r dr d\theta \tag{19b}$$

Unlike the uniform rectangular plate, simply-supported, clamped, and free-edge uniform circular plates have eigenfunctions. Hence, the deflection of the circular plate can be expressed as the combination of eigenfunctions and generalized coordinates.

$$w_c(r,\theta,t) = \sum_{i=1}^{n_c} \Phi_{ci}(r,\theta) q_{ci}(t) = \Phi_c(r,\theta) q_c(t) \tag{20}$$

where $\Phi_{ci}(r,\theta)$ represents the eigenfunction of the uniform circular plate and $q_{ci}(t)$ represents the generalized coordinate. Inserting Eq. (20) into Eq. (19) results in the following.

$$T_C = \frac{1}{2}\dot{q}_c^T M_c \dot{q}_c \,,\ V_C = \frac{1}{2} q_c^{\ T} K_c q_c \tag{21a,b}$$

where

$$M_c = \rho h \pi R^2 I \,,\quad K_c = \frac{\pi D}{R^2} \Lambda_c \tag{22a,b}$$

in which I is an $n_c \times n_c$ identity matrix, Λ_c is an $n_c \times n_c$ diagonal matrix whose diagonals are λ_i^4 . The eigenvalue has the expression, $\lambda^4 = \omega^2 \rho h R^4 / D$.
Since our study is concerned with either a rectangular or a circular hole, we consider only a free-edge circular plate [Itao and Crandall(1979)]. If the eigenfunctions are rearranged in ascending order, we can have

$$\Phi_{c1} = 1\,,\ \Phi_{c2} = \frac{r}{R}\cos\theta\,,\ \Phi_{c3} = \frac{r}{R}\sin\theta \tag{23a-c}$$

$$\Phi_{c(k+3)} = A_k\left[J_{n_k}\left(\lambda_k\frac{r}{R}\right) + C_k I_{n_k}\left(\lambda_k\frac{r}{R}\right)\right]f_k(\theta),\ k = 1,2,... \tag{23d}$$

where J_{n_k} and I_{n_k} are the Bessel functions of the first kind and the modified Bessel functions of order n_k, respectively. The first three modes represent the rigid-body modes and other modes represent the elastic vibration modes. The characteristic values obtained from Eq. (23d) are tabulated in the work of Kwak and Han(2007) by rearranging the values given in reference [Leissa(1993)]. In this case, Λ_c has the following form.

$$\Lambda_c = diag\left(\begin{bmatrix} 0 & 0 & 0 & \lambda_1^4 & \lambda_2^4 & \lambda_3^4 & \cdots & \lambda_{n_c-3}^4 \end{bmatrix}\right) \tag{24}$$

4. Free vibration analysis of rectangular plate with a hole by use of global coordinates

Let us consider a rectangular plate with a rectangular hole, as shown in Figure 1.

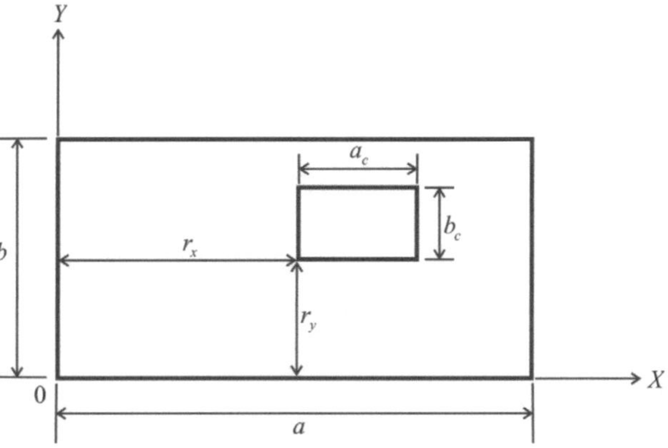

Fig. 1. Rectangular plate with a rectangular hole with global axes.

In this case, the total kinetic and potential energies can be obtained by subtracting the energies belonging to the hole domain from the total energies for the global domain.

$$\begin{aligned} T_{total} &= T_R - T_{RH} = \frac{1}{2}\dot{q}_r^T(M_r - M_{rh}^*)\dot{q}_r = \frac{1}{2}\dot{q}_r^T M_{rrh}\dot{q}_r \\ V_{total} &= V_R - V_{RH} = \frac{1}{2}q_r^T(K_r - K_{rh}^*)q_r = \frac{1}{2}q_r^T K_{rrh}q_r \end{aligned} \tag{25a,b}$$

where

$$M_{rrh} = M_r - M_{rh}^*\,,\ K_{rrh} = K_r - K_{rh}^* \tag{26a,b}$$

in which M_r, K_r are mass and stiffness matrices for the whole rectangular plate, which are given by Eq. (5), and M_{rh}^*, K_{rh}^* reflect the reductions in mass and stiffness matrices due to the hole, which can be also expressed by non-dimensionalized mass and stiffness matrices, respectively.

$$M_{rh}^* = \rho hab\,\bar{M}_{rh}^*\,,\; K_{rh}^* = \frac{Db}{a^3}\bar{K}_{rh}^* \tag{27a,b}$$

where

$$\bar{M}_{rh}^* = \int_{\bar{r}_x}^{\bar{r}_x+\bar{a}_c}\int_{\bar{r}_y}^{\bar{r}_y+\bar{b}_c} \Phi_r^T \Phi_r \, d\xi d\eta \tag{28a}$$

$$\begin{aligned}\bar{K}_{rh}^* = \int_{\bar{r}_x}^{\bar{r}_x+\bar{a}_c}\int_{\bar{r}_y}^{\bar{r}_y+\bar{b}_c} \Bigg[& \frac{\partial^2\Phi_r^T}{\partial\xi^2}\frac{\partial^2\Phi_r}{\partial\xi^2} + \alpha^4 \frac{\partial^2\Phi_r^T}{\partial\eta^2}\frac{\partial^2\Phi_r}{\partial\eta^2} + \nu\alpha^2\left(\frac{\partial^2\Phi_r^T}{\partial\xi^2}\frac{\partial^2\Phi_r}{\partial\eta^2} + \frac{\partial^2\Phi_r^T}{\partial\eta^2}\frac{\partial^2\Phi_r}{\partial\xi^2}\right) \\ & + 2(1-\nu)\alpha^2 \frac{\partial^2\Phi_r^T}{\partial\xi\partial\eta}\frac{\partial^2\Phi_r}{\partial\xi\partial\eta}\Bigg] d\xi d\eta \end{aligned} \tag{28b}$$

in which $\bar{r}_x = r_x / a, \bar{r}_y = r_y / b, \bar{a}_c = a_c / a, \bar{b}_c = b_c / b$ represent various aspect ratios. Hence, the non-dimensionalized eigenvalue problem for the addressed problem can be expressed as:

$$\left(\bar{K}_{rrh} - \bar{\omega}^2 \bar{M}_{rrh}\right) A = 0 \tag{29}$$

where

$$\bar{M}_{rrh} = \bar{M}_r - \bar{M}_{rh}^*\,,\; \bar{K}_{rrh} = \bar{K}_r - \bar{K}_{rh}^* \tag{30a,b}$$

To calculate the non-dimensionalized mass and stiffness matrices for the hole domain given by Eq. (28), we generally resort to numerical integration. However, in the case of a simply-supported rectangular plate with a rectangular hole, the exact expressions exists for the non-dimensionalized mass and stiffness matrices for the hole[Kwak & Han(2007)].

5. Independent coordinate coupling method for a rectangular plate with a rectangular hole

Let us consider again the rectangular plate with a rectangular hole, as shown in Fig. 2. As can be seen from Fig. 2, the local coordinates fixed to the hole domain is introduced. Considering the non-dimensionalized coordinates, $\xi_h = x_h / a_c$, $\eta_h = y_h / b_c$, we can express the displacement inside the hole domain as

$$w_h(\xi_h,\eta_h) = \Phi_h(\xi_h,\eta_h) q_h \tag{31}$$

where $\Phi_h(\xi_h,\eta_h) = [\Phi_{h1}\,\Phi_{h2} \ldots \Phi_{hm_h}]$ is the $1\times m_h$ admissible function matrix, and $q_h(t) = [q_{h1}\,q_{h2} \ldots q_{hm_h}]^T$ is the $m_h \times 1$ generalized coordinate vector. If we apply the separation of variables to the admissible function as we did in Eq. (10), then we have

$$\Phi_{hi}(\xi_h,\eta_h) = \phi_{hi}(\xi_h)\psi_{hi}(\eta_h),\; i = 1,2,\ldots,m_h \tag{32}$$

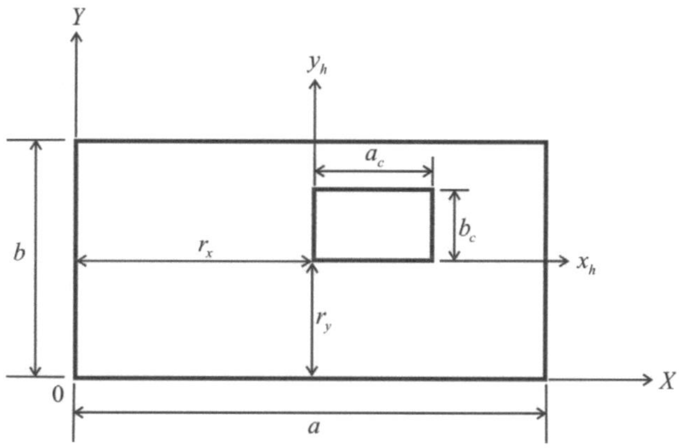

Fig. 2. Rectangular plate with a rectangular hole with local axes.

Using Eqs. (31) and (32), we can express the kinetic and potential energies in the hole domain as

$$T_{RH} = \frac{1}{2}\dot{q}_{rh}^T M_{rh}\dot{q}_{rh}\,,\quad V_{RH} = \frac{1}{2}q_{rh}{}^T K_{rh} q_{rh} \tag{33a,b}$$

Hence, the total kinetic and potential energies can be written as

$$T_{total} = \frac{1}{2}\dot{q}_r^T M_r \dot{q}_r - \frac{1}{2}\dot{q}_{rh}^T M_{rh}\dot{q}_{rh}\,,\quad V_{total} = \frac{1}{2}q_r^T K_r q_r - \frac{1}{2}q_{rh}^T K_{rh} q_{rh} \tag{34a,b}$$

Where M_r, K_r are defined by Eqs. (5) and (6), and

$$M_{rh} = \rho h a_c b_c \bar{M}_{rh}\,,\quad K_{rh} = \frac{D b_c}{a_c^3}\bar{K}_{rh} \tag{35a,b}$$

in which

$$\bar{M}_{rh} = \int_0^1\int_0^1 \Phi_h^T \Phi_h \, d\xi_h d\eta_h \tag{36a}$$

$$\bar{K}_{rh} = \int_0^1\int_0^1 \left[\frac{\partial^2\Phi_h^T}{\partial\xi_h^2}\frac{\partial^2\Phi_h}{\partial\xi_h^2} + \alpha_c^4 \frac{\partial^2\Phi_h^T}{\partial\eta_h^2}\frac{\partial^2\Phi_h}{\partial\eta_h^2} + \nu\alpha_c^2\left(\frac{\partial^2\Phi_h^T}{\partial\xi_h^2}\frac{\partial^2\Phi_h}{\partial\eta_h^2} + \frac{\partial^2\Phi_h^T}{\partial\eta_h^2}\frac{\partial^2\Phi_h}{\partial\xi_h^2} \right) + 2(1-\nu)\alpha_c^2 \frac{\partial^2\Phi_h^T}{\partial\xi_h\partial\eta_h}\frac{\partial^2\Phi_h}{\partial\xi_h\partial\eta_h} \right] d\xi_h d\eta_h \tag{36b}$$

and $\alpha_c = a_c / b_c$. Note that the definite integrals in Eq. (36) has distinctive advantage compared to Eq. (28) because it has an integral limit from 0 to 1 thus permitting closed form expressions. Therefore, we can use the same expression used for the free-edge rectangular plate.

Since the local coordinate system is used for the hole domain, we do not have to carry out integration for the hole domain, as in Eq. (28). However, the displacement matching condition between the global and local coordinates should be satisfied inside the hole domain. The displacement matching condition inside the hole domain can be written as

$$w_{rh}(\xi_h,\eta_h)=w_r(\xi,\eta) \tag{37}$$

The relationship between the non-dimensionalized global and local coordinates can be written as

$$\xi=\frac{r_x}{a}+\frac{a_c}{a}\xi_h,\quad \eta=\frac{r_y}{b}+\frac{b_c}{b}\eta_h \tag{38a,b}$$

Considering Eqs. (3), (10), (31) and (32), and inserting them into Eq. (37), we can derive

$$\sum_{j=1}^{m_h}\Phi_{rhj}(\xi_h,\eta_h)q_{rhj}(t)=\sum_{j=1}^{m_h}\phi_{hj}(\xi_h)\psi_{hj}(\eta_h)q_{rhj}(t)=\sum_{k=1}^{m}\Phi_{rk}(\xi,\eta)q_{rk}(t)=\sum_{k=1}^{m}\phi_k(\xi)\psi_k(\eta)q_{rk}(t) \tag{39}$$

Multiplying Eq. (39) by $\phi_{hi}(\xi_h)\psi_{hi}(\eta_h)$ and performing integration, we can derive

$$\begin{aligned}&\sum_{j=1}^{m_h}\int_0^1\int_0^1\phi_{hi}(\xi_h)\psi_{hi}(\eta_h)\phi_{hj}(\xi_h)\psi_{hj}(\eta_h)\,d\xi_h\,d\eta_h\,q_{rhj}(t)=\\&=\sum_{k=1}^{m}\int_0^1\int_0^1\phi_{hi}(\xi_h)\psi_{hi}(\eta_h)\phi_{rk}(\xi)\psi_{rk}(\eta)\,d\xi_h\,d\eta_h\,q_{rk}(t),\quad i=1,2,\ldots,m_h\end{aligned} \tag{40}$$

Using the orthogonal property of the eigenfunctions of the uniform beam, Eq. (40) can be rewritten as

$$\begin{aligned}q_{rhi}(t)&=\sum_{k=1}^{m}\int_0^1\phi_{hi}(\xi_h)\phi_k(\xi)\,d\xi_h\int_0^1\psi_{hi}(\eta_h)\psi_k(\eta)\,d\eta_h\,q_{rk}(t)\\&=\sum_{k=1}^{m}(T_{rrh})_{ik}\,q_{rk}(t),\ i=1,2,\ldots,m_h\end{aligned} \tag{41}$$

If we express Eq. (41) in the matrix form, we can have

$$q_{rh}=T_{rrh}q_r \tag{42}$$

where T_{rrh} is the $m_h\times m$ transformation matrix between two coordinates. Inserting Eq. (42) into Eq. (34), we can derive

$$T_{total}=\frac{1}{2}\dot{q}_r^T M_r\dot{q}_r-\frac{1}{2}\dot{q}_{rh}^T T_{rrh}^T M_{rh}T_{rrh}\dot{q}_{rh}=\frac{1}{2}\dot{q}_r M_{rrh}\dot{q}_r \tag{43a}$$

$$V_{total}=\frac{1}{2}q_r^T K_r q_r-\frac{1}{2}q_{rh}^T T_{rrh}^T K_{rh}T_{rrh}q_{rh}=\frac{1}{2}q_r^T K_{rrh}q_r \tag{43b}$$

where

$$M_{rrh}=M_r-T_{rrh}^T M_{rh}T_{rrh},\ K_{rrh}=K_r-T_{rrh}^T K_{rh}T_{rrh} \tag{44a,b}$$

Equation (44) can be expressed by means of non-dimensionalized parameters

$$M_{rrh}=\rho hab\,\bar{M}_{rrh},\ K_{rrh}=\frac{Db}{a^3}\bar{K}_{rrh} \tag{45a,b}$$

where

$$\bar{M}_{rrh} = \bar{M}_r - (\bar{a}_c \bar{b}_c) T_{rrh}^T \bar{M}_{rh} T_{rrh} \,, \quad \bar{K}_{rrh} = \bar{K}_r - \frac{\bar{b}_c}{\bar{a}_c^3} T_{rrh}^T \bar{K}_{rh} T_{rrh} \tag{46a,b}$$

Hence, the non-dimensionalized eigenvalue problem can be written in the same form as Eq. (29).

In deriving the mass and stiffness matrices, Eq. (46), for the eigenvalue problem, we only needed the transformation matrix, T_{rrh}. $\bar{M}_r$, $\bar{K}_r$ can be easily computed by Eq. (11) according to the edge boundary conditions and $\bar{M}_{rh}$, $\bar{K}_{rh}$ can be computed from the results of Eq. (11) for the all free-edge rectangular plate. On the other hand, the computation of $\bar{M}_{rh}^*$, $\bar{K}_{rh}^*$ based on the global coordinates is not easy because of integral limits. Compared to the approach based on the global coordinates, the numerical integration for the transformation matrix, T_{rrh}, is easy because the integral limits are 0 and 1. The process represented by Eqs. (42) and (46) is referred to as the ICCM in the study by Kwak and Han(2007). The ICCM enables us to solve the free vibration problem of the rectangular plate with a rectangular hole more easily than the previous approaches based on the global coordinates do. The advantage of the ICCM becomes more apparent when we deal with a circular hole, as will be demonstrated in the next section.

6. Free vibration analysis of rectangular plate with multiple rectangular cutouts by independent coordinate coupling method

As in the case of single rectangular hole, the total energy can be computed by subtracting the energy belonging to holes from the energy of the whole rectangular plate, which is not an easy task when applying the classical Rayleigh-Ritz method. However, the ICCM enables us to formulate the free vibration problem for the rectangular plate with multiple holes more easily than the CRRM.

Let us consider a rectangular plate with n rectangular holes as shown in Fig. 3.

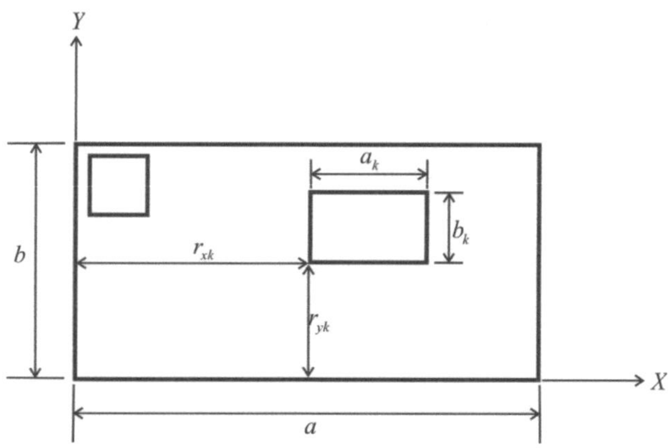

Fig. 3. Rectangular plate with multiple rectangular holes

By employing the same formulation used in the case of a rectangular hole with a single rectangular hole, the non-dimensionalized mass and stiffness matrices can be derived considering a single hole case:

$$\bar{M}_{rrh} = \bar{M}_r - \sum_{k=1}^{n} (\bar{a}_k \bar{b}_k) T_{rrhk}^T \bar{M}_{rh} T_{rrhk} \, , \; \bar{K}_{rrh} = \bar{K}_r - \sum_{k=1}^{n} \frac{\bar{b}_k}{\bar{a}_k^3} T_{rrhk}^T \bar{K}_{rh} T_{rrhk} \tag{47a,b}$$

where the following non-dimensionalized variables are introduced for the analysis

$$\bar{r}_{xk} = r_{xk} / a , \; \bar{r}_{yk} = r_{yk} / b , \; \bar{a}_k = a_k / a , \; \bar{b}_k = b_k / b \tag{48a-d}$$

And the transformation matrix can be expressed by considering Eq. (41)

$$\left(T_{rrhk}\right)_{ij} = \int_0^1 \phi_{hi}(\xi_{hi}) \phi_j(\xi) d\xi_{hi} \int_0^1 \psi_{hi}(\eta_{hi}) \psi_j(\eta) d\eta_{hi} \tag{49}$$

In order to validate the efficacy of the ICCM for the rectangular plate with multiple rectangular holes, the rectangular plate with two square holes as shown in Fig. 4 is considered as a numerical example, in which $\nu = 0.3$. The results of the ICCM are compared to those obtained by the classical Rayleigh-Ritz method.

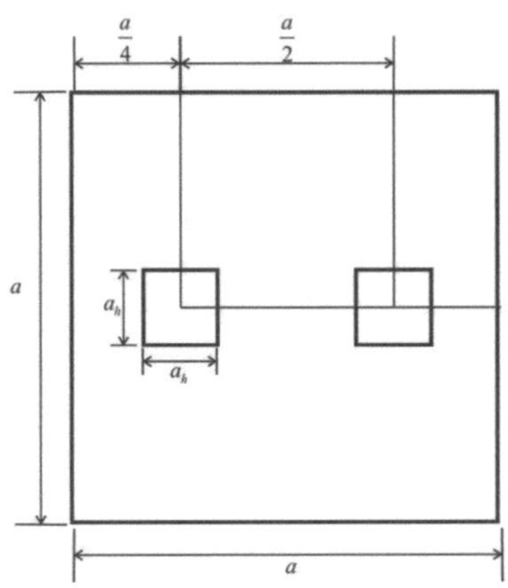

Fig. 4. Square plate with two square holes

Ten admissible functions in each direction were employed, which implies one hundred admissible functions, for both CRRM and ICCM. In the case of the ICCM, the additional admissible functions are necessary for the hole domain. In our study ten admissible functions in each direction of the rectangular hole domain, which implies one hundred admissible functions, were used. The number of admissible functions guaranteeing the convergence are referred to the work of Kwak and Han(2007).

Fig. 5 shows the non-dimensionalized natural frequencies obtained by the CRRM and ICCM for the case that the plate shown in Fig. 4 has all simply-supported boundary conditions, where $\bar{a}_h = a_h / a$. As shown in Fig. 5, the results obtained by the ICCM agree well with the results obtained by the CRRM. The fundamental frequency increases as the size of the hole increases but higher natural frequencies undergo rapid change as the size of the hole increases. This result is similar to the one obtained by Kwak and Han(2007) for a single hole case.

In the case of the simply-supported rectangular plate with a hole, the solutions of integrals can be obtained in a closed form without numerical integral technique. However, in the case of the clamped rectangular plate, the closed-form solution can't be obtained, so the numerical integrations are necessary. Figure 6 shows the advantage of the ICCM over the CRRM regarding the computational time. As can be seen from Fig. 6, the computational time increases enormously in the case of the CRRM compared to the ICCM as the size of the hole increases. Hence, it can be readily recognized that the ICCM has the computational efficiency compared to the CRRM, which was confirmed in the work of Kwak and Han(2007) for a single hole case.

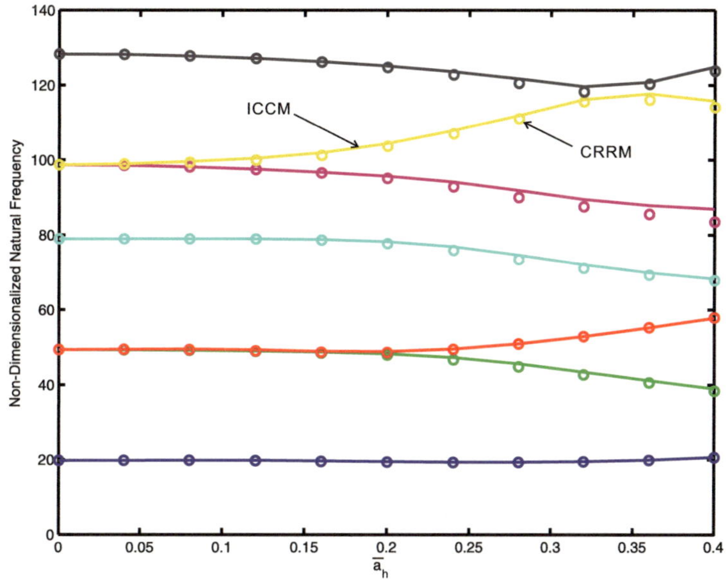

Fig. 5. Simply-supported square plate with two square holes

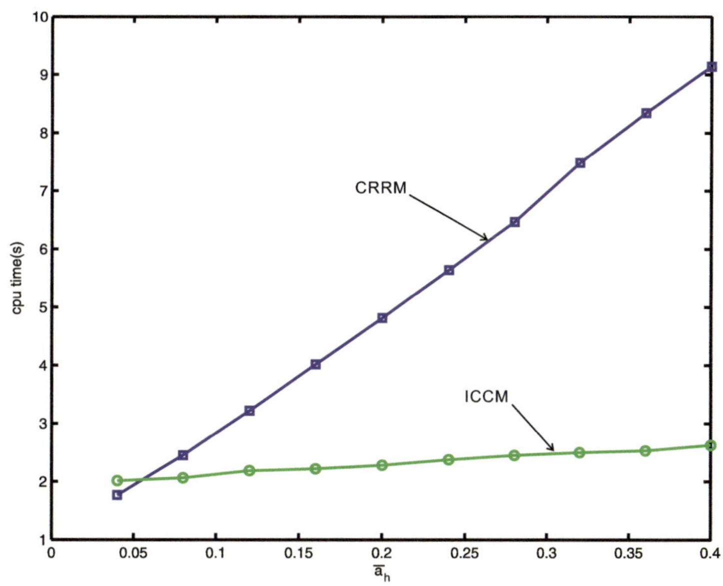

Fig. 6. CPU time vs. hole size

7. Independent coordinate coupling method for a rectangular plate with a circular hole

Let us consider a rectangular plate with a circular hole, as shown in Fig. 7. The global coordinate approach used in Section 4 can be used for this problem but we must resort to numerical integration technique. If we use the ICCM, we can avoid the complex numerical computation and thus simplify the computation as in the case of a rectangular hole.

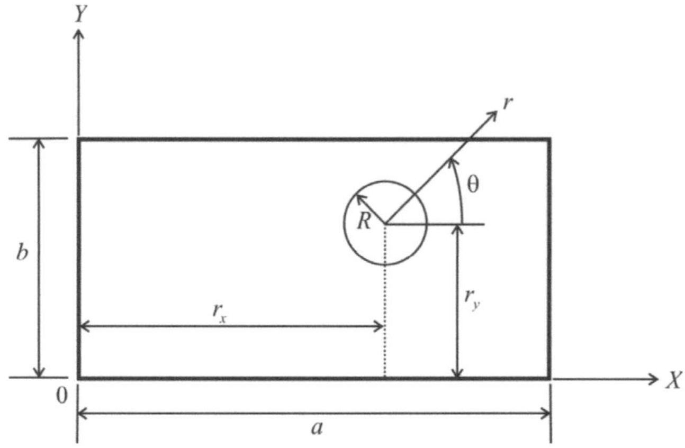

Fig. 7. Rectangular plate with a circular hole.

The total kinetic and potential energies of the rectangular plate with a circular hole are obtained by subtracting the energies of the circular hole domain from the energies of the whole plate, as we did for the case of a rectangular hole. Hence, the following equations can be obtained by using Eqs. (4) and (21).

$$T_{total}=\frac{1}{2}\dot{q}_r^T M_r\dot{q}_r-\frac{1}{2}\dot{q}_{ch}^T M_{ch}\dot{q}_{ch},\quad V_{total}=\frac{1}{2}q_r^T K_r q_r-\frac{1}{2}q_{ch}^T K_{ch}q_{ch} \qquad (50a,b)$$

In order to apply the ICCM, the displacement matching condition should be satisfied. Hence, the following condition should be satisfied inside the circular hole domain.

$$w_c(r,\theta)=w_r(\xi,\eta) \qquad (51)$$

Considering Eqs. (20), (3) and (10), we can obtain.

$$\sum_{j=1}^{m_c}\Phi_{cj}(r,\theta)q_{chj}(t)=\sum_{k=1}^{m}\Phi_{rk}(\xi,\eta)q_{rk}(t)=\sum_{k=1}^{m}\phi_k(\xi)\psi_k(\eta)q_{rk}(t) \qquad (52)$$

Multiplying Eq. (52) by $\Phi_{ci}(r,\theta)$ and performing integration over the circular hole domain result in

$$\sum_{j=1}^{m_c}\int_0^{2\pi}\int_0^R\Phi_{ci}(r,\theta)\Phi_{cj}(r,\theta)r dr\, d\theta\, q_{chj}(t)=\sum_{k=1}^{m}\int_0^{2\pi}\int_0^R\Phi_{ci}(r,\theta)\phi_k(\xi)\psi_k(\eta)r dr\, d\theta\, q_{rk}(t),\quad i=1,2,\ldots,m_c \qquad (53)$$

Using the orthogonal property of $\Phi_{ci}(r,\theta)$, Eq. (53) can be rewritten as

$$q_{chi}(t)=\sum_{k=1}^{m}\int_0^{2\pi}\int_0^{R}\Phi_{ci}(r,\theta)\phi_k(\xi)\psi_k(\eta)rdr\,d\theta\, q_{rk}(t)=\sum_{k=1}^{m}\left(T_{ch}\right)_{ik}q_{rk}(t),\ i=1,2,...,m_c \tag{54}$$

Equation (54) can be expressed in matrix form.

$$q_{ch}=T_{rch}\,q_r \tag{55}$$

where T_{ch} is a $m_c \times m$ transformation matrix. We also need the relationship between the global and local coordinates, which can be expressed as follows.

$$\xi=\frac{r_x}{a}+\frac{r\cos\theta}{a},\quad \eta=\frac{r_y}{b}+\frac{r\sin\theta}{b} \tag{56a,b}$$

Using Eq. (55), the mass and stiffness matrices can be easily derived as in the case of a rectangular hole.

$$M_{rch}=M_r-T_{rch}^T M_{ch}T_{rch},\ K_{rch}=K_r-T_{rch}^T K_{ch}T_{rch} \tag{57a,b}$$

Eq. (57) can be nondimensionalized using Eqs. (5) and (22) as for the rectangular hole. Hence, we obtain

$$M_{rch}=\rho hab\,\bar{M}_{rch},\ K_{rch}=\frac{Db}{a^3}\bar{K}_{rch} \tag{58a,b}$$

where

$$\bar{M}_{rch}=\bar{M}_r-\left(\pi\alpha\beta^2\right)T_{rch}^T T_{rch},\ \bar{K}_{rch}=\bar{K}_r-\left(\frac{\pi\alpha}{\beta^2}\right)T_{rch}^T\Lambda_c T_{rch} \tag{59a,b}$$

in which $\beta = R/a$.
As shown in the process from Eq. (55), (57) and (59), it can be readily seen that the application of the ICCM is very straightforward and the theoretical background is solid. The efficacy of the ICCM are fully demonstrated in the numerical results[Heo and Kwak(2008), Kwak et al.(2005,2006,2007)].

8. Free vibration analysis of rectangular plate with multiple circular cutouts by independent coordinate coupling method

Let us consider a rectangular plate with multiple circular holes as shown in Fig. 8. We can easily extend the formulation developed in the previous section to the case of a rectangular plate with multiple circular holes. The resulting mass and stiffness matrices can be expressed as:

$$\bar{M}_r=\bar{M}-\sum_{k=1}^{n}\pi\alpha\beta_k^2 T_{rchk}^T T_{rchk},\ \bar{K}_{rch}=\bar{K}_r-\sum_{k=1}^{n}\frac{\pi\alpha}{\beta_k^2}T_{rchk}^T\Lambda_c T_{rchk} \tag{60a,b}$$

where T_{rchk} represents the transformation matrix for kth circular hole

$$\left(T_{rchk}\right)_{ij} = \int_0^{2\pi}\int_0^{R_k} \Psi_{ki}\left(r_k,\theta_k\right)\phi_j\left(\xi\right)\psi_j\left(\eta\right) r dr d\theta \tag{61}$$

We also need the relationship between the global and local coordinates, which can be expressed as follows:

$$\xi = \frac{r_{kx}}{a} + \frac{r_k\cos\theta_k}{a}, \quad \eta = \frac{r_{ky}}{b} + \frac{r_k\sin\theta_k}{b} \tag{62a,b}$$

For the numerical study, we considered a square plate with two circular holes as shown in Fig. 9. The results of the ICCM were compared to those obtained by the commercial finite element method, ANSYS. $\nu = 0.3$, $E = 76\,\mathrm{GPa}$, $a = 1\,\mathrm{m}$, $\rho = 7800\ \mathrm{kg/m^3}$ were used and non-dimensionalized frequencies were estimated from the computed natural frequencies. For the ICCM, ten admissible functions were used for each direction of the square plate and fifty admissible functions were used for each circular hole.

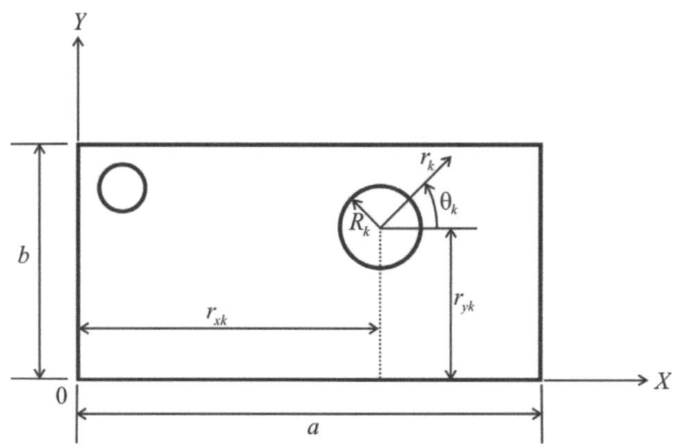

Fig. 8. Rectangular plate with multiple circular holes with local axes

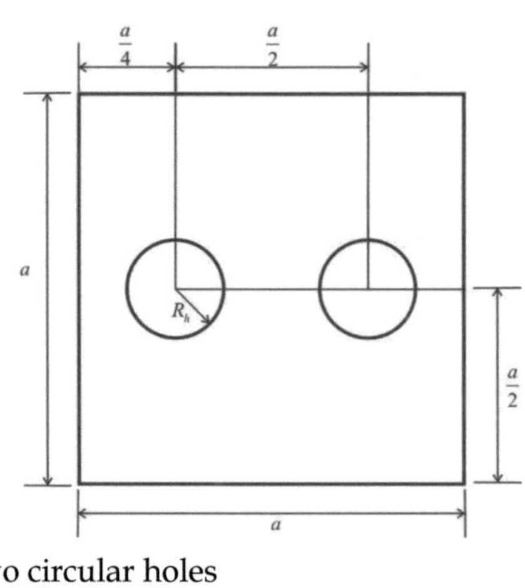

Fig. 9. Square plate with two circular holes

Figure 10 shows the non-dimensionalized natural frequencies obtained by the ICCM and ANSYS for simply-supported square plate with two circular hole, where $\beta = R_h / a$. As shown in the figure, the results obtained by the ICCM are in good agreement with those obtained by ANSYS.

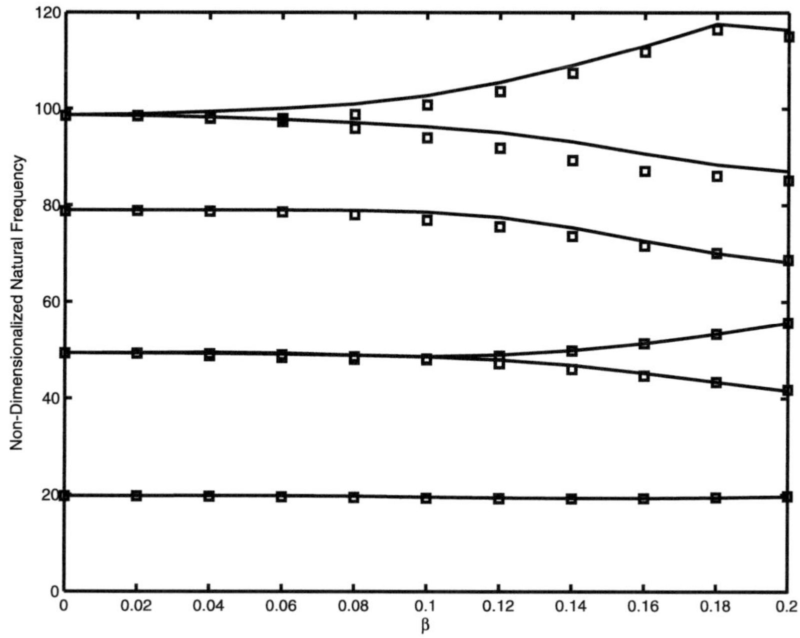

Fig. 10. Simply-supported square plate with two circular holes (—: ICCM, □:ANSYS)

9. Independent coordinate coupling method for a circular plate with an eccentric circular hole

Let us consider a circular plate with an eccentric circular hole as shown in Fig. 11 to demonstrate the efficacy of the ICCM.
The total kinetic and potential energies can be written as

$$T_{total} = T_C - T_{CH} \,,\; V_{total} = V_C - V_{CH} \,. \tag{63a,b}$$

However, it is not easy to express the energies belonging to the eccentric circular hole using the global coordinate system whose origin is fixed to the circular plate since the integral limits cannot be easily established. In addition, the numerical integration for the eccentric circular hole is also not an easy task. These complexities can be avoided with the use of the ICCM [Heo and Kwak(2008)]. Based on the ICCM, the deflection of the circular plate with the eccentric circular hole can be expressed as a combination of eigenfunctions and generalized coordinates, which are based on the local coordinates, r_c, θ_c, as shown in Fig.11. Inserting Eq. (21) into Eq. (63), the total kinetic and potential energies can be written as

$$T_{total} = \frac{1}{2}\dot{q}_c^T M_c \dot{q}_c - \frac{1}{2}\dot{q}_{ch}^T M_{ch} \dot{q}_{ch} \,,\; V_{total} = \frac{1}{2} q_c^T K_c q_c - \frac{1}{2} q_{ch}^T K_{ch} q_{ch} \tag{64a,b}$$

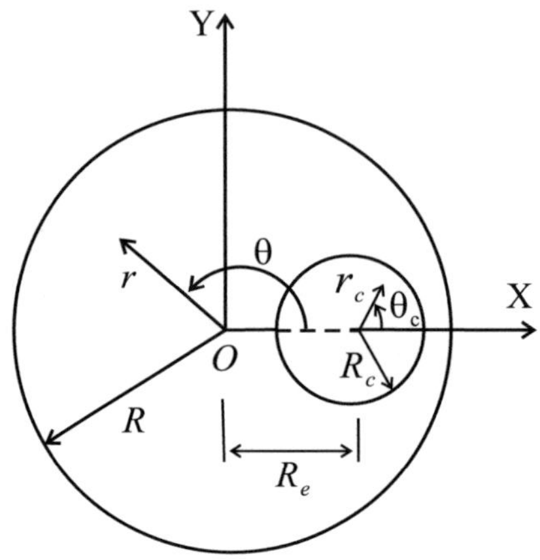

Fig. 11. Circular plate with an eccentric hole for coordinate system

In the next task in the ICCM, the displacement matching condition is satisfied inside the eccentric circular hole domain, i.e.

$$w_{ch}(r_c,\theta_c) = w_c(r,\theta)\,. \tag{65}$$

Inserting Eqs. (20) into (65), we then obtain

$$\sum_{j=1}^{n_c} \Phi_{cj}(r_c,\theta_c) q_{chj}(t) = \sum_{j=1}^{n} \Phi_j(r,\theta) q_{cj}(t)\,. \tag{66}$$

Multiplying Eq. (661) by $\Phi_{ci}(r_c,\theta_c)$ and integrating over the eccentric circular hole domain result in

$$\sum_{j=1}^{n_c} \int_0^{2\pi}\int_0^{R_c} \Phi_{ci}(r_c,\theta_c)\Phi_{cj}(r_c,\theta_c) r_c dr_c\, d\theta_c\, q_{chj}(t) = \sum_{j=1}^{n} \int_0^{2\pi}\int_0^{R_c} \Phi_{ci}(r_c,\theta_c)\Phi_j(r,\theta) r_c dr_c\, d\theta_c\, q_j(t)\,. \quad i = 1,2,\ldots,n_c \tag{67}$$

Using the orthogonal property of $\Phi_{ci}(r_c,\theta_c)$, Eq. (67) can be rewritten as

$$q_{chi}(t) = \sum_{j=1}^{n} \int_0^{2\pi}\int_0^{R_c} \Phi_{ci}(r_c,\theta_c)\Phi_j(r,\theta) r_c dr_c\, d\theta_c\, q_j(t) = \sum_{k=1}^{n} (T_{cch})_{ik}\, q_{jk}(t),\; i = 1,2,\ldots,n_c\,. \tag{68}$$

Equation (68) can be expressed in matrix form

$$q_{ch} = T_{cch} q_c \tag{69}$$

where T_{cch} is a $n_c \times n$ transformation matrix. The relationships between the global and local coordinates are needed to compute each element in the transformation matrix, which can be expressed as follows.

$$r = \sqrt{r_c^2 + R_e^2 - 2 r_c R_e \cos(\pi - \theta_c)},\quad \theta = \tan^{-1}\left(\frac{r_c \sin(\pi - \theta_c)}{R_e + r_c \cos\theta_c}\right). \tag{70a,b}$$

Using Eqs. (69) into (64), we can derive the mass and stiffness matrices as follows:

$$M_{cch} = M_c - T_{cch}^T M_{ch} T_{cch}\,,\ K_{cch} = K_c - T_{cch}^T K_{ch} T_{cch}\,. \tag{71a,b}$$

which can be expressed in terms of the non-dimensionalized mass and stiffness matrices

$$M_{cch} = \rho h \pi R^2 \bar{M}_{cch}\,,\ K_{cch} = \frac{D\pi}{R^2} \bar{K}_{cch} \tag{72a,b}$$

where

$$\bar{M}_{cch} = I - \alpha^2 T_{cch}^T T_{cch}\,,\ \bar{K}_{cch} = \Lambda - \frac{1}{\alpha^2} T_{cch}^T \Lambda_c T_{cch} \tag{73a,b}$$

in which $\alpha = R_c / R$ is the ratio of the radius of the eccentric circular hole to the radius of the circular plate. Hence, the non-dimensionalized eigenvalue problem for the circular plate with an eccentric circular hole can be expressed as

$$\left[\bar{K}_{cch} - \bar{\omega}^2 \bar{M}_{cch}\right] A = 0 \tag{74}$$

where $\bar{\omega} = \omega\sqrt{\rho h R^4 / D}$ is the non-dimensionalized natural frequency.

The finite element commercial code, ANSYS, was used for the calculation of non-dimensionalized natural frequencies of the simply-supported circular plate with an eccentric circular hole, where material constants, $\rho = 2700\,kg/m^3$, $E = 69GPa$, $\nu = 0.3$ and h =2 mm, R =1 m were used. Figure 12 shows the mesh configuration of two cases for $\alpha = 0.25$, $e = 0.4$ and $\alpha = 0.5$, $e = 0.4$, respectively, where the non-dimensionalized eccentric constant, $e = R_e / R$, is introduced. The mesh for the first case consisted of 4261 elements and 4395 nodes and the mesh for the second case consisted of 3197 elements and 3357 nodes.

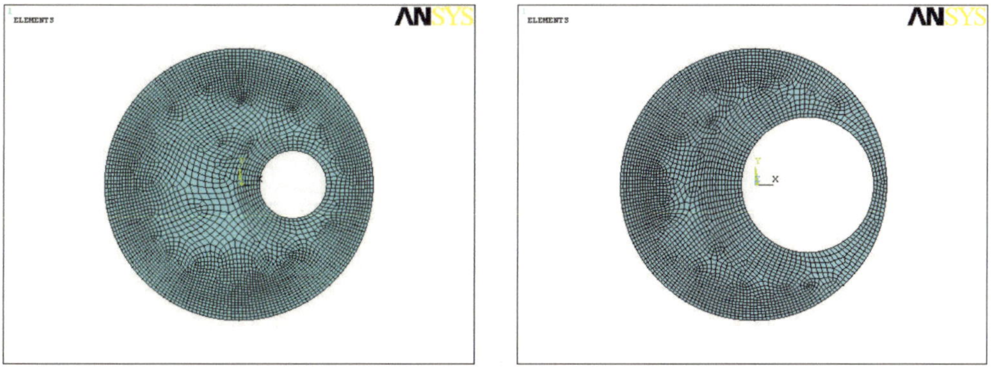

(a) $\alpha = 0.25$, $e = 0.4$ (b) $\alpha = 0.5$, $e = 0.4$

Fig. 12. Mesh Configurations by ANSYS

Figures 13 and 14 show the changes in the non-dimensionalized natural frequencies of the simply-supported circular plate with an eccentric circular hole with respect to the eccentricity when α = 0.25 and 0.5, respectively. Figs. 13 and 14 show the good agreement between the results obtained by the ICCM and the results by ANSYS. Eccentricity had a small effect on the fundamental mode, regardless of the hole size. However, the increases of the hole size and eccentricity had a large effect on higher natural frequencies, which changed unpredictably. Instead of commercial finite element codes, the ICCM can be used as an effective tool for the estimation of natural frequencies of a circular plate with an eccentric circular hole. Different boundary conditions were treated in the work by Heo and Kwak(2008).

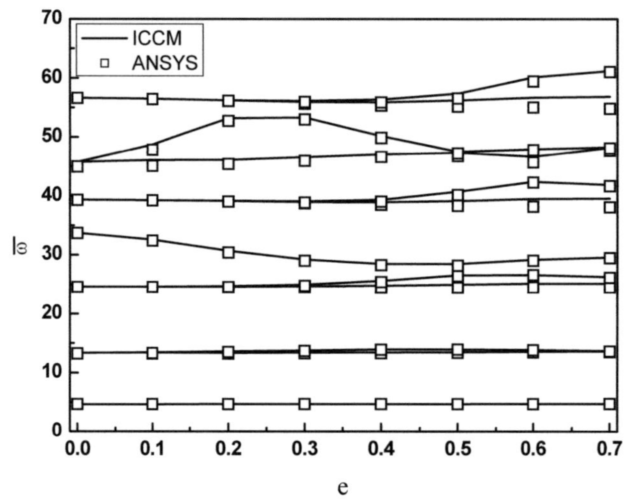

Fig. 13. Non-Dimensionalized Natural Frequency vs. Eccentricity for $\alpha = 0.25$

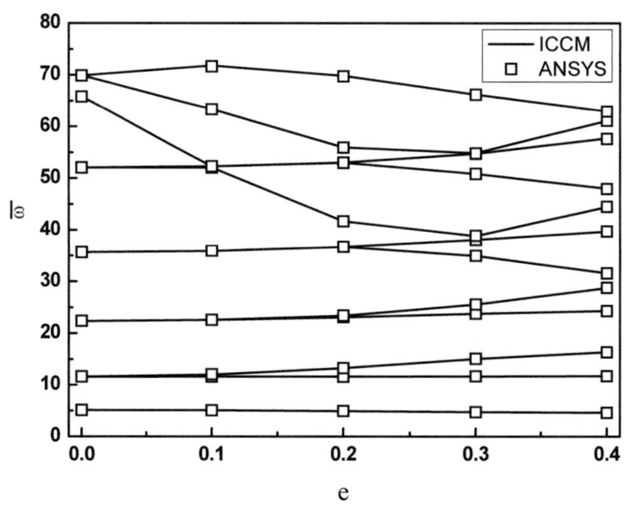

Fig. 14. Non-Dimensionalized Natural Frequency vs. Eccentricity for $\alpha = 0.5$

10. Discussion and conclusions

In general, the free vibration problem of a plate with holes can't be solved analytically. Therefore, we have to resort to numerical approach such as the finite element method. The classical Rayleigh-Ritz method has been popularly used for the analysis of a uniform rectangular plate and the exact solution exists for uniform circular plate. The procedure of the classical Rayleigh-Ritz method was first explained in detail. In applying the classical Rayleigh-Ritz method based on the global coordinates only, the kinetic and potential energies of the rectangular plate with a hole were calculated by subtracting the hole domain in the integrals. However, the Rayleigh-Ritz method can't be effectively used when the plate has holes because the numerical computation of integrals is required. If the plate hole geometry belongs to either rectangular or circular shape, the newly developed method, so called the independent coordinate coupling method (ICCM) can be effectively used. The ICCM has proved its effectiveness in analyzing the free vibration of a rectangular plate with a rectangular hole, a rectangular plate with multiple rectangular holes, a rectangular plate with a circular hole, a rectangular plate with multiple circular holes, and a circular plate with a circular hole. However, the ICCM can be easily extended to a circular plate with a rectangular hole and circular plate with multiple circular holes.

To apply the ICCM to the addressed problem, the global coordinates are set up for the plate and the local coordinates are set up for the hole domain, independently. The kinetic and potential energy expressions for the plate and the inner hole were then derived independently. Since the plate inside the hole domain can be regarded as a virtual free-edge plate, the energies, which are to be subtracted from the total energies, can be easily expressed in closed form. The resulting total energies are expressed in terms of generalized coordinates, which belong to either global or local coordinates. Hence, we need to unify the generalized coordinates. To this end, the relationships between the generalized coordinates belonging to the global and local coordinates were then derived using the displacement matching condition inside the hole domain and the orthogonal property of the admissible functions. In this way, the total kinetic and potential energies can be easily obtained and used for the calculation of the natural frequencies and modes of the circular plate with holes. To verify results of the proposed ICCM, numerical calculations were carried out using the classical Rayleigh-Ritz method based on the global coordinates only and the commercial finite element program. Experiments have been also carried out for the free-edge square plate with a square and circular hole. Both numerical and experimental results showed that good agreement exists between the results by the ICCM and the results obtained by the different algorithms and experiments. Hence, it can be concluded that the proposed ICCM can be effectively used for the free vibration analysis of a plate with holes.

11. References

Aksu, G. & Ali, R. (1976). Determination of Dynamic Characteristics of Rectangular Plates with Cut-outs Using a Finite Difference Formulation. *Journal of Sound and Vibration*, Vol. 44, (147-158), ISSN0022-460X

Ali, R. & Atwal, S. J. (1980). Prediction of Natural Frequencies of Vibration of Rectangular Plates with Rectangular Cutouts. *Computers and Structures*, Vol. 12, No. 9, (819-823), ISSN0045-7949

Avalos, D. R. & Laura, P. A. A. (2003). Transverse Vibrations of Simply Supported Rectangular Plates with Two Rectangular Cutouts. *Journal of Sound and Vibration*, Vol.267, (967-977), ISSN0022-460X

Bhat, R. B. (1985). Natural Frequencies of Rectangular Plates Using Characteristic Orthogonal Polynomials in Rayleigh-Ritz Method. *Journal of Sound and Vibration*, Vol. 102, (493-499), ISSN0022-460X

Bhat, R. B. (1985). Plate Deflections Using Orthogonal Polynomials. *American Society of Civil Engineers, Journal of the Engineering Mechanics Division*, Vol. 111, (1301-1309), ISSN0044-7951

Bhat, R. B. (1990). Numerical Experiments on the Determination of Natural Frequencies of Transverse Vibrations of Rectangular Plates of Nonuniform Thickness. *Journal of Sound and Vibration*, Vol. 138, (205-219), ISSN0022-460X

Cheng, L.; Li, Y. Y. & Yam, L. H. (2003). Vibration Analysis of Annular-Like Plates. *Journal of Sound and Vibration*, Vol. 262, (1153-1170), ISSN0022-460X

Eastep, F.E. & Hemmig, F.G. (1978). Estimation of Fundamental Frequency of Non-Circular Plates with Free, Circular Cutouts. *Journal of Sound and Vibration*, Vol. 56, No. 2, (155-165), ISSN0022-460X

Hegarty, R.F. & Ariman, T. (1975). Elasto-Dynamic Analysis of Rectangular Plates with Circular Holes. *Int. J. Solids Structures*, Vol. 11, (895-906), ISSN0020-7683

Heo, S. & Kwak, M. K. (2008). Free Vibration Analysis of an Annular Plate by the Independent Coordinate Coupling Method. *Transactions of the Korean Society for Noise and Vibration Engineering*, Vol.18 No.5, (564-571), ISSN1598-2785

Heo, S & Kwak, M. K. (2008). Free Vibration Analysis of a Circular Plate with an Eccentric Circular Hole by the Independent Coordinate Coupling Method. *Transactions of the Korean Society for Noise and Vibration Engineering*, Vol.18 No.6, (681-689), ISSN1598-2785

Itao, K. & Crandall, S.H. (1979). Natural Modes and Natural Frequencies of Uniform, Circular, Free-Edge Plates. *Journal of Applied Mechanics*, Vol. 46, (448-453), ISSN0021-8936

Joga-Rao, C. V. & Pickett, G. (1961). Vibrations of Plates of Irregular Shapes and Plates with Holes. *Journal of the Aeronautical Society of India*, Vol. 13, No. 3, (83-88), ISSN0001-9267

Khurasia, H. B. & Rawtani, S. (1978). Vibration Analysis of Circular Plates with Eccentric Hole. *Journal of Applied Mechanics*, Vol.45, No.1, (215~217), ISSN0021-8936

Kim, K.C.; Han, S.Y. & Jung, J.H. (1987). Transverse Vibration of Stiffened Rectangular Plates Having an Inner Cutout. *Journal of the Society of Naval Architects of Korea*, Vol. 24, No. 3, (35-42), ISSN1225-1143

Kumai, T. (1952). The Flexural Vibrations of a Square Plate with a Central Circular Hole. *Proc. 2nd Japan Natl. Congr. Appl. Mech.*, (339-342), ISSN0448-8660

Kwak, M. K. & Han, S. B. (2005). Free Vibration Analysis of Rectangular Plate with a Rectangular Cutout by Independent Coordinate Coupling Method. *Transactions of the Korean Society for Noise and Vibration Engineering*, Vol. 15, No. 12, (1398-1407), ISSN1598-2785

Kwak, M. K. & Han, S. B. (2006). Free Vibration Analysis of Simply-supported Rectangular Plate with a Circular Cutout by Independent Coordinate Coupling Method. *Transactions of the Korean Society for Noise and Vibration Engineering*, Vol. 16, No. 6, (643-650), ISSN1598-2785

Kwak, M. K. & Han, S. B. (2007). Free Vibration Analysis of Rectangular Plate with a Hole by means of Independent Coordinate Coupling Method. *Journal of Sound and Vibration*, Vol. 306, (12-30), ISSN0022-460X

Kwak, M. K. & Song, M. H. (2007). Free Vibration Analysis of Rectangular Plate with Multiple Rectangular Cutouts by Independent Coordinate coupling Method. *Transactions of the Korean Society for Noise and Vibration Engineering*, Vol. 17, No. 9, (881-887), ISSN1598-2785

Kwak, M. K. & Song, M. H. (2007). Free Vibration Analysis of Rectangular Plate with Multiple Circular Cutouts by Independent Coordinate Coupling Method. *Transactions of the Korean Society for Noise and Vibration Engineering*, Vol. 17, No. 11, (1086-1092), ISSN1598-2785

Lam, K. Y.; Hung, K. C. & Chow, S. T. (1989). Vibration Analysis of Plates with Cut-outs by the Modified Rayleigh-Ritz Method. *Applied Acoustics*, Vol. 28, (49-60), ISSN0003-682X

Lam, K. Y. & Hung, K. C. (1990). Vibration Study on Plates with Stiffened Openings Using Orthogonal Polynomials and Partitioning Method. *Computers and Structures*, Vol. 37, (295-301), ISSN0045-7949

Laura, P. A. A.; Romanelli, E. & Rossi, R. E. (1997). Transverse Vibrations of Simply-Supported Rectangular Plates with Rectangular Cutouts. *Journal of Sound and Vibration*, Vol. 202, No. 2, (275-283), ISSN0022-460X

Laura, P. A. A.; Masia, U. & Avalos, D. R. (2006). Small Amplitude, Transverse Vibrations of Circular Plates Elastically Restrained Against Rotation with an Eccentric Circular Perforation with a Free Edge. *Journal of Sound and Vibration*, Vol.292, (1004~1010), ISSN0022-460X

Lee, H.S. & Kim, K.C. (1984). Transverse Vibration of Rectangular Plates Having an Inner Cutout in Water. *Journal of the Society of Naval Architects of Korea*, Vol. 21, No. 1, (21-34), ISSN1225-1143

Lee, W. M.; Chen, J. T. & Lee, Y. T. (2007). Free Vibration analysis of Circular Plates with Multiple Circular Holes using Indirect BIEMs. *Journal of Sound and Vibration*, Vol.304, (811-830), ISSN0022-460X

Lee, Y.-S. & Lee, Y.-B. (1994). Free Vibration Analysis of 4 Edges clamped, Isotropic Square Plates with 2 Collinear Circular Holes. *Transactions of Korean Society for Noise and Vibration Engineering*, Vol.4, No.3, (283-293), ISSN1598-2785

Leissa, A. W. (1993). *Vibration of Plates*, Acoustical Society of America, New York.

Lin, W. H. (1982). Free Transverse Vibrations of Uniform Circular Plates and Membranes with Eccentric Holes. *Journal of Sound and Vibration*, Vol.81, No.3, (425~435), ISSN0022-460X

Monahan, L. J.; Nemergut, P. J. & Maddux, G. E. (1970). Natural Frequencies and Mode Shapes of Plates with Interior Cut-outs. *The Shock and Vibration Bulletin*, Vol. 41, (37-49)

Nagaya, K. (1951). Transverse Vibration of a Plate having an Eccentric Inner Boundary. *Journal of Applied Mechanics*, Vol.18, No.3, (1031~1036), ISSN0021-8936

Nagaya, K. (1980). Transverse Vibration of a Rectangular Plate with an Eccentric Circular Inner Boundary. *Int. J. Solids Structures*, Vol. 16, (1007-1016), ISSN0020-7683

Paramasivam, P. (1973). Free Vibration of Square Plates with Square Opening. *Journal of Sound and Vibration*, Vol. 30, (173-178), ISSN0022-460X

Rajamani, A. & Prabhakaran, R. (1977). Dynamic Response of Composite Plates with Cut-outs, Part I: Simply-Supported Plates. *Journal of Sound and Vibration*, Vol. 54, (549-564), ISSN0022-460X

Rajamani, A. & Prabhakaran, R. (1977). Dynamic Response of Composite Plates with Cut-outs, Part II: Clamped-Clamped Plates. *Journal of Sound and Vibration*, Vol. 54, (565-576), ISSN0022-460X

Sakiyama, T.; Huang, M.; Matsuda, H. & Morita, C. (2003). Free Vibration of Orthotropic Square Plates with a Square Hole. *Journal of Sound and Vibration*, vol. 259, No. 1, (63-80), ISSN0022-460X

Takahashi, S. (1958). Vibration of Rectangular Plates With Circular Holes. *Bulletin of JSME*, Vol. 1, No. 4, (380-385), ISSN0021-3764

Zhong, H. & Yu, T. (2007). Flexural Vibration Analysis of an Eccentric Annular Mindlin Plate. *Archive of Applied Mechanics*, Vol.77, (185~195), ISSN0939-1533

6

Free Vibration of Smart Circular Thin FGM Plate

Farzad Ebrahimi
Mechanical Engineering Department, University of Tehran,
North Kargar St., Tehran, 11365-4563,
Iran

1. Introduction

A new class of materials known as 'functionally graded materials' (FGMs) has emerged recently, in which the material properties are graded but continuous particularly along the thickness direction. In an effort to develop the super heat resistant materials, Koizumi [1] first proposed the concept of FGM. These materials are microscopically heterogeneous and are typically made from isotropic components, such as metals and ceramics.

In the quest for developing lightweight high performing flexible structures, a concept emerged to develop structures with self-controlling and self-monitoring capabilities. Expediently, these capabilities of a structure were achieved by exploiting the converse and direct piezoelectric effects of the piezoelectric materials as distributed actuators or sensors, which are mounted or embedded in the structure [2, 3]. Such structures having built-in mechanisms are customarily known as 'smart structures'. The concept of developing smart structures has been extensively used for active control of flexible structures during the past decade [4].

Recently considerable interest has also been focused on investigating the performance of FG plates integrated with piezoelectric actuators. For example, Ootao and Tanigawa [5] theoretically investigated the simply supported FG plate integrated with a piezoelectric plate subjected to transient thermal loading. A 3-D solution for FG plates coupled with a piezoelectric actuator layer was proposed by Reddy and Cheng [6] using transfer matrix and asymptotic expansion techniques. Wang and Noda [7] analyzed a smart FG compositestructure composed of a layer of metal, a layer of piezoelectric and a FG layer in between, while in [8] a finite element model was developed for studying the shape and vibration control of FG plates integrated with piezoelectric sensors and actuators. Yang et al. [9] investigated the nonlinear thermo-electro-mechanical bending response of FG rectangular plates covered with monolithic piezoelectric actuator layers; most recently, Huang and Shen [10] investigated the dynamics of a FG plate coupled with two monolithic piezoelectric layers undergoing nonlinear vibrations in thermal environments. All the aforementioned studies focused on the rectangular-shaped plate structures.

However, to the authors' best knowledge, no researches dealing with the free vibration characteristics of the circular FGM plate integrated with the piezoelectric layers has been reported. Therefore, the present work attempts to solve the problem of providing analytical solution for free vibration of thin circular FG plates with two full size surface-bonded piezoelectric layers on the top and the bottom of the FG plate. The formulations are based on CPT. A consistent formulation that satisfies the Maxwell static electricity equation is presented so that the full coupling effect of the piezoelectric layer on the dynamic characteristics of the circular FGM plate can be estimated based on the free vibration results. The physical and mechanical properties of the FG substrate plate are assumed to be graded continuously in the thickness direction according to the power-law distribution in terms of the volume fractions of the constituents. The differential equations of motion are solved analytically for clamped edge boundary condition of the plate. By using of some mathematical techniques these differential equations are transformed to a sixth order ordinary differential equation and finally by implementing the operator decomposition method on this equation, three Bessel types of equations are obtained which can easily be solved for the plate deflection and the potential function. The detailed mathematical derivations are presented. In Numerical investigations, the emphasis is placed on investigating the effect of varying the gradient index of FG plate on the free vibration characteristics of the structure. The results are verified by those obtained from 3D finite element analyses.

2. Functionally graded materials

In a FG material made of ceramic and metal mixture, if the volume fraction of the ceramic part is represented by V_c and the metallic part by V_m, we have;

$$V_m + V_c = 1 \tag{1}$$

Based on the power law distribution [11], the variation of V_c vs. thickness coordinate (z) placed at the middle of thickness, can be expressed as;

$$V_c = (z/2h_f + 1/2)^g \quad , g \geq 0 \tag{2}$$

We assume that the inhomogeneous material properties, such as the modulus of elasticity E and the density ρ change in the thickness direction z based on Voigt's rule over the whole range of the volume fraction [12]; while Poisson's ratio υ is assumed to be constant in the thickness direction [13] as;

$$E(z) = (E_c - E_m)V_c(z) + E_m \tag{3a}$$

$$\rho(z) = (\rho_c - \rho_m)V_c(z) + \rho_m \tag{3b}$$

where subscripts m and c refer to the metal and ceramic constituents, respectively. After substituting V_c from Eq. (2) into Eqs. (3), material properties of the FGM plate are determined in the power law form which are the same as those proposed by Reddy et al. [11] i.e.;

$$E(z) = (E_c - E_m)(z/2h_f + 1/2)^g + E_m \tag{4a}$$

$$\rho(z) = (\rho_c - \rho_m)(z/2h_f + 1/2)^g + \rho_m \tag{4b}$$

$$\nu(z) = \nu \tag{4c}$$

3. Piezoelectric materials

For symmetry piezoelectric materials in polar coordinate, the stress - strain - electric field intensity relations based on well-known assumptions of classical plate theory, can be written as [16];

$$\sigma_{rr}^{p} = \bar{C}_{11}^{E}\varepsilon_{rr} + \bar{C}_{12}^{E}\varepsilon_{\theta\theta} - \bar{e}_{31}E_{z} \tag{5}$$

$$\sigma_{\theta\theta}^{p} = \bar{C}_{12}^{E}\varepsilon_{rr} + \bar{C}_{11}^{E}\varepsilon_{\theta\theta} - \bar{e}_{31}E_{z} \tag{6}$$

$$\tau_{r\theta}^{p} = \left(\bar{C}_{11}^{E} - \bar{C}_{12}^{E}\right)\varepsilon_{r\theta} = -z\left(\bar{C}_{11}^{E} - \bar{C}_{12}^{E}\right) \tag{7}$$

in which σ_i , ε_k and e represent the stress and strain components and the permeability constant of piezoelectric material and E_k indicates the components of the electric field and $\bar{C}_{ij}^{E}$ are the components of the symmetric piezoelectric stiffness matrix and $\bar{e}_{31}$ is the reduced permeability constant of piezoelectric material as [13];

$$\bar{C}_{11}^{E} = C_{11}^{E} - \left(C_{13}^{E}\right)^{2} / C_{33}^{E}\,,\ \bar{C}_{12}^{E} = C_{12}^{E} - \left(C_{13}^{E}\right)^{2} / C_{33}^{E}$$

$$\bar{e}_{31}^{E} = e_{31} - C_{13}^{E}e_{33} / C_{33}^{E}$$

4. Constitutive relations

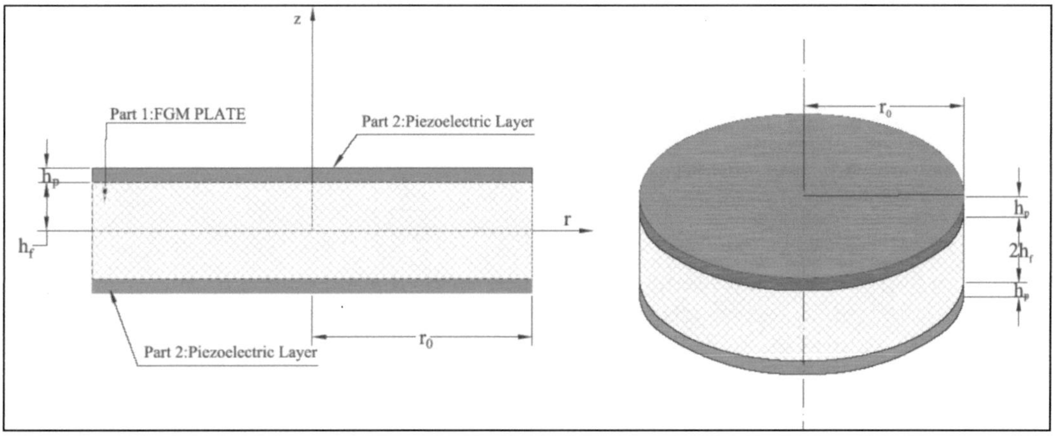

Fig. 1. Schematic representation of the FGM circular plate with two piezoelectric layers mounted on its upper and lower surfaces

The cross section of a circular FGM plate with a piezoelectric layer mounted on its surface is shown in Fig. 1. In most practical applications, the ratio of the radius to the thickness of the plate is more than ten, and the Kirchhoff assumption for thin plates is applicable, whereby the shear deformation and rotary inertia can be omitted. For such a structure, the displacement field is assumed as follows:

$$u_{z} = u_{z}(r,\theta,t) = w(r,\theta,t) \tag{8}$$

$$u_r = u_r(r,\theta,t) = -z\frac{\partial u_z}{\partial r} \quad (9)$$

$$u_\theta = u_\theta(r,\theta,t) = -z\frac{\partial u_z}{r\partial\theta} \quad (10)$$

where u_z, u_r and u_θ are the displacements in the transverse z-direction, radial r-direction, and tangential θ-direction of the plate, respectively.
It is also assumed that the poling direction of the piezoelectric material to be in the z-direction. A differential strain can be induced in case of applying external electric potential across the piezoelectric layer resulting in bending of the plate. The strain of the FGM plate and piezoelectric layer in the radial and tangential directions and the shear component are given by [14]

$$\varepsilon_{rr} = \frac{\partial u_r}{\partial r} = -z\frac{\partial^2 w}{\partial r^2} \quad (11)$$

$$\varepsilon_{\theta\theta} = \frac{\partial u_\theta}{r\partial\theta} + \frac{u_r}{r} = -z(\frac{\partial^2 w}{r^2\partial\theta^2} + \frac{\partial w}{r\partial r}) \quad (12)$$

$$\varepsilon_{r\theta} = \frac{1}{2}(\frac{\partial u_r}{r\partial\theta} + \frac{\partial u_\theta}{\partial r} - \frac{u_\theta}{r}) \quad (13)$$

The stress components in the FGM plate in terms of strains or component of displacement field based on the generalized Hooke's Law are [14];

$$\sigma_{rr}^f = E(z)(\varepsilon_{rr} + \nu\varepsilon_{\theta\theta})/(1-\nu^2) \quad (14)$$

$$\sigma_{\theta\theta}^f = E(z)(\varepsilon_{\theta\theta} + \nu\varepsilon_{rr})/(1-\nu^2) \quad (15)$$

$$\tau_{r\theta}^f = -\frac{zE(z)}{1+\nu}\left(\frac{\partial^2 w}{r\partial r\partial\theta} - \frac{\partial w}{r^2\partial\theta}\right) \quad (16)$$

where the superscript f represents the variable in the FGM structure; Two piezoelectric layers are attached to the FG plate and intended to be used as an actuator or sensor to determine the natural frequencies of a vibrating coupled plate,. There are several different models representing the input electric potential for such a piezoelectric layer. In this paper we decided to adopt the following Wang et al. electric potential function which is appropriate for free vibrations of proposed system [13];

$$\phi = \left[1 - \left(\left(2z - 2h_f - h_p\right)/h_p\right)^2\right]\varphi(r,\theta,t) \quad (17)$$

where $\varphi(r,\theta,t)$ is the electric potential on the mid-surface of the piezoelectric layer.
Based on Eq. (17), the components of electric field intensity E and electric flux density D is written as [15]:

$$D_r = \Xi_{11}E_r = \Xi_{11}(-\frac{\partial\phi}{\partial r}) \quad (18)$$

$$D_\theta = \bar{\Xi}_{11} E_\theta = \bar{\Xi}_{11}(-\frac{\partial \phi}{r\partial \theta}) \tag{19}$$

$$D_z = \bar{\Xi}_{33} E_z + \bar{e}_{31}\left(\varepsilon_{rr} + \varepsilon_{\theta\theta}\right) \tag{20}$$

where $\bar{\Xi}_{11}$, $\bar{\Xi}_{33}$ are the symmetric reduced dielectric constants of piezo layer and given by [17];

$$\bar{\Xi}_{33} = \bar{\Xi}_{33} + (e_{33}^2 / C_{33}^E) \, , \; \bar{\Xi}_{11} = \Xi_{11} \tag{21}$$

in which Ξ_{33}, Ξ_{11} are the dielectric constants.

5. Governing equations

In order to obtain the governing differential equation of the coupled circular plate, we begin with resultant moments components as [16];

$$M_{rr} = \int_{-h_f}^{h_f} z\sigma_{rr}^f dz + 2\int_{h_f}^{h_f+h_p} z\sigma_{rr}^p dz \tag{22}$$

$$M_{\theta\theta} = \int_{-h_f}^{h_f} z\sigma_{\theta\theta}^f dz + 2\int_{h_f}^{h_f+h_p} z\sigma_{\theta\theta}^p dz \tag{23}$$

$$M_{r\theta} = \int_{-h_f}^{h_f} z\tau_{r\theta}^f dz + 2\int_{h_f}^{h_f+h_p} z\tau_{r\theta}^p dz \tag{24}$$

and the resultant shear forces are herein written as

$$q_r = \frac{\partial M_{rr}}{\partial r} + \frac{\partial M_{r\theta}}{r\partial \theta} + \frac{M_{rr} - M_{\theta\theta}}{r} \tag{25}$$

$$q_\theta = \frac{\partial M_{r\theta}}{\partial r} + \frac{\partial M_{\theta\theta}}{r\partial \theta} + \frac{2M_{r\theta}}{r} \tag{26}$$

Substituting Eqs. (11-13) in to Eqs. (14-16) and Eqs. (5-7) and substituting the results in to Eqs. (22-26) and substituting the final results into the governing equation for the Kirchhoff plate,

$$\frac{\partial q_r}{\partial r} + \frac{\partial q_\theta}{r\partial \theta} + \frac{q_r}{r} - \left(\int_{-h_f}^{h_f} \rho_f(z) \frac{\partial^2 u_z}{\partial t^2} dz + 2\int_{h_f}^{h_f+h_p} \rho_p \frac{\partial^2 u_z}{\partial t^2} dz \right) = 0 \tag{27}$$

will result in the equation for the piezoelectric coupled circular FGM plate,

$$\left(D_1 + D_2\right) \Delta\Delta w + \frac{4}{3} h_p \bar{e}_{31} \Delta\varphi + P_0 \frac{\partial^2 w}{\partial t^2} = 0 \tag{28}$$

where Δ is the Laplace operator in polar coordinate and

$$D_1 = \int_{-h_f}^{h_f} \frac{z^2 E(z)}{1-\nu^2} dz$$

$$D_2 = \frac{2}{3} h_p (3h_f^2 + 3h_f h_p + h_p^2)\bar{C}_{11}^E$$

$$\tilde{\rho}_f = \frac{1}{2h_f} \int_{-h_f}^{h_f} \rho_f(z) dz \,, \; P_0 = 2(\tilde{\rho}_f h_f + \tilde{\rho}_p h_p)$$

where ρ_f and ρ_p are material densities of the FGM plate and piezoelectric layer, respectively. Note that all of the electrical variables primarily must satisfy the Maxwell's equation which requires that the divergence of the electric flux density vanishes at any point within the media as [15];

$$\int_{h_f}^{h_f+h_p} \left(\frac{\partial(rD_r)}{r\partial r} + \frac{\partial D_\theta}{r\partial\theta} + \frac{\partial D_z}{\partial z} \right) dz = 0 \tag{29}$$

Now, by substituting Eqs. (18- 20) into the above equation we arrive at;

$$\frac{h_p^2 \bar{\Xi}_{11}}{12\bar{\Xi}_{33}} \Delta\varphi - \varphi + \frac{h_p^2 \bar{e}_{31}}{8\bar{\Xi}_{33}} \Delta w = 0 \tag{30}$$

6. Solution method

Primarily we solve Eqs. (28) and (30) simultaneously by which ϕ can be expressed in terms of w as;

$$\varphi(r,\theta,t) = -\frac{(D_1 + D_2) h_p \bar{\Xi}_{11}}{16\bar{e}_{31}\bar{\Xi}_{33}} \Delta\Delta w + \frac{h_p^2 \bar{e}_{31}}{8\bar{\Xi}_{33}} \Delta w - \frac{h_p(\tilde{\rho}_f h_f + \rho_p h_p)\bar{\Xi}_{11}}{8\bar{e}_{31}\bar{\Xi}_{33}} \frac{\partial^2 w}{\partial t^2} \tag{31}$$

Applying the Laplacian operator to the above equation and substituting the result into equation (28) gives a decoupled sixth-order partial differential equation, namely

$$P_3 \Delta\Delta\Delta w - P_2 \Delta\Delta w + P_1 \Delta\left(\frac{\partial^2 w}{\partial t^2}\right) - P_0 \frac{\partial^2 w}{\partial t^2} = 0 \tag{32}$$

where

$$\begin{aligned} P_1 &= h_p^2 \bar{\Xi}_{11} P_0 / 12\bar{\Xi}_{33} \,, \\ P_2 &= D_1 + D_2 + h_p^3 \bar{e}_{31}^2 / 6\bar{\Xi}_{33} \,, \\ P_3 &= (D_1 + D_2) h_p^2 \bar{\Xi}_{11} / 12\bar{\Xi}_{33} \end{aligned} \tag{33}$$

To solve Eq. (34) for w, we first assume that;

$$w(r,\theta,t)=w_1(r)e^{i(m\theta-\omega t)} \tag{34}$$

where $w_1(r)$ is the displacement amplitude in the z - direction as a function of radial displacement only; ω is the natural angular frequency of the compound plate; and m is the wave number in the circumferential direction. Rewriting Eq. (32) in terms of $w_1(r)$ and using Eq. (34), after canceling the exponential term one would get;

$$P_3\bar{\Delta}\bar{\Delta}\bar{\Delta}w_1 - P_2\bar{\Delta}\bar{\Delta}w_1 - \omega^2 P_1\bar{\Delta}w_1 + \omega^2 P_0 w_1 = 0 \tag{35}$$

where $\bar{\Delta} = d^2/dr^2 + d/rdr - m^2/r^2$

Eq. (35) can be solved by the method of decomposition operator and noting that the w_1 is non-singular at the center of the plate its general solution yields to

$$w_1 = \sum_{n=1}^{3} A_{nm} Z_{nm}(\alpha_n r) \tag{36}$$

here

$$\alpha_1 = \sqrt{|x_1|}\,,\ \alpha_2 = \sqrt{|x_2|}\,,\ \alpha_3 = \sqrt{|x_3|} \tag{37}$$

in which x_1, x_2 and x_3 are the roots of the following cubic characteristic equation,

$$P_3x^3 - P_2x^2 - \omega^2 P_1 x + \omega^2 P_0 = 0 \tag{38}$$

and

$$Z_{im}(\alpha_i r) = Z_{im}(\alpha_i, r) = \begin{cases} J_m(\alpha_i r) & , x_i < 0 \\ I_m(\alpha_i r) & , x_i > 0 \end{cases} \tag{39}$$

here i=(1,2,3) and $J_m(\alpha_i r), I_m(\alpha_i r)$ are the first type and the modified first type Bessel function ,both of them of the order of m. In order to obtain appropriate solution for $\varphi(r,\theta,t)$, we assume;

$$\varphi(r,\theta,t)=\varphi_1(r)e^{i(m\theta-\omega t)} \tag{40}$$

then substituting Eq. (36) in to Eq. (31)we arrive to the following relation for $\varphi(r,\theta,t)$;

$$\begin{aligned}\varphi_1(r) = \left[16\bar{e}_{31}\bar{\Xi}_{33}\right]^{-1}\sum_{n=1}^{3}\Big[A_{nm}h_p(2s_n\alpha_n^2h_p\bar{e}_{31}^2 - \\ (D_1+D_2)\alpha_n^4\bar{\Xi}_{11} + P_0\omega^2\bar{\Xi}_{11})\Big]\times Z_{nm}(\alpha_n r)\end{aligned} \tag{41}$$

7. Case studies, results and discussions

We will solve above the relations in this section; the material parameters and geometries are listed in Table 1.

FGM Plate:	E_c = 205 GPa	ρ_c =*8900* (kg/ m³)
	E_m = 200 GPa	ρ_m =7800
PZT4:	$C_{11}^E = 132$	$C_{33}^E = 115$
$C_{55}^E = 26$ GPa	$C_{13}^E = 73$	$C_{12}^E = 71$
e_{31} =-4.1 (C/m²)	e_{33} =14.1	e_{15} =10.5
Ξ_{11} =7.124 (nF/m)	Ξ_{33} =5.841	ρ_p =7500 (kg/ m³)
Geometry(mm):	r_0=600	h_f =2, h_p =10

Table 1. Material properties and geometric size of the piezoelectric coupled FGM plate [13,17]

7.1 Clamped circular piezo-coupled FGM plate

The boundary condition is given by

$$w_1 = dw_1/dr = d\varphi_1/dr = 0 \quad at\ (r = r_0) \tag{42}$$

and the characteristic equation is

$$\begin{vmatrix} c_{11} & c_{12} & c_{13} \\ c_{21} & c_{22} & c_{23} \\ c_{31} & c_{32} & c_{33} \end{vmatrix} = 0 \tag{43}$$

$$c_{1i} = Z_{im}(\alpha_i r_0), \quad c_{2i} = \alpha_i r_0 Z'_{im}(\alpha_i r_0)$$
$$c_{3i} = \left(\frac{h_p^2 r_0 s_i \alpha_i^3}{8} - \frac{(D_1 + D_2) h_p r_0 \alpha_i^5 \overline{\Xi}_{11}}{16 \overline{e}_{31}^2} + \frac{(D_1 + D_2) h_p \alpha_i \lambda^4 \overline{\Xi}_{11}}{16 \overline{e}_{31}^2 r_0^3} \right) Z'_{im}(\alpha_i r_0) \tag{44}$$

$$\lambda = r_0 \left[\frac{2(\tilde{\rho}_f h_f + \rho_p h_p)\omega^2}{D_1 + D_2} \right]^{\frac{1}{4}} \tag{45}$$

$$\omega = \frac{\lambda^2}{r_0^2} \sqrt{\frac{D_1 + D_2}{2(\tilde{\rho}_f h_f + \rho_p h_p)}} \tag{46}$$

in which the $()'$ symbol indicates the derivative with respect to r and λ is the nondimensional angular natural frequency.

After calculating ω from Eq. (43) and using Eqs. (36, 42) we find the mode shape for w_1 as;

$$w_1(r) = A_{3m} \times \left[\left(\frac{\alpha_3 Z_{2m}(\alpha_2 r_0) Z'_{3m}(\alpha_3 r_0) - \alpha_2 Z_{3m}(\alpha_3 r_0) Z'_{2m}(\alpha_2 r_0)}{\alpha_2 Z_{1m}(\alpha_1 r_0) Z'_{2m}(\alpha_2 r_0) - \alpha_1 Z_{2m}(\alpha_2 r_0) Z'_{1m}(\alpha_1 r_0)} \right) \times Z_{1m}(\alpha_1 r) + \right.$$
$$\left. \left(\frac{\alpha_1 Z_{3m}(\alpha_3 r_0) Z'_{1m}(\alpha_1 r_0) - \alpha_3 Z_{1m}(\alpha_1 r_0) Z'_{3m}(\alpha_3 r_0)}{\alpha_2 Z_{1m}(\alpha_1 r_0) Z'_{2m}(\alpha_2 r_0) - \alpha_1 Z_{2m}(\alpha_2 r_0) Z'_{1m}(\alpha_1 r_0)} \right) \times Z_{2m}(\alpha_2 r) + Z_{3m}(\alpha_3 r) \right] \tag{47}$$

and moreover, by using Eqs. (36, 41, 42) we have the electric potential as;

$$\hat{\varphi}(r) = A_{3m} \times \left[\left(\frac{\alpha_3 Z_{2m}(\alpha_2 r_0) Z'_{3m}(\alpha_3 r_0) - \alpha_2 Z_{3m}(\alpha_3 r_0) Z'_{2m}(\alpha_2 r_0)}{\alpha_2 Z_{1m}(\alpha_1 r_0) Z'_{2m}(\alpha_2 r_0) - \alpha_1 Z_{2m}(\alpha_2 r_0) Z'_{1m}(\alpha_1 r_0)} \right) \times \right.$$
$$\times Z_{1m}(\alpha_1 r) \times \left[h_p (2 s_1 \alpha_1^2 h_p \bar{e}_{31}^2 - (D_1 + D_2)\alpha_1^4 \bar{\Xi}_{11} + P_0 \omega^2 \bar{\Xi}_{11}) \right] \times \left[16 \bar{e}_{31} \bar{\Xi}_{33} \right]^{-1} +$$
$$+ \left(\frac{\alpha_1 Z_{3m}(\alpha_3 r_0) Z'_{1m}(\alpha_1 r_0) - \alpha_3 Z_{1m}(\alpha_1 r_0) Z'_{3m}(\alpha_3 r_0)}{\alpha_2 Z_{1m}(\alpha_1 r_0) Z'_{2m}(\alpha_2 r_0) - \alpha_1 Z_{2m}(\alpha_2 r_0) Z'_{1m}(\alpha_1 r_0)} \right) \times \tag{48}$$
$$\times Z_{2m}(\alpha_2 r) \times \left[h_p (2 s_2 \alpha_2^2 h_p \bar{e}_{31}^2 - (D_1 + D_2)\alpha_2^4 \bar{\Xi}_{11} + P_0 \omega^2 \bar{\Xi}_{11}) \right] \times \left[16 \bar{e}_{31} \bar{\Xi}_{33} \right]^{-1} +$$
$$\left. + \left[h_p (2 s_3 \alpha_3^2 h_p \bar{e}_{31}^2 - (D_1 + D_2)\alpha_3^4 \bar{\Xi}_{11} + P_0 \omega^2 \bar{\Xi}_{11}) \right] \times \left[16 \bar{e}_{31} \bar{\Xi}_{33} \right]^{-1} Z_{3m}(\alpha_3 r) \right]$$

Power Index	Mode no.	FGM plate			
g	m	Present Method	Present (FEM)	Error (%)	Wang et al. [13]
	0	138.42	139.27	0.61	138.48
0	1	288.05	289.70	0.57	288.20
	2	472.55	473.45	0.19	472.79
	0	134.63	135.43	0.59	-
1	1	280.17	281.78	0.57	-
	2	459.62	460.45	0.18	-
	0	132.70	133.63	0.69	-
3	1	276.19	278.04	0.67	-
	2	453.09	454.34	0.28	-
	0	132.12	133.06	0.70	-
5	1	274.96	276.85	0.68	-
	2	451.06	452.39	0.29	-
	0	131.85	132.78	0.70	-
7	1	274.39	276.25	0.67	-
	2	450.13	451.46	0.29	-
	0	131.69	132.70	0.76	-
9	1	274.07	276.09	0.73	-
	2	449.60	450.84	0.28	-
	0	131.64	132.55	0.68	-
10	1	273.96	275.79	0.67	-
	2	449.42	450.66	0.28	-

Table 2. First three resonance frequencies (Hz) of FGM plate

In order to validate the obtained results, we compared our results with those given in the literature [7,9,10].Further as there were no published results for the compound piezoelectric FGM plate, we verify the validity of obtained results with those of FEM results.

Our FEM model for piezo- FG plate comprises: a 3D 8-noded solid element with: number of total nodes 26950, number of total element 24276, 3 DOF per node in the host plate element and 6 DOF per node in the piezoelectric element. Tables 2 and 3 shows the numerical results of our method compared with other references and techniques.
As one can see from Table 2, the obtained results from the analytical method when g=0 (isotropic steel plate) corresponds closely with the results of [7-9] and FEM solution. As it is seen in these tables the maximum estimated error of our solution with FEM is about 1.51%.

Power Index	Mode no.	Coupled Piezo-FGM plate			
g	m	Present Method	Present (FEM)	Error (%)	Wang et al. [13]
	0	143.63	144.69	0.73	143.71
0	1	298.92	300.49	0.52	299.07
	2	490.37	492.62	0.46	490.62
	0	140.26	142.22	1.38	-
1	1	291.89	295.82	1.33	-
	2	478.84	482.09	0.67	-
	0	138.54	140.60	1.46	-
3	1	288.33	292.47	1.42	-
	2	472.99	476.61	0.76	-
	0	138.01	140.07	1.47	-
5	1	287.21	291.39	1.43	-
	2	471.16	474.81	0.77	-
	0	137.76	139.82	1.47	-
7	1	286.69	290.83	1.43	-
	2	470.30	473.95	0.77	-
	0	137.62	139.73	1.51	-
9	1	286.40	290.54	1.43	-
	2	469.83	473.16	0.70	-
	0	137.57	139.61	1.46	-
10	1	286.30	290.41	1.42	-
	2	469.66	473.26	0.76	-

Table 3. First three resonance frequencies (Hz) for piezo-coupled FGM plate for various values of power index

A close inspection of results listed in Tables 2 and 3 indicates that the amount of error between analytical and FEM results for the natural frequencies in FGM plate alone in the all vibration modes and for all values of g are less than the similar results for the compound plate.
The obtained results in Table 3 indicate that by increasing the value of g, the frequency of system decreases in all different vibrational modes. Moreover, this decreasing trend of frequency for smaller values of g is more pronounced, for example by increasing value of g from 1 to 3 (~200%) the frequency of the first mode for the compound plate decreases by

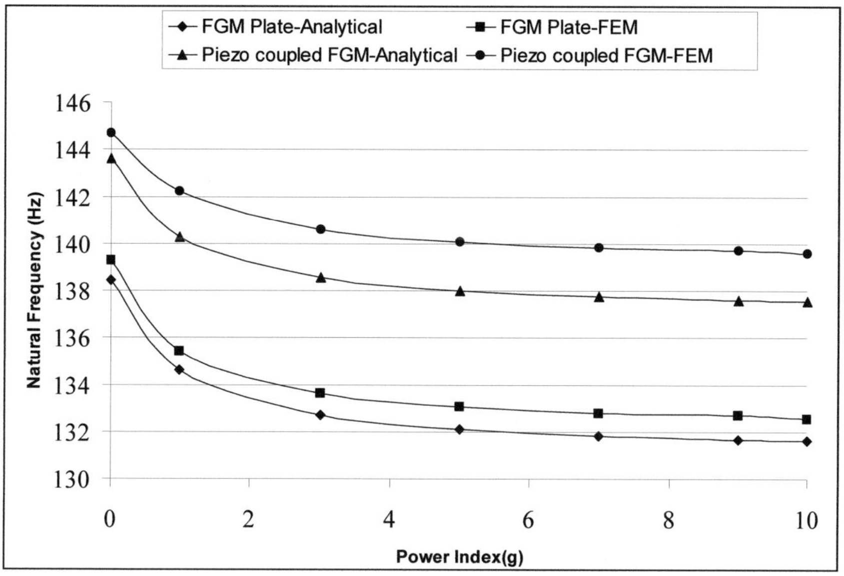

Fig. 2. Effect of power index on the natural frequencies (first mode)

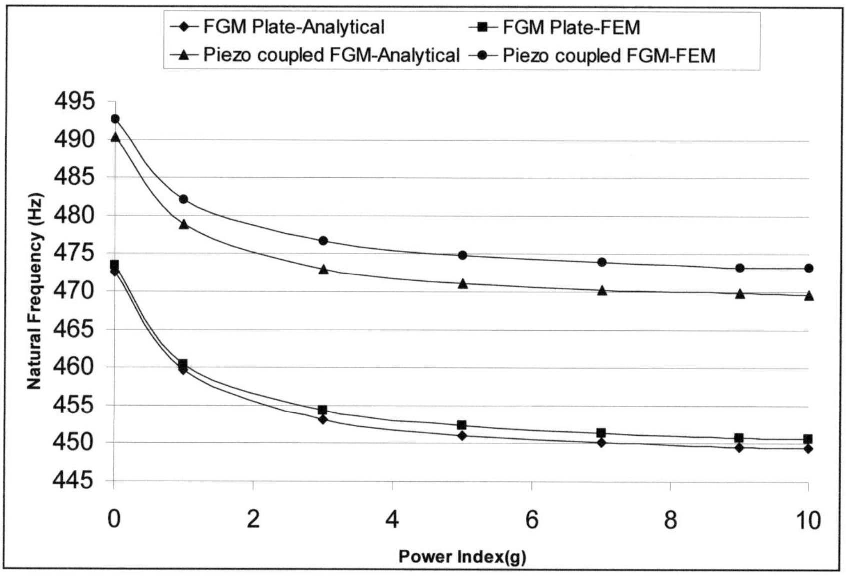

Fig. 3. Effect of power index on the natural frequencies (third mode)

1.23% but by increasing g from 3 to 9 (~ 200%) of the same plate and for the same mode, the frequency decreases by 0.66%. In order to see better the effect of g variations on the natural frequencies of the different plates, Fig. 2 and Fig. 3 also illustrate these variations for the first and third mode shapes.

As it is seen from Figs. 2 and Fig. 3, the behavior of the system follows the same trend in all different cases, i.e. the natural frequencies of the system decrease by increasing of g and stabilizes for g values greater than 7. In fact for g>>1 the FGM plate becomes a ceramic plate and the compound plate transforms to a laminated plate with ceramic core as a host plate.

8. Conclusion

In this paper free vibration of a thin FGM plus piezoelectric laminated circular plate based on CPT is investigated. The properties of FG material changes according to the Reddy's model in direction of thickness of the plate and distribution of electric potential in the piezoelectric layers follows a quadratic function in short circuited form. The validity of the obtained results was done by crossed checking with other references as well as by obtained results from FEM solutions. It is further shown that for vibrating circular compound plates with specified dimensions, one can select a specific piezo-FGM plate which can fulfill the designated natural frequency and indeed this subject has many industrial applications.

9. References

[1] Koizumi M. concept of FGM. Ceram. Trans. 1993; 34: 3–10.
[2] Bailey T, Hubbard JE. Distributed piezoelectric polymer active vibration control of a cantilever beam. J. Guidance, Control Dyn. 1985; 8: 605-11.
[3] Millerand SE, Hubbard JE. Observability of a Bernoulli–Euler beam using PVF2 as a distributed sensor. MIT Draper Laboratory Report, 1987.
[4] Peng F, Ng A, Hu YR. Actuator placement optimization and adaptive vibration control of plate smart structures. J. Intell. Mater. Syst. Struct. 2005; 16: 263–71.
[5] Ootao Y, Tanigawa Y. Three-dimensional transient piezo-thermo-elasticity in functionally graded rectangular plate bonded to a piezoelectric plate. Int. J. Solids Struct. 2000; 37: 4377–401.
[6] Reddy JN, Cheng ZQ. Three-dimensional solutions of smart functionally graded plates. ASME J. Appl. Mech 2001; 68: 234–41.
[7] Wang BL, Noda N. Design of smart functionally graded thermo-piezoelectric composite structure. Smart Mater. Struct. 2001; 10: 189–93.
[8] He XQ, Ng TY, Sivashanker S, Liew KM. Active control of FGM plates with integrated piezoelectric sensors and actuators. Int. J. Solids Struct.2001; 38: 1641–55.
[9] Yang J, Kitipornchai S, Liew KM. Non-linear analysis of thermo-electro-mechanical behavior of shear deformable FGM plates with piezoelectric actuators. Int. J. Numer. Methods Eng. 2004; 59:1605–32.
[10] Huang XL, Shen HS. Vibration and dynamic response of functionally graded plates with piezoelectric actuators in thermal environments. J. Sound Vib. 2006; 289: 25–53.
[11] Reddy JN, Praveen GN. Nonlinear transient thermoelastic analysis of functionally graded ceramic-metal plate. Int. J. Solids Struct 1998; 35: 4457-76.
[12] Wetherhold RC, Wang S. The use of FGM to eliminate or thermal deformation. Composite Sci Tech 1996; 56: 1099-104.
[13] Wang Q, Quek ST, Liu X. Analysis of piezoelectric coupled circular plate. Smart Mater. Struct 2001; 10: 229-39.
[14] Reddy J.N, Theory and analysis of elastic plates, Philadelphia: Taylor and Francis, 1999.
[15] Halliday D, Resniek R. Physics, John Wiley and Sons, 1978.
[16] Brush DO, Almroth BO. Buckling of bars plates and shells. New York: Mac-hill, 1975.
[17] Loy CT, Lam KL, Reddy JN. Vibration of functionally graded cylindrical shells. Int. J Mech Sciences 1999;41: 309-24

7

An Atomistic-based Spring-mass Finite Element Approach for Vibration Analysis of Carbon Nanotube Mass Detectors

S.K. Georgantzinos and N.K. Anifantis
Machine Design Laboratory
Mechanical and Aeronautics Engineering Department
University of Patras, GR 26500
Greece

1. Introduction

Since their discovery in 1991 by Ijima [1], carbon nanotubes (CNT) have received much attention as a new class of nanomaterials revealing a significant potential for use in a diverse range of novel and evolving applications. Much of the interest in CNTs has focused on their particular molecular structures and their unique electronic and mechanical properties. For example, their elastic stiffness is comparable to that of diamond (1000 GPa), while their strength is ten times larger (yield strength 100 GPa). Furthermore, CNTs conduct heat and electricity along their length with very little resistance, and therefore they act as tiny electrical wires or paths for the rapid diffusion of heat. As a result, progressive research activities regarding CNTs have been ongoing in recent years. For more detail on theoretical predictions and experimental measurements of both mechanical and physical properties, see the comprehensive reviews in [2,3].

The combination of an extremely high stiffness and light weight in CNTs results in vibration frequencies on the order of GHz and THz. There is a wide range of applications in which the vibrational characteristics of CNTs are significant. In applications such as oscillators, charge detectors, field emission devices, vibration sensors, and electromechanical resonators, oscillation frequencies are key properties. An representative application is the development of sensors for gaseous molecules, which play significant roles in environmental monitoring, chemical process control, and biomedical applications. Mechanical resonators are widely used as inertial balances to detect small quantities of adsorbed mass through shifts in oscillation frequency. Recently, advances in lithography and materials synthesis have enabled the fabrication of nanoscale mechanical resonators that utilize CNTs [4,5]. The use of a CNT to make the lightest inertial balance ever is essentially a target to make a nanoscale mass spectrometer of ultrahigh resolution. Building such a mass spectrometer that is able to make measurements with atomic mass sensitivity is one of the main goals in the burgeoning field of nanomechanics. An inertial balance relies only on the mass and does not, therefore, require the ionization or acceleration stages that might damage the molecules being measured. This means that a nanoscale inertial balance would be able to measure the mass of macromolecules that might be too fragile to be measured by conventional instruments [5]. Several efforts for the building of CNT-based sensors have been presented in the literature. Mateiu *et al.* [6] described an approach for building a mass sensor based on multi-walled CNTs with an atomic force microscope. Chiu *et al.* [7] demonstrated atomic-scale mass sensing using doubly clamped, suspended CNT resonators in which their single-electron transistor properties allowed the self-detection of nanotube vibration. They used the detection of shifts in the resonance frequency of the nanotubes to sense and determine the inertial mass of atoms as well as the mass of the nanotube itself. Commonly, multi-walled CNTs are less sensitive than single-walled CNTs. However, multi-walled carbon nanotubes are easier to manipulate and more economical to be produced, since they are both longer and have larger diameters than single-walled CNTs [8].

Hence, it is important to develop accurate theoretical models for evaluation of natural frequencies and mode shapes of CNTs. An excellent review article was recently published by Gibson *et al.* [9] that presents related scientific efforts in dealing with the vibrational behavior of CNTs and their composites, including both theoretical and experimental studies. Controlled experiments performed at nanoscale dimensions remain both difficult and expensive. Despite of this fact, Garcia-Sanchez *et al.* [10] have recently presented a mechanical method for detecting CNT resonator vibrations using a novel scanning force microscopy method. The comparison between experimental and theoretical methods pre-require the complete definition of all parameters such as the length of the vibrating nanotube, the nanotube type and other conditions that influence the vibrational behavior such as the slack phenomenon, nature of the support condition, environmental conditions and other influences.

In an attempt to approach the vibration behavior of CNTs, various theoretical methods have been reported in literature. Molecular dynamics (MD) and molecular mechanics, as well as elastic continuum mechanics, are considered efficient because they can accurately and cost-effectively produce results that closely approximate the behavior of CNTs. . Each of the previously mentioned approaches offers different advantages, but also certain drawbacks. MD is an accurate method capable of simulating the full mechanical CNT performance. However, it carries a high computational cost that may be prohibitive for large-scale problems, especially in the context of vibration analysis. Molecular mechanics-based techniques, such as those in [11-13], have been used for vibration analysis of CNTs and have been shown to be accurate and also more computationally cost-effective than MD. Nevertheless, under such approaches, the modeling of atomic interactions requires special attention because the mechanical equivalent used to simulate the carbon-carbon bond deformations must be efficient for the studied problem. Generally, typical elements of classical mechanics, such as rods [14], beams [15, 16], springs [17-19] and cells [20] have been proposed including appropriate stiffness parameters, thus their strain energies are equivalent to the potential energies of each interatomic interaction. Furthemore, elastic continuum mechanics methods based on well-known beam theories have also been successfully used to evaluate the vibration characteristics of CNTs under typical boundary conditions [21-24]. Xu *et al.* [25] studied the free vibration of double-walled CNTs modeled as two individual beams interacting with each other taking van der Waals forces into account and supported with different boundary conditions between the inner and outer tubes. These methods have the lowest computational cost; however, they can compute only a subset (mainly the bending modes) of the vibrational modes and natural frequencies.

In terms of CNT mass detector function, the principle of mass detection using CNT-based resonators is based on the fact that the vibrational behavior of the resonator is sensitive to changes in its mass due to attached particles. The change of the resonator mass due to an added mass causes frequency shifts. The key challenge in mass detection is quantifying the changes in the resonant frequencies due to added masses. Based on this principle, the usage of computational tools, as presented in prevous paragraph, capable of simulating the vibrational behavior of CNT-based mass detectors is important for two reasons. First, they can cost-effectively predict the mass sensing characteristics of different nanoresonator types, thereby allowing the optimal design of detectors with a specific sensing range. Second, their cooperation with experimental measurements can improve the detection abilities of the nanodevice. With respect to theoretical studies on CNT-based sensors, Li and Chou [26] examined the potential of nanobalances using individual single-walled CNTs in a cantilevered or bridged configuration. Wu *et al.* [27] explored the resonant frequency shift of a fixed-free single-walled CNT caused by the addition of a nanoscale particle to the beam tip. This was done to explore the suitability of a single-walled CNT as a mass detector device in a micro-scale situation via a continuum mechanics-based finite element method simulation using a beam-bending model. Chowdhury *et al.* [28] examined the potential of single-walled CNTs as biosensors using a continuum mechanics-based approach and derived a closed-form expression to calculate the mass of biological objects from the frequency shift.

In this chapter, an atomistic spring-mass based finite element approach appropriate to simulate the vibration characteristics of single-walled and multi-walled CNTs is presented. The method uses spring-mass finite elements that simulate the interatomic interactions and the inertia effects in CNTs. It uses a special technique for simulating the bending between adjacent bonds, distinguishing it from other mechanics-based methods. This method utilizes the exact microstructure of the CNTs while allowing the straightforward input of the force constants that molecular theory provides. In addition, spring-like elements are formulated in order to simulate the interlayer van der Waals interactions. These elements connect all atoms between different CNT layers at a distance smaller than the limit below which the van der Waals potential tends to zero. The related stiffness is a function of this distance. The resulting dynamic equilibrium equations can be used to generate new results. Results available in the literature were used to validate the proposed method. Parametric analyses are performed reporting the natural frequencies as well as the mode shapes of various multi-walled CNTs for different aspect geometric characteristics. Furthermore, the principle of mass detection using resonators is based on the fact that the resonant frequency is sensitive to the resonator mass, which includes the self-mass of the resonator and the attached mass. The change of the attached mass on the resonator causes a shift to the resonant frequency. Since, the key issue of mass detection is in quantifying the change in the resonant frequency due to the added mass, the effect of added mass to the vibration signature of CNTs is investigated for the clamped-free and clamped-clamped support conditions. And different design parameters. Additionally, the frequency shifts of single- and multi-walled CNTs were compared.

2. CNTs geometry

A planar layer of carbon atoms forms a periodic structure called the graphene sheet. Pencil lead consists of a stack of overlaying graphene sheets that easily separate upon shearing in writing. A perfect graphene sheet in the xy-plane consists of a doubly periodic hexagonal lattice defined by two base vectors,

$$\mathbf{v_1} = a(1,0) \text{ and } \mathbf{v_2} = a\left(\cos 60^o, \sin 60^o\right) = a\left(\frac{1}{2}, \frac{\sqrt{3}}{2}\right) \tag{1}$$

where α is equal to $\sqrt{3}r_h$ and r_h is the radius of the hexagonal cell. Note that the lengths of these vectors are equal. Any point of plane $P = (x, y)$ is uniquely defined as a linear combination of these two vectors,

$$\mathbf{P} = \mathbf{v}_0 + a\mathbf{v}_1 + b\mathbf{v_2}\,, \tag{2}$$

where a and b are integers, provided that $\mathbf{v}_0$ is the center of a hexagon.
Knowing the geometry of graphene, a single-walled CNT can be geometrically generated by rolling a single-layer graphene sheet, which is ideally cut, to make a cylinder. The graphene sheet must be rolled up in the direction of the chiral vector $\mathbf{C_h}$ defined as (see Figure 1):

$$\mathbf{C_h} = n\,\mathbf{a}_1 + m\,\mathbf{a}_2 \tag{3}$$

where $\mathbf{a}_1$ and $\mathbf{a}_2$ are the basis vectors of the honeycomb lattice and integers (n , m) are the number of steps along the zigzag carbon bonds and generally are used to name a nanotube.

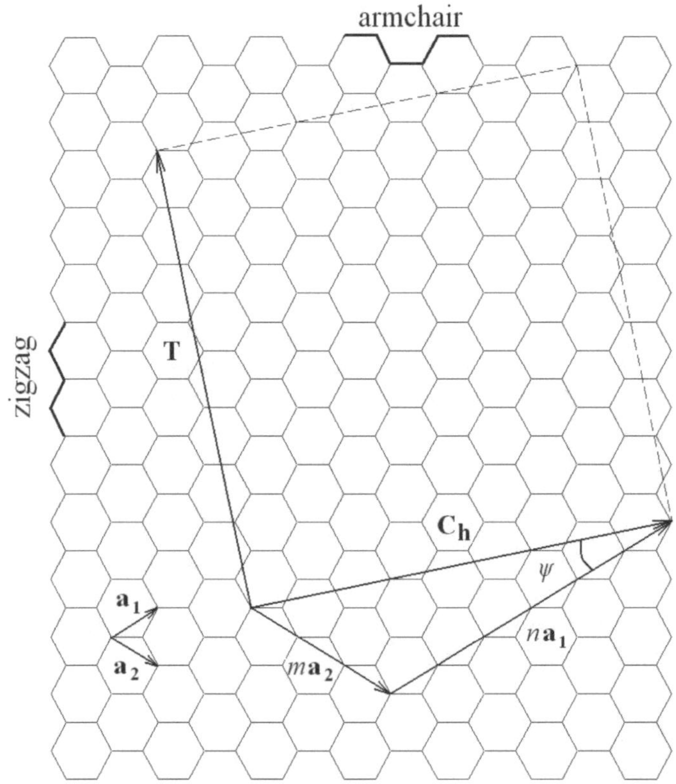

Fig. 1. Generation of a SWCNT from a graphene sheet.

A nanotube (n, n) is usually named as armchair (Figure 2(a)) while the nanotube $(n, 0)$ is usually named zigzag (Figure 2(b)). The chiral angle ψ $(0 \le \psi \le 30^{\circ})$ is defined as:

$$\tan\psi = \frac{\sqrt{3}m}{(2n+m)} \tag{4}$$

It is obvious that for an armchair nanotube $\psi = 30^{\circ}$ while for a zigzag $\psi = 0^{\circ}$. The nanotube's diameter D is given by the following equation:

$$D = \frac{a_{\mathrm{c-c}}\sqrt{3(n^2+nm+m^2)}}{\pi} \tag{5}$$

where $a_{\mathrm{c-c}}$ is distance between two neighbor carbon atoms and is equal to 0.1421 nm. Chiral vector $\mathbf{C_h}$ and the following translational vector $\mathbf{T}$ define the ideal rectangular cutting area of graphene sheet:

$$\mathbf{T} = \left(\frac{2m+n}{W}\right)\mathbf{a}_1 + \left(-\frac{2n+m}{W}\right)\mathbf{a}_2 \tag{6}$$

where W defines the higher common divisor of $2m+n$ and $2n+m$.

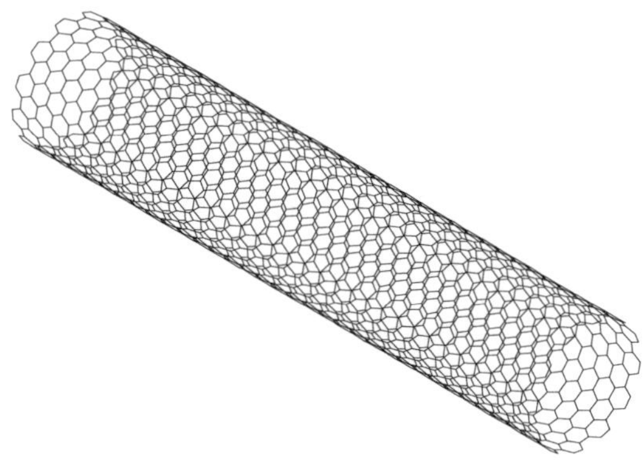

(a)

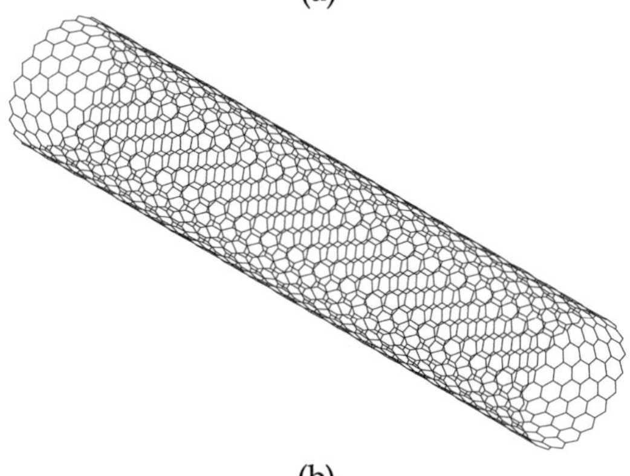

(b)

Fig. 2. Geometry of an (a) armchair and (b) zigzag single-walled CNTs

For simplicity, the original coordinate system of the graphene sheet (x', y') is transformed into a new system (x, y, z) of the nanotube such that $\mathbf{T}$ is along y'-axis. Then, the graphene atomic coordinates are converted to those of the nanotube according to the equation (Kołoczek et al. [29]):

$$(x, y, z) = \left(R\cos\left(\frac{x'}{R}\right),\ r\sin\left(\frac{x'}{R}\right),\ y' \right) \tag{7}$$

where R is the nanotube's radius.

A multi-walled CNT consists of multiple layers of graphene rolled in on themselves to form a tube shape. In other words, every multi-walled CNT consists of more than one coaxial single-walled CNTs. Since single-walled CNTs are parts of multi-walled CNTs, the layers of multi-walled CNTs have similar geometric characteristics. Given that the interlayer distance is 0.34nm, as has been observed in [1], the difference between diameters of neighbouring layers, where the diameter of every layer can be calculated using the Equation (5), is ΔD=0.68nm. Knowing that this equation is a function only of chirality indexes of the two neighboring nanotubes in a multi-walled CNT, someone can calculate the convenient types of single-walled CNTs able to apart the multi-walled CNT of

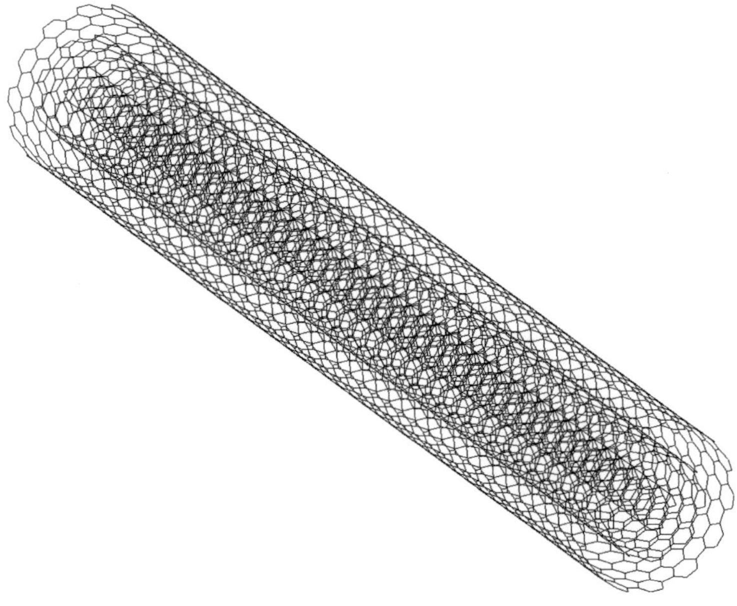

Fig. 3. Geometry of a multi-walled CNT.

specific number of layers and outer diameter. If (n_1, m_1) and (n_2, m_2) are the types of the inner and outer neighboring layers respectively, it is observed that for zig-zag nanotubes, the chirality indexes are $n_2 = n_1 + 9$ and $m_2 = m_1 = 0$. Correspondingly, if the neighboring nanotubes are armchair then $n_2 = m_2 = n_1 + 5$ and $m_1 = n_1$. The type of one MWCNT, here, is declared as the sequence of the types of all layers $(n_{in}, m_{in}) - \ldots - (n_{out}, m_{out})$, starting from the type of the innermost tube and finishing to the type of the outermost tube. A representative example of multi-walled CNT geometry consists of three layers is depicted in Figure 3.

3. Computational model

3.1 Force field

The total potential energy, omitting non-bonded interactions, i.e. the electrostatic energy and the energy due to van der Waals interaction, is a sum of energies caused by the bonded interatomic interactions, which are depicted in Figure 4(a), and may be expressed by the following equation (Rappe et al. [30]):

$$U = \sum U_r + \sum U_\theta + \sum U_\phi + \sum U_\omega \tag{8}$$

where U_r represents the energy due to bond stretching, U_θ the energy due to bond angle bending, U_ϕ the energy due to dihedral angle torsion and U_ω the energy due to out of plane torsion.

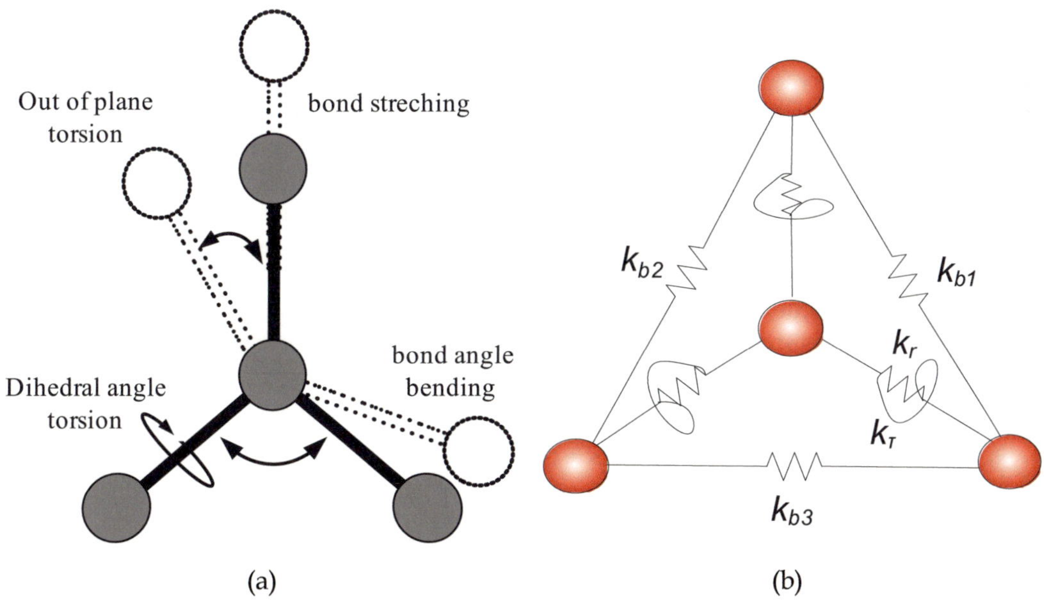

Fig. 4. Force field in neighboring atoms (a) interatomic interactions and (b) spring model.

Under the assumption of small deformations, the harmonic approximation is adequate for describing potential energy (Gelin [31]) and therefore the force field. By adopting the simplest harmonic forms and combining the dihedral angle, torsion with the out of plane torsion into a single equivalent term then the following terms can describe the total potential energy [17]:

$$U_r = \frac{1}{2}k_r(\Delta r)^2, \quad \frac{d^2U_r}{d\Delta r^2} = k_r \tag{9}$$

$$U_\theta = \frac{1}{2}k_\theta(\Delta\theta)^2, \quad \frac{d^2U_\theta}{d\Delta\theta^2} = k_\theta \tag{10}$$

$$U_\tau = U_\varphi + U_\omega = \frac{1}{2}k_\tau(\Delta\phi)^2, \quad \frac{d^2U_\tau}{d\Delta\tau^2} = k_\tau \tag{11}$$

where k_r , k_θ and k_τ are the bond stretching, bond angle bending, and torsional resistance force constants, respectively, while Δr , $\Delta\theta$ and $\Delta\phi$ represent the bond length, bond angle and twisting bond angle variations, respectively.
The second derivatives of the potential energy terms appearing in equations (7), (8) and (9) with respect to bond stretch, bond angle and twisting bond angle variations, respectively, produce spring stiffness coefficients k_r , k_θ and k_τ. Thus, here, axial and torsional springs that straightforwardly introduce the physical constants are utilized (Figure 4(b)) in order to describe the force field. The springs interconnect the nodes that have been extracted from the eq. (3). The bond angle bending interaction is simulated by axial springs, which have stiffness

$$k_{bi} = \left(\frac{1}{a_{c-c} \cos(90^\circ - \gamma)} \right)^2 k_\theta \tag{12}$$

as has been described in [17], where $\gamma = 30^\circ$ in the hexagonal lattice of the graphene. This angle may vary for each C-C-C microstructure in a CNT according to its type and radius due to its cylindrical shape. In the case of chiral nanotubes, the stiffness of the three different bending springs (Figure 4(b)) varies $k_{b1} \neq k_{b2} \neq k_{b3}$. In the cases of armchair and zigzag nanotubes, two of the three bending spring stiffnesses are equal due to the same angle γ . In the other hand, because of the planar shape of the graphene sheets, all the bond angle bending springs have the same stiffness, i.e. $k_{b1} = k_{b2} = k_{b3}$.
The interlayer interactions between the walls of a multi-walled CNT is caused by the van der Waals forces and can described through the Lennard-Jones pair potential [32,33]

$$U(R) = 4\varepsilon \left[\left(\frac{\sigma}{R} \right)^{12} - \left(\frac{\sigma}{R} \right)^{6} \right] \tag{13}$$

where R is the distance between the interacting atoms, ε is the depth of the potential and σ is a parameter that is determined by the equilibrium distance. The van der Waals force F is obtained by taking the derivative of the Lennard-Jones pair potential, i.e.,

$$F(R) = -\frac{dU(R)}{dr} = -4\varepsilon \left(6\frac{\sigma^6}{R^7} - 12\frac{\sigma^{12}}{R^{13}} \right) \tag{14}$$

It should be noted that the initial pressure exerted on a sheet is negligible at the equilibrium distance, and thus the van der Waals force can be estimated by the Taylor expansion to the first order around the equilibrium position, i.e.,

$$F(R) = F(R_0) + \frac{dF(R_0)}{dR}(R - R_0) = -24\varepsilon \left(26\frac{\sigma^{12}}{R_0^{13}} - 7\frac{\sigma^6}{R_0^{7}} \right)(R - R_0) \tag{15}$$

Where $R_0 = \sqrt{(X_j - X_i)^2 + (Y_j - Y_i)^2 + (Z_j - Z_i)^2}$ is the initial distance between the atoms of the different layers.

3.2 CNT modeling

In order to evaluate the vibrational characteristics of CNTs, we must develop equations that describe the dynamic equilibrium of the entire model. The elemental equations must be constructed first before the global stiffness and mass matrices can be assembled.

The elemental equation for the a_i-element, as defined and developed in [18] to represent the bond stretching as well as twisting bond angle interactions, is

$$\begin{bmatrix} k_r\mathbf{I}_{20} & 0 \\ 0 & k_\tau\mathbf{I}_{20} \end{bmatrix}\begin{bmatrix} \mathbf{u} \\ \boldsymbol{\theta} \end{bmatrix}+\begin{bmatrix} m_r\mathbf{I}_{21} & 0 \\ 0 & 0 \end{bmatrix}\begin{bmatrix} \ddot{\mathbf{u}} \\ \ddot{\boldsymbol{\theta}} \end{bmatrix}=\begin{bmatrix} \mathbf{F} \\ \mathbf{T} \end{bmatrix} \quad \text{or} \quad \mathbf{k_a}\mathbf{U}+\mathbf{m_a}\ddot{\mathbf{U}}=\mathbf{P}, \tag{16}$$

where

$$\mathbf{I}_{20}=\begin{bmatrix} 1 & -1 \\ -1 & 1 \end{bmatrix},\ \mathbf{I}_{21}=\begin{bmatrix} 0 & -1 \\ 0 & 1 \end{bmatrix}, \tag{17}$$

m_r is the concentrated mass equal to the half or whole mass of the carbon nuclei [18], $\mathbf{F}$ represents the forces applied to nodes 1 and 2 of the element, $\mathbf{u}$ is the vector of nodal displacements ($\mathbf{u}=[u_1 \quad u_2]^{\mathrm{T}}$, $\mathbf{F}=[F_1 \quad F_2]^{\mathrm{T}}$), $\boldsymbol{\theta}$ is the vector of nodal rotations, $\mathbf{T}$ is the vector of the applied torsional moments ($\boldsymbol{\theta}=[\theta_1 \quad \theta_2]^{\mathrm{T}}$, $\mathbf{T}=[T_1 \quad T_2]^{\mathrm{T}}$), $\mathbf{U}$ is the vector of nodal translations and rotations, $\ddot{\mathbf{U}}$ is the vector of nodal accelerations, $\mathbf{P}$ is the column vector of loads, and finally, $\mathbf{k_a}$ and $\mathbf{m_a}$ are the elemental stiffness and mass matrix, respectively. Similarly, the equation for the b_i-element, which describes the bond angle interaction in the hexagonal lattice, is

$$\begin{bmatrix} k_b\mathbf{I}_{20} & 0 \\ 0 & 0 \end{bmatrix}\begin{bmatrix} \mathbf{u} \\ \boldsymbol{\theta} \end{bmatrix}+\begin{bmatrix} 0 & 0 \\ 0 & 0 \end{bmatrix}\begin{bmatrix} \ddot{\mathbf{u}} \\ \ddot{\boldsymbol{\theta}} \end{bmatrix}=\begin{bmatrix} \mathbf{F} \\ \mathbf{T} \end{bmatrix} \quad \text{or} \quad \mathbf{k_b}\mathbf{U}+\mathbf{m_b}\ddot{\mathbf{U}}=\mathbf{P}, \tag{18}$$

where

$$k_b=\begin{cases} k_{b1}, & \text{when } b\text{-element is straight in respect to the hexagonal cell} \\ k_{b2}, & \text{when } b\text{-element is slant in respect to the hexagonal cell} \end{cases}$$

is the stiffness coefficient, as described in [18], and $\mathbf{k_b}$ and $\mathbf{m_b}$ are the corresponding elemental stiffness and mass matrix, respectively.

Moreover, we must derive the elemental equation for the van der Waals nanosprings (*vdw* elements). Because this spring is only translational, we can write the elemental equation as follows:

$$\begin{bmatrix} k_{vdw}\mathbf{I}_{20} & 0 \\ 0 & 0 \end{bmatrix}\begin{bmatrix} \mathbf{u} \\ \boldsymbol{\theta} \end{bmatrix}+\begin{bmatrix} 0 & 0 \\ 0 & 0 \end{bmatrix}\begin{bmatrix} \ddot{\mathbf{u}} \\ \ddot{\boldsymbol{\theta}} \end{bmatrix}=\begin{bmatrix} \mathbf{F} \\ \mathbf{T} \end{bmatrix} \quad \text{or} \quad \mathbf{k_{vdw}}\mathbf{U}+\mathbf{m_{vdw}}\ddot{\mathbf{U}}=\mathbf{P}, \tag{19}$$

where k_{vdw} is the stiffness as derived by Equation (15) [19]. Note that the mass matrix $\mathbf{m_{vdw}}$ is a null matrix because all of the inertia effects are included in the previously defined elements.

To express the stiffness matrix of the elements in the global coordinate system, a transformation matrix must be used. Let (local) nodes 1 and 2 of the axial spring correspond to nodes *i* and *j*, respectively, of the global system. The local displacements u_1 and u_2 can be resolved into the respective components u_{x1}, u_{y1}, u_{z1} and u_{x2}, u_{y2}, u_{z2}. These groups of components are parallel to the global X, Y, Z axes, respectively. Then, the two sets of displacements are related as

$$\mathbf{u} = \mathbf{T_1 u_{xyz}} \tag{20}$$

where $\mathbf{u_{xyz}}$ is the vector of nodal displacements of the axial spring expressed in the global coordinate system, and the transformation matrix $\mathbf{T}$ is given by

$$\mathbf{T_1} = \begin{bmatrix} c_{xx} & c_{yy} & c_{zz} & 0 & 0 & 0 \\ 0 & 0 & 0 & c_{xx} & c_{yy} & c_{zz} \end{bmatrix} \tag{21}$$

Here, c_{xx}, c_{yy}, and c_{zz} are the direction cosines of the angles between the line ij and the directions OX, OY, and OZ, respectively. The direction cosines can be expressed with respect to the global coordinates of nodes i and j as

$$\begin{aligned} c_{xx} &= \cos(x, X) = \frac{X_j - X_i}{l_e} \\ c_{yy} &= \cos(x, Y) = \frac{Y_j - Y_i}{l_e} \\ c_{zz} &= \cos(x, Z) = \frac{Z_j - Z_i}{l_e} \end{aligned} \tag{22}$$

where the nanospring length is

$$l_e = \sqrt{\left(X_j - X_i\right)^2 + \left(Y_j - Y_i\right)^2 + \left(Z_j - Z_i\right)^2}\ . \tag{23}$$

The transformation matrix of the rotations for the corresponding rotational spring is similar to the one for the axial spring. The transformation matrix for a nanospring has dimension 4 x 12,

$$\mathbf{T} = \begin{bmatrix} \mathbf{T_1} & \mathbf{0} \\ \mathbf{0} & \mathbf{T_1} \end{bmatrix} \tag{24}$$

Hence, the elemental matrices expressed in the global coordinate system are

$$\begin{aligned} \mathbf{K_a} &= \mathbf{T^T k_a T} \\ \mathbf{M_a} &= \mathbf{T^T m_a T} \\ \mathbf{K_b} &= \mathbf{T^T k_b T} \\ \mathbf{K_{vdw}} &= \mathbf{T^T k_{vdw} T} \end{aligned} \tag{25}$$

The displacements and rotations as well as the loads are related by the equation

$$\mathbf{U} = \mathbf{TD}\,,\, \mathbf{R} = \mathbf{T^T P}\,, \tag{26}$$

where $\mathbf{D}$ is the vector of displacements and rotations and $\mathbf{R}$ is the vector of loads with respect to the global coordinates. The superscript T in the above equations denotes matrix transposition. Finally, the elemental equations in the global system become

$$\mathbf{K_a D} + \mathbf{M_a \ddot{D}} = \mathbf{R} \tag{27}$$

$$\mathbf{K_b D} + \mathbf{M_b \ddot{D}} = \mathbf{R} \tag{28}$$

$$\mathbf{K_{vdw} D} + \mathbf{M_{vdw} \ddot{D}} = \mathbf{R}\,, \tag{29}$$

We assemble the global stiffness **K** and global mass matrices **M** from the above elemental matrices. Considering undamped free vibration for the tubes, the equation of motion becomes:

$$\mathbf{M\ddot{X}} + \mathbf{KX} = \mathbf{0} \tag{30}$$

After applying the CNT support conditions, the eigenvalue problem can be solved using common finite element procedures. The solution to the eigenvalue problem reveals the natural frequencies of vibration and the corresponding mode shapes.

3.3 CNT with added mass modeling

In the case that a nanoparticle is attached to a CNT mass detector, the nano-particle mass is considered to be located at a node of the CNT model and has a value m. Thus, for one a_i-element that includes this node, the value of the additional mass is added in Equation (16). The location of the nano-particle can be considered at the nanotube tip or at an intermediate position (Figure 5).

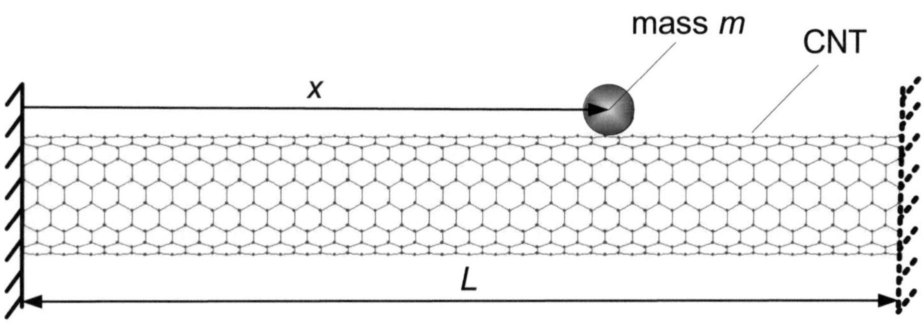

Fig. 5. CNT mass resonant sensor with an attached mass at an intermediate position.

As previously mentioned, the principle of mass detection using CNT-based detectors is based on resonant frequency shifts of a CNT due to a change in mass. The key challenge of mass detection and measuring is quantifying the changes in the resonant frequencies due to the added mass. The frequency shift is defined as

$$\textit{Frequency}\ \text{Shift} = f_0 - f\,, \tag{31}$$

where f_0 is the frequency a CNT without an attached mass, and f is the frequency of the same tube with an attached mass.

4. Results and discussion

4.1 Validation of model for vibration analysis

First, in order to validate the proposed method, we compare the results obtained from the present method with outputs from other theoretical approaches based on molecular or continuum mechanics, as shown in Table 1. The comparison is limited in terms of different sequence of vibration modes. This is mainly because of differences in

D_i (nm)	D_o (nm)	L (nm)	Support Condition of inner CNT	Support Condition of outer CNT	Fundamental Frequency (THz)	
					Present	Other Studies
0.4	1.1	4.1	C-C	C-C	0.9478	0.9276 [13]
0.4	1.1	5.5	C-C	C-C	0.6410	0.7355 [13]
0.4	1.1	8.0	C-C	C-C	0.3551	0.3323 [13]
0.7	1.4	14	Free	C-C	0.1582	0.1665 [25]
0.7	1.4	14	Free	C-F	0.0288	0.0270 [25]
0.7	1.4	14	C-C	C-C	0.1661	0.1718 [25]
0.7	1.4	20	C-C	C-C	0.04	~0.03 [24]

Table 1. Comparison of fundamental frequencies of MWCNTs resulting from different theoretical approaches. fundamental frequencies because the other methods obtain a formulation as well as the presence of new modes that are not reported in the other methods. Very good agreement is shown between results for different support conditions.

4.2 Modes of vibration of pure CNTs

The solution of the eigenvalue problem reveals numerous natural frequencies and corresponding mode shapes. The lengths of studied nanotubes are appropriate for possible future miniaturized products based on nanoscale structures. Figure 6 illustrates the modes of vibration for an armchair (12, 12) nanotube with length L = 10.95 nm subjected to clamped-clamped support condition. Figure 6(a) depicts the first radial breathing mode shape in which the atoms at the half-length of the tube have the largest displacement due to the fixed-end boundary constraints. The second radial breathing mode has a vase-like shape. On the other hand, Figure 6(b) shows the first bending-like mode that simultaneously exhibits radial breathing. As a result, a triangle-like shape is observed at every cross-section of the tube. For this reason, hereafter, modes with a similar shape will be called triangular mode shapes. Furthermore, in Figure 6(c), the deformation of the nanotube has a shape in which every cross-section has a cross-like form.

For this reason, modes with this type of shape will be called as cross mode shapes of vibration. The modal analysis for the specific nanotube also reveals bending, axial, and twisting modes. The first bending mode has a half-sine shape, whereas the second has a full-sine shape, as observed in a beam. In the axial modes of vibration, the movement of the atoms is parallel to the longitudinal axis of the tube. The first axial mode is accompanied by a simultaneous movement of all atoms along the longitudinal direction. In the second one, longitudinal movement of the vectors towards the center of the tube is observed. Another type of mode is the twisting (torsional) mode, as referred to in [18]. The atoms move in the circumferential direction with a simultaneous increase in the radius of the single-walled CNT. In the first twisting mode, all atoms have the same circumferential movement and the maximum increase in the radius is observed on the half length of the tube.

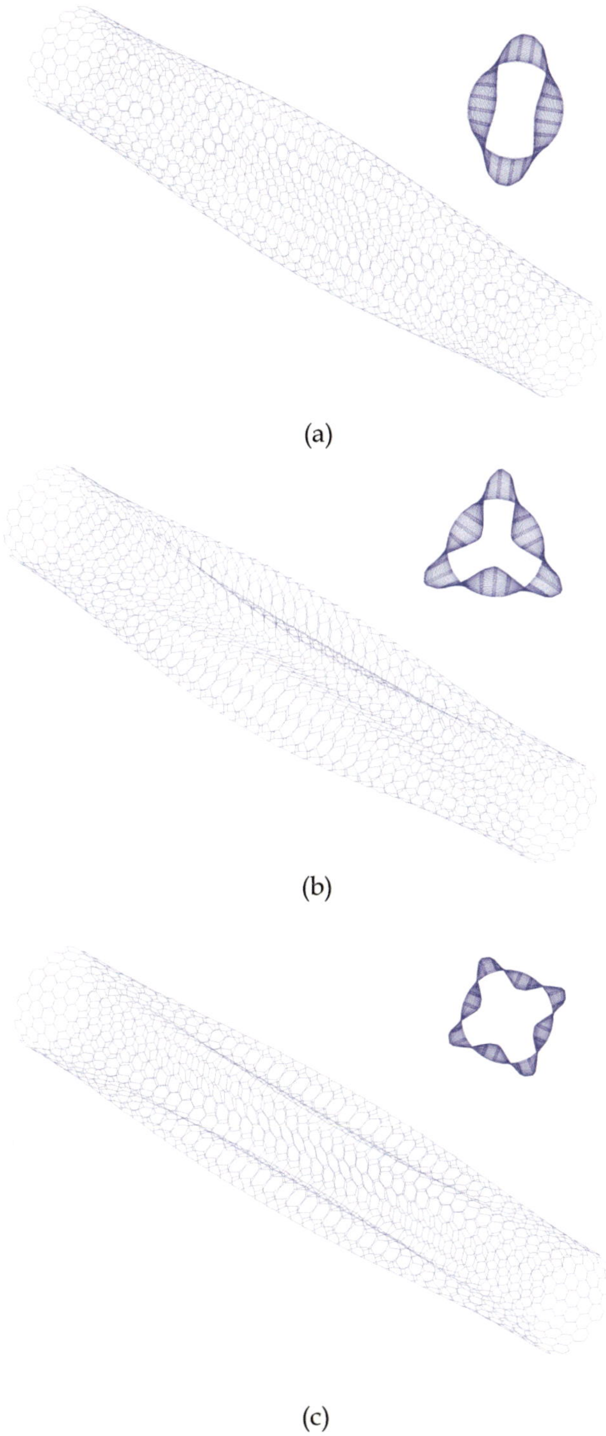

Fig. 6. Radial-like modes of vibration of clamped-clamped supported single-walled CNTs: (a) radial breathing mode, (b) triangular mode, and (c) cross mode.

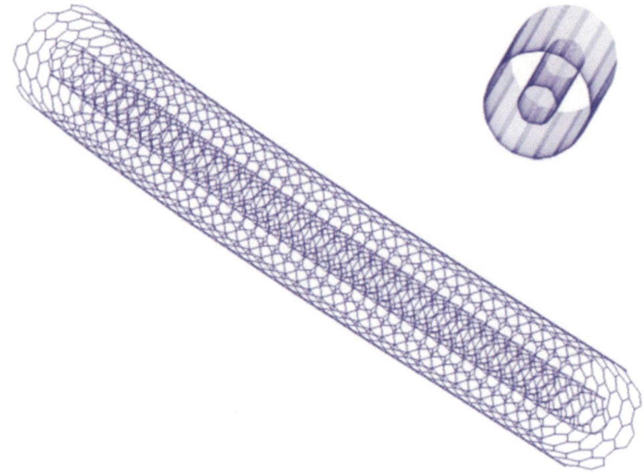

(a)

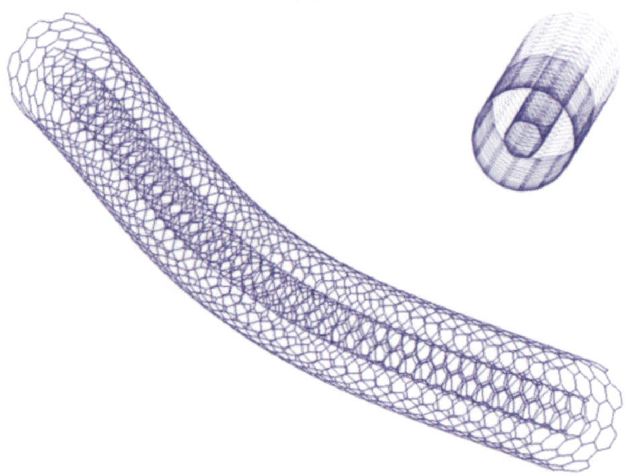

(b)

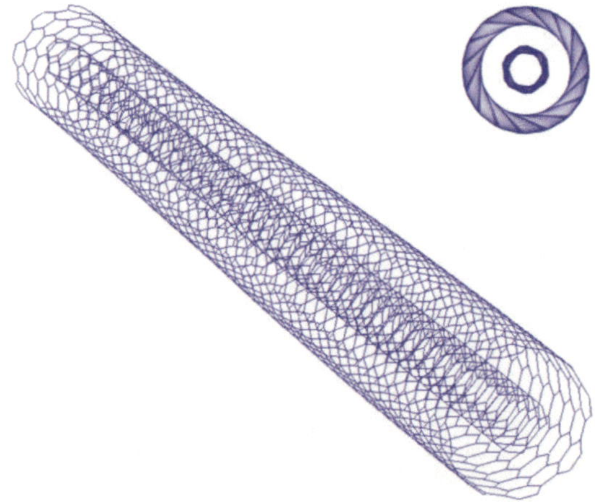

(c)

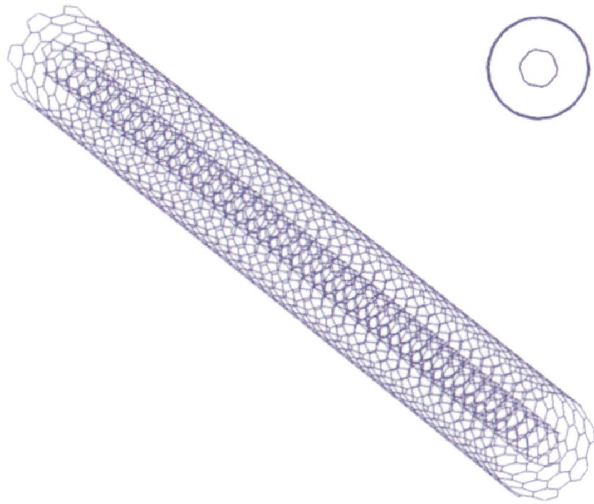

(d)

Fig. 7. Beam-like modes of vibration of clamped-free supported MWCNTs: (a) first bending, (b) second bending, (c) first twisting, and (d) first axial modes.

For the clamped-free support condition, the basic mode shapes of vibration will be described for a zigzag (5,0)-(14,0) double-walled CNT with length L = 15 nm and using a higher aspect ratio than the previous example. Figure 7 correspondingly depicts modes that are basic for large aspect ratios. Figures 7a and 7b illustrate the first and second bending mode shapes. These are similar to the bending modes presented in a macro-scale cantilever beam. The twisting modes (Figure 7c) exhibit a behavior similar to the clamped-clamped case, but the shapes differ because of the free end. The first axial mode (Figure 7d) is accompanied by a simultaneous movement of all atoms in the longitudinal direction and changing tube length during the vibration.

4.3 Effect of layers on CNT vibration

In order to investigate the influence of the number of layers on the vibration characteristics of a nanotube, CNTs of the same aspect ratio (i.e., the same length and outer diameter) were chosen for analysis with the proposed technique. Figure 8a depicts how the natural frequencies change for armchair tubes of length L = 17 nm and outer diameter D_o = 2.45 nmwhen subjected to a clamped-clamped support condition. It is observed that the lower the number of layers, the lower the frequency of the first radial breathing, second radial breathing, first triangular, and first cross modes. In contrast, the higher the number of layers, the lower the frequency of the first bending, second bending, first twisting, and first axial modes. As the number of layers increases, the tube tends to behave more like a beam. Note that for tubes with only one layer (single-walled CNTs), the basic modes are the first triangular, the first cross, and the first radial breathing shapes. Figure 8b illustrates similar variations for the clamped-free support condition. Here, the tube has length L = 15 nm and outer diameter D_o = 2.5 nm. In this case, the frequencies are certainly lower, as expected with the less strict support condition. We also note that non-coaxial mode shapes are revealed, together with modes for which the inner layers exhibit different shapes from the outer ones. These are not analytically described because they are not basics modes of vibration.

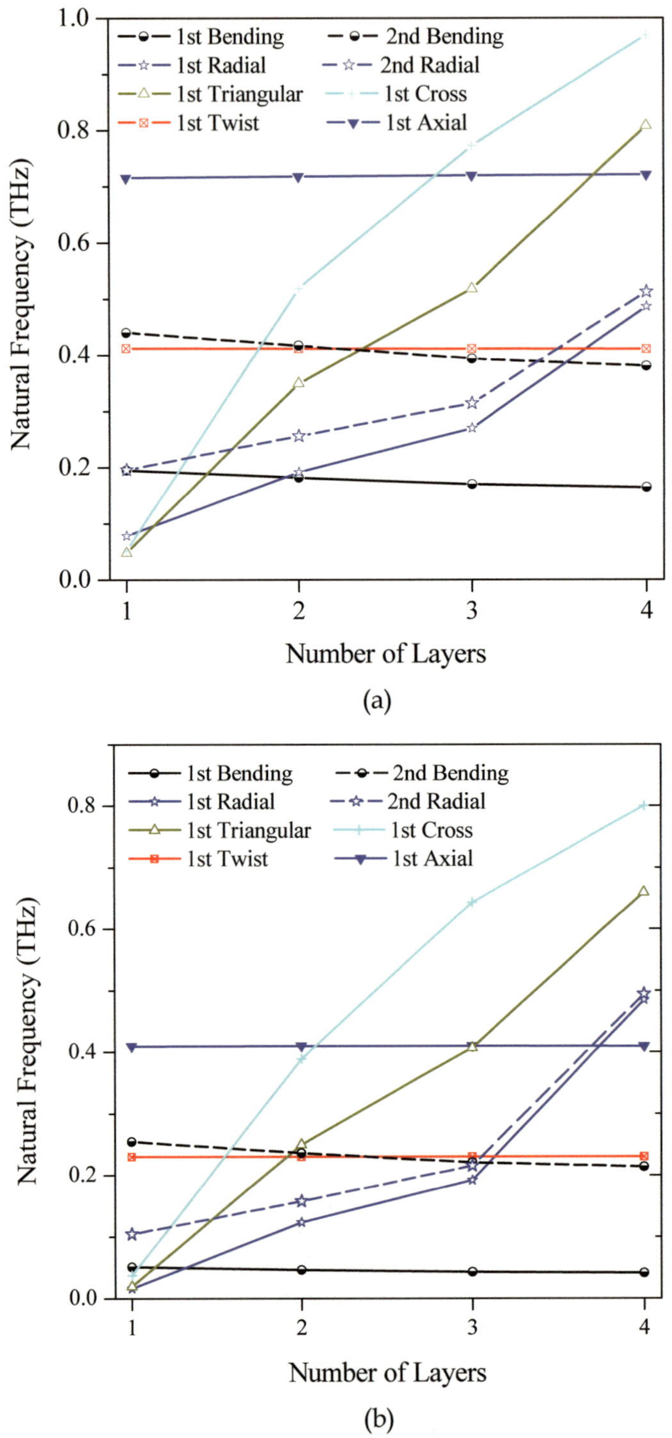

Fig. 8. Natural frequencies of (a) a clamped-clamped armchair CNT (D_o = 2.45nm, L = 17nm), and (b) a clamped-free zigzag CNT (D_o = 2.5nm, L = 15nm) vs number of layers.

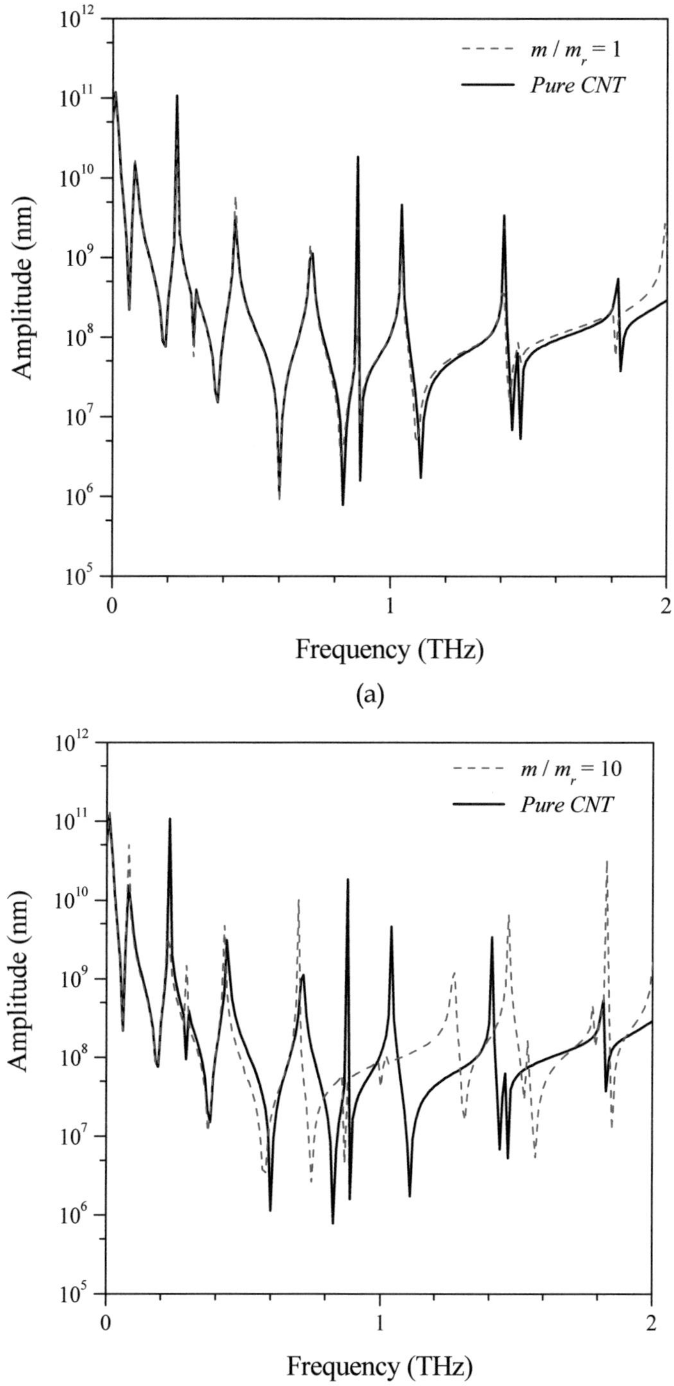

Fig. 9. Vibration spectra of clamped-free supported (5,0) CNT with a mass (a) $m/m_r = 1$ and (b) $m/m_r = 10$ attached on the CNT tip.

4.4 Vibration signature of CNT mass detectors

Also, the Fast Fourier Transform (FFT) of CNT mass detector is resembled for various cases in order to be depicted the changes occurred in the spectrum due to the magnitude and location of the added mass on CNT. In Figure 9, the spectra of a simply supported zigzag (5,0) with and without an added mass equal to one (Figure 9(a)) or ten carbon atoms (Figure 9(b)) attached on the tip are illustrated. The length of the CNT is L = 10.95 nm. The continuous red line gives the response of the CNT with the added mass, while the dashed blue line gives the corresponding signature of the same, however, pure CNT.

Despite the very small value of the added mass, important changes are observed in frequencies of the fundamental and higher order modes of vibration. Here, it has to be noticed that, most of the available studies in the literature investigate mainly the fundamental frequency shift. Nevertheless, significant changes in frequencies of higher order modes are also obtained here, as in [34] also reported. The results demonstrate that the higher order frequency shifts may be very helpful in the design and function of CNT mass detectors. Similar results are obtained also for the case of clamped-clamped support condition for the same CNT, when the added mass attached on the center of the CNT length (Figure 10).

An important question, which should be answered before a practical CNT mass detector may be made, is the problem of determining where the mass lands on the nanotube. Because some parts of the tube vibrate much more than the base, a mass that lands near the fast-moving region has a much greater effect on the resonant frequency and equivalent to a much greater mass arriving near to the base. Moreover, in addition to the fundamental mode of vibration, higher-order modes in which the maximum displacement occurs at two or more positions along the nanotube are possible as have been previously seen. These higher-order vibration modes could be exploited in experiments because the change in the resonant frequency also depends on how much the nanotube moves at the absorbing point [5]. This means that measurements made with higher-order modes could potentially allow both the mass and landing position to be determined. In literature, the majority of theoretical models are focused to sense the shift of the fundamental frequency due to an added mass. However, this suffices to sense a mass addition to the CNT, but not to measure the amount of mass because the magnitude of the shift is influenced by the mass and its location on the CNT. Hence, the sensing and measurement of the mass require investigation of the frequency shifts of the fundamental vibration mode as well as higher-order modes of vibration [5,35]. Based on this concept, the spectra of the CNT-mass system for different positions of the added mass on the CNT as well as different support conditions are depicted in Figure 11 and Figure 12. Here, we assume a lager mass (m/m_r = 100) than in previous results. In all cases, there are a lot of additional modes of vibration in comparison with the pure CNT case (see Figure 9). This can be explained by the asymmetry of the system existing due to the large mass value. Extremely different behaviors for different mass positions on CNT are observed in frequencies of the fundamental and higher order modes of vibration.

The previous mentioned results demonstrate that CNTs reveal a very different vibration signature even if a very small mass is added on its body. Moreover, they indicate that the position of the added mass noticeably influences the CNTs vibration spectra. The present methodology is capable to quantify these effects and thus can be used as a helpful computational tool for the design of CNT mass detectors.

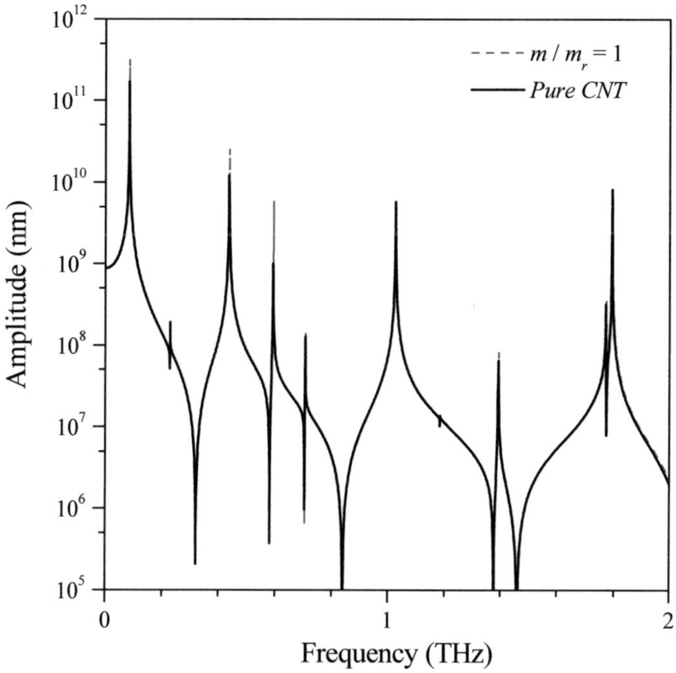

(a)

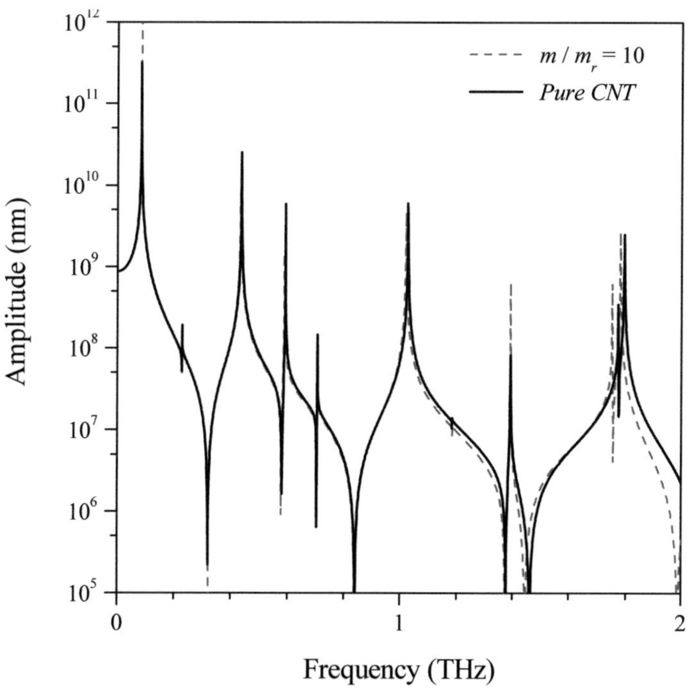

(b)

Fig. 10. Vibration spectra of clamped-clamped supported (5,0) CNT with a mass (a) $m/m_r = 1$ and (b) $m/m_r = 10$ attached on the middle of CNT length.

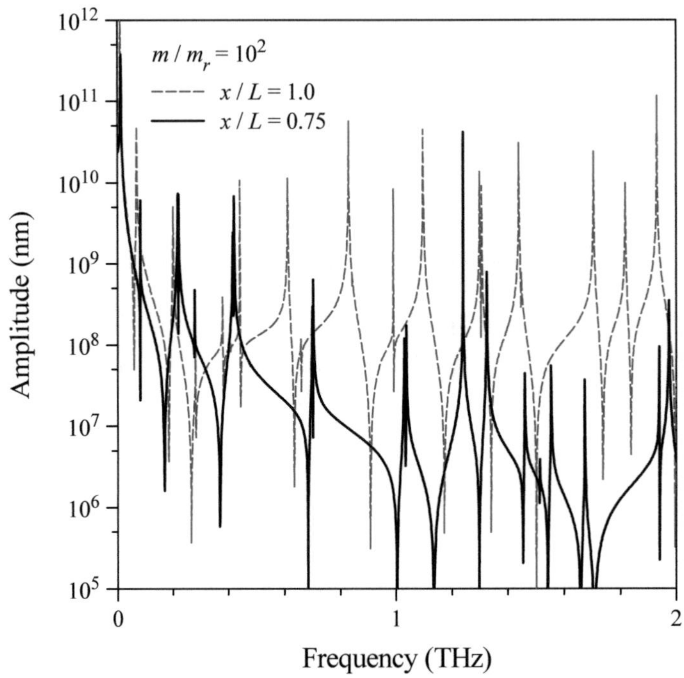

(a)

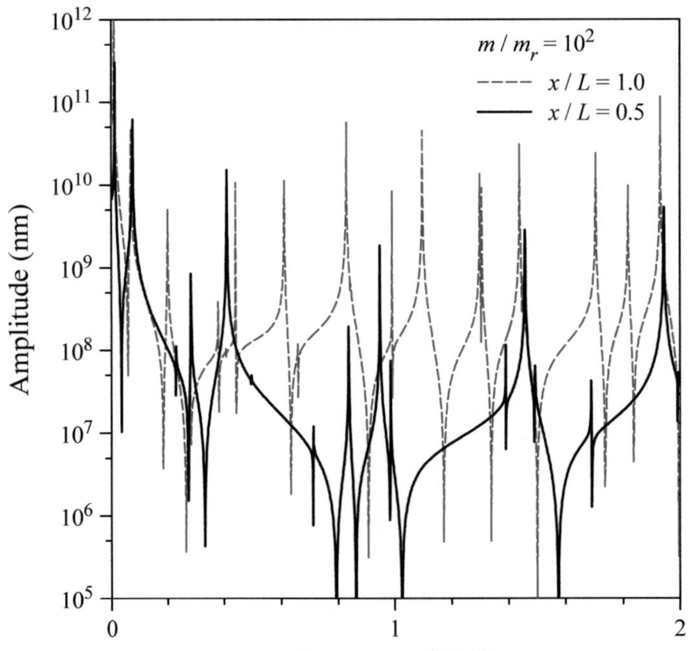

(b)

Fig. 11. Vibration spectra of clamped-free supported (5,0) CNT with a mass $m/m_r = 100$ attached on the (a) $x/L = 0.75$ and (b) $x/L = 0.5$ position compared with CNT tip case.

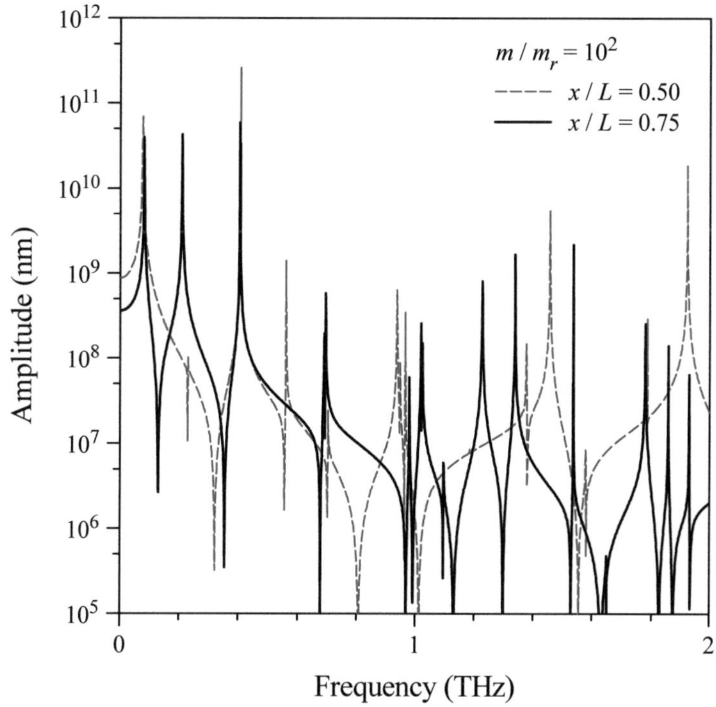

Fig. 12. Vibration spectra of clamped-clamped supported (5,0) CNT with a mass $m/m_r = 100$ attached on the $x/L = 0.75$ position compared with $x/L = 0.5$ case.

L (nm)	D_{in} (nm)	D_{out} (nm)	Layers	Support condition	Additional mass Position	Frequency Shift (MHz)			
						$m/mr = 10^0$	$m/mr = 10^1$	$m/mr = 10^2$	$m/mr = 10^3$
17	-	2.44	1	C-F	Free End	10	98	1208	7880.9
	1.76		2			8	80	793	7693
	1.09		3			6	58	579	5630
	0.41		4			5	52	518	5021
17	-	2.44	1	C-C	Middle	21	228	3907	26907
	1.76		2			20	260	6940	30420
	1.09		3			20	190	2800	19390
	0.41		4			20	170	2320	16700

Table 2. Comparison of the frequency shift between single- and multi-walled CNTs with the same lengths and diameters.

4.5 Comparison between single and multi-walled CNT mass detectors

The main issue to examine is whether multi-walled CNTs are more or less sensitive to an added mass in terms of frequency shift than single-walled CNTs with the same geometry, i.e. same outer diameter and length. Based on the data included in Table 2, it can be observed that a significant difference in frequency shift exists for all of the different masses attached to the CNTs. Furthermore, a higher frequency shift was observed for CNTs with fewer layers apparently because the fractional change in mass is larger. There are cases where the frequency shift of single-walled CNTs is more than 100% higher than the corresponding shift present in a multi-walled CNT. These observations suggest that single-walled CNTs are more suitable as mass resonant detectors than multi-walled CNTs.

5. Conclusions

The present chapter presented an efficient method for vibration analysis of CNTs mass detectors. The numerical method uses the CNT atomistic microstructure in order to assemble the dynamic equilibrium equation utilizing appropriate spring-mass based finite elements, which simulate the interatomic interactions and the inertia effects of CNTs. The analysis revealed the basic mode shapes of vibration and the corresponding natural frequencies of CNTs. Assuming the presence of an added mass attached on CNT, the method was utilized to resemble the FFT of CNT mass detector. The results demonstrate a significantly different vibration signature for different values and locations of the added mass in terms of frequency shift of fundamental and higher order modes. The method is capable to quantify these effects and hence could be used and facilitate the optimal design of CNT mass detectors of ultrahigh sensitivity. Finally, a parametric analysis demonstrated that single-walled carbon nanotubes are more sensitive than multi-walled CNTs and hence more suitable as mass resonant detectors.

6. References

[1] S. Iijima, Helical microtubes of graphite carbon, Nature 354 (1991) 56–58.

[2] R.S. Ruoff, D. Qian, W.K. Liu, Mechanical properties of carbon nanotubes: theoretical predictions and experimental measurements, *C. R. Physique* 4 (2003) 993–1008.

[3] M.S. Dresselhaus, G. Dresselhaus, J.C. Charlier, E. Hernandez, Electronic, thermal and mechanical properties of carbon nanotubes, *Philosophical Transactions of the Royal Society A: Mathematical, Physical and Engineering Sciences* 362 (2004) 2065-2098.

[4] K. Jensen, K. Kim, D.A. Zettl, An atomic-resolution nanomechanical mass sensor, Nature Nanotechnology 3 (2008) 533-537.

[5] R.G. Knobel, Weighing single atoms with a nanotube, Nature Nanotechnology 3 (2008) 525-526.

[6] R. Mateiu, A. Kuhle, R. Marie, A. Boisen, Building a multi-walled carbon nanotube-based mass sensor with the atomic force microscope, Ultramicroscopy 105 (2005) 233–237.

[7] H.-Y. Chiu, P. Hung, H.W.Ch. Postma, M. Bockrath, Atomic-Scale Mass Sensing Using Carbon Nanotube Resonators, *Nano Leters* 8 (2008) 4342–4346.

[8] R. Mateiu, Z.J. Davis, D.N. Madsen, K. Mølhave, P. Bøggild, A-M. Rassmusen, M. Brorson, C.J. H. Jacobsen, A Boisen, An approach to a multi-walled carbon nanotube based mass sensor, *Microelectronic Engineering* 73-74 (2004) 670-674.

[9] R.F. Gibson, E.O. Ayorinde, Y. Weng, Vibrations of carbon nanotubes and their composites: A review, *Composites Science and Technology*, 67 (2007) 1-28.

[10] B. Liu, Y. Huand, H. Jiang, S. Qu, K.C. Hwang, The atomic-scale finite element method, *Computer methods in applied mechanics and engineering*, 193 (2004) 1849-1864.
[11] B. Liu, Y. Huand, H. Jiang, S. Qu, K.C. Hwang, The atomic-scale finite element method, *Computer methods in applied mechanics and engineering*, 193 (2004) 1849-1864.
[12] M. Mir, A. Hosseini, G.H. Majzoobi, A numerical study of vibrational properties of single-walled carbon nanotubes, *Computational Materials Science*, 43 (2008) 540-548.
[13] C. Li, T.-W. Chou, Vibrational behaviors of multiwalled-carbon-nanotube-based nanomechanical resonator, *Applied Physics Letters*, 84 (2004) 121-123.
[14] G.M. Odegard, T.S. Gates, L.M. Nicholson, K.E. Wise, Equivalent-continuum modeling of nano-structured materials, *Composites Science and Technology* 62 (2002) 1869-1880.
[15] C. Li, T.-W. Chou, A structural mechanics approach for the analysis of carbon nanotubes, *International Journal of Solids and Structures* 40 (2003) 2487-2499.
[16] K.I. Tserpes, P. Papanikos, Finite element modeling of single-walled carbon nanotubes, Composites Part B: Engineering 36 (2005) 468-477.
[17] G.I. Giannopoulos, P.A. Kakavas, , N.K. Anifantis, Evaluation of the effective mechanical properties of single walled carbon nanotubes using a spring based finite element approach, *Computational Materials Science* 41 (2008) 561-569.
[18] S.K. Georgantzinos, G.I. Giannopoulos, N.K. Anifantis, An efficient numerical model for vibration analysis of single-walled carbon nanotubes, *Computational Mechanics* 43 (2009) 731-741.
[19] S.K. Georgantzinos, N.K. Anifantis, Vibration analysis of multi-walled carbon nanotubes using a spring-mass based finite element model, *Computational Materials Science* 47 (2009) 168-177.
[20] T.C. Theodosiou, D.A. Saravanos, Molecular mechanics based finite element for carbon nanotube modeling, *CMES - Computer Modeling in Engineering and Sciences* 19(2) (2007) 121-134.
[21] C.M. Wang, V.B.C. Tan, Y.Y. Zhang, Timoshenko beam model for vibration analysis of multi-walled carbon nanotubes, *Journal of Sound and Vibration,* 294 (2006) 1060-1072.
[22] J. Yoon, C.Q. Ru, A. Mioduchowski, Terahertz vibration of short carbon nanotubes modeled as Timoshenko beams, *Journal of Applied Mechanics, Transactions ASME,* 72 (2005) 10-17.
[23] K.Y. Xu, X.N. Guo, C.Q. Ru, Vibration of a double-walled carbon nanotube aroused by nonlinear intertube van der Waals forces, *Journal of Applied Physics,* 99 (2006) 064303.
[24] T. Natsuki, Q.Q. Ni, M. Endo, Analysis of the vibration characteristics of double-walled carbon nanotubes, *Carbon,* 46 (2008) 1570–1573.
[25] K.-Y. Xu, E.C. Aifantis, Y.-H. Yan, Vibrations of double-walled carbon nanotubes with different boundary conditions between inner and outer tubes, *Journal of Applied Mechanics, Transactions ASME* 75 (2008) 0210131-0210139.
[26] C. Li, T.-W. Chou, Mass detection using carbon nanotube-based nanomechanical resonators, *Applied Physics Letters,* 84 (25) (2004) 5246-5248.
[27] D.H. Wu, W.T. Chien, C.S. Chen, H.H. Chen, Resonant frequency analysis of fixed-free single-walled carbon nanotube-based mass sensor, *Sensors and Actuators A* 126 (2006) 117–121.
[28] R. Chowdhury, S. Adhikari, J. Mitchell, Vibrating carbon nanotube based bio-sensors, *Physica E,* doi:10.1016/j.physe.2009.09.007.

[29] J. Kołoczek, K. Young-Kyun, A. Burian, Characterization of spatial correlations in carbon nanotubes-modelling studies, *Journal of Alloys and Compounds*, 28 (2001) 222–225.

[30] A.K. Rappe, C.J. Casewit, K.S. Colwell, W.A. Goddard, W.M. Skiff, UFF, A full periodic table force-field for molecular mechanics and molecular dynamics simulations, *Journal of American Chemical Society*, 114 (1992) 10024–10035.

[31] B.R. Gelin, Molecular modeling of polymer structures and properties. Cincinnati: Hanser/Gardner Publishers, 1994.

[32] J.E. Lennard-Jones, The determination of molecular fields: from the variation of the viscosity of a gas with temperature, *Proceedings of the Royal Society of London Series A* 106 (1924) 441.

[33] L.A. Girifalco, R.A. Lad, Energy of cohesion, compressibility, and the potential energy functions of the graphite system, *The Journal of Chemical Physics*, 25 (1956) 693–697.

[34] S.K. Georgantzinos, N.K. Anifantis, Carbon nanotube-based resonant nanomechanical sensors:a computational investigation of their behavior, *Physica E-Low-Dimensional Systems & Nanostructures*, 42 (2010) 1795–1801.

[35] S. Dohn, W. Svendsen, A. Boisen, O. Hansen, Mass and position determination of attached particles on cantilever based mass sensors, *Review of Scientific Instruments* 78 (2007) 103303.

8

B-spline Shell Finite Element Updating by Means of Vibration Measurements

Antonio Carminelli and Giuseppe Catania
DIEM, Dept. of Mechanical Design, University of Bologna,
viale Risorgimento 2, 40136 Bologna,
Italy

1. Introduction

Within the context of structural dynamics, Finite Element (FE) models are commonly used to predict the system response. Theoretically derived mathematical models may often be inaccurate, in particular when dealing with complex structures. Several papers on FE models based on B-spline shape functions have been published in recent years (Kagan & Fischer, 2000; Hughes et al, 2005). Some papers showed the superior accuracy of B-spline FE models compared with classic polynomial FE models, especially when dealing with vibration problems (Hughes et al, 2009). This result may be useful in applications such as FE updating.

Estimated data from measurements on a real system, such as frequency response functions (FRFs) or modal parameters, can be used to update the FE model. Although there are many papers in the literature dealing with FE updating, several open problems still exist. Updating techniques employing modal data require a previous identification process that can introduce errors, exceeding the level of accuracy required to update FE models (D'ambrogio & Fregolent, 2000). The number of modal parameters employed can usually be smaller than that of the parameters involved in the updating process, resulting in ill-defined formulations that require the use of regularization methods (Friswell et al., 2001; Zapico et al.,2003). Moreover, correlations of analytical and experimental modes are commonly needed for mode shapes pairing. Compared with updating methods using modal parameters as input, methods using FRFs as input present several advantages (Esfandiari et al., 2009; Lin & Zhu, 2006), since several frequency data are available to set an over-determined system of equations, and no correlation analysis for mode pairing is necessary in general.

Nevertheless there are some issues concerning the use of FRF residues, such as the number of measurement degrees of freedom (dofs), the selection of frequency data and the ill-conditioning of the resulting system of equations. In addition, common to many FRF updating techniques is the incompatibility between the measurement dofs and the FE model dofs. Such incompatibility is usually considered from a dof number point of view only, measured dofs being a subset of the FE dofs. Reduction or expansion techniques are a common way to treat this kind of incompatibility (Friswell & Mottershead, 1995). A more general approach should also take into account the adoption of different dofs in the two models. As a matter of result, the adoption of B-spline functions as shape functions in a FE

model leads to non-physical dofs, and the treatment of this kind of coordinate incompatibility must be addressed.
In this paper a B-spline based FE model updating procedure is proposed. The approach is based on the least squares minimization of an objective function dealing with residues, defined as the difference between the model based response and the experimental measured response, at the same frequency. A proper variable transformation is proposed to constrain the updated parameters to lie in a compact domain without using additional variables. A B-spline FE model is adopted to limit the number of dofs. The incompatibility between the measured dofs and the B-spline FE model dofs is also dealt with.
An example dealing with a railway bridge deck is reported, considering the effect of both the number of measurement dofs and the presence on random noise. Results are critically discussed.

2. B-spline shell finite element model

2.1 B-spline shell model

A shell geometry can be efficiently described by means of B-spline functions mapping the parametric domain (ξ,η,τ) $(with\ 0 \le \xi,\eta,\tau \le 1)$ into the tridimensional Euclidean space (x,y,z). The position vector of a single B-spline surface patch, with respect to a Cartesian fixed, global reference frame **O**, {x,y,z}, is usually defined by a tensor product of B-spline functions (Piegl & Tiller, 1997):

$$\mathbf{r}(\xi,\eta) = \begin{Bmatrix} r_x \\ r_y \\ r_z \end{Bmatrix} = \sum_{i=1}^{m}\sum_{j=1}^{n} B_i^p(\xi) \cdot B_j^q(\eta) \cdot \mathbf{P_{ij}}, \tag{1}$$

involving the following parameters:

- a control net of $m \times n$ Control Points (CPs) $\mathbf{P_{ij}}$;
- the uni-variate normalized B-spline functions $B_i^p(\xi)$ of degree *p*, defined with respect to the curvilinear coordinate ξ by means of the knot vector:

$$\mathbf{U} = \left\{\xi_1,\ldots,\xi_{m+p+1}\right\} = \left\{\underbrace{0,\ldots,0}_{p+1},\xi_{p+1},\ldots,\xi_m,\underbrace{1,\ldots,1}_{p+1}\right\};$$

- the uni-variate normalized B-spline functions $B_j^q(\eta)$ of degree *q*, defined with respect to the curvilinear coordinate η by means of the knot vector:

$$\mathbf{V} = \left\{\eta_1,\ldots,\eta_{n+q+1}\right\} = \left\{\underbrace{0,\ldots,0}_{q+1},\eta_{q+1},\ldots,\eta_n,\underbrace{1,\ldots,1}_{q+1}\right\}.$$

The degenerate shell model is a standard in FE software because of its simple formulation (Cook et al., 1989). The position vector of the solid shell can be expressed as:

$$\mathbf{s}(\xi,\eta,\tau) = \sum_{i=1}^{m}\sum_{j=1}^{n} B_i^p(\xi) \cdot B_j^q(\eta) \cdot \left[\mathbf{P_{ij}} + t_{ij}\left(\tau - \frac{1}{2}\right)\mathbf{v}_{ij}^3\right], \tag{2}$$

where the versors $\mathbf{v}^3_{ij}$ and the thickness values t_{ij} can be calculated from the interpolation process proposed in (Carminelli & Catania, 2009).
The displacement field can be defined by following the isoparametric approach and enforcing the fiber inextensibility in the thickness direction (Cook et al., 1989):

$$\mathbf{d}(\xi,\eta,\tau)=\begin{Bmatrix} d_x \\ d_y \\ d_z \end{Bmatrix}=\sum_{i=1}^{m}\sum_{j=1}^{n}B_i^p(\xi)\cdot B_j^q(\eta)\cdot\left(\begin{Bmatrix} \mathrm{u_{ij}} \\ \mathrm{v_{ij}} \\ \mathrm{w_{ij}} \end{Bmatrix}+\mathrm{t}_{ij}\left(\tau-\frac{1}{2}\right)[\textbf{-v}_{ij}^2 \quad \mathbf{v}_{ij}^1]\begin{Bmatrix} \alpha_{\mathrm{ij}} \\ \beta_{\mathrm{ij}} \end{Bmatrix}\right)=$$

$$=\sum_{i=1}^{m}\sum_{j=1}^{n}B_i^p(\xi)\cdot B_j^q(\eta)\cdot\begin{bmatrix} 1 & 0 & 0 & & \\ 0 & 1 & 0 & \mathrm{t}_{ij}\left(\tau-\frac{1}{2}\right)[\textbf{-v}_{ij}^2 & \mathbf{v}_{ij}^1] \\ 0 & 0 & 1 & & \end{bmatrix}\cdot\begin{Bmatrix} \mathrm{u_{ij}} \\ \mathrm{v_{ij}} \\ \mathrm{w_{ij}} \\ \alpha_{\mathrm{ij}} \\ \beta_{\mathrm{ij}} \end{Bmatrix}= \tag{3}$$

$$=\begin{bmatrix} \mathbf{N_u} \\ \mathbf{N_v} \\ \mathbf{N_w} \end{bmatrix}\cdot\boldsymbol{\delta}=\mathbf{N}\cdot\boldsymbol{\delta}\,,$$

where $\boldsymbol{\delta}$ is the vector collecting the $(5\cdot m\cdot m)$ generalized dofs:

$$\boldsymbol{\delta}^T=\{u_{11} \quad v_{11} \quad w_{11} \quad \alpha_{11} \quad \beta_{11} \quad \cdots \quad u_{mn} \quad v_{mn} \quad w_{mn} \quad \alpha_{mn} \quad \beta_{mn}\}\,, \tag{4}$$

$\left(\mathbf{v}_{\mathrm{ij}}^1,\mathbf{v}_{\mathrm{ij}}^2,\mathbf{v}_{\mathrm{ij}}^3\right)$ refer to orthonormal sets defined on $\mathbf{P}_{ij}$ starting from the vector $\mathbf{v}_{\mathrm{ij}}^3$ (Carminelli & Catania, 2007), u_{ij}, v_{ij} and w_{ij} are translational dofs, α_{ij} and β_{ij} are rotational dofs.
The strains can be obtained from displacements in accordance with the standard positions assumed in three-dimensional linear elasticity theory (small displacements and small deformations), and can be expressed as:

$$\boldsymbol{\varepsilon}=\left\{\varepsilon_x \quad \varepsilon_y \quad \varepsilon_z \quad \gamma_{xy} \quad \gamma_{yz} \quad \gamma_{xz}\right\}^T=\mathbf{L}\cdot\mathbf{N}\cdot\boldsymbol{\delta}=\mathbf{D}\cdot\boldsymbol{\delta}\,, \tag{5}$$

where $\mathbf{D}=\mathbf{L}\cdot\mathbf{N}$ and $\mathbf{L}$ is the linear operator:

$$\mathbf{L}=\begin{bmatrix} \frac{\partial}{\partial x} & 0 & 0 & \frac{\partial}{\partial y} & 0 & \frac{\partial}{\partial z} \\ 0 & \frac{\partial}{\partial y} & 0 & \frac{\partial}{\partial x} & \frac{\partial}{\partial z} & 0 \\ 0 & 0 & \frac{\partial}{\partial z} & 0 & \frac{\partial}{\partial y} & \frac{\partial}{\partial x} \end{bmatrix}^T. \tag{6}$$

The stress tensor $\boldsymbol{\sigma}$ and strain $\boldsymbol{\varepsilon}$ are related by the material constitutive relationship:

$$\boldsymbol{\sigma}=\left\{\sigma_x \quad \sigma_y \quad \sigma_z \quad \tau_{xy} \quad \tau_{yz} \quad \tau_{xz}\right\}^T=\mathbf{E}\cdot\boldsymbol{\varepsilon}\,, \tag{7}$$

where **E** is the plane stress constitutive matrix obtained according to the Mindlin theory. **T** is the transformation matrix from the local material reference frame (1,2,3) to the global reference frame (x,y,z) (Cook et al., 1989):

$$\mathbf{E} = \mathbf{T}^{\mathrm{T}} \cdot \mathbf{E}^{'} \cdot \mathbf{T}, \tag{8}$$

and $\mathbf{E}^{'}$ is the plane stress constitutive matrix in the local material reference frame:

$$\mathbf{E}^{'} = \begin{bmatrix} \dfrac{E_1}{(1-\nu_{12}\nu_{21})} & \dfrac{\nu_{12}E_2}{(1-\nu_{12}\nu_{21})} & 0 & 0 & 0 & 0 \\ \dfrac{\nu_{12}E_2}{(1-\nu_{12}\nu_{21})} & \dfrac{E_2}{(1-\nu_{12}\nu_{21})} & 0 & 0 & 0 & 0 \\ 0 & 0 & 0 & 0 & 0 & 0 \\ 0 & 0 & 0 & G_{12} & 0 & 0 \\ 0 & 0 & 0 & 0 & G_{23} & 0 \\ 0 & 0 & 0 & 0 & 0 & G_{13} \end{bmatrix}, \tag{9}$$

where E_{ij} are Young modulus, G_{ij} are shear modulus and υ_{ij} are Poisson's ratios in the material reference frame.
The expressions of the elasticity, inertia matrices and of the force vector can be obtained by means of the principle of minimum total potential energy:

$$\Pi = U + W \rightarrow \min, \tag{10}$$

where U is the potential of the strain energy of the system:

$$U = \frac{1}{2}\int_{\Omega} \boldsymbol{\varepsilon}^{\mathrm{T}} \cdot \boldsymbol{\sigma}\, d\Omega, \tag{11}$$

and W is the potential of the body force **f** and of the surface pressure Q, and includes the potential W_i of the inertial forces:

$$W = -\int_{\Omega} \mathbf{d}^{\mathrm{T}} \cdot \mathbf{f} \cdot d\Omega - \int_{S} \mathbf{d}^{\mathrm{T}} \cdot \mathbf{Q} \cdot dS + W_i, \tag{12}$$

where:

$$W_i = \int_{\Omega} \rho \cdot \mathbf{d}^{\mathrm{T}} \cdot \ddot{\mathbf{d}} \cdot d\Omega. \tag{13}$$

The introduction of the displacement function (Eq.3) in the functional Π (Eq.10), imposing the stationarity of the potential energy:

$$\nabla_{\boldsymbol{\delta}}(\Pi) = 0, \tag{14}$$

yields the equations of motion:

$$\mathbf{M} \cdot \ddot{\boldsymbol{\delta}} + \mathbf{K_f} \cdot \boldsymbol{\delta} = \mathbf{F}, \tag{15}$$

where the unconstrained stiffness matrix is:

$$\mathbf{K_f} = \int_{\Omega} \mathbf{D}^{\mathrm{T}} \cdot \mathbf{E} \cdot \mathbf{D} \, d\Omega \,, \tag{16}$$

the mass matrix is:

$$\mathbf{M} = \int_{\Omega} \rho \cdot \mathbf{N}^{\mathrm{T}} \cdot \mathbf{N} \, d\Omega \,, \tag{17}$$

and the force vector is:

$$\mathbf{F} = \int_{\Omega} \mathbf{N}^{\mathrm{T}} \cdot \mathbf{f} \, d\Omega + \int_{S} \mathbf{N}^{\mathrm{T}} \cdot \mathbf{Q} \, dS \,, \tag{18}$$

where ρ is the mass density, Ω being the solid structure under analysis and S the external surface of solid Ω.

2.2 Constraint modeling

Distributed elastic constraints are taken into account by including an additional term $\Delta\mathrm{W}$ in the functional of the total potential energy. The additional term $\Delta\mathrm{W}$ takes into account the potential energy of the constraint force per unit surface area Q_C, assumed as being applied on the external surface of the shell model:

$$\mathbf{Q_C} = -\mathbf{R} \cdot \mathbf{d} \,, \tag{19}$$

where **R** is the matrix containing the stiffness coefficients r_{ab} of a distributed elastic constraint, modeled by means of B-spline functions:

$$r_{ab} = \sum_{i=1}^{m^{ab}} \sum_{j=1}^{n^{ab}} B_i^{p^{ab}} \cdot B_j^{q^{ab}} \cdot \kappa_{ij}^{ab} \,, \tag{20}$$

where $B_i^{p^{ab}}$ and $B_j^{q^{ab}}$ are the uni-variate normalized B-spline functions defined by means of the knot vectors, respectively, $\mathbf{U^{ab}}$ and $\mathbf{V^{ab}}$:

$$\Delta\mathrm{W} = -\frac{1}{2} \int_{S} (\mathbf{d}^{T} \cdot \mathbf{Q_C}) dS = \frac{1}{2} \boldsymbol{\delta}^{\mathrm{T}} \cdot \int_{S} \left(\mathbf{N}^{\mathrm{T}} \cdot \mathbf{R} \cdot \mathbf{N} \right) dS \cdot \boldsymbol{\delta} \,. \tag{21}$$

The stiffness matrix due to the constraint forces is

$$\mathbf{\Delta K} = \int_{S} \left(\mathbf{N}^{\mathrm{T}} \cdot \mathbf{R} \cdot \mathbf{N} \right) dS \,. \tag{22}$$

The introduction of $\Delta\mathrm{W}$ this last term in the total potential energy Π yields the equation of motion:

$$\mathbf{M} \cdot \ddot{\boldsymbol{\delta}} + \left(\mathbf{K_f} + \mathbf{\Delta K} \right) \cdot \boldsymbol{\delta} = \mathbf{F} \,. \tag{23}$$

2.3 Damping modelling

For lightly damped structures, effective results may be obtained by imposing the real damping assumption (real modeshapes).

The real damping assumption is imposed by adding a viscous term in the equation of motion:

$$\mathbf{M}\cdot\ddot{\boldsymbol{\delta}}+\mathbf{C}\cdot\dot{\boldsymbol{\delta}}+\left(\mathbf{K}_{\mathrm{f}}+\Delta\mathbf{K}\right)\cdot\boldsymbol{\delta}=\mathbf{F}, \tag{24}$$

where the damping matrix $\mathbf{C}$ is:

$$\mathbf{C}=\boldsymbol{\Phi}^{-T}\cdot\mathbf{diag}(2\zeta\omega)\cdot\boldsymbol{\Phi}^{-1}, \tag{25}$$

and

$$\mathbf{diag}(2\zeta\omega)=\begin{bmatrix} 2\zeta_1\omega_1 & 0 & \cdots & 0 \\ 0 & 2\zeta_2\omega_2 & & \vdots \\ \vdots & & \ddots & 0 \\ 0 & \cdots & 0 & 2\zeta_N\omega_N \end{bmatrix}, \tag{26}$$

where $\boldsymbol{\Phi}$ is the matrix of the eigen-modes $\boldsymbol{\Phi}_i$ obtained by solving the eigen-problem:

$$\left(\mathbf{K}-\omega_i^2\mathbf{M}\right)\boldsymbol{\Phi}_i=\mathbf{0}, \tag{27}$$

and ω_i^2 is the i-th eigen-value of Eq.(27). Modal damping ratios ζ_i can be evaluated from:

$$\zeta_i=\zeta\left(f_i\right)=\zeta\left(2\pi\cdot\omega_i\right), \tag{28}$$

where the damping $\zeta(f)$ is defined by means of control coefficients γ_z and B-spline functions B_z defined on a uniformly spaced knot vector:

$$\zeta(f)=\zeta(f(u))=\sum_{z=1}^{n^z}B_z(u)\cdot\gamma_z;\quad f=f_{ST}+u\cdot\left(f_{FI}-f_{ST}\right);\quad u\in[0,1], \tag{29}$$

where f_{ST} and f_{FI} are, respectively, the lower and upper bound of the frequency interval in which the spline based damping model is defined.

3. Updating procedure

The parametrization adopted for the elastic constraints and for the damping model is employed in an updating procedure based on Frequency Response Functions (FRFs) experimental measurements.

The ℓ measured FRFs $H_b^X(\omega)$, with $b=1,\ldots,\ell$, are collected in a vector $\mathbf{h}_X(\omega)$:

$$\mathbf{h}_X(\omega)=\begin{Bmatrix} H_1^X(\omega) \\ \vdots \\ H_\ell^X(\omega) \end{Bmatrix}. \tag{30}$$

The dynamic equilibrium equation in the frequency domain, for the spline-based finite element model, can be defined by Fourier transforming Eq.(24), where $\mathrm{F}(\ )=\tilde{(\ )}$:

$$\left(-\omega^2\mathbf{M} + j\omega\mathbf{C} + \mathbf{K_f} + \Delta\mathbf{K}\right)\cdot\tilde{\boldsymbol{\delta}} = \mathbf{Z}(\omega)\cdot\tilde{\boldsymbol{\delta}} = \mathbf{H}^{-1}(\omega)\cdot\tilde{\boldsymbol{\delta}} = \tilde{\mathbf{F}}, \tag{31}$$

where $\mathbf{Z}(\omega)$ is the dynamic impedance matrix and $\mathbf{H}(\omega) = (\mathbf{Z}(\omega))^{-1}$ is the receptance matrix. Since the vector $\tilde{\boldsymbol{\delta}}$ contains non-physical displacements and rotations, the elements of the matrix $\mathbf{H}(\omega)$ cannot be directly compared with the measured FRFs $H_q^X(\omega)$. The analytical FRFs related to physical dofs of the model can be obtained by means of the FE shape functions. Starting from the input force applied and measured on the point $\mathbf{P^i} = \mathbf{s}(\xi_i, \eta_i, \tau_i)$ along a direction $\boldsymbol{\varphi}$ and the response measured on the point $\mathbf{P^r} = \mathbf{s}(\xi_r, \eta_r, \tau_r)$ along the direction $\boldsymbol{\psi}$, the corresponding analytical FRF is:

$$H_{\boldsymbol{\psi},\boldsymbol{\varphi}}^{r,i}(\omega) = \mathbf{N}_{\boldsymbol{\psi}}(\xi_r, \eta_r, \tau_r)\cdot\mathbf{H}(\omega)\cdot\mathbf{N}_{\boldsymbol{\varphi}}^T(\xi_i, \eta_i, \tau_i), \tag{32}$$

where $\boldsymbol{\varphi}$ and $\boldsymbol{\psi}$ can assume a value among **u**, **v** or **w** (Eq.3).
The sensitivity of the FRF $H_{\boldsymbol{\psi},\boldsymbol{\varphi}}^{r,i}$ with respect to a generic parameter p_k is:

$$\begin{aligned}\frac{\partial H_{\boldsymbol{\varphi},\boldsymbol{\psi}}^{i,s}(\omega,\mathbf{p})}{\partial p_k} &= \mathbf{N}_{\boldsymbol{\psi}}(\varsigma_r, \eta_r, \gamma_r)\cdot\frac{\partial\mathbf{H}(\omega,\mathbf{p})}{\partial p_k}\cdot\mathbf{N}_{\boldsymbol{\varphi}}^T(\varsigma_i, \eta_i, \gamma_i) = \\ &= -\mathbf{N}_{\boldsymbol{\psi}}(\xi_r, \eta_r, \gamma_r)\cdot\mathbf{H}(\omega,\mathbf{p})\cdot\frac{\partial\mathbf{Z}(\omega,\mathbf{p})}{\partial p_k}\cdot\mathbf{H}(\omega,\mathbf{p})\cdot\mathbf{N}_{\boldsymbol{\varphi}}^T(\xi_i, \eta_i, \gamma_i),\end{aligned} \tag{33}$$

where $\mathbf{p} = \left\{p_1 \quad \cdots \quad p_{n^p}\right\}^T$ is the vector containing the updating parameters p_k.
Since each measured FRF $H_b^X(\omega)$ refers to a well-defined set $\{i, r, \boldsymbol{\varphi}, \boldsymbol{\psi}\}$, it is possible to collect, with respect to each measured FRF, the analytical FRFs in the vector:

$$\mathbf{h_a}(\omega,\mathbf{p}) = \begin{Bmatrix} H_{\boldsymbol{\varphi},\boldsymbol{\psi}}^{i,s}(\omega,\mathbf{p}) \\ \vdots \\ H_{\boldsymbol{\theta},\boldsymbol{\sigma}}^{\ell,t}(\omega,\mathbf{p}) \end{Bmatrix}. \tag{34}$$

The elements of $\mathbf{h_a}(\omega,\mathbf{p})$ are generally nonlinear functions of **p**. The problem can be linearized, for a given angular frequency ω_i, by expanding $\mathbf{h_a}(\omega,\mathbf{p})$ in a truncated Taylor series around $\mathbf{p}=\mathbf{p_0}$:

$$\mathbf{h_a}(\omega_i,\mathbf{p_o}) + \sum_{k=1}^{n^p}\frac{\partial\mathbf{h_a}(\omega_i,\mathbf{p_o})}{\partial p_k}\Delta p_k = \mathbf{h_x}(\omega_i), \tag{35}$$

in matrix form:

$$\left[\frac{\partial\mathbf{h_a}(\omega_i,\mathbf{p_o})}{\partial p_1},\ldots,\frac{\partial\mathbf{h_a}(\omega_i,\mathbf{p_o})}{\partial p_k},\ldots,\frac{\partial\mathbf{h_a}(\omega_i,\mathbf{p_o})}{\partial p_{np}}\right]\begin{bmatrix}\Delta p_1 \\ \vdots \\ \Delta p_k \\ \vdots \\ \Delta p_{np}\end{bmatrix} = \mathbf{h_x}(\omega_i) - \mathbf{h_a}(\omega_i,\mathbf{p_o}), \tag{36}$$

or:

$$\mathbf{S}_i \cdot \mathbf{\Delta p} = \mathbf{\Delta h}_i\,, \tag{37}$$

where $\mathbf{S}_i$ is the sensitivity matrix for the *i-th* angular frequency value ω_i.
It is possible to obtain a least squares estimation of the n^p parameters p_k, by defining the error function **e**:

$$\mathbf{e} = \sum_{i=1}^{n^f} \mathbf{S}_i \cdot \mathbf{\Delta p} - \mathbf{\Delta h}_i\,, \qquad n^f \gg n^p\,, \tag{38}$$

and by minimizing the objective function g:

$$g = \left(\mathbf{e}^T \cdot \mathbf{e}\right) \rightarrow \min\,. \tag{39}$$

Since the updating parameters p_k belong to different ranges of value, ill-conditioned updating equations may result. A normalization of the variables was employed to prevent ill-conditioning of the sensitivity matrix:

$$p_k = p_{0_k} \cdot \left(1 + x_k\right) \quad ; k=1,\ldots,n^p\,, \tag{40}$$

where p_{0_k} is a proper normalization value for the parameter p_k.
Moreover, to avoid updating parameters assuming non-physical values during the iterative procedure, a proper variable transformation is proposed to constrain the parameters in a compact domain without using additional variables:

$$x_{k_{min}} \le x_k \le x_{k_{max}}\,, \quad \left(x_{k_{min}} = \frac{p_{k_{min}}}{p_{0_k}} - 1, \quad x_{k_{max}} = \frac{p_{k_{max}}}{p_{0_k}} - 1\right), \tag{41}$$

where $p_{k_{max}}$ and $p_{k_{min}}$ are, respectively, the maximum and minimum values allowed for the parameter p_k. The transformation is:

$$\begin{aligned} p_k &= p_{0_k} \cdot \left(1 + 0.5 \cdot \left(x_{k_{min}} + x_{k_{max}} + \left(x_{k_{max}} - x_{k_{min}}\right) \cdot \sin\left(y_k\right)\right)\right) = \\ &= p_{0_k} + 0.5 \cdot \left(p_{k_{min}} + p_{k_{max}} - 2 \cdot p_{0_k} + \left(p_{k_{max}} - p_{k_{min}}\right) \cdot \sin\left(y_k\right)\right). \end{aligned} \tag{42}$$

The sensitivity matrices were derived with respect to the new variables y_k:

$$\frac{\partial \mathbf{h}_a}{\partial y_k} = \frac{\partial \mathbf{h}_a}{\partial p_k} \cdot \frac{\partial p_k}{\partial y_k} = 0.5 \cdot \left(p_{k_{max}} - p_{k_{min}}\right) \cdot \cos\left(y_k\right) \cdot \frac{\partial \mathbf{h}_a}{\partial p_k}\,, \tag{43}$$

which are allowed to take real values ($-\infty \le y_k \le \infty$) during the updating procedure.
Since FRF data available from measurement are usually large in quantity, a least squares estimation of the parameters can be obtained by adopting various FRF data at different frequencies. The proposed technique is iterative because a first order approximation was made during derivation of Eq.(35). At each step the updated global variables p_k can be obtained by means of Eq.(42).

4. Applications

The numerical example concerns the deck of the "Sinello" railway bridge (Fig.1). It is a reinforced concrete bridge located between Termoli and Vasto, Italy. It has been studied by several authors (Gabriele et al., 2009; Garibaldi et al., 2005) and design data and dynamical simulations are available.
The second deck span is a simply supported grillage with five longitudinal and five transverse beams. The grillage and the slab were modeled with an equivalent orthotropic plate, with fourth degree B-spline functions and 13x5 CPs (blue dot in Fig.2), for which the equivalent material properties were estimated by means of the design project:

$$E_1 = 5.5 \cdot 10^9\, Pa, \quad E_2 = 9.6 \cdot 10^8\, Pa, \quad G = 4.3 \cdot 10^8\, Pa,$$
$$\rho = 975\, Kg/m^3, \quad \nu_{12} = 0.3.$$

Because of FRF experimental measurement unavailability, two sets of experimental measurements were simulated assuming the input force applied on point 1 along z direction (Fig. 2). Twelve response dofs (along z direction) were used in the first set (red squares in Fig.2), while the second set contains only four measurement response dofs (red squares 1-4 in Fig. 2), in the frequency range [0, 80] Hz.
The simply supported constraint was modelled as a distributed stiffness acting on a portion of the bottom surface of the plate ($\tau = 0$):

$$\Delta\mathbf{K} = \int_S \left(\mathbf{N}^T \cdot \mathbf{R} \cdot \mathbf{N}\right) \cdot dS\,, \tag{44}$$

where $\mathbf{R}$ is the matrix containing the stiffness of distributed spring acting only in vertical direction z:

$$\mathbf{R} = \begin{bmatrix} 0 & 0 & 0 \\ 0 & 0 & 0 \\ 0 & 0 & r_{33}(\xi,\eta) \end{bmatrix}. \tag{45}$$

The distributed stiffness r_{33} is modelled by means of B-spline functions:

$$r_{33} = \sum_{i=1}^{1}\sum_{j=1}^{4} B'^0_i(\xi) \cdot B'^2_j(\eta) \cdot \kappa'_{ij} + \sum_{i=1}^{1}\sum_{j=1}^{4} B''^0_i(\xi) \cdot B''^2_j(\eta) \cdot \kappa''_{ij}\,, \tag{46}$$

where:

- $\boldsymbol{\kappa}' = 10^9 \cdot [0.4 \quad 1.5 \quad 1.8 \quad 0.6]\, N/m^3$, and the associated B-spline functions are defined on the knot vectors $\mathbf{U}' = \{0, 0.03\}$ and $\mathbf{V}' = \{0,0,0,0.5,1,1,1\}$;
- $\boldsymbol{\kappa}'' = 10^9 \cdot [1.5 \quad 0.4 \quad 0.5 \quad 1.8]\, N/m^3$, and the associated B-spline functions are defined on the knot vectors $\mathbf{U}'' = \{0.97, 1\}$ and $\mathbf{V}'' = \{0,0,0,0.5,1,1,1\}$.

The distribution of the spring stiffness is plotted in Fig.3. In order to simplify the presentation of the numerical results, the stiffness coefficients are collected in the vector $\boldsymbol{\kappa}$ as follows:

$$\boldsymbol{\kappa} = [\boldsymbol{\kappa}' \quad \boldsymbol{\kappa}''] = [\kappa_1 \quad \cdots \quad \kappa_j \quad \cdots \quad \kappa_8] = 10^9 \cdot [0.4 \quad 1.5 \quad 1.8 \quad 0.6 \quad 1.5 \quad 0.4 \quad 0.5 \quad 1.8]\, N/m^3\,. \tag{47}$$

The modal damping ratio values reported in Fig.4 were employed for the first 30 eigen-modes.

Fig. 1. Sinello railway bridge (Garibaldi et al., 2005).

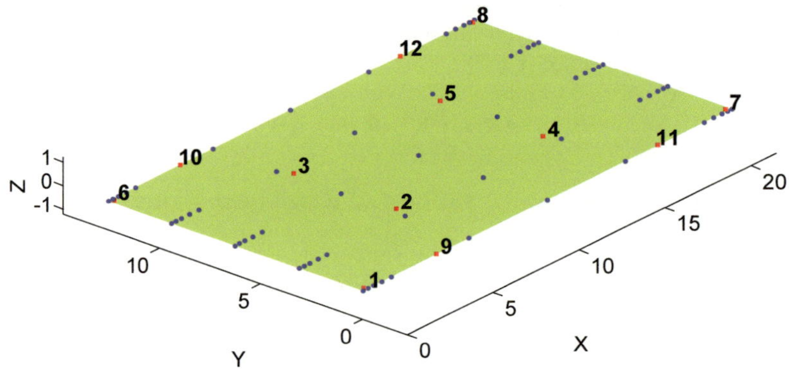

Fig. 2. The B-spline FE model with the 13x5 pdc (blue dot) and the 12 measurement response dofs (red squares).

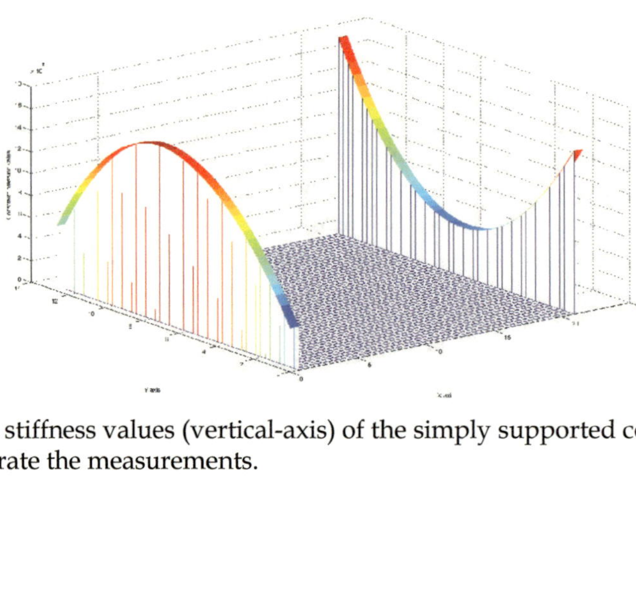

Fig. 3. Distributed stiffness values (vertical-axis) of the simply supported constraint employed to generate the measurements.

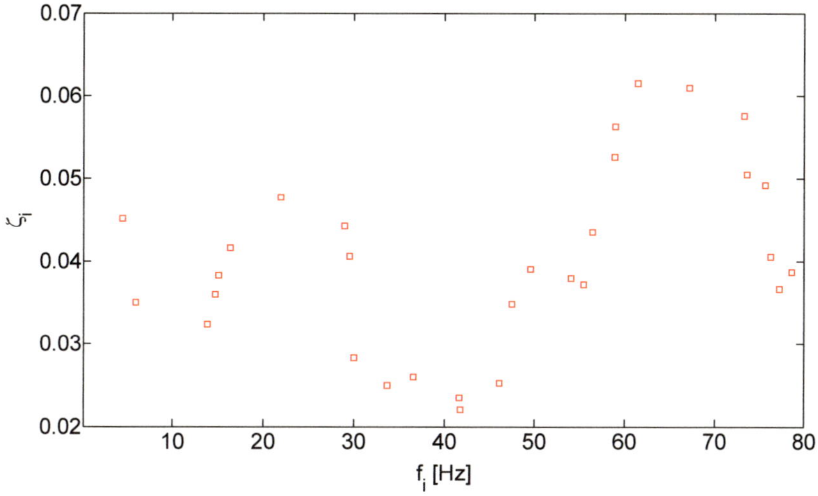

Fig. 4. Modal damping ratio values adopted to simulate the measurements. The values refer to the first 30 modes in the frequency range [0,80] Hz.

4.1 Numerical simulation without noise and with 12 measurement response dofs

Coefficients in vector $\mathbf{\kappa}$ and damping coefficients γ_z (quadratic B-spline functions, n^z=7, f_{ST}=0 Hz and f_{FI}=80 Hz in Eq.28) are assumed as the updating identification variables. The updating procedure is started by setting all of the coefficients in $\mathbf{\kappa}$ to 10^9 N/m^3 and all of the damping coefficients to 0.01. The comparison of the resulting FRFs is reported in Fig.5. The gradient of $\mathbf{C}$ with respect to the stiffness parameters is disregarded, i.e. $\frac{\partial \mathbf{C}}{\partial p_k} \simeq 0$ if $p_k \neq \gamma_z$. All twelve measurements dofs (Fig. 2) are considered as input. The value of the identification parameters at each step, adopting the proposed procedure, is reported in Fig.6 for the stiffness coefficients, and in Fig.7 for the γ_z coefficients; Fig.8 refers to the comparison of the modal damping ratio values used to simulate the measurements (red squares) and the identified curve (black line). The negative values of some parameters can lead to non physical stiffness matrix $\Delta\mathbf{K}$ so that instabilities may occur during the updating procedure. The proposed variable transformation does not allow stiffness coefficients to assume negative values. The comparison of theoretical and input FRF after updating is reported in Fig.9.

4.2 Numerical simulation without noise and with 4 measurement response dofs

The second simulation deals with the same updating parameters adopted in the previous example and with the same starting values, but only four measurement response dofs (dofs from 1 to 4 in Fig. 2) are considered.

The value of the identification parameters at each step, adopting the proposed procedure, is reported in Fig.10 for the stiffness coefficients, and in Fig.11 for the γ_z damping coefficients; Fig.12 refers to the comparison of the modal damping ratio values used to simulate the measurements (red squares) and the identified curve (black line). Fig.13 refers to the comparison of the FRFs after updating.

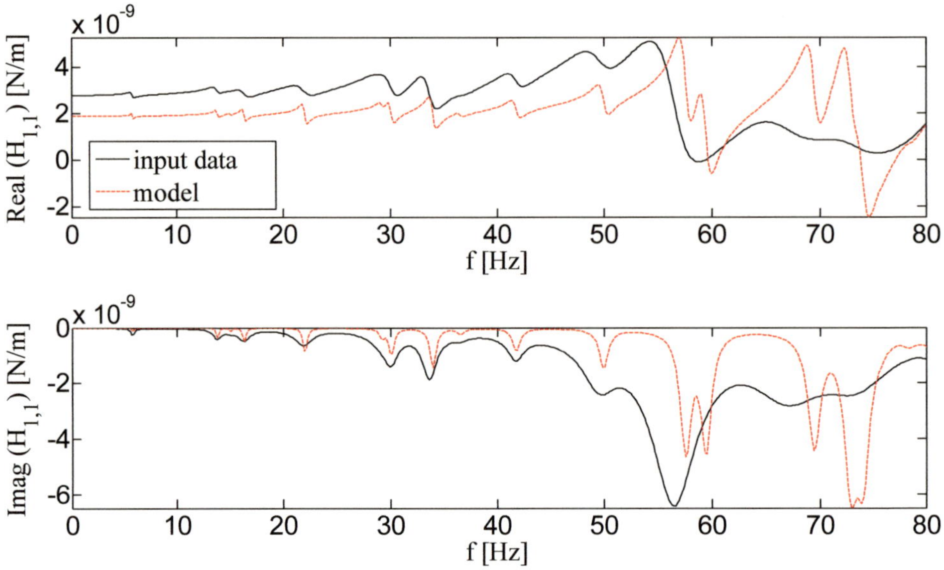

Fig. 5. Comparison of (input in dof 1; output in dof 1) FRF before updating: the input data (black continuous line) and the model (red dotted line).

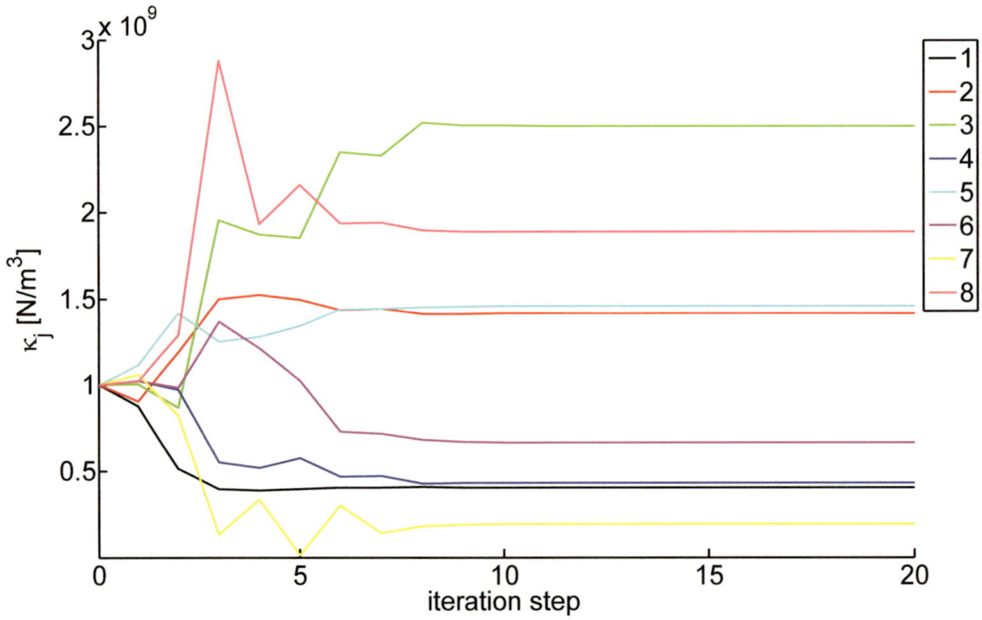

Fig. 6. Evolution of the stiffness parameters κ_j (j=1,...,8 in the legend) during iterations by adopting the proposed updating procedure. Example with 12 measurement response dofs and without noise.

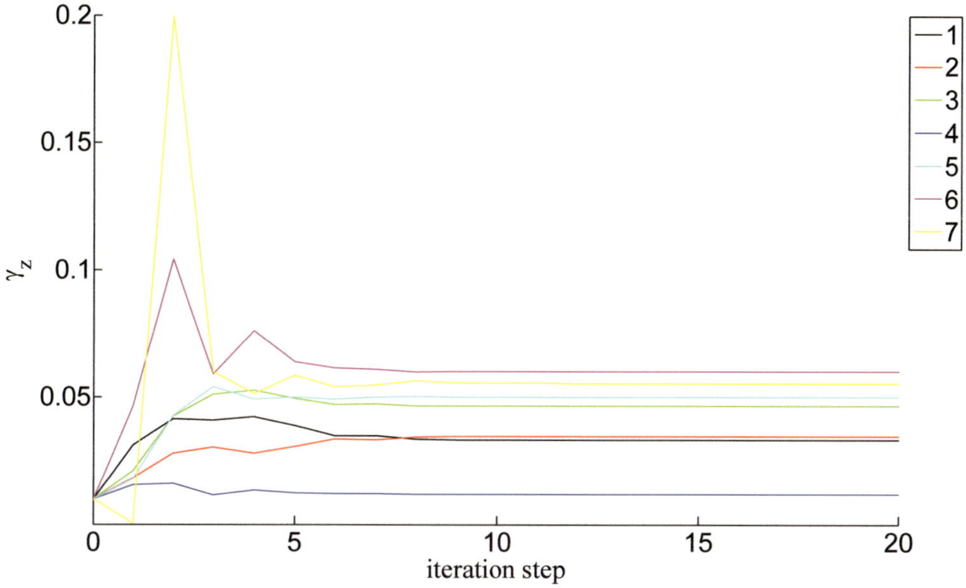

Fig. 7. Evolution of the damping parameters γ_z (z=1,...,7 in the legend) during iterations by adopting the proposed updating procedure. Example with 12 measurement response dofs and without noise.

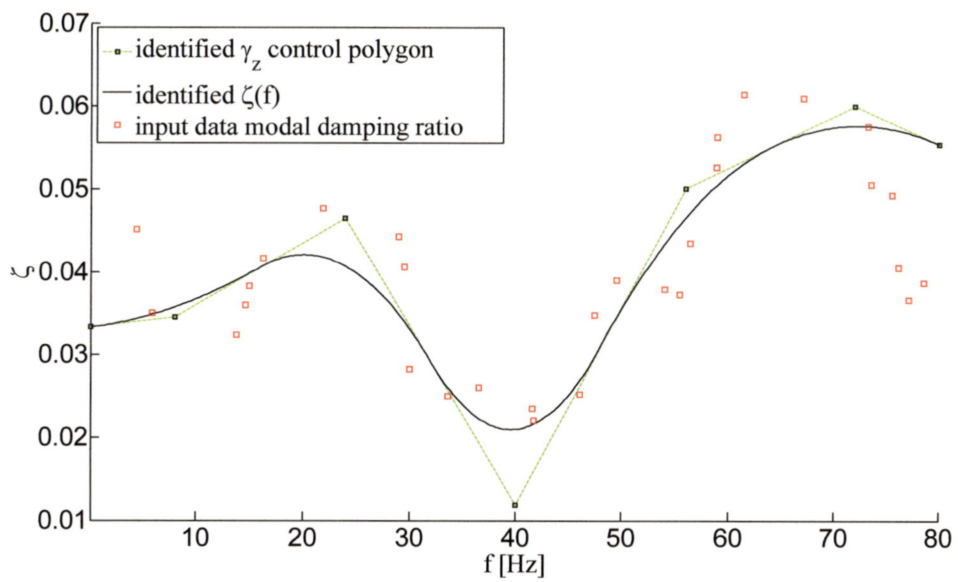

Fig. 8. Comparison of the modal damping ratio used to simulate the measurements (red squares) and the identified $\zeta(f)$ (black line; green filled squares refer to B-spline curve control coefficients). Example with 12 measurement response dofs and without noise.

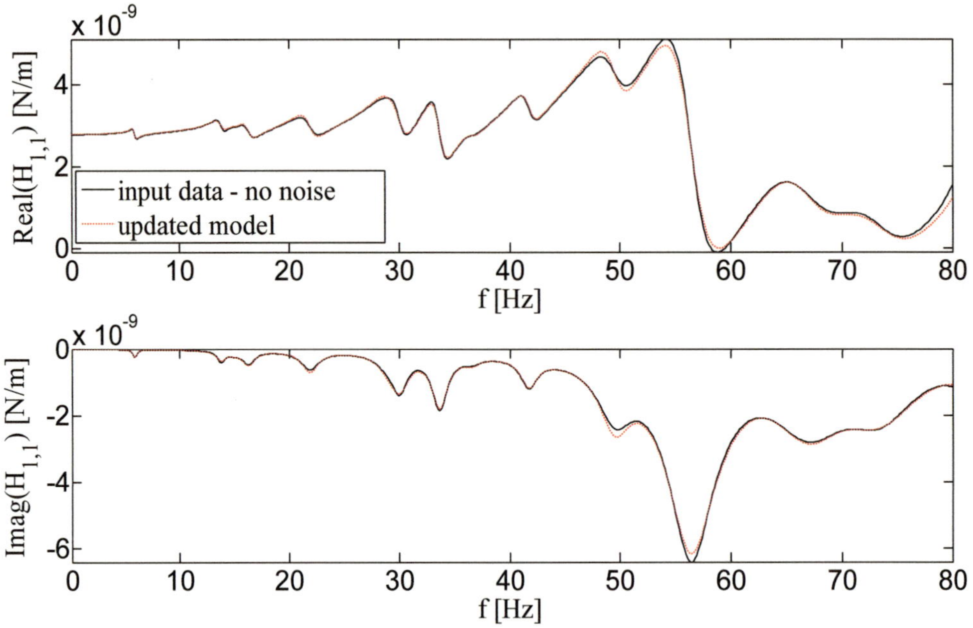

Fig. 9. Comparison of (input in point 1; output in point 1) FRF after updating (example with 12 measurement response dofs without noise): the input data (black continuous line) and the updated model (red dotted line).

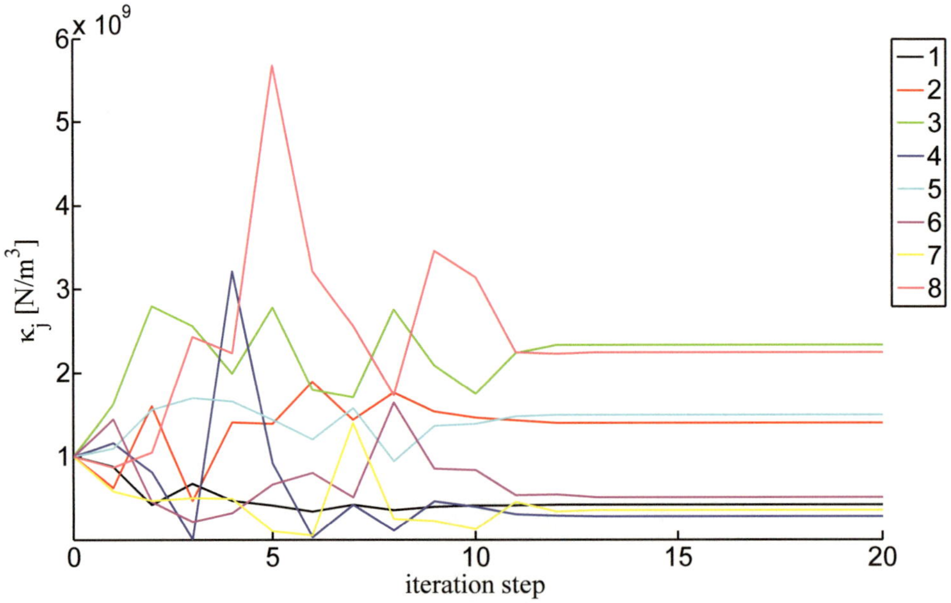

Fig. 10. Evolution of stiffness parameters κ_j (j=1,...,8 in the legend) during iterations by adopting the proposed updating procedure. Example with 4 measurement response dofs and without noise.

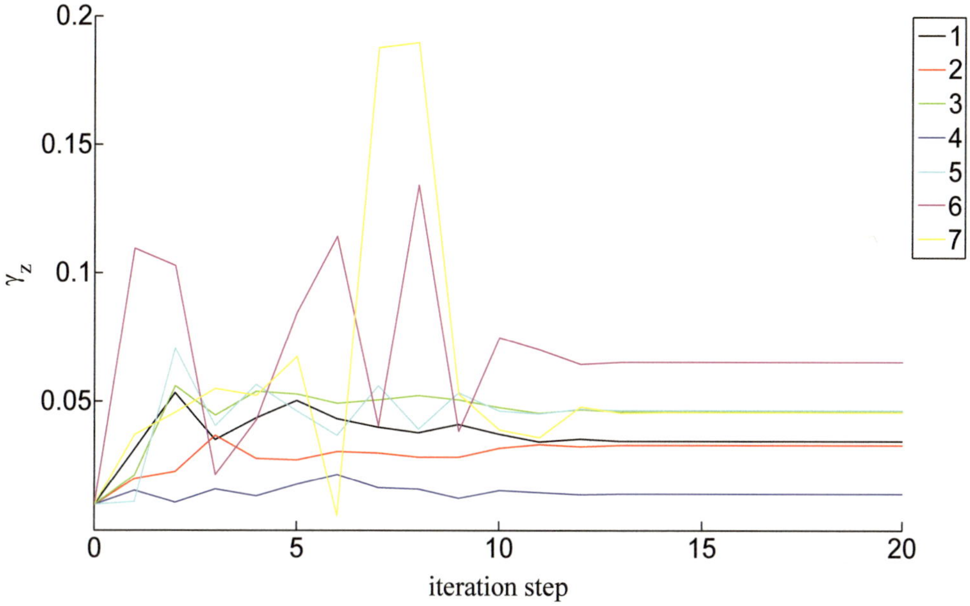

Fig. 11. Evolution of the damping parameters γ_z (z=1,...,7 in the legend) during iterations by adopting the proposed updating procedure. Example with 4 measurement response dofs and without noise.

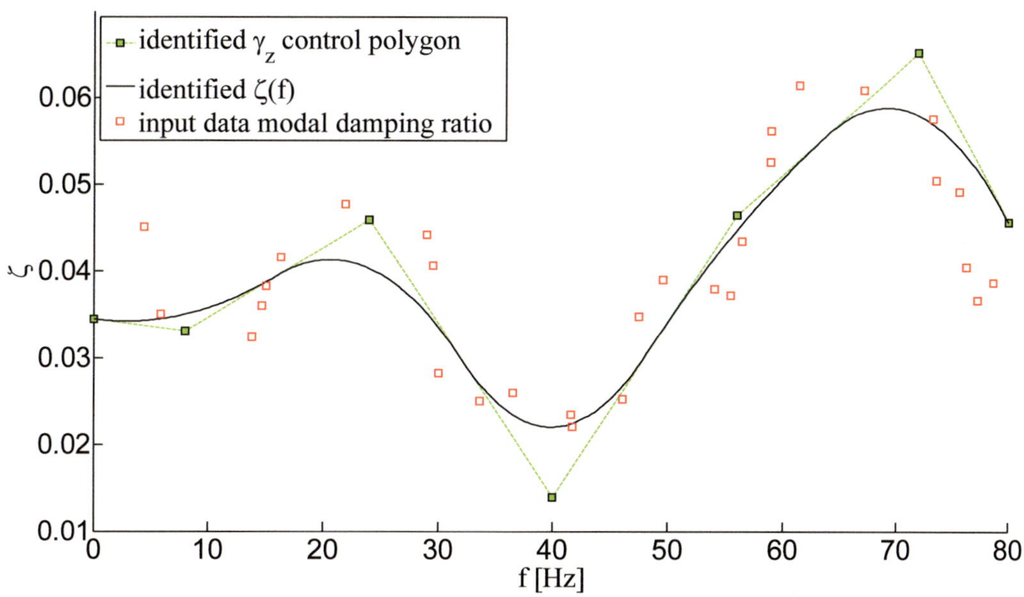

Fig. 12. Comparison of the modal damping ratio used to simulate the measurements (red squares) and the identified $\zeta(f)$ (black line; green filled squares refer to B-spline curve control coefficients). Example with 4 measurement response dofs and without noise.

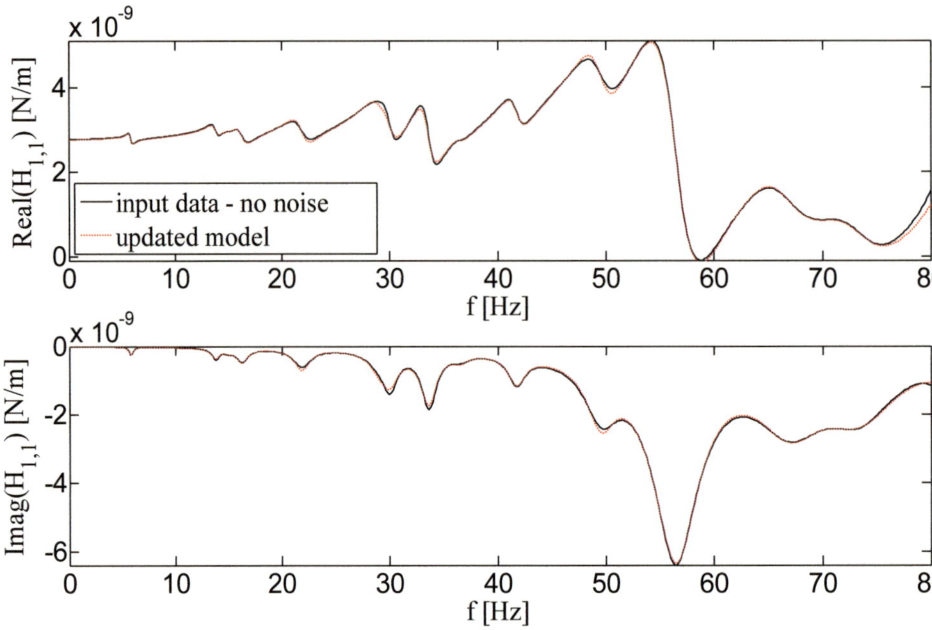

Fig. 13. Comparison of (input in point 1; output in point 1) FRF after updating (example with 4 measurement response dofs, without noise): the input data (black continuous line) and the updated model (red dotted line).

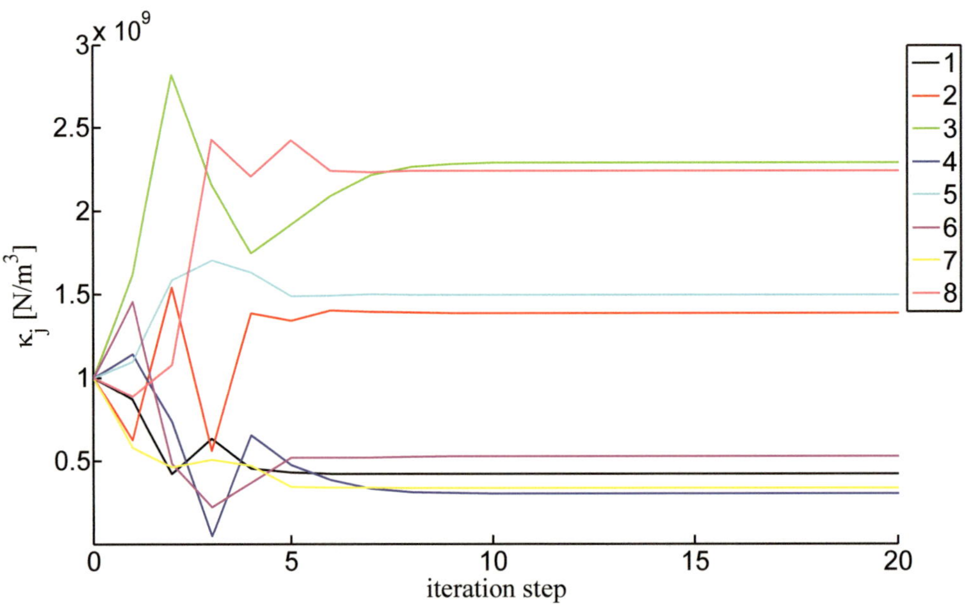

Fig. 14. Evolution of stiffness parameters κ_j (j=1,...,8 in the legend) during iterations by adopting the proposed updating procedure. Example with 4 measurement response dofs and with 3% noise.

4.3 Numerical simulations with noise

In these two simulations, the same updating parameters of the previous examples are considered with the same starting values. A random noise is added in FRFs, by considering a normal distribution with a standard deviation set to 3% and 10% of the signal RMS value. Four FRFs data (dofs from 1 to 4, Fig.2) are employed in the updating process.

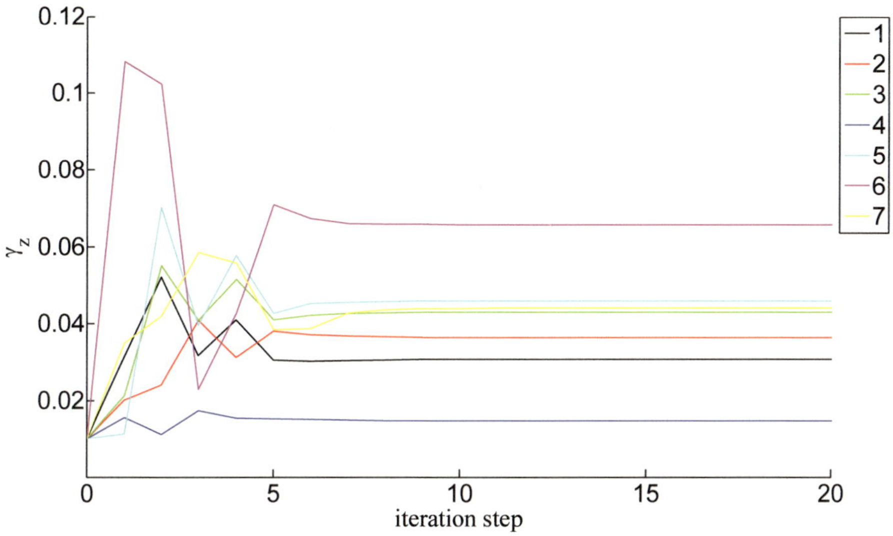

Fig. 15. Evolution of the damping parameters γ_z(z=1,...,7 in the legend) during iterations by adopting the proposed updating procedure. Example with 4 measurement response dofs and with 3% noise.

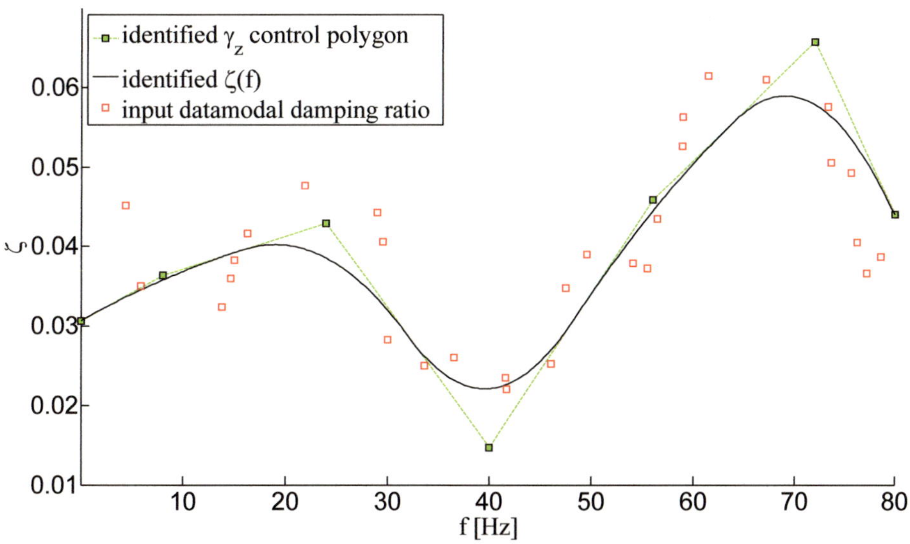

Fig. 16. Comparison of the modal damping ratio used to simulate the measurements (red squares) and the identified $\zeta(f)$ (black line; green filled squares refer to B-spline curve control coefficients). Example with 4 measurement response dofs and with 3% noise.

When 3% noise is added, the value of the identification parameters at each step, adopting the proposed procedure, is reported in Fig.14 for the stiffness coefficients, and in Fig.15 for the γ_z damping coefficients; Fig.16 refers to the comparison of the modal damping ratio used to simulate the measurements (red squares) and the identified curve (black line) where the green filled squares are the B-spline control coefficient γ_z. Fig.17 refers to the comparison of the input and updated FRFs.

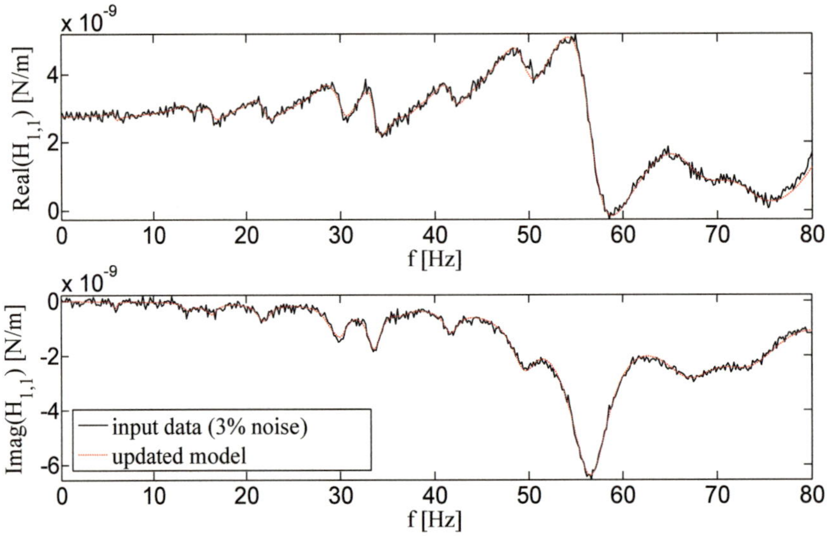

Fig. 17. Comparison of (input point 1; output point 1) FRF considering noise (3% case) after updating (4 measurement response dofs): the input data (black line) and the updated model (red line).

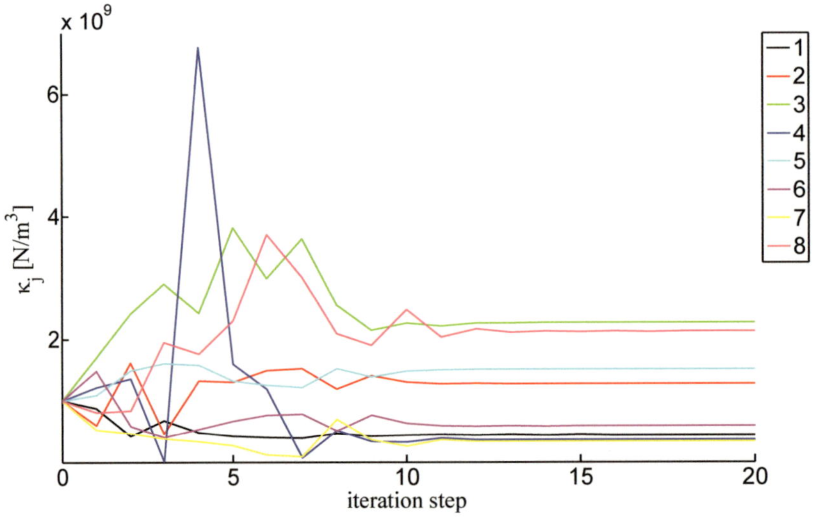

Fig. 18. Evolution of stiffness parameters κ_j (j=1,...,8 in the legend) during iterations by adopting the proposed updating procedure. Example with 4 measurement response dofs and with 10% noise.

For the simulation considering the 10% noise case, Fig.18 and Fig.19 show the evolution during iteration for, respectively, the stiffness coefficients and the γ_z damping coefficients; Fig.20 refers to the comparison of the modal damping ratio values used to simulate the measurements and the identified function. Fig.21 and Fig.22 refer to the comparison of the input and updated FRFs.

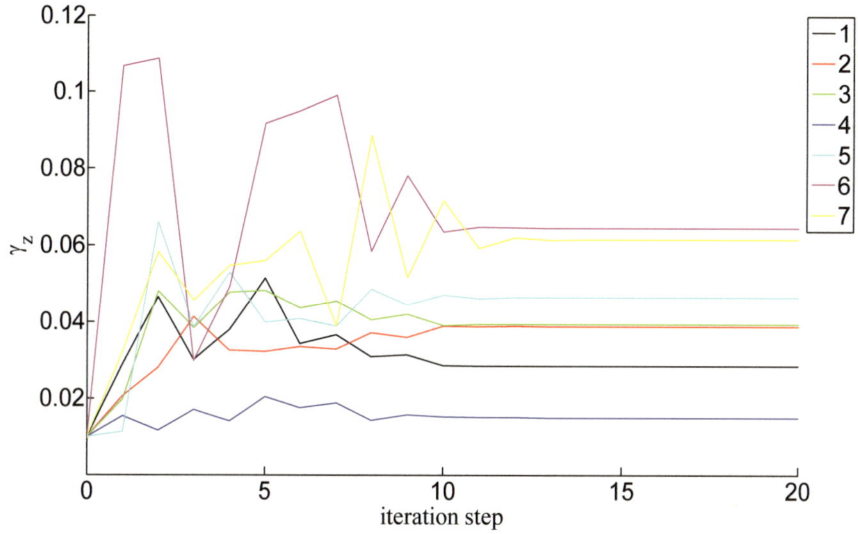

Fig. 19. Evolution of the damping parameters γ_z (z=1,...,7 in the legend) during iterations by adopting the proposed updating procedure. Example with 4 measurement response dofs and with 10% noise.

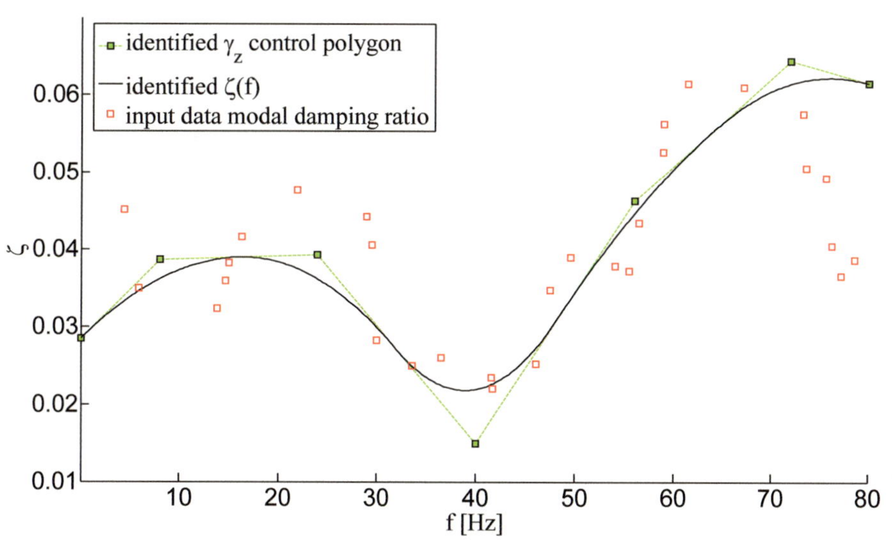

Fig. 20. Comparison of the modal damping ratio ζ used to simulate the measurements (red squares) with the identified $\zeta(f)$ (black line; green filled squares refer to B-spline curve control coefficients). Example with 4 measurement response dofs and with 10% noise.

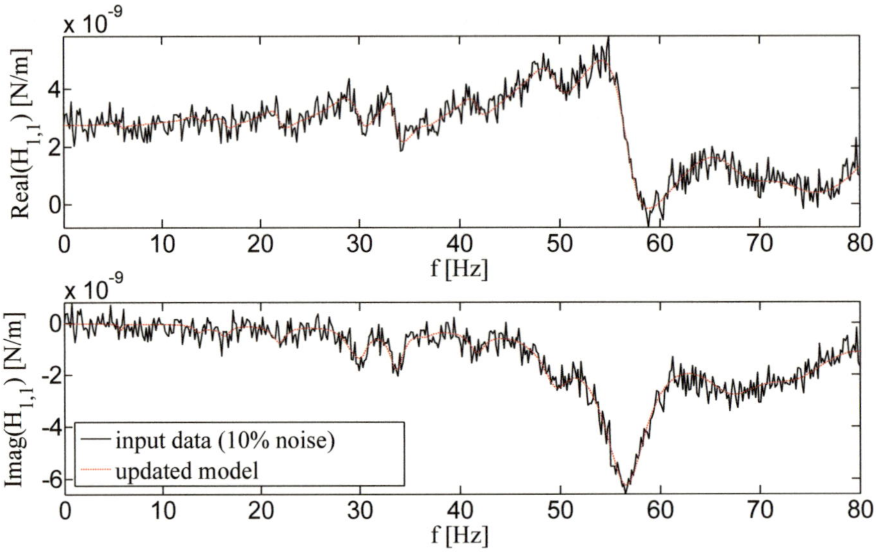

Fig. 21. Comparison of (input point 1; output point 1) FRF considering noise (10% case) after updating (4 measurement response dofs): the input data (black line) and the updated model (red line).

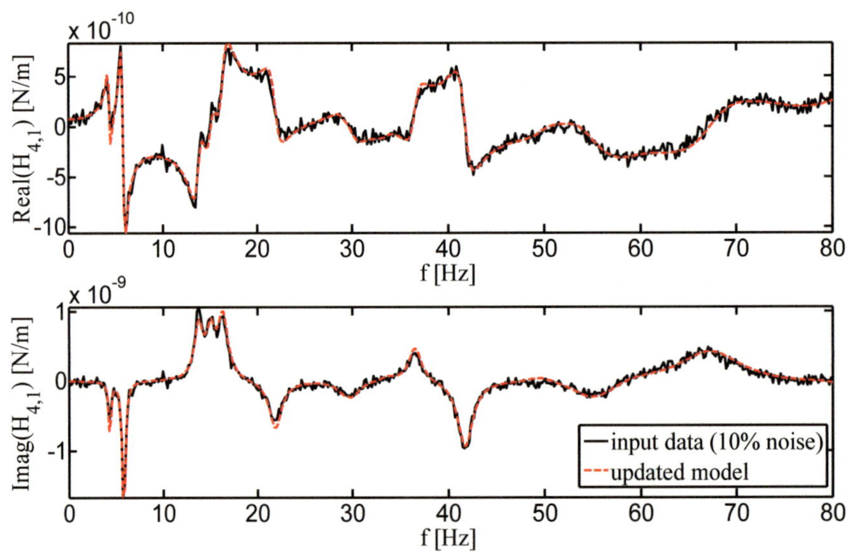

Fig. 22. Comparison of (input point 1; output point 4) FRF considering noise (10% case) after updating (4 measurement response dofs): the input data (black line) and the updated model (red line).

5. Discussion

Experimental measurement data were simulated by adopting the same B-spline analytical model used as the updating model. Numerical results showed good matching of the FRFs

after the updating process with both twelve and four measurement dofs, when noise is not considered. However, when only four measurement dofs are employed, more iterations were necessary to make updating parameter values become stable, with respect to the case in which twelve measurement dofs were adopted. The updated FRFs showed a good matching with the input FRFs even with the adoption of four measurement dofs and noisy data as input in the updating procedure: in the 10% noise case, the procedure required more iterations than in the 3% noise case example, but a moderately fast convergence was obtained anyway. A transformation of the updating variables was proposed to constrain the updated parameters to lie in a compact domain without using additional variables. This transformation ensured physical values to be assumed for all of the parameters during the iteration steps, and convergence was effectively and efficiently obtained in all of the cases under study.
The approach needs to be tested by adopting true measurement data as input. However, the experimental estimate of input-output FRFs for big structures like bridges can be difficult and can also be affected by experimental model errors, mainly due to input force placement, spatial distribution and measurement estimate. A technique employing output-only measured data need to be considered in future studies.

6. Conclusions

An updating procedure of a B-spline FE model of a railway bridge deck was proposed, the updating parameters being the coefficients of a distributed constraint stiffness model and the damping ratios, both modeled by means of B-spline functions. The optimization objective function was defined by considering the difference between the measured (numerically synthesised) FRFs and the linearized analytical FRFs. The incompatibility between the measured dofs and the non-physical B-spline FE model dofs was overcome by employing the same B-spline shape functions, thus adding a small computational cost.
A transformation of the updating variables was proposed to constrain the updated parameters to lie in a compact domain without using additional variables. Some test cases were investigated by simulating the experimental measurements by model based numerical simulations. Results are shown and critically discussed. Future applications will be addressed towards the development of a model updating technique employing output-only vibrational measured data.

7. Acknowledgments

The present study was developed within the MAM-CIRI, with the contribution of the Regione Emilia-Romagna, Progetto Tecnopoli. Support from the Italian Ministero dell'Università e della Ricerca (MIUR), under the "Progetti di Interesse Nazionale" (PRIN07) framework is also kindly acknowledged.

8. References

Carminelli, A. & Catania, G. (2007). Free vibration analysis of double curvature thin walled structures by a B-spline finite element approach. *Proceedings of ASME IMECE 2007*, pp. 1-7, Seattle (Washington), USA, 11-15 November 2007.

Carminelli, A. & Catania, G. (2009). PB-spline hybrid surface fitting technique. *Proceedings of ASME IDETC/CIE 2009*, pp.1-7, San Diego, California, USA, August 30-September 2, 2009.

Cook, R.D.; Malkus, D.S.; Plesha, M.E. & Witt, R.J. (1989). Concepts and applications of finite element analysis, J. Wiley & Sons, ISBN 0-471-35605-0, New York, NY, USA.

D'ambrogio W. & Fregolent A. (2000). Robust dynamic model updating using point antiresonances. *Proceedings of the 18th International Modal Analysis Conference*, pp. 1503-1512, San Antonio, Texas.

Esfandiari, A.; Bakhtiari-Nejad, F.; Rahai, A. & Sanayei, M. (2009). Structural model updating using frequency response function and quasi-linear sensitivity equation. *Journal of Sound and Vibration*, Vol. 326, 3-5, pp. 557-573, ISSN 0022-460X.

Friswell, M. I. & Mottershead, J. E. (1995) *Finite element modal updating in structural dynamics*, Kluwer Academic Publisher, ISBN 0-7923-3431-0, Dordrecht, Netherlands.

Friswell, M.I.; Mottershead, J.E. & Ahmadian, H. (2001). Finite-Element Model Updating Using Experimental Test Data: Parametrization and Regularization. *Philosophical Transactions: Mathematical, Physical and Engineering Sciences*, 359, 1778, Experimental Modal Analysis (Jan. 2001), pp. 169-186.

Gabriele S.; Valente, C. & Brancaleoni, F. (2009). Model calibration by interval analysis. *Proceedings of XIX AIMETA CONFERENCE*, Ancona, Italy, September 14-17, 2009.

Garibaldi, L.; Catania, G.; Brancaleoni, F.; Valente, C. & Bregant, L. (2005). Railway Bridges Identification Techniques. *Proceedings of IDETC2005: The 20th ASME Biennial Conference on Mechanical Vibration and Noise*, Long Beach, CA, USA, September 24-28, 2005.

Hughes, T.J.R.; Cottrell, J.A. & Bazilevs, Y. (2005). Isogeometric analysis: CAD, finite elements, NURBS, exact geometry, and mesh refinement. *Computer Methods in Applied Mechanics and Engineering*, 194, pp. 4135–4195, 2005.

Hughes, T.J.R.; Reali, A. & Sangalli, G. (2009). Isogeometric methods in structural dynamics and wave propagation, *Proceedings of COMPDYN 2009 - Computational Methods in Structural Dynamics and Earthquake Engineering*, Rhodes, Greece, 22-24 June 2009.

Lin, R.M. & Zhu, J. (2006). Model updating of damped structures using FRF data. *Mechanical Systems andSignal Processing*, 20, pp. 2200-2218.

Kagan, P. & Fischer, A. (2000). Integrated mechanically based CAE system using B-spline finite elements. Computer Aided Design, 32, pp. 539-552.

Piegl L. & Tiller, W. (1997). *The NURBS Book, 2nd Edition*. Springer-Verlag, ISBN 3-540-61545-8, New York, NY, USA.

Zapico, J.L.; Gonzalez, M.P.; Friswell, M.I.; Taylor, C.A. & Crewe, A.J. (2003). Finite element model updating of a small scale bridge. *Journal of Sound and Vibrations*, 268, pp. 993-1012.

9

Dynamic Analysis of a Spinning Laminated Composite-Material Shaft Using the *hp*- version of the Finite Element Method

Abdelkrim Boukhalfa
Department of Mechanical Engineering, Faculty of Technology
University of Tlemcen
Algeria

1. Introduction

Because of their high strength, high stiffness, and low density characteristics, composite materials are now used widely for the design of rotating mechanical components such as, for example, driveshafts for helicopters, cars and jet engines, or centrifugal separator cylindrical tubes. The interest of composites for rotordynamic applications has been demonstrated both numerically and experimentally. Accompanied by the development of many new advanced composite materials, various mathematical models of spinning composite shafts were also developed by researchers.

Zinberg and Symonds (Zinberg & Symonds, 1970) investigated the critical speeds for rotating anisotropic shafts and their experiments affirmed the advantages of composite shafts over aluminum alloy shafts. Using Donell's thin shell theory, Reis et al. (Dos Reis et al., 1987) applied finite element method to evaluate critical speeds of thin-walled laminated composite shafts. They concluded that the lay-up of a composite shaft strongly influences the dynamic behavior of this shaft.

Kim and Bert (Kim & Bert, 1993) utilized Sanders' best first approximation shell theory to determine critical speeds of a rotating shaft containing layers of arbitrarily laminated composite materials. Both the thin- and thick-shell models, including the Coriolis effect, were presented. Bert (Bert, 1992), as well as Bert and Kim (Bert & Kim, 1995a), examined critical speeds of composite shafts using Bernoulli-Euler beam theory and Bresse-Timoshenko beam model, respectively. Conventional beam model approaches used to date are Equivalent Modules Beam Theory (EMBT). In another study, Bert and Kim (Bert & Kim, 1995b) have analysed the dynamic instability of a composite drive shaft subjected to fluctuating torque and/or rotational speed by using various thin shell theories. The rotational effects include centrifugal and Coriolis forces. Dynamic instability regions for a long span simply supported shaft are presented.

M- Y. Chang et al (Chang et al., 2004a) published the vibration behaviours of the rotating composite shafts. In the model the transverse shear deformation, rotary inertia and gyroscopic effects, as well as the coupling effect due to the lamination of composite layers have been incorporated. The model based on a first order shear deformable beam theory (continuum- based Timoshenko beam theory). M- Y. Chang et al (Chang et al., 2004b) published the vibration analysis of rotating composite shafts containing randomly oriented reinforcements. The Mori-Tanaka mean-field theory is adopted here to account for the interaction at the finite concentrations of reinforcements in the composite material.

Additional recent work on composite shafts dealing with both the theoretical and experimental aspects was reported by Singh (Singh, 1992), Gupta and Singh (Gupta & Singh, 1996) and Singh and Gupta (Singh & Gupta, 1994a). Rotordynamic formulation based on equivalent modulus beam theory was developed for a composite rotor with a number of lumped masses, and supported on general eight coefficient bearings. A Layerwise Beam Theory (LBT) was derived by Singh and Gupta (Gupta & Singh, 1996) from an available shell theory, with a layerwise displacement field, and was then extended to solve a general composite rotordynamic problem. The conventional rotor dynamic parameters as well as critical speeds, natural frequencies, damping factors, unbalance response and threshold of stability were analyzed in detail and results from the formulations based on the two theories, namely, the equivalent modulus beam theory (EMBT) and layerwise beam theory (LBT) were compared (Singh & Gupta, 1994a). The experimental rotordynamic studies carried by Singh and Gupta (Singh & Gupta, 1995-1996) were conducted on two filament wound carbon/epoxy shafts with constant winding angles (±45° and ±60°). Progressive balancing had to be carried out to enable the shaft to traverse through the first critical speed. Inspire of the very different shaft configurations used, the authors' have shown that bending-stretching coupling and shear-normal coupling effects change with stacking sequence, and alter the frequency values. Some practical aspects such as effect of shaft disk angular misalignment, interaction between shaft bow, which is common in composite shafts and rotor unbalance, and an unsuccessful operation of a composite rotor with an external damper were discussed and reported by Singh and Gupta (Singh & Gupta, 1995). The Bode and cascade plots were generated and orbital analysis at various operating speeds was performed. The experimental critical speeds showed good correlation with the theoretical prediction.

Mastering vibratory behavior requires knowledge of the characteristics of the composite material spinning shafts, the prediction of this knowledge is fundamental in the design of the rotating machinery in order to provide a precise idea of the safe intervals in terms of spinning speeds. Within the framework of this idea, our work concerns to the study of the vibratory behavior of the spinning composite material shafts, and more precisely, their behavior in rotation by taking into account the effects of the transverse shear deformation, rotary inertia and gyroscopic effects, as well as the coupling effect due to the lamination of composite layers, the effect of the elastic bearings and external damping and the effect of disk. In the presented composite shaft model, the Timoshenko theory will be adopted. An *hp*- version of the finite element method (combination between the conventional version of the finite element method (*h*- version) and the hierarchical finite element method (*p*-version) with trigonometric shape functions (Boukhalfa et al., 2008-2010) is used to model the structure. A hierarchical finite element of beam type with six degrees of freedom per node is developed. The assembly is made same manner as the standard version of the finite element method for several elements. The theoretical study allows the establishment of the kinetic energy and the strain energy of the system (shaft, disk and bearings) necessary to determine the motion equations. A program is elaborated to calculate the Eigen-frequencies and the critical speeds of the system. The results obtained are compared with those available in the literature and show the speed of convergence, the precision and the effectiveness of the method used. Several examples are treated, and a discussion is established to determine the influence of the various parameters and boundary conditions. In the *hp*- version of the finite element method, the error in the solution is controlled by both the number of elements h and the polynomial order p ((Babuska & Guo, 1986); (Demkowicz et al., 1989)). The *hp*-version of the finite element method has been exploited in a few areas including plate vibrations (Bardell et al., 1995) and beam statics (Bardell, 1996) and has been shown to offer considerable savings in computational effort when compared with the standard *h*-version of the finite element method.

2. Equations of motion

2.1 Kinetic and strain energy expressions of the shaft

The shaft is modeled as a Timoshenko beam, that is, first-order shear deformation theory with rotary inertia and gyroscopic effect is used. The shaft rotates at constant speed about its longitudinal axis. Due to the presence of fibers oriented than axially or circumferentially, coupling is made between bending and twisting. The shaft has a uniform, circular cross section.

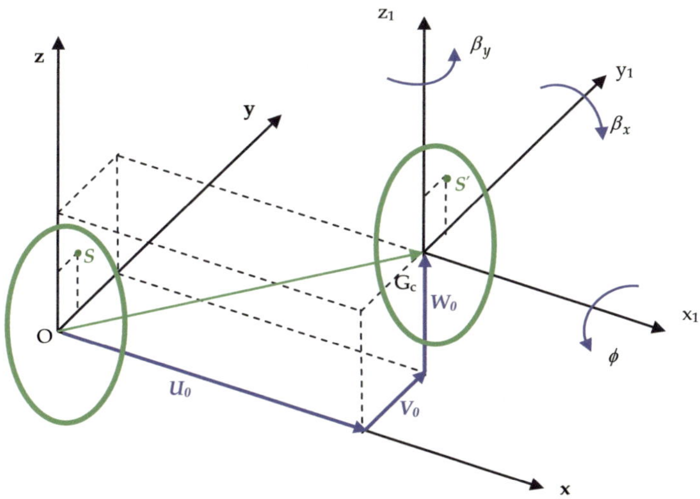

Fig. 1. The elastic displacements of a typical cross-section of the shaft

The following displacement field of a spinning shaft (one beam element) is assumed by choosing the coordinate axis x to coincide with the shaft axis:

$$\begin{cases} U(x,y,z,t) = U_0(x,t) + z\beta_x(x,t) - y\beta_y(x,t) \\ V(x,y,z,t) = V_0(x,t) - z\phi(x,t) \\ W(x,y,z,t) = W_0(x,t) + y\phi(x,t) \end{cases} \quad (1)$$

Where U, V and W are the flexural displacements of any point on the cross-section of the shaft in the x, y and z directions respectively, the variables U_0, V_0 and W_0 are the flexural displacements of the shaft's axis, while β_x and β_y are the rotation angles of the cross-section, about the y and z axis respectively. The ϕ is the angular displacement of the cross-section due to the torsion deformation of the shaft (see figure 1).

The strain components in the cylindrical coordinate system (As shown in figure 2-3) can be written in terms of the displacement variables defined earlier as

$$\begin{cases} \varepsilon_{xx} = \dfrac{\partial U_0}{\partial x} + r\sin\theta \dfrac{\partial \beta_x}{\partial x} - r\cos\theta \dfrac{\partial \beta_y}{\partial x} \\ \varepsilon_{rr} = \varepsilon_{\theta\theta} = \varepsilon_{r\theta} = 0 \\ \varepsilon_{x\theta} = \varepsilon_{\theta x} = \dfrac{1}{2}(\beta_y \sin\theta + \beta_x \cos\theta - \sin\theta \dfrac{\partial V_0}{\partial x} + \cos\theta \dfrac{\partial W_0}{\partial x} + r\dfrac{\partial \phi}{\partial x}) \\ \varepsilon_{xr} = \varepsilon_{rx} = \dfrac{1}{2}(\beta_x \sin\theta - \beta_y \cos\theta - \sin\theta \dfrac{\partial W_0}{\partial x} + \cos\theta \dfrac{\partial V_0}{\partial x}) \end{cases} \tag{2}$$

Let us consider a composite shaft consists of k layered (see figure 3) of fiber inclusion reinforced laminate. The constitutive relations for each layer are described by

$$\begin{Bmatrix} \sigma_{xx} \\ \sigma_{\theta\theta} \\ \sigma_{rr} \\ \tau_{r\theta} \\ \tau_{xr} \\ \tau_{x\theta} \end{Bmatrix} = \begin{bmatrix} C'_{11} & C'_{12} & C'_{13} & 0 & 0 & C'_{16} \\ C'_{12} & C'_{22} & C'_{23} & 0 & 0 & C'_{26} \\ C'_{13} & C'_{23} & C'_{33} & 0 & 0 & C'_{36} \\ 0 & 0 & 0 & C'_{44} & C'_{45} & 0 \\ 0 & 0 & 0 & C'_{45} & C'_{55} & 0 \\ C'_{16} & C'_{26} & C'_{36} & 0 & 0 & C'_{66} \end{bmatrix} \begin{Bmatrix} \varepsilon_{xx} \\ \varepsilon_{\theta\theta} \\ \varepsilon_{rr} \\ \gamma_{r\theta} \\ \gamma_{xr} \\ \gamma_{x\theta} \end{Bmatrix} \tag{3}$$

Where C_{ij}' are the effective elastic constants, they are related to lamination angle η (as shown in figure 4-5) and the elastic constants of principal axes.

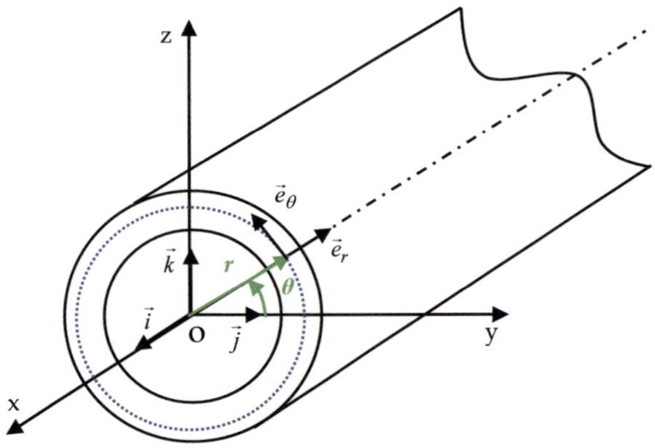

Fig. 2. The cylindrical coordinate System

The stress-strain relations of the nth layer expressed in the cylindrical coordinate system (see figure 6) can be expressed as

$$\begin{cases} \sigma_{xx} = C'_{11n}\varepsilon_{xx} + k_s C'_{16n}\gamma_{x\theta} \\ \tau_{x\theta} = \tau_{\theta x} = k_s C'_{16n}\varepsilon_{xx} + k_s C'_{66n}\gamma_{x\theta} \\ \tau_{xr} = \tau_{rx} = k_s C'_{55n}\gamma_{xr} \end{cases} \tag{4}$$

Where k_s is the transverse shear correction factor.

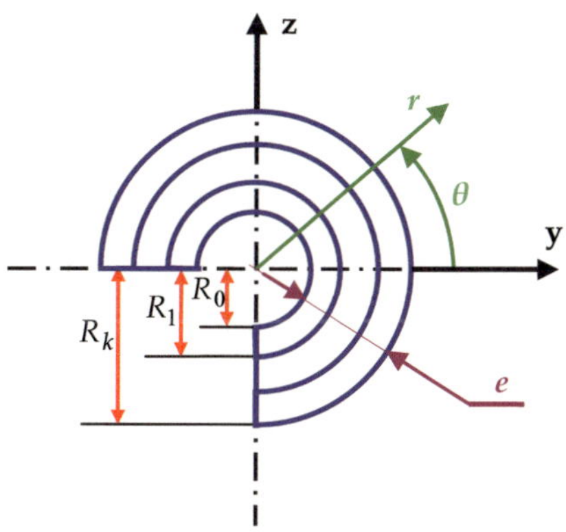

Fig. 3. k -layers of composite shaft

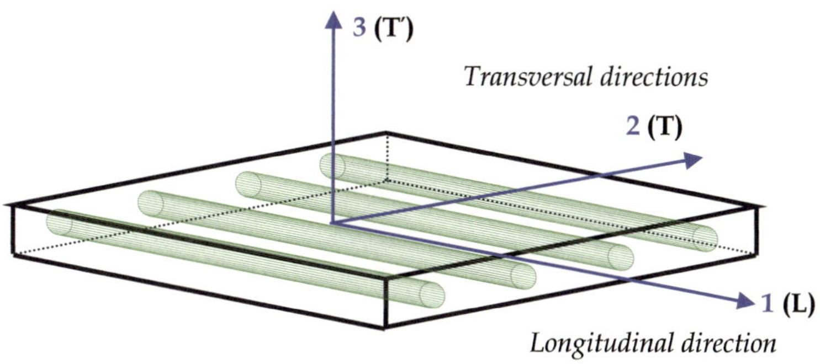

Fig. 4. A typical composite lamina and its principal axes

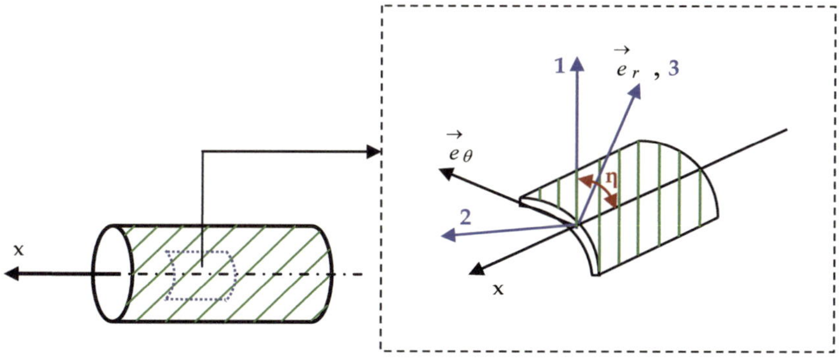

Fig. 5. The definitions of the principal coordinate axes on an arbitrary layer of the composite

The formula of the strain energy is

$$E_d = \frac{1}{2}\int_V (\sigma_{xx}\varepsilon_{xx} + 2\tau_{xr}\varepsilon_{xr} + 2\tau_{x\theta}\varepsilon_{x\theta})\, dV \tag{5}$$

The various components of strain energy of the shaft are presented as follow (one beam element)

$$E_{da}=\frac{1}{2}A_{11}\int_0^L\left(\frac{\partial U_0}{\partial x}\right)^2dx+\frac{1}{2}B_{11}\left[\int_0^L\left(\frac{\partial \beta_x}{\partial x}\right)^2dx+\int_0^L\left(\frac{\partial \beta_y}{\partial x}\right)^2dx\right]+\frac{1}{2}k_sB_{66}\int_0^L\left(\frac{\partial \phi}{\partial x}\right)^2dx+$$
$$\frac{1}{2}k_sA_{16}\left[2\int_0^L\frac{\partial \phi}{\partial x}\frac{\partial U_0}{\partial x}dx+\int_0^L\beta_y\frac{\partial \beta_x}{\partial x}dx-\int_0^L\beta_x\frac{\partial \beta_y}{\partial x}dx-\int_0^L\frac{\partial V_0}{\partial x}\frac{\partial \beta_x}{\partial x}dx-\int_0^L\frac{\partial W_0}{\partial x}\frac{\partial \beta_y}{\partial x}dx\right]+ \tag{6}$$
$$\frac{1}{2}k_s(A_{55}+A_{66})\left[\int_0^L\left(\frac{\partial V_0}{\partial x}\right)^2dx+\int_0^L\left(\frac{\partial W_0}{\partial x}\right)^2dx+\int_0^L\beta_x^2dx+\int_0^L\beta_y^2dx+2\int_0^L\beta_x\frac{\partial W_0}{\partial x}dx-2\int_0^L\beta_y\frac{\partial V_0}{\partial x}dx\right]$$

where A_{ij} and B_{ij} are given in Appendix.
The kinetic energy of the spinning composite shaft (one beam element) (Boukhalfa et al., 2008), including the effects of translatory and rotary inertia, can be written as

$$E_{ca}=\frac{1}{2}\int_0^L\Big[I_m\left(\dot{U}_0^{\,2}+\dot{V}_0^{\,2}+\dot{W}_0^{\,2}\right)+I_d\left(\dot{\beta}_x^{\,2}+\dot{\beta}_y^{\,2}\right)$$
$$-2\Omega I_p\beta_x\dot{\beta}_y+2\Omega I_p\dot{\phi}+I_p\dot{\phi}^2+\Omega^2I_p+\Omega^2I_d\left(\beta_x^{\,2}+\beta_y^{\,2}\right)\Big]dx \tag{7}$$

where Ω is the rotating speed of the shaft which is assumed constant, L is the length of the shaft, the $2\Omega I_p\beta_x\dot{\beta}_y$ term accounts for the gyroscopic effect, and $I_d\left(\dot{\beta}_x^{\,2}+\dot{\beta}_y^{\,2}\right)$ represent the rotary inertia effect. The mass moments of inertia I_m, the diametrical mass moments of inertia I_d and polar mass moment of inertia I_p of spinning shaft per unit length are defined in the appendix. As the $\Omega^2I_d\left(\beta_x^{\,2}+\beta_y^{\,2}\right)$ term is far smaller than Ω^2I_p, it will be neglected in further analysis.

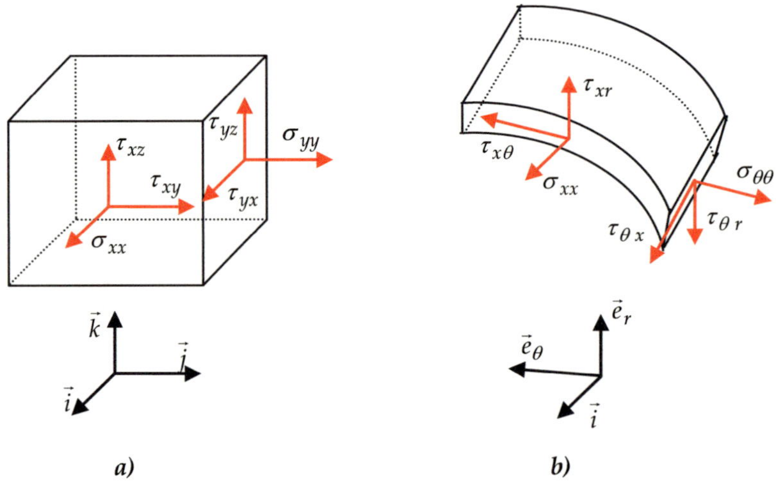

Fig. 6. The stress components; ***a)*** in the coordinate axes (x, y, z) - ***b)*** in the coordinate axes $(x\,,r\,,\theta)$

2.2 Kinetic energy of the disk

The disk fixed to the composite shaft (see figure 7) is assumed rigid and made of isotropic material. According to Equation (7) the kinetic energy of the disk can be expressed as

$$E_{cD} = \frac{1}{2}\Big[I_m^D\left(\dot{U}_0^{\,2} + \dot{V}_0^{\,2} + \dot{W}_0^{\,2}\right) + I_d^D\left(\dot{\beta}_x^{\,2} + \dot{\beta}_y^{\,2}\right) \\ -2\Omega I_p^D \beta_x \dot{\beta}_y + 2\Omega I_p^D \dot{\phi} + I_p^D \dot{\phi}^2 + \Omega^2 I_p^D + \Omega^2 I_d^D\left(\beta_x^{\,2} + \beta_y^{\,2}\right)\Big] \tag{8}$$

where I_m, I_d and I_p are the mass, the diametrical mass moment of inertia and the polar mass moment of inertia of the disk. As the $\Omega^2 I_p^D\left(\beta_x^{\,2} + \beta_y^{\,2}\right)$ term is far smaller than $\Omega^2 I_p^D$, it will be neglected in further analysis.

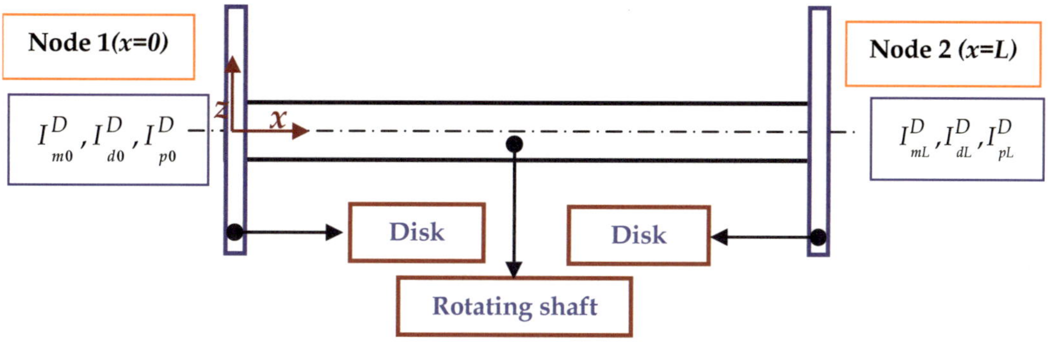

Fig. 7. Various positions of the disk on the spinning shaft (one element).

2.3 Virtual work of the bearings

The bearings are characterized by values of stiffness and viscous damping following the y and z directions and the cross terms (see Figures 8 and 9). The stiffness and damping effects of the bearings are modeled using springs and viscous dampers.

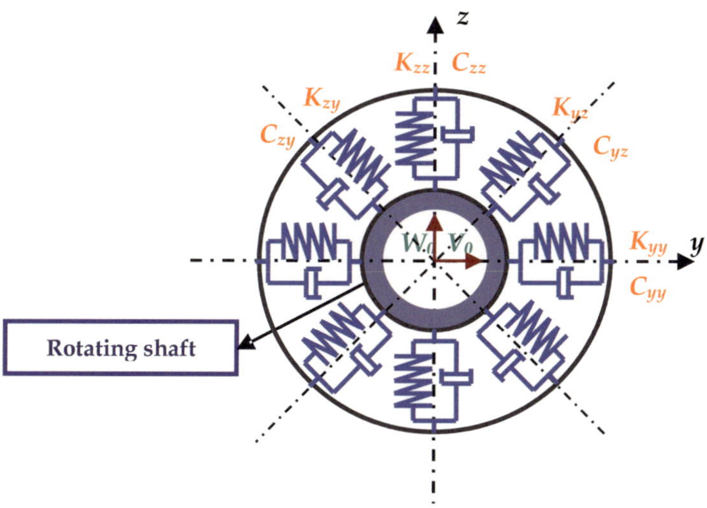

Fig. 8. Model of bearings

The virtual work δA done by these external forces can be written as

$$\delta A = F_{V_0}\delta_{V_0} + F_{W_0}\delta_{W_0} \tag{9}$$

where F_{V_0} and F_{w_0} are the generalized forces expressed by

$$\begin{Bmatrix} F_{V0} \\ F_{W_0} \end{Bmatrix} = -\begin{bmatrix} C_{yy} & C_{yz} \\ C_{zy} & C_{zz} \end{bmatrix} \begin{Bmatrix} \dot{V}_0 \\ \dot{W}_0 \end{Bmatrix} - \begin{bmatrix} K_{yy} & K_{yz} \\ K_{zy} & K_{zz} \end{bmatrix} \begin{Bmatrix} V_0 \\ W_0 \end{Bmatrix} \tag{10}$$

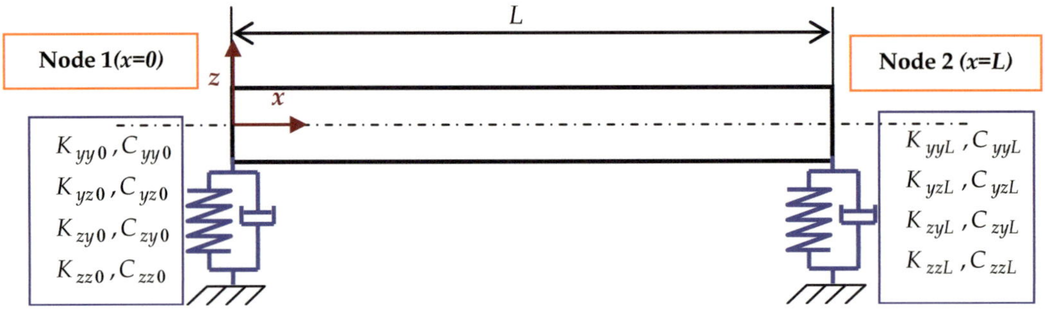

Fig. 9. Spinning shaft (one element) supported by two bearings

2.4 Hierarchical Beam element formulation

The spinning flexible beam is descretised by hierarchical beam elements. Each element with two nodes 1 and 2 is shown in figure 10. In the case of a staged shaft, several elements can be used (see figure 11). The element's nodal d.o.f. at each node are U_0 , V_0 , W_0 , β_x , β_y and ϕ . The local and non-dimensional co-ordinates are related by

$$\xi = x/L \qquad \text{With } (0 \le \xi \le 1) \tag{11}$$

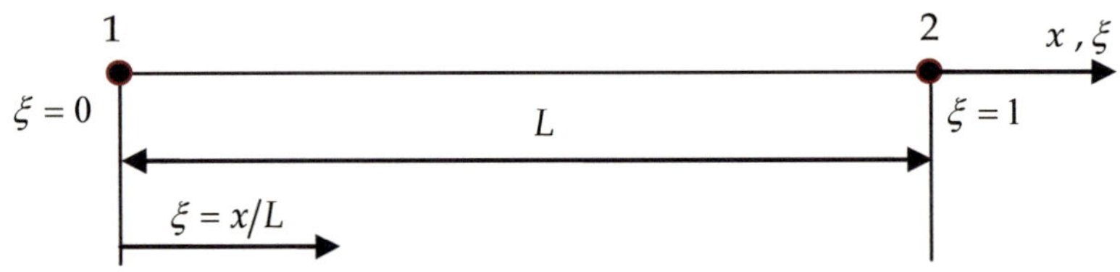

Fig. 10. 3D Beam element with two nodes

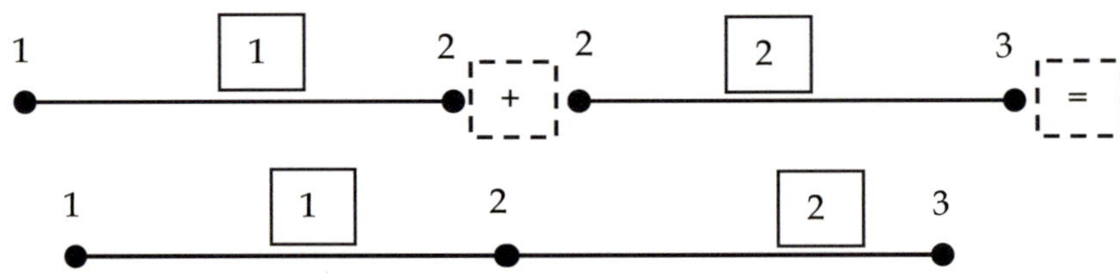

Fig. 11. Assembly between two *p*- elements

The vector displacement formed by the variables U_0 , V_0 , W_0 , β_x , β_y and ϕ can be written as

$$\begin{cases} U_0 = [N_U]\{q_U\} = \sum_{m=1}^{p_U} x_m(t) \cdot f_m(\xi) \\ V_0 = [N_V]\{q_V\} = \sum_{m=1}^{p_V} y_m(t) \cdot f_m(\xi) \\ W_0 = [N_W]\{q_W\} = \sum_{m=1}^{p_W} z_m(t) \cdot f_m(\xi) \\ \beta_x = [N_{\beta_x}]\{q_{\beta_x}\} = \sum_{m=1}^{p_{\beta_x}} \beta_{xm}(t) \cdot f_m(\xi) \\ \beta_y = [N_{\beta_y}]\{q_{\beta_y}\} = \sum_{m=1}^{p_{\beta_y}} \beta_{ym}(t) \cdot f_m(\xi) \\ \phi = [N_\phi]\{q_\phi\} = \sum_{m=1}^{p_\phi} \phi_m(t) \cdot f_m(\xi) \end{cases} \tag{12}$$

where $[N]$ is the matrix of the shape functions, given by

$$\left[N_{U,V,W,\beta_x,\beta_y,\phi}\right] = \left[f_1\ f_2 \cdots\cdots f_{p_U,p_V,p_W,p_{\beta_x},p_{\beta_y},p_\phi}\right] \tag{13}$$

where $p_U, p_V, p_W, p_{\beta_x}, p_{\beta_y}$ and p_ϕ are the numbers of hierarchical terms of displacements (are the numbers of shape functions of displacements). In this work, $p_U = p_V = p_W = p_{\beta_x} = p_{\beta_y} = p_\phi = p$

The vector of generalized coordinates given by

$$\{q\} = \left\{ q_U, q_V, q_W, q_{\beta_x}, q_{\beta_y}, q_\phi \right\}^T \tag{14}$$

where

$$\left\{ \begin{array}{l} \{q_U\} = \{x_1, x_2, x_3, \ldots, x_{p_U}\}^T \exp(j\omega t); \{q_V\} = \{y_1, y_2, y_3, \ldots, y_{p_V}\}^T \exp(j\omega t); \\ \{q_W\} = \{z_1, z_2, z_3, \ldots, y_{p_W}\}^T \exp(j\omega t); \{q_{\beta_x}\} = \left\{\beta_{x_1}, \beta_{x_2}, \beta_{x_3}, \ldots, \beta_{x_{p_{\beta_x}}}\right\}^T \exp(j\omega t); \\ \{q_{\beta_y}\} = \left\{\beta_{y_1}, \beta_{y_2}, \beta_{y_3}, \ldots, \beta_{y_{p_{\beta_y}}}\right\}^T \exp(j\omega t); \{q_\phi\} = \{\phi_1, \phi_2, \phi_3, \ldots, \phi_{p_\phi}\}^T \exp(j\omega t) \end{array} \right\} \tag{15}$$

The group of the shape functions used in this study is given by

$$\left\{ (f_1 = 1 - \xi)(f_2 = \xi)\left(f_{r+2} = \sin(\delta_r\, \xi), \delta_r = r\pi\ ;\ r = 1, 2, 3, \ldots\right) \right\} \tag{16}$$

The functions (f_1, f_2) are those of the finite element method necessary to describe the nodal displacements of the element; whereas the trigonometric functions f_{r+2} contribute only to the internal field of displacement and do not affect nodal displacements. The most attractive particularity of the trigonometric functions is that they offer great numerical stability. The shaft is modeled by elements called hierarchical finite elements with p shape functions for each element. The assembly of these elements is done by the *h*- version of the finite element method.

After modelling the spinning composite shaft using the *hp*- version of the finite element method and applying the Euler-Lagrange equations, the motion's equations of free vibration of spinning flexible shaft can be obtained.

$$[M]\{\ddot{q}\}+\left[[G]+\left[C_p\right]\right]\{\dot{q}\}+[K]\{q\}=\{0\} \tag{17}$$

$[M]$ and $[K]$ are the mass and stiffness matrix respectively, $[G]$ is the gyroscopic matrix and $[C_p]$ is the damping matrix of the bearing (the different matrices of the equation (17) are given in the appendix).

3. Results

A program based on the formulation proposed to resolve the resolution of the equation (17).

3.1 Convergence

First, the mechanical properties of boron-epoxy are listed in table 1, and the geometric parameters are L =2.47 m, D =12.69 cm, e =1.321 mm, 10 layers of equal thickness (90°, 45°,-45°,0°$_6$, 90°). The shear correction factor k_s =0.503 and the rotating speed Ω =0. In this example, the boron -epoxy spinning shaft is modeled by one element of length L, then by two elements of equal length $L/2$.

	Graphite-epoxy	Boron-epoxy
E_{11} (GPa)	139.0	211.0
E_{22} (GPa)	11.0	24.1
G_{12} (GPa)	6.05	6.9
G_{23} (GPa)	3.78	6.9
ν_{12}	0.313	0.36
ρ (kg/m^3)	1578.0	1967.0

Table 1. Properties of composite materials (Bert & Kim, 1995a)

The results of the five bending modes for various boundary conditions of the composite shaft as a function of the number of hierarchical terms p are shown in figure 12. Figure clearly shows that rapid convergence from above to the exact values occurs when the number of hierarchical terms increased. The bending modes are the same for a number of hierarchical finite elements, equal 1 then 2. This shows the exactitude of the method even with one element and a reduced number of the shape functions. It is noticeable in the case of low frequencies, a very small p is needed (p=4 sufficient), whereas in the case of the high frequencies, and in order to have a good convergence, p should be increased.

3.2 Validation

In the following example, the critical speeds of composite shaft are analyzed and compared with those available in the literature to verify the present model. In this example, the composite hollow shafts made of boron-epoxy laminae, which are considered by Bert and

Kim (Bert & Kim, 1995a), are investigated. The properties of material are listed in table1. The shaft has a total length of 2.47 m. The mean diameter D and the wall thickness of the shaft are 12.69 cm and 1.321 mm respectively. The lay-up is [90°/45°/-45°/$0°_6$/90°] starting from the inside surface of the hollow shaft. A shear correction factor of 0.503 is also used. The shaft is modeled by one element. The shaft is simply-supported at the ends. In this validation, p =10.

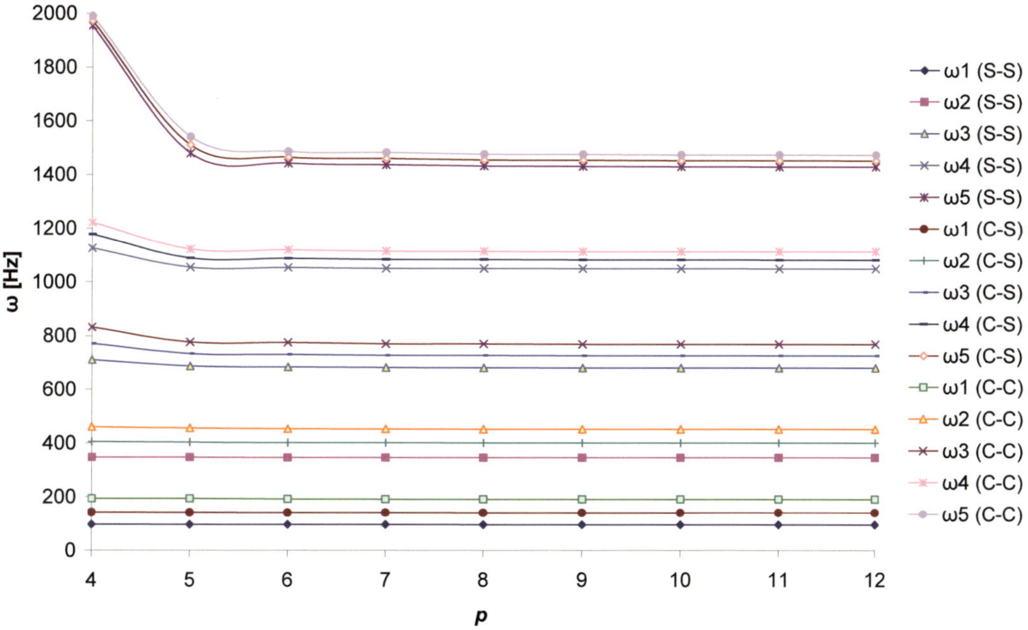

Fig. 12. Convergence of the frequency ω for the 5 bending modes of the composite shaft for different boundary conditions (S: simply-supported; C: clamped) as a function of the number of hierarchical terms p

The result obtained using the present model is shown in table 2 together with those of referenced papers. As can be seen from the table our results are close to those predicted by other beam theories. Since in the studied example the wall of the shaft is relatively thin, models based on shell theories (Kim & Bert, 1993) are expected to yield more accurate results. In the present example, the critical speed measured from the experiment however is still underestimated by using the Sander shell theory while overestimated by the Donnell shallow shell theory. In this case, the result from the present model is compatible to that of the Continuum based Timoshenko beam theory of M-Y. Chang et al (Chang et al., 2004a). In this reference the supports are flexible but in our application the supports are rigid.

In our work, the shaft is modeled by one element with two nodes, but in the model of the reference (Chang et al., 2004a) the shaft is modeled by 20 finite elements of equal length (*h*-version). The rapid convergence while taking one element and a reduced number of shape functions shows the advantage of the method used. One should stress here that the present model is not only applicable to the thin-walled composite shafts as studied above, but also to the thick-walled shafts as well as to the solid ones.

L=2.47 m, D =12.69 cm, e =1.321 mm, 10 layers of equal thickness (90°, 45°,-45°,0°$_6$,90°)		
	Theory or Method	Ω_{cr1} (rpm)
Zinberg & Symonds, 1970	Measured experimentally EMBT	6000 5780
Dos Reis et al., 1987	Bernoulli–Euler beam theory with stiffness determined by shell finite elements	4942
Kim & Bert, 1993	Sanders shell theory Donnell shallow shell theory	5872 6399
Bert, 1992	Bernoulli–Euler beam theory	5919
Bert & Kim, 1995a	Bresse–Timoshenko beam theory	5788
Singh & Gupta, 1996	EMBT LBT	5747 5620
Chang et al., 2004a	Continuum based Timoshenko beam theory	5762
Present	Timoshenko beam theory by the *hp*- version of the FEM.	5760

Table 2. The first critical speed of the boron-epoxy composite shaft

The first eigen-frequency of the boron-epoxy spinning shaft calculated by our program in the stationary case is 96.0594 Hz on rigid supports and 96.0575 Hz on two elastic supports of stiffness 1740 GN/m. In the reference (Chatelet et al., 2002), they used the shell's theory for the same shaft studied in our case and on rigid supports; the frequency is 96 Hz. In this example, is not noticeable the difference between shaft bi-supported on rigid supports or elastic supports because the stiffness of the supports are very large.

3.3 Results and interpretations

In this study, the results obtained for various applications are presented. Convergence towards the exact solutions is studied by increasing the numbers of hierarchical shape functions for two elements. The influence of the mechanical and geometrical parameters and the boundary conditions on the eigen-frequencies and the critical speeds of the embarked spinning composite shafts are studied. In this study, p = 10.

3.3.1 Influence of the gyroscopic effect on the eigen-frequencies

In the following example, the frequencies of a graphite- epoxy spinning shaft are analyzed. The mechanical properties of shaft are shown in table 1, with k_s = 0.503. The ply angles in the various layers and the geometrical properties are the same as those in the first example.
Figure 13 shows the variation of the bending fundamental frequency ω as a function of rotating speed Ω for different boundary conditions. The gyroscopic effect inherent to rotating structures induces a precession motion. When the rotating speed increase, the forward modes (1F) increase, whereas the backward modes (1B) decrease. The gyroscopic effect causes a coupling of orthogonal displacements to the axis of rotation, and by consequence separate the frequencies in two branches: backward precession mode and forward precession mode. In all cases, the forward modes increase with increasing rotating speed however the backward modes decrease.

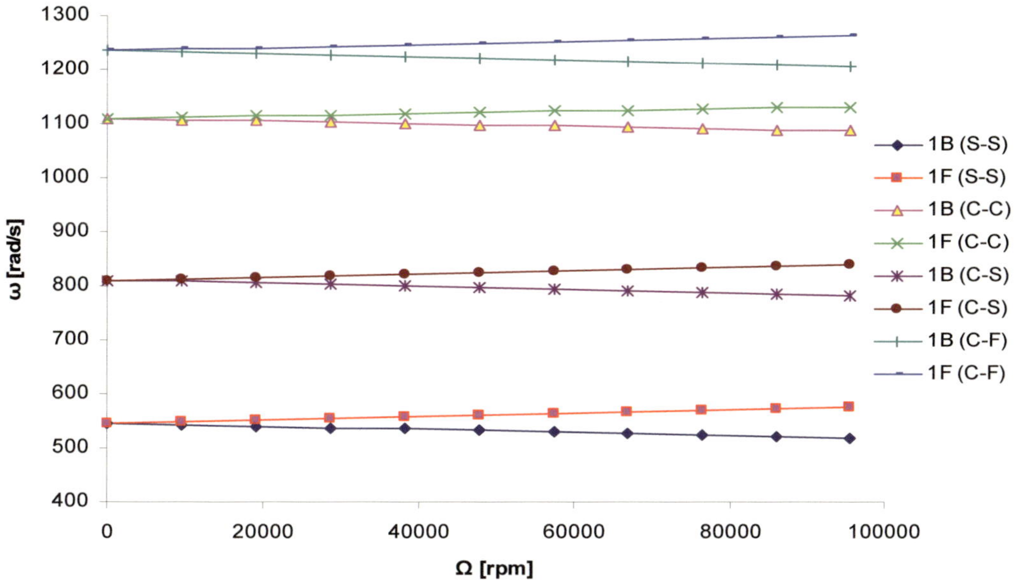

Fig. 13. The first backward (1B) and forward (1F) bending mode of a graphite- epoxy shaft for different boundary conditions and different rotating speeds (S: simply-supported; C: clamped; F: free)

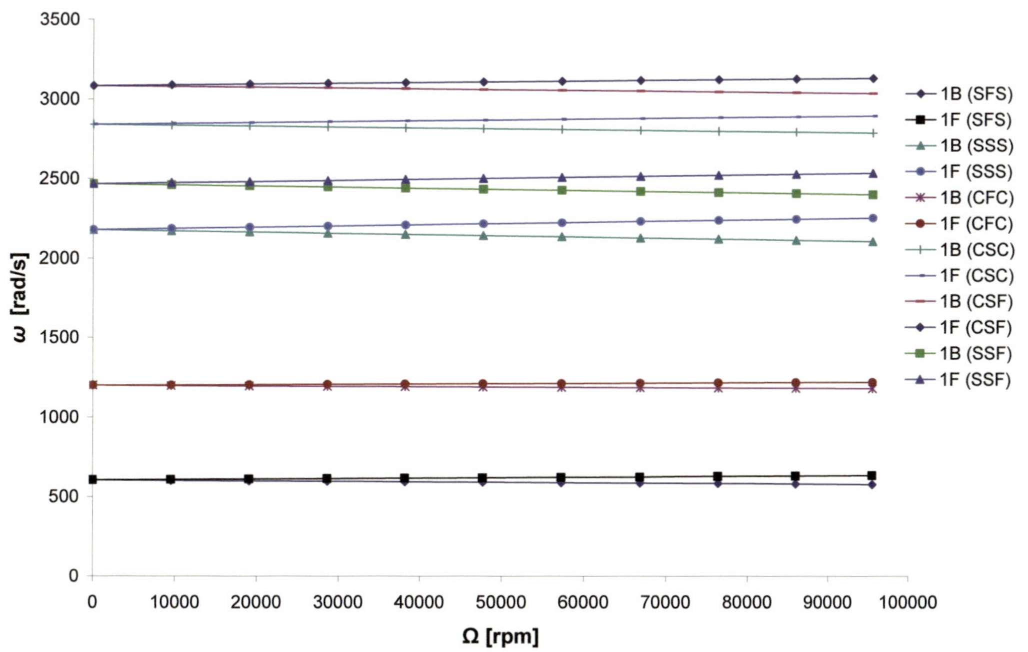

Fig. 14. The first backward (1B) and forward (1F) bending mode of a boron- epoxy shaft for different boundary conditions and different rotating speeds. *L =2.47 m, D =12.69 cm, e =1.321 mm, 10 layers of equal thickness (90°, 45°,-45°,0°$_6$, 90°)*

3.3.2 Influence of the boundary conditions on the eigen-frequencies

In the following example, the boron-epoxy shaft is modeled by two elements of equal length $L/2$. The frequencies of the spinning shaft are analyzed. The mechanical properties of shaft are shown in table 1, with $k_s = 0.503$. The ply angles in the various layers and the geometrical properties are the same as those in the preceding example.

Figure 14 shows the variation of the bending fundamental frequency ω according to the rotating speeds Ω for various boundary conditions. According to these found results, it is noticed that, the boundary conditions have a very significant influence on the eigen-frequencies of a spinning composite shaft. For example, by adding a support to the mid-span of the spinning shaft, the rigidity of the shaft increases which implies the increase in the eigen-frequencies.

3.3.3 Influence of the lamination angle on the eigen-frequencies

By considering the same preceding example, the lamination angles have been varied in order to see their influences on the eigen-frequencies of the spinning composite shaft.

Figure 15 shows the variation of the bending fundamental frequency ω according to the rotating speeds Ω (Campbell diagram) for various ply angles. According to these results, the bending frequencies of the composite shaft decrease when the ply angle increases and vice versa.

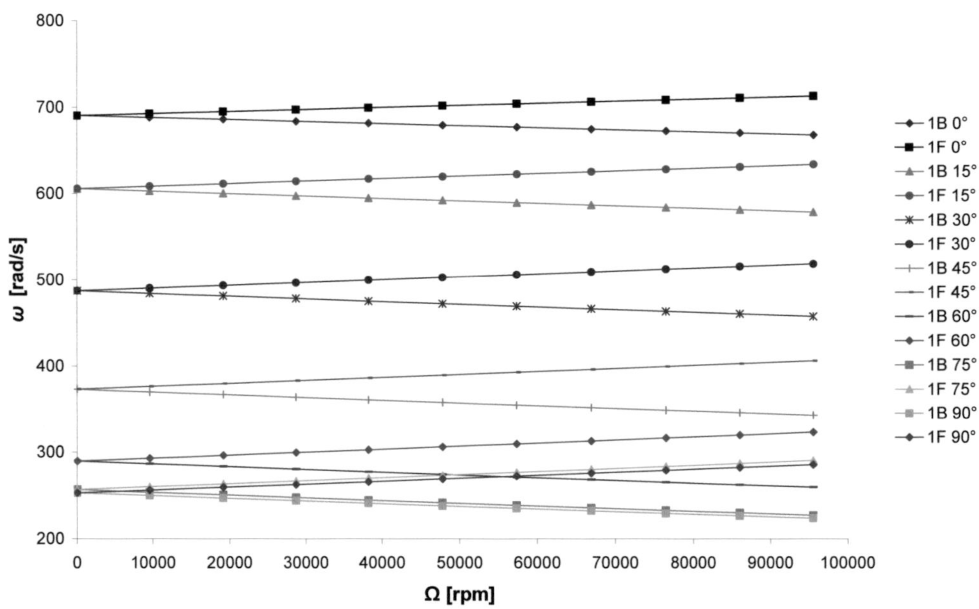

Fig. 15. The first backward (1B) and forward (1F) bending mode of a boron- epoxy shaft (S-S) for different lamination angles and different rotating speeds. *L =2.47 m, D =12.69 cm, e =1.321 mm, 10 layers of equal thickness*

3.3.4 Influence of the ratios L/D, e/D and η on the critical speeds and rigidity

The intersection point of the line ($\Omega = \omega$) with the bending frequency curves (diagram of Campbell) indicate the speed at which the shaft will vibrate violently (i.e., the critical speed Ω_{cr}).

In figure 16, the first critical speeds of the graphite-epoxy composite shaft (the properties are listed in table 1, with k_s =0.503) are plotted according to the lamination angle for various ratios *L/D* and various boundary conditions (S-S, C-C). From figure 16, the first critical speed of shaft bi-simply supported (S-S) has the maximum value at η = 0° for a ratio *L/D* = 10, 15 and 20, and at η = 15° for a ratio *L/D* = 5. For the case of a shaft bi-clamped (C-C), the maximum critical speed is at η = 0° for a ratio *L/D* = 20 and at η = 15° for a ratio *L/D* = 10 and 15, and at η = 30° for a ratio *L/D* = 5.
Above results can be explained as follows. The bending rigidity reaches maximum at η = 0° and reduces when the lamination angle increases; in addition, the shear rigidity reaches maximum at η = 30° and minimum with η = 0° and η = 90°. A situation in which the bending rigidity effect predominates causes the maximum to be η = 0°. However, as described by Singh ad Gupta (Singh & Gupta, 1994b), the maximum value shifts toward a higher lamination angle when the shear rigidity effect increases. Therefore, while comparing the phenomena of figure 16, the constraint from boundary conditions would raise the rigidity effect. A similar is observed for short shafts.

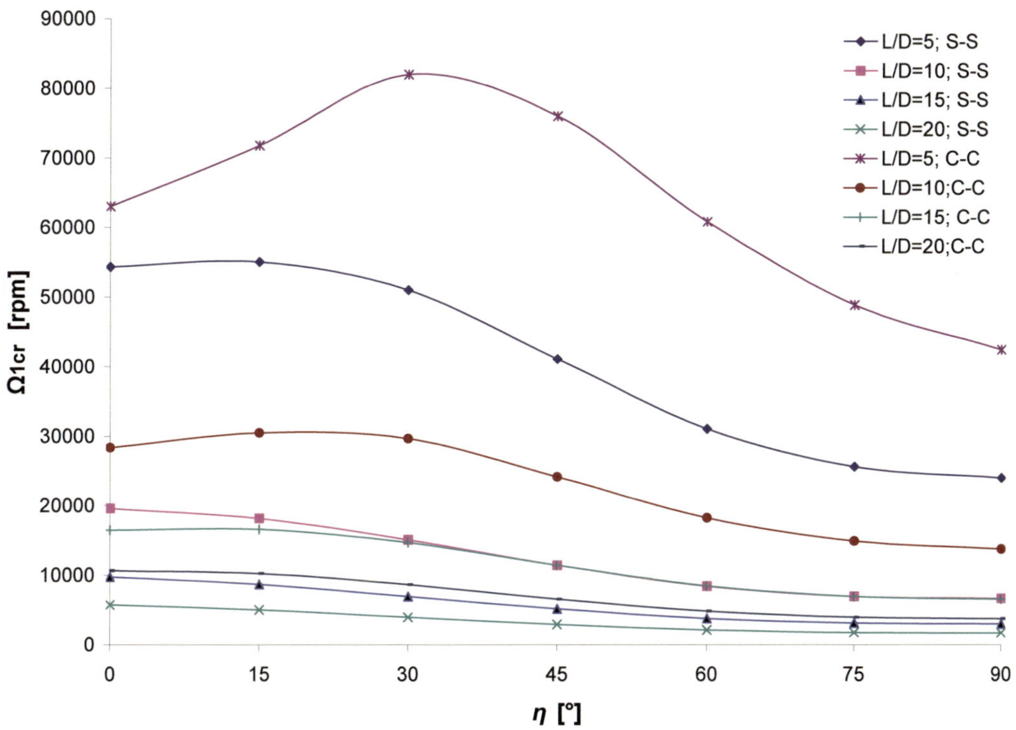

Fig. 16. The first critical speed Ω_{1cr} of spinning composite shaft according to the lamination angle η for various ratios *L/D* and various boundary conditions (S-S, C-C)

In figures 17 and 18, the first critical speeds according to ratio *L/D* of the same graphite-epoxy shaft bi-simply supported (S-S) and the same graphite-epoxy shaft bi- clamped (C-C) for various lamination angles. It is noticeable, if ratio *L/D* increases, the critical speed decreases and vice versa.

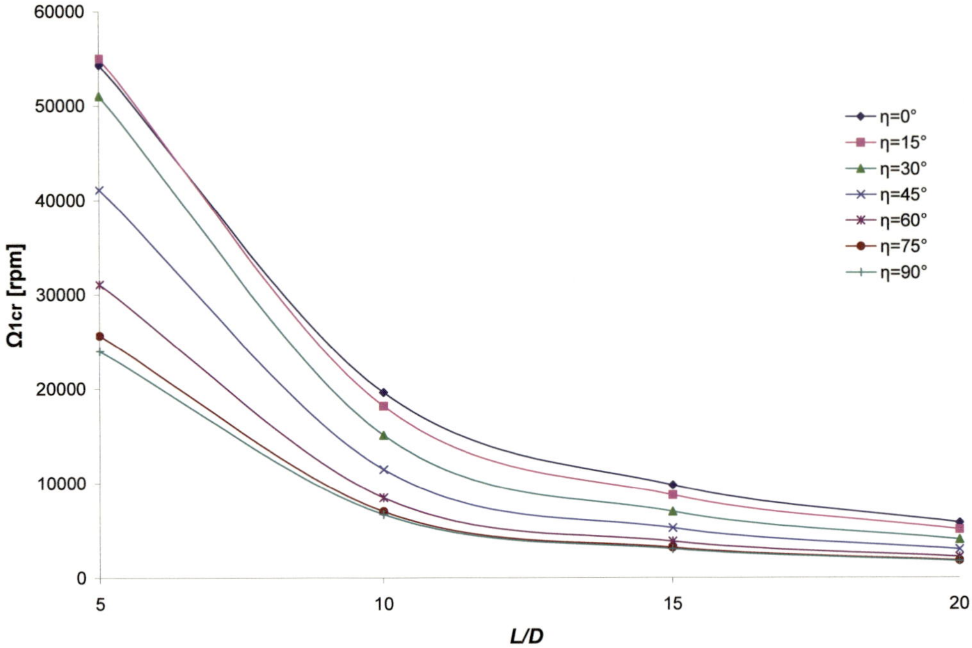

Fig. 17. The first critical speed Ω_{1cr} of spinning composite shaft bi- simply supported (S-S) according to ratio *L/D* for various lamination angles η

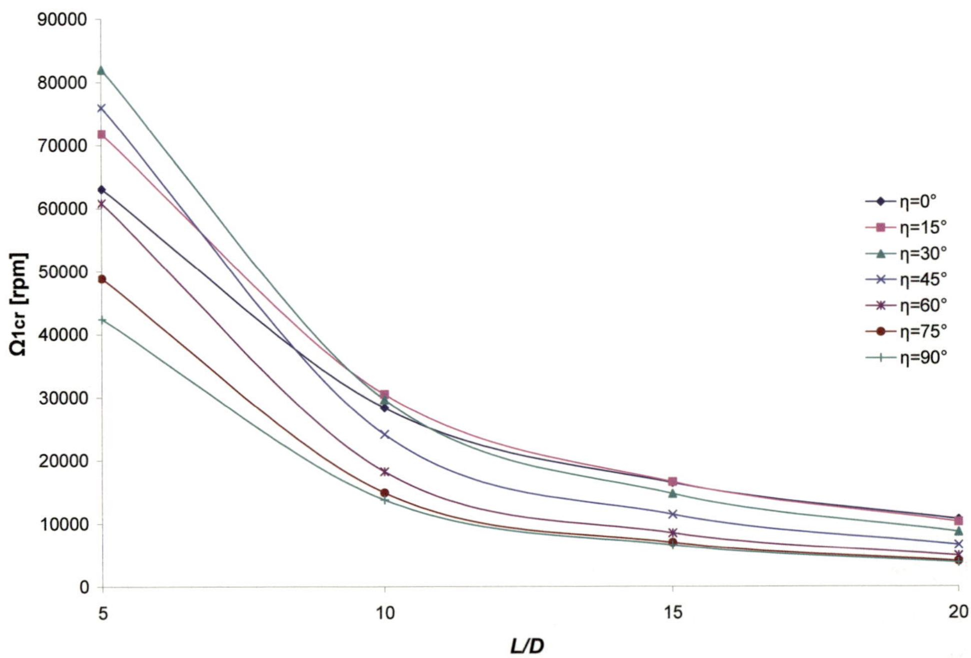

Fig. 18. The first critical speed Ω_{1cr} of spinning composite shaft bi- clamped (C-C) according to ratio *L/D* for various lamination angles η

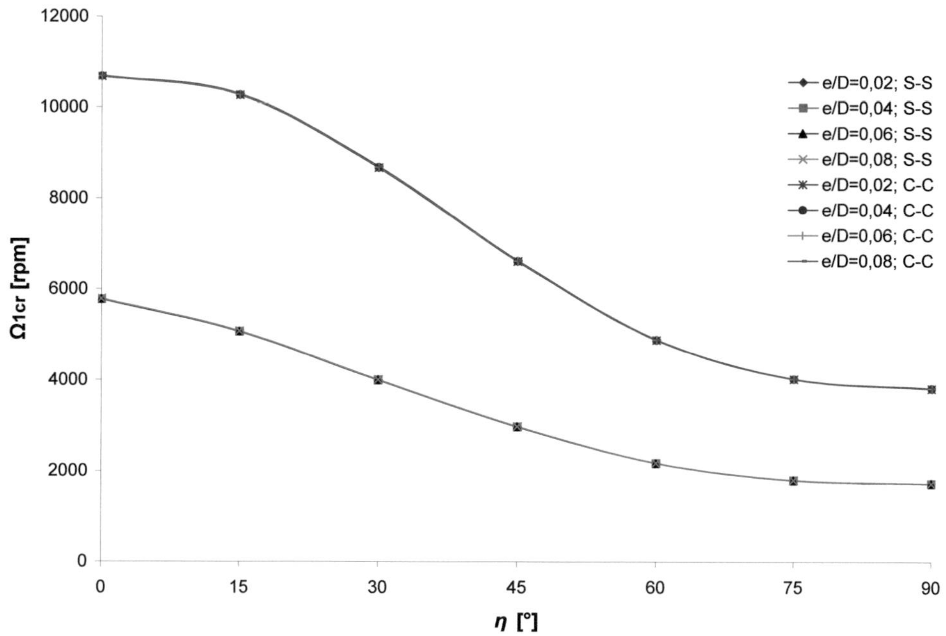

Fig. 19. The first critical speed Ω_{1cr} of spinning composite shaft according to the lamination angle η for various ratios *e/D* and various boundary conditions (S-S, C-C); (*L/D* = 20)

Figure 19 plots the variation of first critical speeds of the same graphite-epoxy composite shaft with ratio *L/D* = 20 according to the lamination angle for various *e/D* ratios and various boundary conditions. It is noticed the influence of the *e/D* ratio on the critical speed is almost negligible; the curves are almost identical for the various *e/D* ratios of each boundary condition. This is due to the deformation of the cross section is negligible, and thus the critical speed of the thin-walled shaft would approximately independent of thickness ratio *e/D*. According to above results, while predicting which stacking sequence of the spinning composite shaft having the maximum critical speed, we should consider *L/D* ratio and the type of the boundary conditions. I.e., the maximum critical speed of a spinning composite shaft is not forever at ply angle equalizes zero degree, but it depends on the *L/D* ratio and the type the boundary conditions.

3.3.5 Influence of the stacking sequence on the eigen-frequencies

In order to show the effects of the stacking sequence on the eigen-frequencies, a spinning carbon- epoxy shaft is mounted on two rigid supports; the mechanical and geometrical properties of this shaft are (Singh & Gupta, 1996):

E_{11} = 130 GPa, E_{22} = 10 GPa, G_{12} = G_{23} = 7 GPa, ν_{12} = 0.25, ρ = 1500 Kg/m^3

L =1.0 m, D = 0.1 m, e = 4 mm, 4 layers of equal thickness, k_s = 0.503

A four-layered scheme was considered with two layers of 0° and two of 90° fibre angle. The flexural frequencies have been obtained for different combinations (both symmetric and unsymmetric) of 0° and 90° orientations (see figure 20). This figure plots the Campbell diagram of the first eigen-frequency of a spinning shaft for various stacking sequences. It can be observed from this figure that, for symmetric configurations, the frequency values of the spinning composite shaft are very close, and do have a slight dependence on the relative positioning of the 0° and 90° layers.

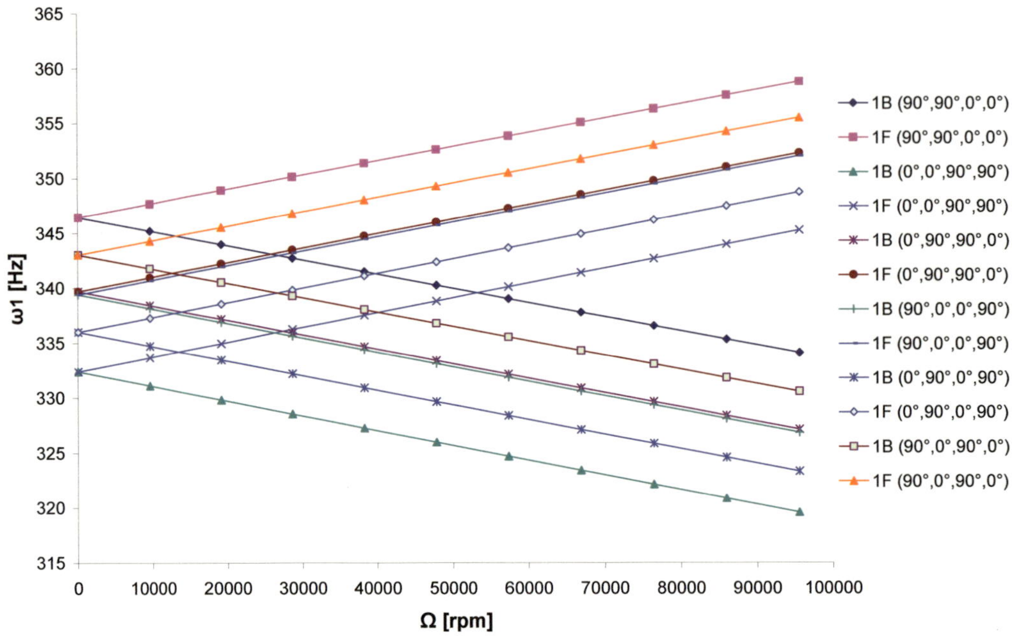

Fig. 20. First bending eigen-frequency of the spinning carbon- epoxy shaft bi- simply supported (S-S) for various stacking sequences according to the rotating speed

3.3.6 Influence of the disk's position according to the spinning shaft on on the eigen-frequencies

By considering another example, the eigen-frequencies of a graphite-epoxy shaft system are analyzed. The material properties are those listed in table 1. The lamination scheme remains the same as example 1, while its geometric properties, the properties of a uniform rigid disk are listed in table 3. The disk is placed at the mid-span of the shaft. The shaft system is shown in figure 21. For the finite element analysis, the shaft is modeled into two elements of equal lengths. The first element is simply-supported - free (S-F) and the second element is free- simply-supported (F-S). The disk is placed at the free boundary (F).

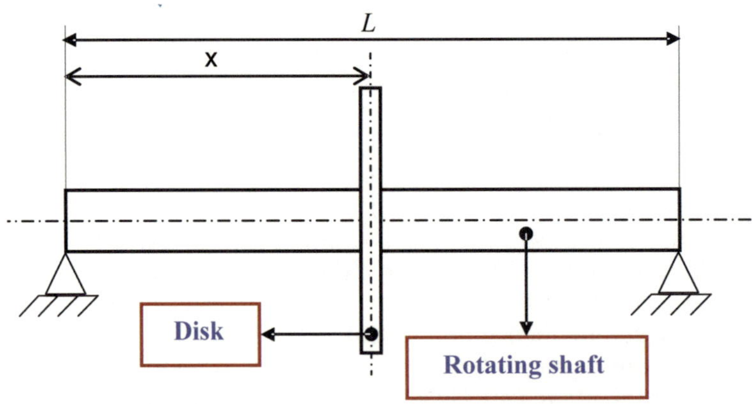

Fig. 21. System; embarked hollow spinning shaft.

The Campbell diagram containing the frequencies of the second pairs of bending whirling modes of the above composite system is shown in figure 22. Denote the ratio of the whirling bending frequency and the rotation speed of shaft as γ. The intersection point of the line (γ=1) with the whirling frequency curves indicate the speed at which the shaft will vibrate violently (i.e., the critical speed). In figure 22 the second pair of the forward and backward whirling frequencies falls more wide apart in contrast to other pairs of whirling modes. This might be due to the coupling of the pitching motion of the disk with the transverse vibration of shaft. Note that the disk is located at the mid-span of the shaft, while the second whirling forward and backward bending modes are skew-symmetric with respect to the mid-span of the shaft. Figure 23 shows the Campbell diagram of the first two bending frequencies of the embarked graphite- epoxy shaft for various disk's positions (x) according to the shaft (see figure 21). It is noted that when the disk approaches the support, the first bending frequency decreases and the second bending frequency increases and vice versa.

Properties	Shaft	Disk
L (m)	0.72	
Interior ray (m)	0.028	
external ray (m)	0.048	
k_s	0.56	
I_m (kg)		2.4364
I_d (kg m^2)		0.1901
I_p (kg m^2)		0.3778

Table 3. Properties of the system (shaft + disk)

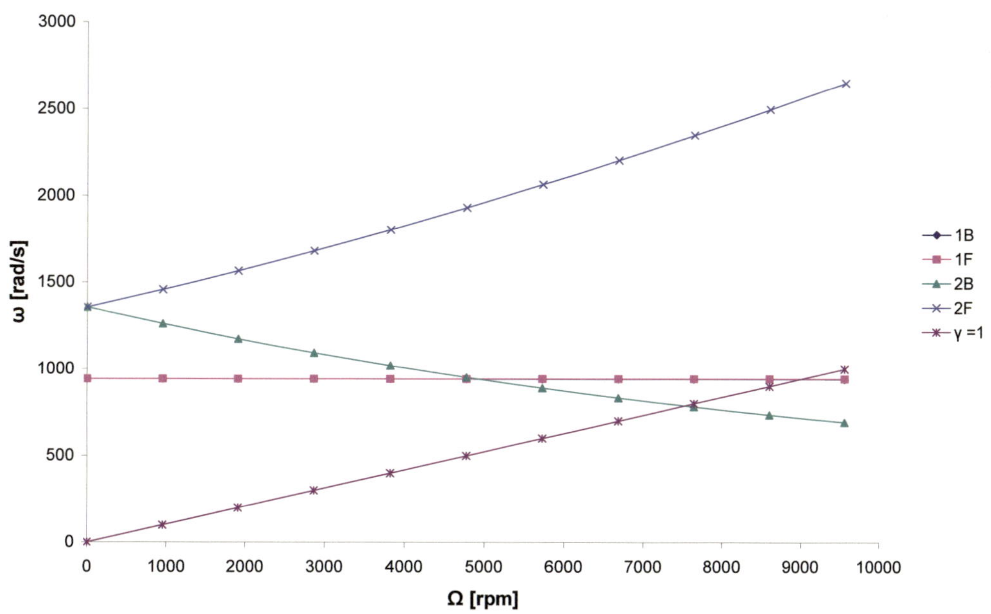

Fig. 22. Campbell diagram of the first two bending frequencies of the embarked graphite-epoxy shaft

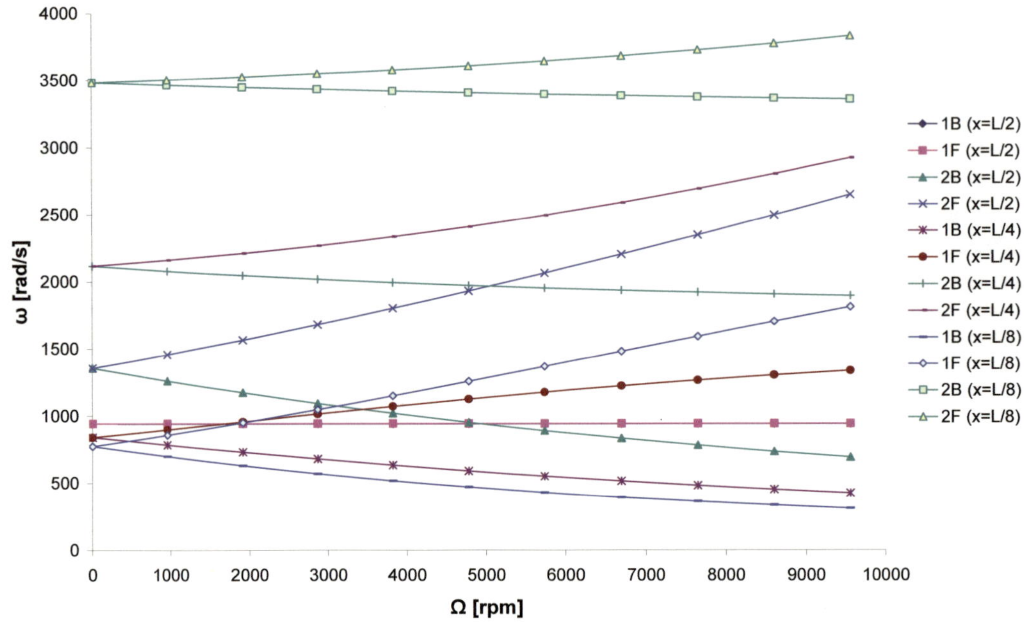

Fig. 23. Campbell diagram of the first two bending frequencies of the graphite-epoxy shaft for various disk's positions (x) according to the shaft

4. Conclusion

The analysis of the free vibrations of the spinning composite shafts using the *hp*-version of the finite element method (hierarchical finite element method (*p*-version) with trigonometric shape functions combined with the standard finite element method (*h*-version)), is presented in this work. The results obtained agree with those available in the literature. Several examples were treated to determine the influence of the various geometrical and physical parameters of the embarked spinning shafts. This work enabled us to arrive at the following conclusions:

a. Monotonous and uniform convergence is checked by increasing the number of the shape functions p, and the number of the hierarchical finite elements h. The convergence of the solutions is ensured by the element beam with two nodes. The results agree with the solutions found in the literature.
b. The gyroscopic effect causes a coupling of orthogonal displacements to the axis of rotation, and by consequence separates the frequencies in two branches, backward and forward precession modes. In all cases the forward modes increase with increasing rotating speed however the backward modes decrease. This effect has a significant influence on the behaviours of the spinning shafts.
c. The dynamic characteristics and in particular the eigen-frequencies, the critical speeds and the bending and shear rigidity of the spinning composite shafts are influenced appreciably by changing the ply angle, the stacking sequence, the length, the mean diameter, the materials, the rotating speed and the boundary conditions.
d. The critical speed of the thin-walled spinning composite shaft is approximately independent of the thickness ratio and mean diameter of the spinning shaft.
e. The dynamic characteristics of the system (shaft + disk + support) are influenced appreciably by changing disk's positions according to the shaft.

Prospects for future studies can be undertaken following this work: a study which takes into account damping interns in the case of a functionally graded material rotor with flexible disks, supported by supports with oil and subjected to disturbing forces like the air pockets or seisms, etc.

5. Nomenclature

U(x, y, z)	Displacement in x direction.
V(x, y, z)	Displacement in y direction.
W(x, y, z)	Displacement in z direction.
β_x	Rotation angles of the cross-section about the y axis.
β_y	Rotation angles of the cross-section, about the z axis.
ϕ	Angular displacement of the cross-section due to the torsion deformation of the shaft.
E	Young modulus.
G	Shear modulus.
(1, 2, 3)	Principal axes of a layer of laminate
(x, y, z)	Cartesian coordinates.
(x, r, θ)	Cylindrical coordinates.
G_c	Centre of the cross-section.
(O, x, y, z)	Inertial reference frame.
(G_c, x_1, y_1, z_1)	Local reference frame is located in the centre of the cross-section.
C_{ij}'	Elastic constants.
k_s	Shear correction factor.
ν	Poisson coefficient.
ρ	Masse density.
L	Length of the shaft.
D	Mean radius of the shaft.
e	Wall thickness of the shaft.
R_n	The nth layer inner radius of the composite shaft.
R_{n+1}	The nth layer outer radius of the composite shaft.
k	Number of the layer of the composite shaft.
η	Lamination (ply) angle.
θ	Circumferential coordinate.
ξ	Local and non-dimensional co-ordinates.
ω	Frequency, eigen-value.
Ω	Rotating speed.
[N]	Matrix of the shape functions.
f (ξ)	Shape functions.
p	Number of the shape functions or number of hierarchical terms.
t	Time.
E_c	Kinetic energy.
E_d	Strain energy.
$\{q_i\}$	Generalized coordinates, with (i = U, V, W, β_x, β_y, ϕ)
[M]	Masse matrix.
[K]	Stiffness matrix.
[G]	Gyroscopic matrix.

$[C_P]$	Damping matrix.
$K_{yy0}, K_{yz0}, K_{zy0}, K_{zz0}$	Bearing stiffness coefficients in x = 0.
$K_{yyL}, K_{yzL}, K_{zyL}, K_{zzL}$	Bearing stiffness coefficients in x = L.
$C_{yy0}, C_{yz0}, C_{zy0}, C_{zz0}$	Bearing damping coefficients in x = 0.
$C_{yyL}, C_{yzL}, C_{zyL}, C_{zzL}$	Bearing damping coefficients in x = L.

6. Appendix

The terms A_{ij}, B_{ij} of the equation (6) and I_m, I_d, I_p of the equation (7) are given as follows:

$$\begin{cases} A_{11} = \pi \sum_{n=0}^{k} C'_{11n}(R_{n+1}^2 - R_n^2)\,;\; A_{55} = \frac{\pi}{2} \sum_{n=0}^{k} C'_{55n}(R_{n+1}^2 - R_n^2) \\ A_{66} = \frac{\pi}{2} \sum_{n=0}^{k} C'_{66n}(R_{n+1}^2 - R_n^2)\,;\; A_{16} = \frac{2\pi}{3} \sum_{n=0}^{k} C'_{16n}(R_{n+1}^3 - R_n^3)\,; \\ B_{11} = \frac{\pi}{4} \sum_{n=0}^{k} C'_{11n}(R_{n+1}^4 - R_n^4)\,;\; B_{66} = \frac{\pi}{2} \sum_{n=0}^{k} C'_{66n}(R_{n+1}^4 - R_n^4) \end{cases} \begin{cases} I_m = \pi \sum_{n=0}^{k} \rho_n \left(R_{n+1}^2 - R_n^2\right) \\ I_d = \frac{\pi}{4} \sum_{n=0}^{k} \rho_n \left(R_{n+1}^4 - R_n^4\right) \\ I_p = \frac{\pi}{2} \sum_{n=0}^{k} \rho_n \left(R_{n+1}^4 - R_n^4\right) \end{cases} \quad \text{A1-2}$$

where k is the number of the layer, R_{n-1} is the nth layer inner radius of the composite shaft and R_n it is the nth layer outer of the composite shaft. L is the length of the composite shaft and ρ_n is the density of the nth layer of the composite shaft.

The indices used in the matrix forms are as follows:

a: shaft; **D**: disk; **e**: element; **P**: bearing (support)

The various matrices of the equation (13) which assemble the elementary matrices of the system as follows

- Shaft

$$\left[M_a^e\right] = \begin{bmatrix} [M_U] & 0 & 0 & 0 & 0 & 0 \\ 0 & [M_V] & 0 & 0 & 0 & 0 \\ 0 & 0 & [M_W] & 0 & 0 & 0 \\ 0 & 0 & 0 & \left[M_{\beta_x}\right] & 0 & 0 \\ 0 & 0 & 0 & 0 & \left[M_{\beta_y}\right] & 0 \\ 0 & 0 & 0 & 0 & 0 & [M_\phi] \end{bmatrix} \quad \text{A3}$$

$$\left[K_a^e\right] = \begin{bmatrix} [K_U] & 0 & 0 & 0 & 0 & [K_1] \\ 0 & [K_V] & 0 & [K_2] & [K_3] & 0 \\ 0 & 0 & [K_W] & [K_4] & [K_5] & 0 \\ 0 & [K_2]^T & [K_4]^T & \left[K_{\beta_x}\right] & [K_6] & 0 \\ 0 & [K_3]^T & [K_5]^T & [K_6]^T & \left[K_{\beta_y}\right] & 0 \\ [K_1]^T & 0 & 0 & 0 & 0 & [K_\phi] \end{bmatrix} \quad \text{A4}$$

$$\left[G_a^e\right]=\begin{bmatrix}0&0&0&0&0&0\\0&0&0&0&0&0\\0&0&0&0&0&0\\0&0&0&0&[G_1]&0\\0&0&0&-[G_1]^T&0&0\\0&0&0&0&0&0\end{bmatrix} \quad \text{A5}$$

$$[M_U]=I_mL\int_0^1[N_U]^T[N_U]\,d\xi\ ,\quad [M_V]=I_mL\int_0^1[N_V]^T[N_V]\,d\xi \quad \text{A6-7}$$

$$[M_W]=I_mL\int_0^1[N_W]^T[N_W]\,d\xi\ ,\quad \left[M_{\beta_x}\right]=I_dL\int_0^1\left[N_{\beta_x}\right]^T\left[N_{\beta_x}\right]d\xi \quad \text{A8-9}$$

$$\left[M_{\beta_y}\right]=I_dL\int_0^1\left[N_{\beta_y}\right]^T\left[N_{\beta_y}\right]d\xi\ ,\quad \left[M_{\phi}\right]=I_pL\int_0^1\left[N_{\phi}\right]^T\left[N_{\phi}\right]d\xi \quad \text{A10-11}$$

$$[K_U]=\frac{1}{L}A_{11}\int_0^1[N'_U]^T[N'_U]\,d\xi\ ,\quad [K_V]=\frac{1}{L}k_s(A_{55}+A_{66})\int_0^1[N'_V]^T[N'_V]\,d\xi \quad \text{A12-13}$$

$$[K_W]=\frac{1}{L}k_s(A_{55}+A_{66})\int_0^1[N'_W]^T[N'_W]\,d\xi\ ,\ [K_1]=\frac{1}{L}k_sA_{16}\int_0^1\left[N'_{\phi}\right]^T[N'_U]\,d\xi \quad \text{A14-15}$$

$$[K_2]=-\frac{1}{2L}k_sA_{16}\int_0^1[N'_V]^T\left[N'_{\beta_x}\right]d\xi\ ,\ [K_3]=-k_s(A_{55}+A_{66})\int_0^1\left[N_{\beta_y}\right]^T[N'_V]\,d\xi \quad \text{A16-17}$$

$$[K_4]=k_s(A_{55}+A_{66})\int_0^1\left[N_{\beta_x}\right]^T[N'_W]\,d\xi\ ,\ [K_5]=-\frac{1}{2L}k_sA_{16}\int_0^1[N'_W]^T\left[N'_{\beta_y}\right]d\xi \quad \text{A18-19}$$

$$[K_6]=\left[\frac{1}{2}k_sA_{16}\int_0^1\left[N_{\beta_y}\right]^T\left[N'_{\beta_x}\right]d\xi\right]-\left[\frac{1}{2}k_sA_{16}\int_0^1\left[N_{\beta_x}\right]^T\left[N'_{\beta_y}\right]d\xi\right] \quad \text{A20}$$

$$\left[K_{\beta_x}\right]=\left[\frac{1}{L}B_{11}\int_0^1\left[N'_{\beta_x}\right]^T\left[N'_{\beta_x}\right]d\xi\right]+\left[Lk_s(A_{55}+A_{66})\int_0^1\left[N_{\beta_x}\right]^T\left[N_{\beta_x}\right]\,d\xi\right] \quad \text{A21}$$

$$\left[K_{\beta_y}\right]=\left[\frac{1}{L}B_{11}\int_0^1\left[N'_{\beta_y}\right]^T\left[N'_{\beta_y}\right]d\xi\right]+\left[Lk_s(A_{55}+A_{66})\int_0^1\left[N_{\beta_y}\right]^T\left[N_{\beta_y}\right]\,d\xi\right] \quad \text{A22}$$

$$\left[K_\phi\right]=\frac{1}{L}B_{66}\int_0^1\left[N'_\phi\right]^T\left[N'_\phi\right]d\xi\ ,\ \left[G_1\right]=\Omega I_p L\int_0^1\left[N_{\beta_x}\right]^T\left[N_{\beta_y}\right]d\xi \qquad \text{A23-24}$$

Where $\left[N'_i\right]=\dfrac{\partial\left[N_i\right]}{\partial\xi}$, with ($i = U, V, W, \beta_x, \beta_y, \phi$)

- Disk

$$\left[M_D^e\right]=\begin{bmatrix}\left[I_m^D\right] & 0 & 0 & 0 & 0 & 0\\ 0 & \left[I_m^D\right] & 0 & 0 & 0 & 0\\ 0 & 0 & \left[I_m^D\right] & 0 & 0 & 0\\ 0 & 0 & 0 & \left[I_d^D\right] & 0 & 0\\ 0 & 0 & 0 & 0 & \left[I_d^D\right] & 0\\ 0 & 0 & 0 & 0 & 0 & \left[I_p^D\right]\end{bmatrix}, \qquad \text{A25}$$

$$\left[G_D^e\right]=\begin{bmatrix}0 & 0 & 0 & 0 & 0 & 0\\ 0 & 0 & 0 & 0 & 0 & 0\\ 0 & 0 & 0 & 0 & 0 & 0\\ 0 & 0 & 0 & 0 & \Omega\left[I_p^D\right] & 0\\ 0 & 0 & 0 & -\Omega\left[I_p^D\right]^T & 0 & 0\\ 0 & 0 & 0 & 0 & 0 & 0\end{bmatrix} \qquad \text{A26}$$

- Bearings (Supports)

$$\left[K_P^e\right]=\begin{bmatrix}0 & 0 & 0 & 0 & 0 & 0\\ 0 & \left[K_{yy}\right] & \left[K_{yz}\right] & 0 & 0 & 0\\ 0 & \left[K_{zy}\right] & \left[K_{zz}\right] & 0 & 0 & 0\\ 0 & 0 & 0 & 0 & 0 & 0\\ 0 & 0 & 0 & 0 & 0 & 0\\ 0 & 0 & 0 & 0 & 0 & 0\end{bmatrix}, \qquad \text{A27}$$

$$\left[C_P^e\right]=\begin{bmatrix}0 & 0 & 0 & 0 & 0 & 0\\ 0 & \left[C_{yy}\right] & \left[C_{yz}\right] & 0 & 0 & 0\\ 0 & \left[C_{zy}\right] & \left[C_{zz}\right] & 0 & 0 & 0\\ 0 & 0 & 0 & 0 & 0 & 0\\ 0 & 0 & 0 & 0 & 0 & 0\\ 0 & 0 & 0 & 0 & 0 & 0\end{bmatrix} \qquad \text{A28}$$

The elementary matrices of the system are

$$\begin{cases} \left[M^e\right]=\left[M_a^e\right]+\left[M_D^e\right] \\ \left[G^e\right]=\left[G_a^e\right]+\left[G_D^e\right] \\ \left[K^e\right]=\left[K_a^e\right]+\left[K_P^e\right] \\ \left[C_P^e\right] \end{cases} \quad \text{A29}$$

The various matrices (globally matrices) which assemble the elementary matrices, according to the boundary conditions as follows

$$\begin{cases} [M]=[M_a]+[M_D] \\ [G]=[G_a]+[G_D] \\ [K]=[K_a]+[K_P] \\ [C_P] \end{cases} \quad \text{A30}$$

The terms of the matrices are a function of the integrals: $J_{mn}^{\alpha\beta} = \int_0^1 f_m^{\alpha}(\xi)\, f_n^{\beta}(\xi) \quad d\xi$;

(m, n) indicate the number of the shape functions used, and (α,β) is the order of derivation.

7. References

Babuska, I. & Guo, B. (1986). The *h–p* version of the finite element method, Part I: the basic approximation results. *Computational Mechanics*, Vol. 1 , page numbers (21–41)

Bardell, N.S. (1996). An engineering application of the *h–p* version of the finite element method to the static analysis of a Euler-Bernoulli beam. Computers and Structures, Vol. 59, page numbers (195–211)

Bardell, N.S.; Dunsdon, J.M., & Langley, R.S. (1995). Free vibration analysis of thin rectangular laminated plate assemblies using the *h–p* version of the finite element method. *Composite Structures*, Vol. 32, page numbers (237–246)

Bert, C.W. & Kim, C.D. (1995a). Whirling of composite-material driveshafts including bending, twisting coupling and transverse shear deformation. *Journal of Vibration and Acoustics*, vol. 117, page numbers (17-21)

Bert, C.W. & Kim, C.D. (1995b). Dynamic instability of composite-material drive shaft subjected to fluctuating torque and/or rotational speed. *Dynamics and Stability of Systems*, Vol. 2, page numbers (125-147).

Bert, C.W. (1992). The effect of bending–twisting coupling on the critical speed of a driveshafts. *In: Proceedings. 6th Japan-US Conference on Composites Materials*, pp. 29-36, Orlando. FL. Technomic. Lancaster. PA

Boukhalfa, A. & Hadjoui, A. (2010). Free vibration analysis of an embarked rotating composite shaft using the *hp-* version of the FEM. *Latin American Journal of Solids and Structures*, Vol. 7, No. 2, page numbers (105-141)

Boukhalfa, A.; Hadjoui, A. & Hamza Cherif, S.M. (2008). Free vibration analysis of a rotating composite shaft using the *p*-version of the finite element method. *International Journal of Rotating Machinery*. Article ID 752062. 10 pages, Vol. 2008

Chang, M.Y.; Chang, M.Y. & Huang, J.H. (2004b). Vibration analysis of rotating composite shafts containing randomly oriented reinforcements. *Composite Structures*, Vol. 63, page numbers (21-32)

Chang, M.Y.; Chen, J.K. & Chang, C.Y. (2004a). A simple spinning laminated composite shaft model. *International Journal of Solids and Structures*, Vol. 41, page numbers (637–662)

Chatelet, E.; Lornage, D. & Jacquet-richardet, G. (2002). A three dimensional modeling of the dynamic behavior of composite rotors. *International Journal of Rotating Machinery*, Vol. 8, No. 3, page numbers (185-192)

Demkowicz, L.; Oden, J.T.; Rachowicz, W. & Hardy, O. (1989). Toward a universal *h–p* adaptive finite element strategy, Part I: constrained approximation and data structure. *Computational Methods in Applied Mechanics and Engineering*, Vol. 77, page numbers (79–112)

Dos Reis, H.L.M.; Goldman, R.B. & Verstrate, P.H. (1987). Thin walled laminated composite cylindrical tubes: Part III- Critical Speed Analysis. *Journal of Composites Technology and Research*, Vol. 9, page numbers (58–62)

Gupta, K. & Singh, S.E. (1996). Dynamics of composite rotors, *Proceedings of indo-us symposium on emerging trends in vibration and noise engineering*, pp. 59-70, New Delhi. India

Kim, C.D. & Bert, C.W. (1993). Critical speed analysis of laminated composite hollow drive shafts. *Composites Engineering*, Vol. 3, page numbers (633–643)

Singh, S.E. & Gupta, K. (1994b). Free damped flexural vibration analysis of composite cylindrical tubes using beam and shell theories. *Journal of Sound and Vibration*, Vol. 172, page numbers (171-190)

Singh, S.E. & Gupta, K. (1995). Experimental studies on composite shafts, *Proceedings of the International Conference on Advances in Mechanical Engineering.* , pp. 1205-1221, Bangalore. India

Singh, S.E. & Gupta, K. (1996). Composite shaft rotordynamic analysis using a layerwise theory. *Journal of Sound and Vibration*, Vol. 191, No. 5, page numbers (739–756)

Singh, S.P. & Gupta, K. (1994a). Dynamic analysis of composite rotors. *5th International Symposium on Rotating Machinery (ISROMAC-5)*. Also *International Journal of Rotating Machinery*, vol. 2, page numbers (179-186)

Singh, S.P. (1992). Some studies on dynamics of composite shafts. *Ph.D Thesis. Mechanical Engineering Department. IIT*, Delhi, India

Zinberg, H. & Symonds, M.F. (1970). The development of an advanced composite tail rotor driveshaft, *26th Annual Forum of The American Helicopter Society*, Washington. DC, June 1970

10

The Generalized Finite Element Method Applied to Free Vibration of Framed Structures

Marcos Arndt[1], Roberto Dalledone Machado[2] and Adriano Scremin[2]
[1]*Positivo University,*
[2]*Federal University of Paraná*
Brazil

1. Introduction

The vibration analysis is an important stage in the design of mechanical systems and buildings subject to dynamic loads like wind and earthquake. The dynamic characteristics of these structures are obtained by the free vibration analysis.

The Finite Element Method (FEM) is commonly used in vibration analysis and its approximated solution can be improved using two refinement techniques: *h* and *p*-versions. The *h*-version consists of the refinement of element mesh; the *p*-version may be understood as the increase in the number of shape functions in the element domain without any change in the mesh. The conventional *p*-version of FEM consists of increasing the polynomial degree in the solution. The *h*-version of FEM gives good results for the lowest frequencies but demands great computational cost to work up the accuracy for the higher frequencies. The accuracy of the FEM can be improved applying the polynomial *p* refinement.

Some enriched methods based on the FEM have been developed in last 20 years seeking to increase the accuracy of the solutions for the higher frequencies with lower computational cost. Engels (1992) and Ganesan & Engels (1992) present the Assumed Mode Method (AMM) which is obtained adding to the FEM shape functions set some interface restrained assumed modes. The Composite Element Method (CEM) (Zeng, 1998a and 1998b) is obtained by enrichment of the conventional FEM local solution space with non-polynomial functions obtained from analytical solutions of simple vibration problems. A modified CEM applied to analysis of beams is proposed by Lu & Law (2007). The use of products between polynomials and Fourier series instead of polynomials alone in the element shape functions is recommended by Leung & Chan (1998). They develop the Fourier *p*-element applied to the vibration analysis of bars, beams and plates. These three methods have the same characteristics and they will be called enriched methods in this chapter. The main features of the enriched methods are: (a) the introduction of boundary conditions follows the standard finite element procedure; (b) hierarchical *p* refinements are easily implemented and (c) they are more accurate than conventional *h* version of FEM.

At the same time, the Generalized Finite Element Method (GFEM) was independently proposed by Babuska and colleagues (Melenk & Babuska, 1996; Babuska et al., 2004; Duarte et al., 2000) and by Duarte & Oden (Duarte & Oden, 1996; Oden et al., 1998) under the following names: Special Finite Element Method, Generalized Finite Element Method, Finite Element Partition of Unity Method, *hp* Clouds and Cloud-Based *hp* Finite Element Method.

Actually, several meshless methods recently proposed may be considered special cases of this method. Strouboulis et al. (2006b) define otherwise the subclass of methods developed from the Partition of Unity Method including *hp* Cloud Method of Oden & Duarte (Duarte & Oden, 1996; Oden et al., 1998), the eXtended Finite Element Method (XFEM) of Belytschko and co-workers (Sukumar et al, 2000 and 2001), the Generalized Finite Element Method (GFEM) of Strouboulis et al. (2000 and 2001), the Method of Finite Spheres of De & Bathe (2001), and the Particle-Partition of Unity Method of Griebel & Schweitzer (Schweitzer, 2009). The GFEM, which was conceived on the basis of the Partition of Unity Method, allows the inclusion of a priori knowledge about the fundamental solution of the governing differential equation. This approach ensures accurate local and global approximations. Recently several studies have indicated the efficiency of the GFEM and others methods based on the Partition of Unity Method in problems such as analysis of cracks (Xiao & Karihaloo, 2007; Abdelaziz & Hamouine, 2008; Duarte & Kim, 2008; Nistor et al., 2008), dislocations based on interior discontinuities (Gracie et al., 2007), large deformation of solid mechanics (Khoei et al., 2008) and Helmholtz equation (Strouboulis et al., 2006a; Strouboulis et al., 2008). In structural dynamics, the Partition of Unity Method was applied by De Bel et al. (2005), Hazard & Bouillard (2007) to numerical vibration analysis of plates and by Arndt et al. (2010) to free vibration analysis of bars and trusses. Among the main challenges in developing the GFEM to a specific problem are: (a) choosing the appropriate space of functions to be used as local approximation and (b) the imposition of essential boundary conditions, since the degrees of freedom used in GFEM generally do not correspond to the nodal ones. In most cases the imposition of boundary conditions is achieved by the degeneration of the approximation space or applying penalty or Lagrange multipliers methods.

The purpose of this chapter is to present a formulation of the GFEM to free vibration analysis of framed structures. The proposed method combines the best features of GFEM and enriched methods: (a) efficiency, (b) hierarchical refinements and (c) the introduction of boundary conditions following the standard finite element procedure. In addition the enrichment functions are easily obtained. The GFEM elements presented can be used in plane free vibration analysis of rods, shafts, Euler Bernoulli beams, trusses and frames. These elements can be simply extended to spatial analysis of framed structures. The main features of the GFEM are discussed and the partition of unity functions and the local approximation spaces are presented. The GFEM solution can be improved using three refinement techniques: *h*, *p* and adaptive versions. In the adaptive GFEM, trigonometric and exponential enrichment functions depending on geometric and mechanical properties of the elements are added to the conventional Finite Element Method shape functions by the partition of unity approach. This technique allows an accurate adaptive process that converges very fast and is able to refine the frequency related to a specific vibration mode even for a coarse discretization scheme.

In this chapter the efficiency and convergence of the proposed method for vibration analysis of framed structures are checked. The frequencies obtained by the GFEM are compared with those obtained by the analytical solution, the CEM and the *h* and *p* versions of the Finite Element Method.

The chapter is structured as follows. Section 2 describes the variational form of the free vibration problems of bars and Euler-Bernoulli beams. The enriched methods proposed for free vibration analysis of bars and beams are discussed in Section 3. In Section 4 the main features of the GFEM and the formulation of C^0 and C^1 elements are discussed. Section 5 presents some applications of the proposed GFEM. Section 6 concludes the chapter.

2. Structural free vibration problem

Generally the structural free vibration problems are linear eigenvalue problems that can be described by: find a pair (λ, u) so that

$$Tu = \lambda Qu \text{ on } \Omega, \text{ with} \tag{1}$$

$$Pu = 0 \text{ on } \partial\Omega \tag{2}$$

where T, Q and P are linear operators and $\partial\Omega$ corresponds to the boundary of domain Ω. The vibration of bars, stationary shafts and Euler-Bernoulli beams are mathematically modeled by elliptic boundary value problems, so T is a linear elliptic operator of order $2m$ and P is a consistent boundary operator of order m. Moreover, as the structural free vibration problems are derived from conservative laws, the operator T is formally assumed self-adjoint (Carey & Oden, 1983).

According to Carey & Oden (1984), in order to obtain the variational form of a time dependent problem, one should consider the time t as a real parameter and develop a family of variational problems in t. This consists in selecting test functions w, independent of t, and applying the weighted-residual method.

By this technique the structural free vibration problem becomes an eigenvalue problem with variational statement: find a pair (λ, u), with $u \in H(\Omega)$ and $\lambda \in \mathbf{R}$, so that

$$B(u,w) = \lambda F(u,w), \ \forall w \in H \tag{3}$$

where $B: H \times H \mapsto \mathbf{R}$ and $F: H \times H \mapsto \mathbf{R}$ are bilinear forms.

In numerical methods, finite dimensional subspaces of approximation $H^h \subset H(\Omega)$ are chosen and the variational statement becomes: find $\lambda_h \in \mathbf{R}$ and $u_h \in H^h(\Omega)$ so that

$$B(u_h,w) = \lambda_h F(u_h,w), \ \forall w \in H^h . \tag{4}$$

Established an overview of the problem, in what follows the specific features of the free vibration problems of bars and beams are presented.

2.1 Axial vibration of a straight bar

The bar consists of a straight rod with axial strain (Fig. 1). The basic hypotheses concerning physical modeling of bar vibration are (Craig, 1981): (a) the cross sections which are straight and normal to the axis of the bar before deformation remain straight and normal after deformation; and (b) the material is elastic, linear and homogeneous.

The momentum equation that governs this problem is the partial differential equation

$$\rho A(x)\frac{\partial^2 \bar{u}}{\partial t^2} - \frac{\partial}{\partial x}\left(EA(x)\frac{\partial \bar{u}}{\partial x}\right) = p(x,t) \tag{5}$$

where $A(x)$ is the cross section area, E is the Young modulus, ρ is the specific mass, p is the externally applied axial force per unit length and t is the time. The problem of free vibration is stated as: find the axial displacement $\bar{u} = \bar{u}(x,t)$ which satisfies Eq. (5) when $p(x,t) = 0$.

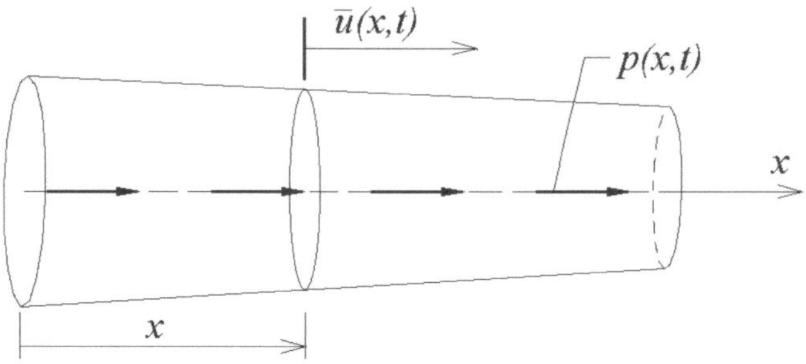

Fig. 1. Straight bar

Assuming periodic solutions $\bar{u}(x,t) = e^{i\omega t}u(x)$, where ω is the natural frequency, the free vibration of a bar becomes an eigenvalue problem with variational statement: find a pair (λ, u), with $u \in H^1(0,L)$ and $\lambda \in \mathbf{R}$, which satisfies Eq. (3) when H space is $H^1(0,L)$, $\lambda = \omega^2$ and L is the bar length.
The bilinear forms B and F in Eq. (3) for Dirichlet and Neumann boundary conditions are

$$B(u,w) = \int_0^L EA \frac{du}{dx}\frac{dw}{dx}dx \tag{6}$$

$$F(u,w) = \int_0^L \rho A u w dx \tag{7}$$

Similarly the bilinear forms for general natural boundary conditions are

$$B(u,w) = \int_0^L EA \frac{du}{dx}\frac{dw}{dx}dx + k_L u(0)w(0) + k_R u(L)w(L) \tag{8}$$

$$F(u,w) = \int_0^L \rho A u w dx + m_L u(0)w(0) + m_R u(L)w(L) \tag{9}$$

where k_L and k_R are the spring stiffness at left and right bar ends, respectively, and m_L and m_R are the masses at left and right bar ends, respectively.
The torsional free vibration of a circular shaft is mathematically identical to the axial free vibration of a straight bar so the variational forms of these problems are the same.

2.2 Transversal vibration of an Euler-Bernoulli beam

Consider a straight beam with lateral displacements, as illustrated in Fig. 2. The basic hypotheses concerning physical modeling of Euler-Bernoulli beam vibration are: (a) there is a neutral axis undergoing no extension or contraction; (b) cross sections in the undeformed beam remain plane and perpendicular to the deformed neutral axis, that is, transverse shear deformation is neglected; (c) the material is linearly elastic and the beam is homogeneous at any cross section; (d) normal stresses σ_y and σ_z are negligible compared to the axial stress σ_x ; and (e) the beam rotary inertia is neglected.
The momentum equation governing this problem is the partial differential equation

$$\frac{\partial^2}{\partial x^2}\left(EI\frac{\partial^2\overline{v}}{\partial x^2}\right)+\rho A\frac{\partial^2\overline{v}}{\partial t^2}=p(x,t) \tag{10}$$

where $I(x)$ is the second moment of area, $A(x)$ is the cross section area, E is the Young modulus, ρ is the specific mass, p is the externally applied transversal force per unit length and t is the time. The free vibration problem consists in finding the lateral displacement $\overline{v}=\overline{v}(x,t)$ which satisfies Eq. (10) when $p(x,t)=0$.
Assuming periodic solutions $\overline{v}(x,t)=e^{i\omega t}v(x)$, where ω is the natural frequency, the free vibration of a beam becomes an eigenvalue problem with variational statement: find a pair (λ,v), with $v\in H^2(0,L)$ and $\lambda\in\mathbf{R}$, which satisfies Eq. (3) when H space is $H^2(0,L)$, $\lambda=\omega^2$, $u=v$ and L is the beam length.

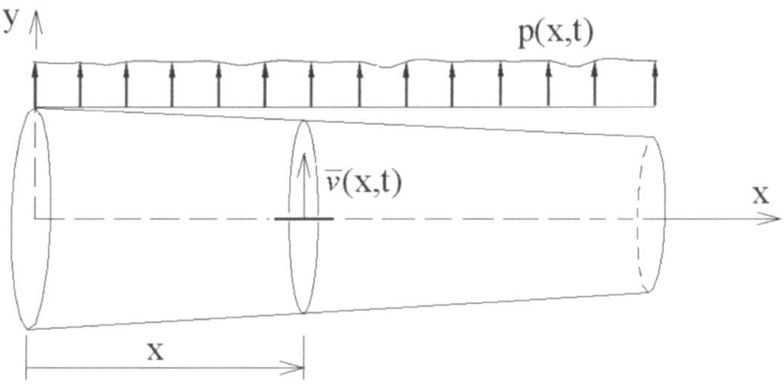

Fig. 2. Straight Euler-Bernoulli beam

For Dirichlet and Neumann boundary conditions the bilinear forms B and F in Eq. (3) are obtained from

$$B(v,w)=\int_0^L EI\frac{d^2v}{dx^2}\frac{d^2w}{dx^2}dx \tag{11}$$

$$F(v,w)=\int_0^L \rho Avw dx\,. \tag{12}$$

Similarly the bilinear forms for general natural boundary conditions are

$$B(v,w)=\int_0^L EI\frac{d^2v}{dx^2}\frac{d^2w}{dx^2}dx+k_{TL}v(0)w(0)+k_{TR}v(L)w(L)+k_{RL}\left.\frac{dv}{dx}\right|_{x=0}\left.\frac{dw}{dx}\right|_{x=0}+ \\ +k_{RR}\left.\frac{dv}{dx}\right|_{x=L}\left.\frac{dw}{dx}\right|_{x=L} \tag{13}$$

$$F(v,w)=\int_0^L \rho Avw dx+m_Lv(0)w(0)+m_Rv(L)w(L)+I_{mL}\left.\frac{dv}{dx}\right|_{x=0}\left.\frac{dw}{dx}\right|_{x=0}+ \\ +I_{mR}\left.\frac{dv}{dx}\right|_{x=L}\left.\frac{dw}{dx}\right|_{x=L} \tag{14}$$

where k_{TL}, k_{RL}, m_L and I_{mL} are translational stiffness, rotational stiffness, concentrated mass and moment of inertia of the attached mass at the left beam end, respectively, and k_{TR}, k_{RR}, m_R and I_{mR} are translational stiffness, rotational stiffness, concentrated mass and moment of inertia of the attached mass at the right beam end, respectively.

3. Enriched methods

Several methods found in the literature have as main feature the enrichment of the shape functions space of the classical FEM by adding other non polynomial functions. In this chapter such methods will be called enriched methods. Actually the Assumed Mode Method (AMM) of Ganesan & Engels (1992), the Composite Element Method (CEM) of Zeng (1998a, b and c) and the Fourier *p*-element of Leung & Chan (1998) are enriched methods. Their main features are: (a) the introduction of boundary conditions follows the standard finite element procedure; (b) hierarchical *p* refinements are easily implemented and (c) they present more accurate results than conventional *h*-version of FEM.
The approximated solution of the enriched methods, in the element domain, is obtained by:

$$u_h^e = u_{FEM}^e + u_{ENRICHED}^e \tag{15}$$

or in matrix shape

$$u_h^e = \mathbf{N}^T \mathbf{q} + \boldsymbol{\varnothing}^T \overline{\mathbf{q}} \tag{16}$$

where u_{FEM}^e is the FEM displacement field based on nodal degrees of freedom, $u_{ENRICHED}^e$ is the enriched displacement field based on field degrees of freedom, $\mathbf{q}$ is the conventional FEM degrees of freedom vector, the vector $\mathbf{N}$ contains the classical FEM shape functions and the vectors $\boldsymbol{\varnothing}$ and $\overline{\mathbf{q}}$ contain the enrichment functions and the field degrees of freedom, respectively. The vectors $\boldsymbol{\varnothing}$ and $\overline{\mathbf{q}}$ can be defined by:

$$\overline{\boldsymbol{\varnothing}}^{T}(\xi) = \begin{bmatrix} F_1 & F_2 & \ldots & F_r & \ldots & F_n \end{bmatrix} \tag{17}$$

$$\overline{\mathbf{q}}^T = \begin{bmatrix} c_1 & c_2 & \cdots & c_n \end{bmatrix} \tag{18}$$

$$\xi = \frac{x}{L_e} \tag{19}$$

where F_r are the enrichment functions, c_r are the field degrees of freedom and L_e is the element length. Different sets of enrichment functions produce different enriched methods. The enrichment functions spaces of the main enriched methods are described as follows.

3.1 Enriched C^0 elements

C^0 elements are used in free vibration analysis of bars and shafts. In this section the enriched C^0 elements are described. In all these enriched methods the FEM displacement field corresponds to the classical FEM with two node elements and linear Lagrangian shape functions. Only the enrichment functions are different.
In the AMM proposed by Engels (1992) the enrichment functions are the normalized analytical solutions of the free vibration problem of a fixed-fixed bar in the form

$$F_r = C\sin(r\pi\xi), \quad r = 1,2,\ldots \tag{20}$$

where C is the mass normalization constant given by

$$C = \sqrt{\frac{2}{\rho A L_e}}\,. \tag{21}$$

The CEM enrichment functions proposed by Zeng (1998a) are trigonometric functions in the form

$$F_r = \sin(r\pi\xi), \quad r = 1,2,\ldots \tag{22}$$

They differ from those of AMM just by the normalization.
The enrichment functions used by Leung & Chan (1998) in the bar Fourier p-element and by Zeng (1998a) in the CEM are the same.
It is noteworthy that all these functions vanish at element nodes. This feature allows the introduction of boundary conditions following the standard finite element procedure.

3.2 Enriched C^1 elements

C^1 elements are used in free vibration analysis of Euler-Bernoulli beams. In this section the enriched C^1 elements are described. The FEM displacement field in these enriched methods corresponds to the classical FEM with two node elements and cubic Hermitian shape functions. The enrichment functions are described below.
In the AMM three different enrichment functions are proposed. Engels (1992) uses analytical free vibration normal modes of a clamped-clamped beam in the classical form

$$F_r = C_r\left\{\sinh(\lambda_r\xi) - \sin(\lambda_r\xi) - \alpha_r\left[\cosh(\lambda_r\xi) - \cos(\lambda_r\xi)\right]\right\} \tag{23}$$

$$C_r = \frac{1}{\sqrt{\rho A L_e \alpha_r^2}} \tag{24}$$

$$\alpha_r = \frac{\sinh(\lambda_r) - \sin(\lambda_r)}{\cosh(\lambda_r) - \cos(\lambda_r)} \tag{25}$$

where C_r is the mass normalization constant for the rth mode and λ_r are the eigenvalues associated to the analytical solution obtained by the following characteristic equation

$$\cos(\lambda_r)\cosh(\lambda_r) - 1 = 0 \tag{26}$$

Alternatively, Ganesan & Engels (1992) propose enrichment functions based on the same analytical solution but in the form presented by Gartner & Olgac (1982) given by

$$F_r = \frac{1}{\sqrt{\rho A L_e}}\left[\cos(\lambda_r\xi) - \frac{1+(-1)^r e^{-\lambda_r}}{1-(-1)^r e^{-\lambda_r}}\sin(\lambda_r\xi) - \frac{e^{-\lambda_r\xi} - (-1)^r e^{-\lambda_r(1-\xi)}}{1-(-1)^r e^{-\lambda_r}}\right] \tag{27}$$

where λ_r are the eigenvalues obtained by solving the equation

$$\cos(\lambda_r) - \frac{2e^{-\lambda_r}}{1+e^{-2\lambda_r}} = 0 \tag{28}$$

Ganesan & Engels (1992) also propose trigonometric enrichment functions in the following form:

$$F_r = \cos\left[(r-1)\pi\xi\right] - \cos\left[(r+1)\pi\xi\right] \tag{29}$$

The Composite Element Method (CEM), proposed by Zeng (1998b), uses enrichment functions given by:

$$F_r = \sin(\lambda_r\xi) - \sinh(\lambda_r\xi) - \frac{\sin\lambda_r - \sinh\lambda_r}{\cos\lambda_r - \cosh\lambda_r}\left[\cos(\lambda_r\xi) - \cosh(\lambda_r\xi)\right] \tag{30}$$

corresponding to the clamped-clamped beam free vibration solution where λ_r are the eigenvalues obtained by the solution of Eq. (26).
Leung & Chan (1998) propose two types of enrichment functions based on the Fourier series: the cosine version

$$F_r = 1 - \cos(r\pi\xi) \tag{31}$$

and the sine version

$$F_r = \xi(1-\xi)\sin(r\pi\xi)\,. \tag{32}$$

The cosine version is the simplest but is not recommended when modeling a free of shear forces structure with only one element. Leung & Chan (1998) also note that the cosine version fails to predict the clamped-hinged and clamped-clamped modes of beams.
It is noteworthy that all these functions and their first derivatives vanish at element nodes. Again this feature allows the introduction of boundary conditions following the standard finite element procedure.

4. Generalized finite element method

The Generalized Finite Element Method (GFEM) is a Galerkin method whose main goal is the construction of a finite dimensional subspace of approximating functions using local knowledge about the solution that ensures accurate local and global results. The GFEM local enrichment in the approximation subspace is incorporated by the partition of unity approach.

4.1 Partition on unity

The Partition of Unity Method is defined as follows.
Let $u \in H^1(\Omega)$ be the function to be approximated and $\{\Omega_i\}$ be an open cover of domain Ω (Fig. 3) satisfying an overlap condition:

$$\exists M_S \in \mathbf{N} \text{ so that } \forall x \in \Omega \quad card\{\, i \,|\, x \in \Omega_i \} \le M_S\,. \tag{33}$$

A partition of unity subordinate to the cover $\{\Omega_i\}$ is the set of functions $\{\eta_i\}$ satisfying the conditions:

$$\text{supp}\ (\eta_i) = \{\ x \in \Omega\ |\eta_i(x) \neq 0\} \subset [\Omega_i],\quad \forall i \tag{34}$$

$$\sum_i \eta_i \equiv 1 \text{ on } \Omega \tag{35}$$

where supp (η_i) denotes the support of definition of the function η_i and $[\Omega_i]$ is the closure of the patch Ω_i.

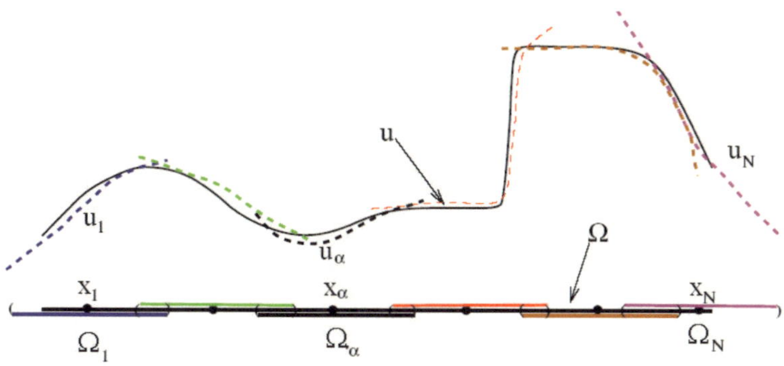

Fig. 3. Open cover $\{\Omega_i\}$ of domain Ω (Duarte et al., 2000)

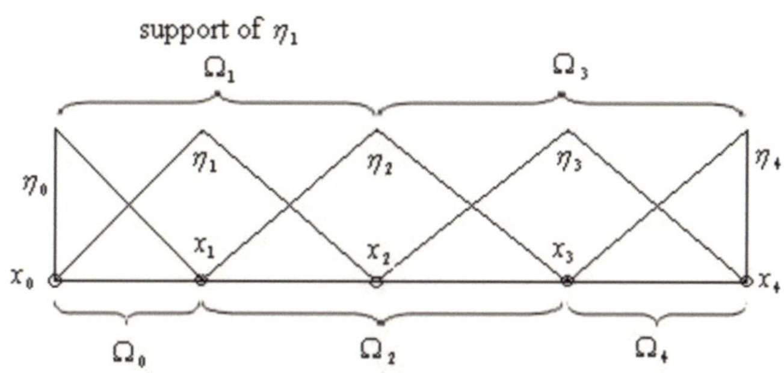

Fig. 4. Patchs and partition of unity set for one-dimensional GFEM finite element mesh

The partition of unity set $\{\eta_i\}$ allows to obtain an enriched set of approximating functions. Let $S_i \subset H^1(\Omega_i \cap \Omega)$ be a set of functions that locally well represents *u*:

$$S_i = \left\{s_i^j\right\}_{j=1}^{m} \tag{36}$$

Then the enriched set is formed by multiplying each partition of unity function η_i by the corresponding s_i^j, i.e.,

$$S := \sum_i \eta_i\ S_i = \left\{\sum_i \eta_i\ s_i^j\ |s_i^j \in S_i\right\} \subset H^1(\Omega) \tag{37}$$

Accordingly, the function *u* can be approximated by the enriched set as:

$$u_h(x) = \sum_i \sum_{s_i^j \in S_i} \eta_i s_i^j(x)\ a_{ij} \tag{38}$$

where a_{ij} are the degrees of freedom.
The proposed C^0 and C^1 generalized elements for free vibration analysis of framed structures are described below. The *h, p* and adaptive refinements of these elements are discussed.
In the proposed GFEM, the cover $\{\Omega_i\}$ corresponds to the finite element mesh and each patch Ω_i corresponds to the sub domain of Ω formed by the union of elements that contain the node x_i (Fig. 4).

4.2 Generalized C^0 elements

The generalized C^0 elements use the classical linear FEM shape functions as the partition of unity, i.e.:

$$\eta_i = \begin{cases} 1 + \dfrac{x - x_i}{x_i - x_{i-1}} & \text{if } x \in (x_{i-1}, x_i) \\ 1 - \dfrac{x - x_i}{x_{i+1} - x_i} & \text{if } x \in (x_i, x_{i+1}) \end{cases} \tag{39}$$

in the patch $\Omega_i = (x_{i-1}, x_{i+1})$.
The proposed local approximation space in the patch $\Omega_i = (x_{i-1}, x_{i+1})$ takes the form:

$$S_i = span\{\, 1 \;\; \gamma_{1j} \;\; \gamma_{2j} \;\; \phi_{1j} \;\; \phi_{2j} \;\; \ldots \}, \; j = 1,2,\ldots,n_l \tag{40}$$

$$\gamma_{1j} = \begin{cases} 0 \;\; if \;\; x \in (x_{i-1}, x_i) \\ \sin\left[\beta_{Rj}(x - x_i)\right] \;\; if \;\; x \in (x_i, x_{i+1}) \end{cases} \tag{41}$$

$$\gamma_{2j} = \begin{cases} \sin\left[\beta_{Lj}(x - x_i)\right] \;\; if \;\; x \in (x_{i-1}, x_i) \\ 0 \;\; if \;\; x \in (x_i, x_{i+1}) \end{cases} \tag{42}$$

$$\phi_{1j} = \begin{cases} 0 \;\; if \;\; x \in (x_{i-1}, x_i) \\ \cos\left[\beta_{Rj}(x - x_i)\right] - 1 \;\; if \;\; x \in (x_i, x_{i+1}) \end{cases} \tag{43}$$

$$\phi_{2j} = \begin{cases} \cos\left[\beta_{Lj}(x - x_i)\right] - 1 \;\; if \;\; x \in (x_{i-1}, x_i) \\ 0 \;\; if \;\; x \in (x_i, x_{i+1}) \end{cases} \tag{44}$$

$$\beta_{Rj} = \sqrt{\frac{\rho_R}{E_R}}\mu_j \tag{45}$$

$$\beta_{Lj} = \sqrt{\frac{\rho_L}{E_L}}\mu_j \tag{46}$$

where E_R and ρ_R are the Young modulus and specific mass on sub domain (x_i, x_{i+1}), E_L and ρ_L are the Young modulus and specific mass on sub domain (x_{i-1}, x_i), and μ_j is a frequency related to the enrichment level *j*.

The enriched functions, so proposed, vanish at element nodes, which allows the imposition of boundary conditions in the same fashion of the finite element procedure.

This C^0 element can be applied to the free vibration analysis of shafts, bars and trusses. Different frequencies μ_j produce different enriched elements. The increase in the number of elements in the mesh with only one level of enrichment (j = 1) and a fixed parameter $\beta_1 = \beta_{R1} = \beta_{L1}$, $\beta_1 = \pi$ for example, produces an h refinement. Otherwise the increase in the number of levels of enrichment, with a different parameter $\beta_j = \beta_{Rj} = \beta_{Lj}$ each, $\beta_j = j\pi$ for example, produces a hierarchical p refinement. Another possible refinement in the proposed GFEM is the adaptive one, which is presented below.

The adaptive GFEM is an iterative approach presented first by Arndt et al. (2010) whose main goal is to increase the accuracy of the frequency (eigenvalue) related to a chosen vibration mode with order denoted by "target order". The flowchart with blocks A to H presented in Fig. 5 represents the adaptive process.

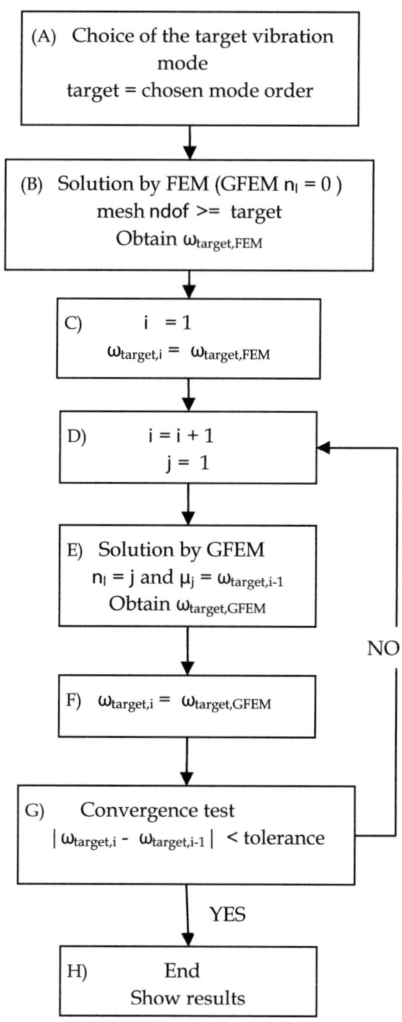

Fig. 5. Flowchart of the adaptive GFEM

In this flowchart, ω_{target} corresponds to the frequency related to the target mode. The first step of the adaptive GFEM process (blocks A to C) consists in obtaining an approximation of the target frequency by the standard FEM (GFEM with $n_l = 0$) with a coarse mesh. The finite element mesh used in the analysis has to be as coarse as necessary to capture a first approximation of the target frequency. The subsequent steps (blocks D to G) consist in applying the GFEM with only one enrichment level ($n_l = 1$) to the same finite element mesh assuming the frequency μ_j ($j = 1$, blocks D and E) of the enrichment functions (Eqs. 41-46) as the target frequency obtained in the last step. Thus, no mesh refinement is necessary along the iterative process.

Both the standard FEM and the adaptive GFEM allow as many frequencies as the total number of degrees of freedom to be obtained. However, in the latter, only the precision of the target frequency is effectively improved by the iterative process. The other frequencies present errors similar to those obtained by the standard FEM with the same mesh. In order to improve the precision of another frequency, it is necessary to perform a new adaptive GFEM analysis, taking this new one as the target frequency.

4.3 Generalized C[1] elements

The generalized C^1 elements also use the classical linear FEM form functions as partition of unity (Eq. (39)). The proposed local approximation space in the patch $\Omega_i = (x_{i-1}, x_{i+1})$ takes the form:

$$S_i = span\{\ \varphi_1 \quad \varphi_2 \quad \gamma_{1j} \quad \gamma_{2j} \quad \dots\},\ j = 1,2,\dots,n_l \tag{47}$$

$$\varphi_1 = \begin{cases} 3\dfrac{x - x_{i-1}}{x_i - x_{i-1}} - 2\left(\dfrac{x - x_{i-1}}{x_i - x_{i-1}}\right)^2 & if \quad x \in (x_{i-1}, x_i) \\ 1 + \dfrac{x - x_i}{x_{i+1} - x_i} - 2\left(\dfrac{x - x_i}{x_{i+1} - x_i}\right)^2 & if \quad x \in (x_i, x_{i+1}) \end{cases} \tag{48}$$

$$\varphi_2 = \begin{cases} \left[\left(\dfrac{x - x_{i-1}}{x_i - x_{i-1}}\right)^2 - \dfrac{x - x_{i-1}}{x_i - x_{i-1}}\right](x_i - x_{i-1}) & if \quad x \in (x_{i-1}, x_i) \\ \left[\dfrac{x - x_i}{x_{i+1} - x_i} - \left(\dfrac{x - x_i}{x_{i+1} - x_i}\right)^2\right](x_{i+1} - x_i) & if \quad x \in (x_i, x_{i+1}) \end{cases} \tag{49}$$

$$\gamma_{1j} = \begin{cases} \cos(\lambda_j z_1) - \dfrac{1 + (-1)^j e^{-\lambda_j}}{1 - (-1)^j e^{-\lambda_j}} \sin(\lambda_j z_1) - \dfrac{e^{-\lambda_j z_1} - (-1)^j e^{-\lambda_j (1 - z_1)}}{1 - (-1)^j e^{-\lambda_j}} & if \quad x \in (x_{i-1}, x_i) \\ 0 \quad if \quad x \in (x_i, x_{i+1}) \end{cases} \tag{50}$$

$$z_1 = \frac{x - x_{i-1}}{x_i - x_{i-1}}$$

$$\gamma_{2j} = \begin{cases} 0 \quad if \quad x \in (x_{i-1}, x_i) \\ \cos(\lambda_j z_2) - \dfrac{1+(-1)^j e^{-\lambda_j}}{1-(-1)^j e^{-\lambda_j}} \sin(\lambda_j z_2) - \dfrac{e^{-\lambda_j z_2} - (-1)^j e^{-\lambda_j (1-z_2)}}{1-(-1)^j e^{-\lambda_j}} \quad if \quad x \in (x_i, x_{i+1}) \end{cases} \tag{51}$$

$$z_2 = \frac{x - x_i}{x_{i+1} - x_i}$$

where λ_j are the eigenvalues obtained by the solution of Eq. (28).

Such partition of unity functions and local approximation space produce the cubic FEM approximation space enriched by functions that represent the local behavior of the differential equation solution. The enriched functions and their first derivatives vanish at element nodes. Hence, the imposition of boundary conditions follows the finite element procedure. This C^1 element is suited to apply to the free vibration analysis of Euler-Bernoulli beams.

Again the increase in the number of elements in the mesh with only one level of enrichment ($j = 1$) and a fixed eigenvalue λ_1 produces the h refinement of GFEM. Otherwise the increase in the number of levels of enrichment, each of one with a different frequency λ_j, produces a hierarchical p refinement. An adaptive GFEM refinement for free vibration analysis of Euler-Bernoulli beams is straight forward, as can be easily seen. However it will not be discussed here.

5. Applications

Numerical solutions for two bars, a beam and a truss are given below to illustrate the application of the GFEM. To check the efficiency of this method the results are compared to those obtained by the *h* and *p*-versions of FEM and the *c*-version of CEM.

The number of degrees of freedom (ndof) considered in each analysis is the total number of effective degrees of freedom after introduction of boundary conditions. As an intrinsic imposition of the adaptive method, each target frequency is obtained by a new iterative analysis. The mesh used in each adaptive analysis is the coarser one, that is, just as coarse as necessary to capture a first approximation of the target frequency.

5.1 Uniform fixed-free bar

The axial free vibration of a fixed-free bar (Fig. 6) with length L, elasticity modulus E, mass density ρ and uniform cross section area A, has exact natural frequencies (ω_r) given by (Craig, 1981):

$$\omega_r = \frac{(2r-1)\pi}{2L}\sqrt{\frac{E}{\rho}} \quad , \; r = 1, 2, \dots . \tag{52}$$

In order to compare the exact solution with the approximated ones, in this example it is used a non-dimensional eigenvalue χ_r given by:

$$\chi_r = \frac{\rho L^2 \omega_r^{\,2}}{E} . \tag{53}$$

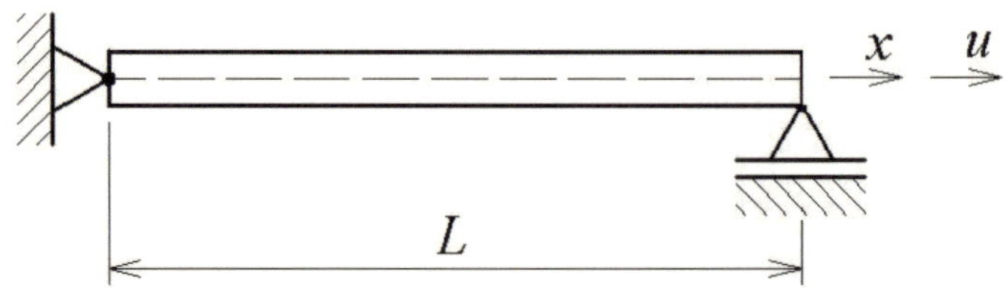

Fig. 6. Uniform fixed-free bar

a) *h* refinement

First the proposed problem is analyzed by a series of *h* refinements of FEM (linear and cubic), CEM and GFEM (C^0 element). A uniform mesh is used in all methods. Only one enrichment function is used in each element of the *h*-version of CEM. One level of enrichment (n_l = 1) with $\beta_1 = \pi$ is used in the *h*-version of GFEM. The evolution of relative error of the *h* refinements for the six earliest eigenvalues in logarithmic scale is presented in Figs. 7-9.
The results show that the *h*-version of GFEM exhibits greater convergence rates than the *h* refinements of FEM and CEM for all analyzed eigenvalues.

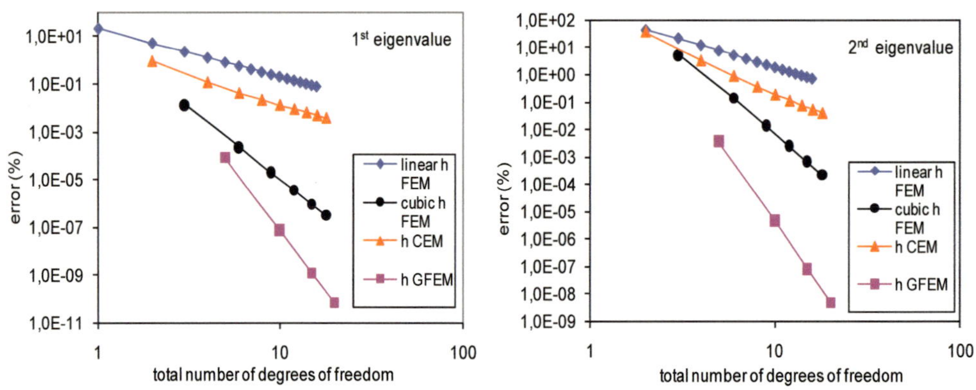

Fig. 7. Relative error (%) for the 1st and 2nd fixed-free bar eigenvalues - *h* refinements

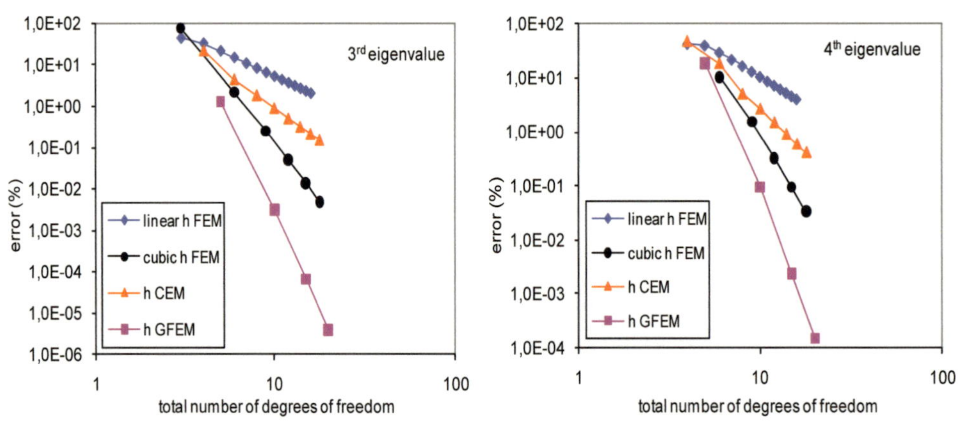

Fig. 8. Relative error (%) for the 3rd and 4th fixed-free bar eigenvalues - *h* refinements

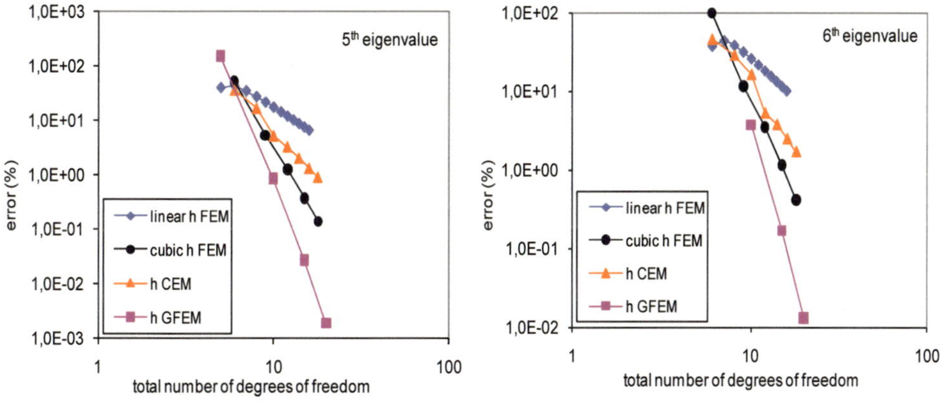

Fig. 9. Relative error (%) for the 5th and 6th fixed-free bar eigenvalues - *h* refinements

b) *p* refinement

The *p* refinement of GFEM is now compared to the hierarchical *p*-version of FEM and the *c*-version of CEM. The *p*-version of GFEM consists in a progressive increase of levels of enrichment with parameter $\beta_j = j\pi$.

The evolution of relative error of the *p* refinements for the six earliest eigenvalues in logarithmic scale is presented in Figs. 10-12.

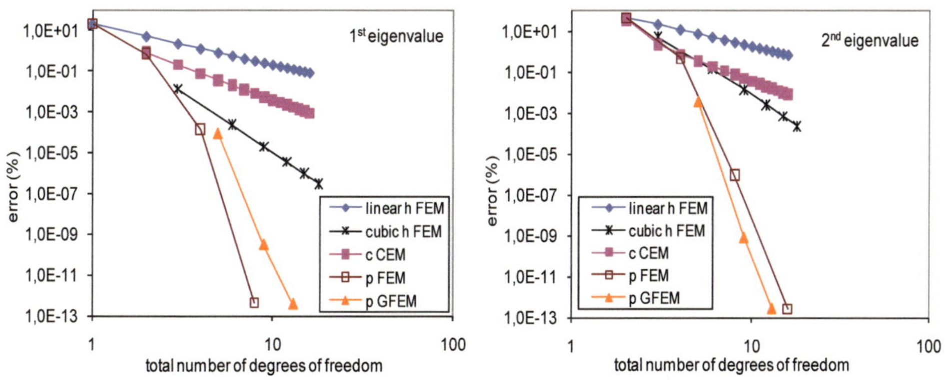

Fig. 10. Relative error (%) for the 1st and 2nd fixed-free bar eigenvalues - *p* refinements

The fixed-free bar results show that the *p*-version of GFEM presents greater convergence rates than the *h* refinements of FEM and the *c*-version of CEM. The hierarchical *p* refinement of FEM only overcomes the results obtained by *p*-version of GFEM for the first eigenvalue. For the other eigenvalues the GFEM presents more precise results and greater convergence rates.

c) adaptive refinement

Four different adaptive GFEM analyses are performed in order to obtain the first four frequencies. The behavior of the relative error in each analysis is presented in Fig. 13.

In order to capture an initial approximation of the target vibration frequency, for the first frequency, the finite element mesh must have at least one bar element (one effective degree of freedom), for the second frequency, it must have at least two bar elements (two effective degrees of freedom), and so on.

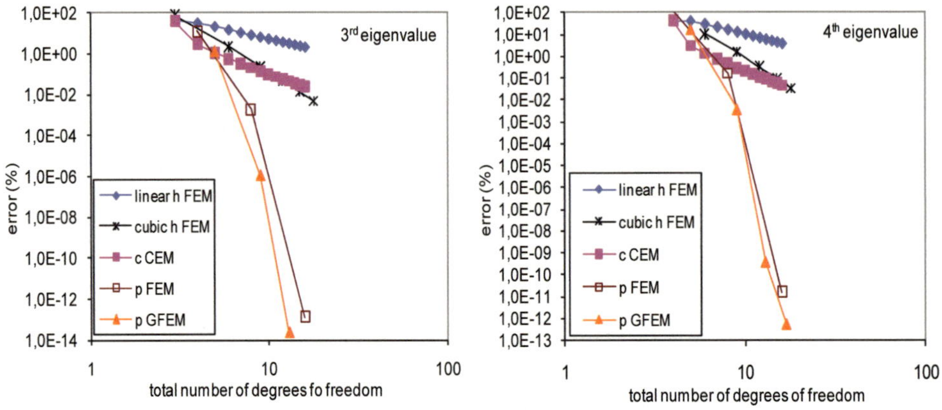

Fig. 11. Relative error (%) for the 3rd and 4th fixed-free bar eigenvalues - *p* refinements

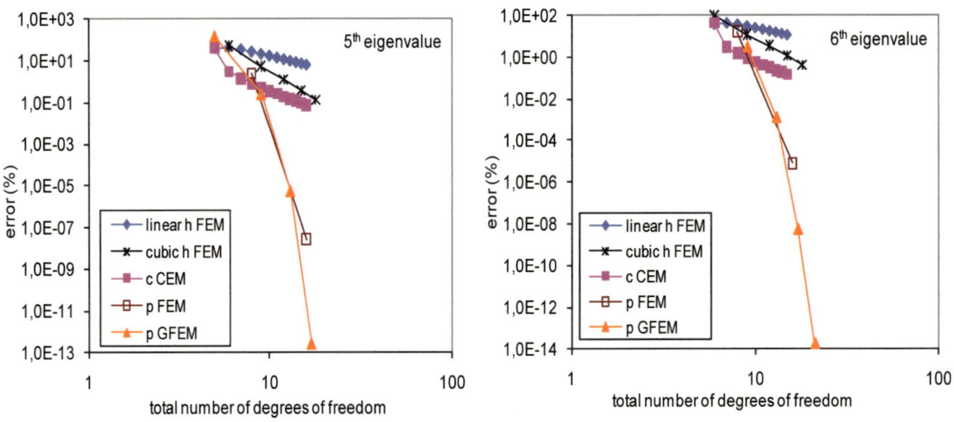

Fig. 12. Relative error (%) for the 5th and 6th fixed-free bar eigenvalues - *p* refinements

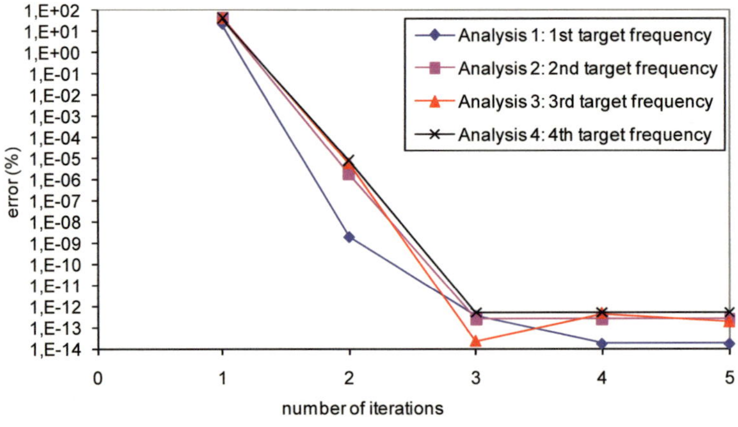

Fig. 13. Error in the adaptive GFEM analyses of fixed-free uniform bar

Table 1 presents the relative errors obtained by the numerical methods. The linear FEM solution is obtained with 100 elements, that is, 100 effective degrees of freedom (dof). The cubic FEM solution is obtained with 20 elements, that is, 60 effective degrees of freedom. The CEM solution is obtained with one element and 15 enrichment functions corresponding to one nodal degree of freedom and 15 field degrees of freedom resulting in 16 effective degrees of freedom. The hierarchical p FEM solution is obtained with a 17-node element corresponding to 16 effective degrees of freedom. The analyses by the adaptive GFEM have no more than 20 degrees of freedom in each iteration. For example, the fourth frequency is obtained taking 4 degrees of freedom in the first iteration and 20 degrees of freedom in the two subsequent ones.

Eigenvalue	linear h FEM (100e) ndof = 100	cubic h FEM (20e) ndof = 60	p FEM (1e 17n) ndof = 16	c CEM (1e 15c) ndof =16	Adaptive GFEM (after 3 iterations)	
	error (%)	error (%)	error (%)	error (%)	error (%)	ndof in iterations
1	2,056 e-3	8,564 e-10	3,780 e-13	8,936 e-4	3,780 e-13	1x 1 dof + 2x 5 dof
2	1,851 e-2	1,694 e-7	2,560 e-13	8,188 e-3	2,560 e-13	1x 2 dof + 2x 10 dof
3	5,141 e-2	3,619 e-6	1,382 e-13	2,299 e-2	2,304 e-14	1x 3 dof + 2x 15 dof
4	1,008 e-1	2,711 e-5	1,602 e-11	4,579 e-2	5,289 e-13	1x 4 dof + 2x 20 dof

Table 1. Results to free vibration of uniform fixed-free bar

The adaptive process converges rapidly, requiring three iterations in order to achieve each target frequency with precision of the 10^{-13} order. For the uniform fixed-free bar, one notes that the adaptive GFEM reaches greater precision than the h versions of FEM and the c-version of CEM. The p-version of FEM is as precise as the adaptive GFEM only for the first two eigenvalues. After this, the precision of the adaptive GFEM prevails among the others. For the sake of comparison, the standard FEM software Ansys© employing 410 truss elements (LINK8) reaches the same precision for the first four frequencies.

5.2 Fixed-fixed bar with sinusoidal variation of cross section area

In order to analyze the efficiency of the adaptive GFEM for non-uniform bars, the longitudinal free vibration of a fixed-fixed bar with sinusoidal variation of cross section area, length L, elasticity modulus E and mass density ρ is analyzed. The boundary conditions are $\bar{u}(0,t)=0$ and $\bar{u}(L,t)=0$, and the cross section area varies as

$$A(x) = A_0 \sin^2\left(\frac{x}{L}+1\right) \tag{54}$$

where A_0 is a reference cross section area.

Kumar & Sujith (1997) presented exact analytical solutions for longitudinal free vibration of bars with sinusoidal and polynomial area variations.

This problem is analyzed by the h and p versions of FEM and the adaptive GFEM. Six adaptive analyses are performed in order to obtain each of the first six frequencies. The behavior of the relative error of the target frequency in each analysis is presented in Fig. 14.

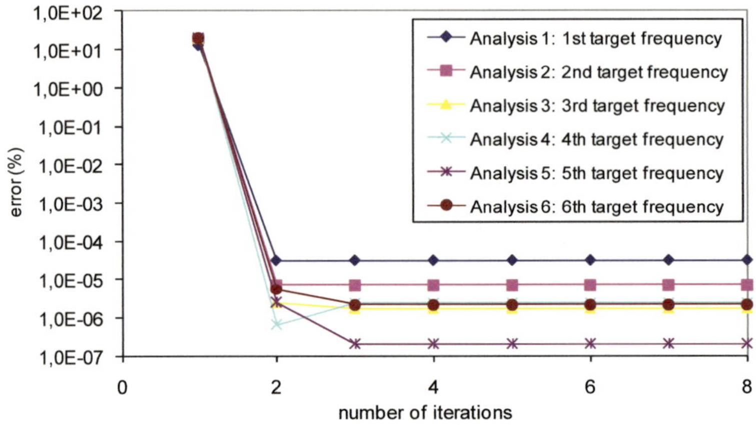

Fig. 14. Error in the adaptive GFEM analyses of fixed-fixed non-uniform bar

Table 2 shows the first six non-dimensional eigenvalues ($\beta_r = \omega_r L\sqrt{\rho/E}$) and their relative errors obtained by these methods. The linear *h* FEM solution is obtained with 100 elements, that is, 99 effective degrees of freedom after introduction of boundary conditions. The cubic *h* FEM solution is obtained with 12 cubic elements, that is, 35 effective degrees of freedom. The *p* FEM solution is obtained with one hierarchical 33-node element, that is, 31 effective degrees of freedom. The analyses by the adaptive GFEM have maximum number of degrees of freedom in each iteration ranging from 9 to 34.

r	Analytical solution (Kumar & Sujith, 1997)	linear h FEM (100e) ndof = 99	cubic h FEM (12e) ndof = 35	hierarchical p FEM (1e 33n) ndof = 31	Adaptive GFEM (after 3 iterations)	
	χ_r	error (%)	error (%)	error (%)	error (%)	ndof in iterations
1	2,978189	4,737 e-3	2,577 e-5	2,998 e-5	2,997 e-5	1x 1 dof + 2x 9 dof
2	6,203097	1,699 e-2	1,901 e-4	6,774 e-6	6,871 e-6	1x 2 dof + 2x 14 dof
3	9,371576	3,753 e-2	3,065 e-4	1,643 e-6	1,731 e-6	1x 3 dof + 2x 19 dof
4	12,526519	6,632 e-2	7,312 e-4	2,498 e-6	2,441 e-6	1x 4 dof + 2x 24 dof
5	15,676100	1,033 e-1	2,332 e-3	2,407 e-7	2,044 e-7	1x 5 dof + 2x 29 dof
6	18,823011	1,486 e-1	6,787 e-3	2,163 e-6	2,187 e-6	1x 6 dof + 2x 34 dof

Table 2. Results to free vibration of non-uniform fixed-fixed bar

The adaptive GFEM exhibits more accuracy than the *h*-versions of FEM with even less degrees of freedom. The precision reached for all calculated frequencies by the adaptive process is similar to the *p*-version of FEM with 31 degrees of freedom. The errors are greater than those from the uniform bars because the analytical vibration modes of non-uniform bars cannot be exactly represented by the trigonometric functions used as enrichment functions; however, the precision is acceptable for engineering applications. Each analysis by the adaptive GFEM is able to refine the target frequency until the exhaustion of the approximation capacity of the enriched subspace.

5.3 Uniform clamped-free beam

The free vibration of an uniform clamped-free beam (Fig. 15) in lateral motion, with length L, second moment of area I, elasticity modulus E, mass density ρ and cross section area A, is analyzed in order to demonstrate the application of the proposed method. The analytical natural frequencies (ω_r) are the roots of the equation:

$$\cos(\kappa_r L)\cosh(\kappa_r L)+1=0\,,\ r=1,2,\ldots \tag{55}$$

$$\kappa_r=\sqrt[4]{\frac{\omega_r{}^2\rho A}{EI}} \tag{56}$$

To check the efficiency of the proposed generalized C^1 element the results are compared to those obtained by the h and p versions of FEM and by the c refinement of CEM. The eigenvalue $\chi_r=\kappa_r.L$ is used to compare the analytical solution with the approximated ones.

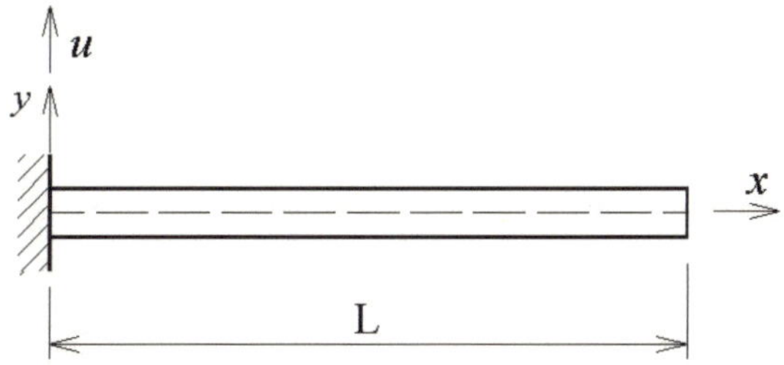

Fig. 15. Uniform clamped-free beam

a) *h* refinement

First this problem is analyzed by the h refinement of FEM, CEM and GFEM. A uniform mesh is used in all methods. Only one enrichment function is used in each element of the h-version of CEM. One level of enrichment ($n_l = 1$) is used in the h-version of GFEM.

The evolution of the relative error of the h refinements for the four earliest eigenvalues in logarithmic scale is presented in Figs. 16 and 17.

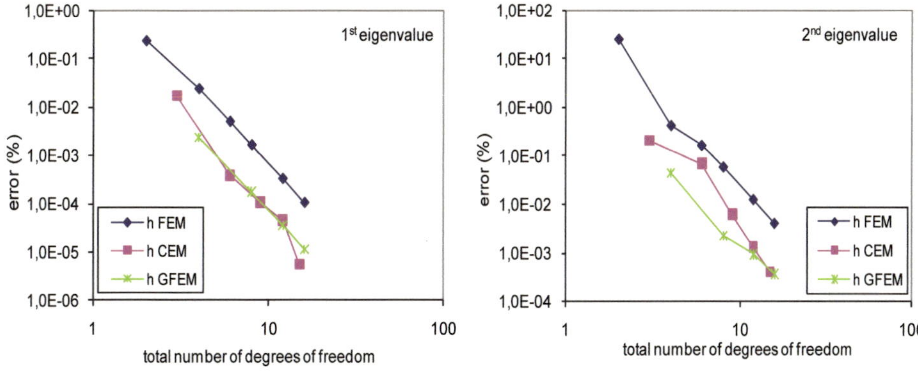

Fig. 16. Relative error (%) for the 1st and 2nd clamped-free beam eigenvalues - h refinements

The results show that the h-version of GFEM presents greater convergence rates than the h refinement of FEM. The results of h-version of CEM for the first two eigenvalues resemble those obtained by the h-version of GFEM. However the results of h-version of GFEM for higher eigenvalues are more accurate.

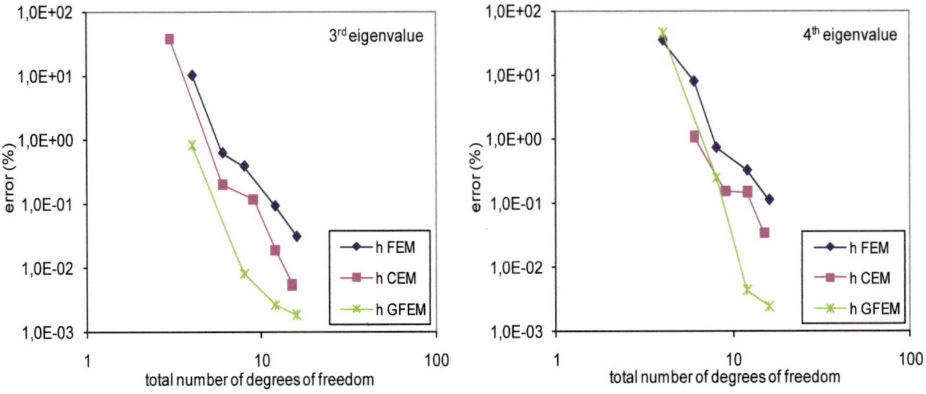

Fig. 17. Relative error (%) for the 3rd and 4th clamped-free beam eigenvalues - h refinements

b) p refinement

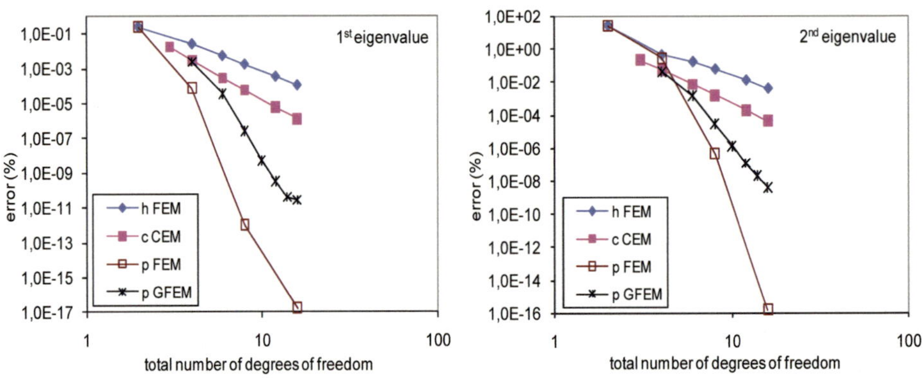

Fig. 18. Relative error (%) for the 1st and 2nd clamped-free beam eigenvalues - p refinements

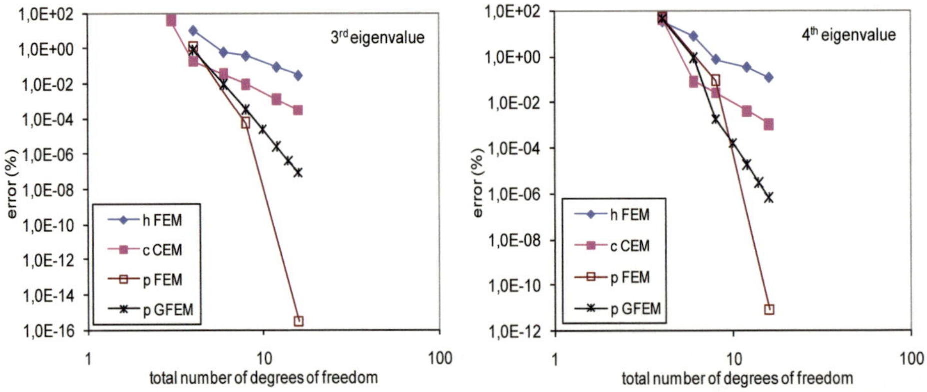

Fig. 19. Relative error (%) for the 3rd and 4th clamped-free beam eigenvalues - p refinements

The p refinement of GFEM is now compared to the hierarchical p-version of FEM and the c-version of CEM. The p-version of GFEM consists in a progressive increase of levels of enrichment. The relative error evolution of the p refinements for the first eight eigenvalues in logarithmic scale is presented in Figs. 18-21.
The results of the p-version of GFEM converge more rapidly than those obtained by the h-version of FEM and the c-version of CEM. The hierarchical p-version of FEM overcomes the precision and convergence rates obtained by the p-version of GFEM for the first six eigenvalues. However the p-version of GFEM is more precise for higher eigenvalues.

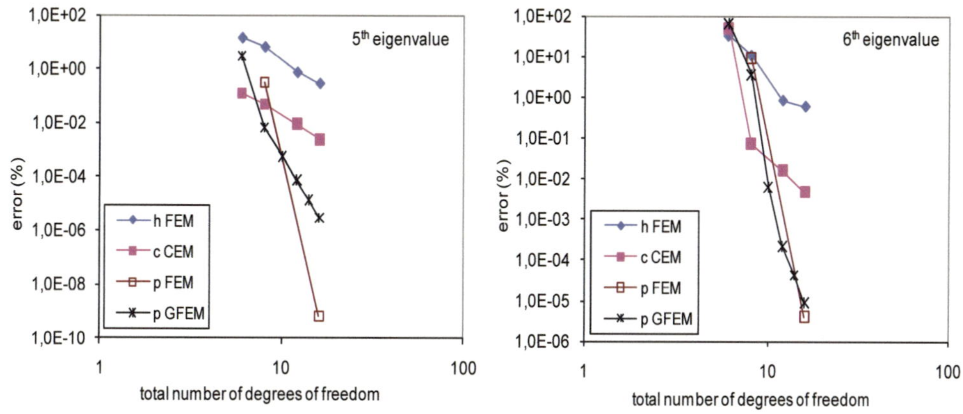

Fig. 20. Relative error (%) for the 5th and 6th clamped-free beam eigenvalues - p refinements

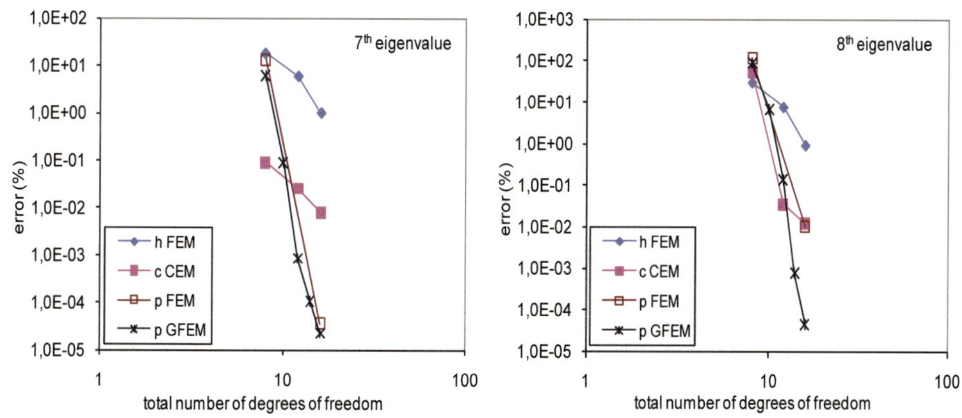

Fig. 21. Relative error (%) for the 7th and 8th clamped-free beam eigenvalues - p refinements

5.4 Seven bar truss

The free axial vibration of a truss formed by seven straight bars is analyzed to illustrate the application of the adaptive GFEM in structures formed by bars. This problem is proposed by Zeng (1998a) in order to check the CEM. The geometry of the truss is presented in Fig. 22. All bars in the truss have cross section area A = 0,001 m^2, mass density $\rho = 8000$ kg m^{-3} and elasticity modulus $E = 2{,}1\ 10^{11}$ N m^{-2}.

All analyses use seven element mesh, the minimum number of C^0 type elements necessary to represent the truss geometry. The linear FEM, the *c*-version of CEM and the *h*-version of GFEM with $n_l = 1$ and $\beta_1 = \pi$ are applied. Six analyses by the adaptive GFEM are performed in order to improve the accuracy of each of the first six natural frequencies. The frequencies obtained by each analysis are presented in Table 3.

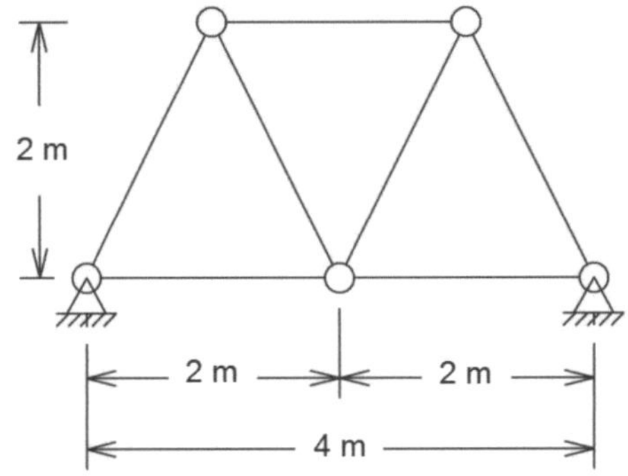

Fig. 22. Seven bar truss

	FEM (7e) ndof = 6	CEM (7e 1c) ndof = 13	CEM (7e 2c) ndof = 20	CEM (7e 5c) ndof = 41	*h* GFEM (7e) $n_l = 1$, $\beta_1 = \pi$ ndof = 34	Adaptive GFEM (7e 3i) 1x 6 dof + 2x 34 dof
i	ω_i (rad/s)	ω_i (rad/s)	ω_i (rad/s)	ω_i (rad/s)	ω_i (rad/s)	ω_i (rad/s)
1	1683,521413	1648,516148	1648,258910	1647,811939	1647,785439	1647,784428
2	1776,278483	1741,661466	1741,319206	1740,868779	1740,840343	1740,839797
3	3341,375203	3119,123132	3113,835167	3111,525066	3111,326191	3111,322715
4	5174,353866	4600,595156	4567,688849	4562,562379	4561,819768	4561,817307
5	5678,184561	4870,575795	4829,702095	4824,125665	4823,253509	4823,248678
6	8315,400602	7380,832845	7379,960217	7379,515018	7379,482416	7379,482322

Table 3. Results to free vibration of seven bar truss

The FEM solution is obtained with seven linear elements, that is, six effective degrees of freedom after introduction of boundary conditions. The *c*-version of the CEM solution is obtained with seven elements and one, two and five enrichment functions corresponding to six nodal degrees of freedom and seven, 14 and 35 field degrees of freedom respectively. All analyses by the adaptive GFEM have six degrees of freedom in the first iteration and 34 degrees of freedom in the following two.

This special case is not well suited to the *h*-version of FEM since it demands the adoption of restraints at each internal bar node in order to avoid modeling instability. The FEM analysis of this truss can be improved by applying bar elements of higher order (*p*-version) or beam elements. The results show that both the *c*-version of CEM and the adaptive GFEM converges to the same frequencies.

6. Conclusion

The main contribution of the present study consists in formulating and investigating the performance of the Generalized Finite Element Method (GFEM) for vibration analysis of framed structures. The proposed generalized C^0 and C^1 elements allow to apply boundary conditions as in the standard finite element procedure. In some of the recently proposed methods such as the modified CEM (Lu & Law, 2007), it is necessary to change the set of shape functions depending on the boundary conditions of the problem. In others, like the Partition of Unity used by De Bel et al. (2005) and Hazard & Bouillard (2007), the boundary conditions are applied under a penalty approach. In addition the GFEM enrichment functions require less effort to be obtained than the FEM shape functions in a hierarchical *p* refinement.

The GFEM results were compared with those obtained by the *h* and *p* versions of FEM and the *c*-version of CEM. The *h*-version of GFEM for C^0 elements exhibits more accuracy than *h* refinements of FEM and CEM. The C^1 *h*-version of GFEM presents more accurate results than *h*-version of FEM for all beam eigenvalues. The results of *h*-version of CEM for the first beam eigenvalues are alike those obtained by the *h*-version of GFEM. However the higher beam eigenvalues obtained by the *h*-version of GFEM are more precise.

The *p*-version of GFEM is quite accurate and its convergence rates are higher than those obtained by the *h*-versions of FEM and the *c*-version of CEM in free vibration analysis of bars and beams. It is observed however that the last eigenvalues obtained in each analysis of *p*-version of GFEM did not show good accuracy, but this deficiency is also found in the other enriched methods, such as the CEM. Although the *p* refinement of GFEM has produced excellent results and convergence rates, the adaptive GFEM exhibits special skills to reach accurately a specific frequency.

In most of the free vibration analysis it is virtually impossible to get all the natural frequencies. However, in practical analysis it is sufficient to work with a set of frequencies in a range (or band), or with those which have more significant participation in the analysis. The adaptive GFEM allows to find a specific natural frequency with accuracy and computational efficiency. It may be used in repeated analyses in order to find all the frequency in the range of interest.

In the C^0 adaptive GFEM, trigonometric enrichment functions depending on geometric and mechanical properties of the elements are added to the linear FEM shape functions by the partition of unity approach. This technique allows an accurate adaptive process that converges very fast and is able to refine the frequency related to a specific vibration mode. The adaptive GFEM shows fast convergence and remains stable after the third iteration with quite precise results for the target frequency.

The results have shown that the adaptive GFEM is more accurate than the *h* refinement of FEM and the *c* refinement of CEM, both employing a larger number of degrees of freedom. The adaptive GFEM in free vibration analysis of bars has exhibited similar accuracy, in some cases even better, to those obtained by the *p* refinement of FEM.

Thus the adaptive GFEM has shown to be efficient in the analysis of longitudinal vibration of bars, so that it can be applied, even for a coarse discretization scheme, in complex practical problems. Future research will extend this adaptive method to other structural elements like beams, plates and shells.

7. References

Abdelaziz, Y. & Hamouine, A. (2008). A survey of the extended finite element. *Computers and Structures*, Vol. 86, 1141-1151.

Arndt, M.; Machado, R.D. & Scremin A. (2010). An adaptive generalized finite element method applied to free vibration analysis of straight bars and trusses. *Journal of Sound and Vibration*, Vol. 329, 659–672.

Babuska, I. ; Banerjee, U. & Osborn, J.E. (2004). Generalized finite element methods: main ideas, results, and perspective. *Technical Report 04-08*, TICAM, University of Texas at Austin.

Carey, G.F. & Oden, J.T. (1983). *Finite elements: A second course*, Vol. 2, Prentice -Hall, New Jersey.

Carey, G.F. & Oden, J.T. (1984). *Finite elements: Computational aspects*, Vol. 3, Prentice -Hall, New Jersey.

Craig, R.R. (1981). *Structural dynamics: an introduction to computer methods*, John Wiley, New York.

De, S. & Bathe, K.J. (2001). The method of finite spheres with improved numerical integration. *Computers and Structures*, Vol. 79, 2183-2196.

De Bel, E.; Villon, P. & Bouillard, Ph. (2005). Forced vibrations in the medium frequency range solved by a partition of unity method with local information. *International Journal for Numerical Methods in Engineering*, Vol. 62, 1105-1126.

Duarte, C.A.; Babuska, I. & Oden, J.T. (2000). Generalized finite element methods for three-dimensional structural mechanics problems. *Computers and Structures*, Vol. 77, 215-232.

Duarte, C.A. & Kim, D-J. (2008). Analysis and applications of a generalized finite element method with global-local enrichment functions. *Computer Methods in Applied Mechanics and Engineering*, Vol. 197, 487-504.

Duarte, C.A. & Oden, J.T. (1996). An *h-p* adaptive method using clouds. *Computer Methods in Applied Mechanics and Engineering*, Vol. 139, 237-262.

Engels, R.C. (1992). Finite element modeling of dynamic behavior of some basic structural members. *Journal of Vibration and Acoustics*, Vol. 114, 3-9.

Ganesan, N. & Engels, R.C. (1992). Hierarchical Bernoulli-Euller beam finite elements. *Computers & Structures*, Vol. 43, No. 2, 297-304.

Gartner, J.R. & Olgac, N. (1982). Improved numerical computation of uniform beam characteristic values and characteristic functions. *Journal of Sound and Vibration*, Vol. 84, No. 4, 481-489.

Gracie, R.; Ventura, G. & Belytschko, T. (2007) A new fast finite element method for dislocations based on interior discontinuities. *International Journal for Numerical Methods in Engineering*, Vol. 69, 423-441.

Hazard, L. & Bouillard, P. (2007). Structural dynamics of viscoelastic sandwich plates by the partition of unity finite element method. *Computer Methods in Applied Mechanics and Engineering*, Vol. 196, 4101-4116.

Khoei, A.R.; Anahid, M. & Shahim, K. (2008). An extended arbitrary Lagrangian-Eulerian finite element method for large deformation of solid mechanics. *Finite Elements in Analysis and Design*, Vol. 44, 401-416.

Kumar, B.M. & Sujith, R.I. (1997). Exact solutions for the longitudinal vibration of non-uniform rods. *Journal of Sound and Vibration*, Vol. 207, 721-729.

Leung, A.Y.T. & Chan, J.K.W. (1998). Fourier *p*-element for the analysis of beams and plates. *Journal of Sound and Vibration*, Vol. 212, No. 1, 179-185.

Lu, Z.R. & Law, S.S. (2007). Discussions on composite element method for vibration analysis of structure. *Journal of Sound and Vibration*, Vol. 305, 357-361.

Melenk, J.M. & Babuska, I. (1996). The partition of unity finite element method: basic theory and applications. *Computer Methods in Applied Mechanics and Engineering*, Vol. 139, No. 1-4, 289-314.

Nistor, I.; Pantalé, O. & Caperaa, S. (2008). Numerical implementation of the extended finite element method for dynamic crack analysis. *Advances in Engineering Software*, Vol. 39, 573-587.

Oden, J.T.; Duarte, C.A.M. & Zienkiewicz, O.C. (1998). A new cloud-based hp finite element method. *Computer Methods in Applied Mechanics and Engineering*, Vol. 153, 117-126.

Schweitzer, M.A. (2009). An adaptive hp-version of the multilevel particle-partition of unity method. *Computer Methods in Applied Mechanics and Engineering*, Vol. 198, No. 13-14, 1260-1272.

Sukumar, N.; Chopp, D.L.; Moes, N. & Belytschko, T. (2001). Modeling holes and inclusions by level sets in the extended finite-element method. *Computer Methods in Applied Mechanics and Engineering*, Vol. 190, 6183-6200.

Sukumar, N.; Moes, N.; Moran, B. & Belytschko, T. (2000). Extended finite element method for three-dimensional crack modeling. *International Journal for Numerical Methods in Engineering*, Vol. 48, 1549-1570.

Strouboulis, T.; Babuska, I. & Copps, K. (2000). The design and analysis of the generalized finite element method. *Computer Methods in Applied Mechanics and Engineering*, Vol. 181, 43-69.

Strouboulis, T.; Babuska, I. & Hidajat, R. (2006a). The generalized finite element method for Helmholtz equation: theory, computation and open problems. *Computer Methods in Applied Mechanics and Engineering*, Vol. 195, 4711-4731.

Strouboulis, T.; Copps, K. & Babuska, I. (2001). The generalized finite element method. *Computer Methods in Applied Mechanics and Engineering*, Vol. 190, 4081-4193.

Strouboulis, T.; Hidajat, R. & Babuska, I. (2008). The generalized finite element method for Helmholtz equation. Part II: effect of choice of handbook functions, error due to absorbing boundary conditions and its assessment. *Computer Methods in Applied Mechanics and Engineering*, Vol. 197, 364-380.

Strouboulis, T.; Zhang, L.; Wang, D. & Babuska, I. (2006b). A posteriori error estimation for generalized finite element methods. *Computer Methods in Applied Mechanics and Engineering*, Vol. 195, 852-879.

Xiao, Q.Z. & Karihaloo, B.L. (2007). Implementation of hybrid crack element on a general finite element mesh and in combination with XFEM. *Computer Methods in Applied Mechanics and Engineering*, Vol. 196, 1864-1873.

Zeng, P. (1998a). Composite element method for vibration analysis of structures, part I: principle and C^0 element (bar). *Journal of Sound and Vibration*, Vol. 218, No. 4, 619-658.

Zeng, P. (1998b). Composite element method for vibration analysis of structures, part II: C^1 element (beam). *Journal of Sound and Vibration*, Vol. 218, No. 4, 659-696.

Zeng, P. (1998c). Introduction to composite element method for structural analysis in engineering. *Key Engineering Materials*, Vol. 145-149, 185-190.

11

Dynamic Characterization of Ancient Masonry Structures

Annamaria Pau and Fabrizio Vestroni
Università di Roma La Sapienza
Italy

1. Introduction

The analysis of the dynamic response induced in a structure by ambient vibrations is important for two reasons. On the one hand, the environmental impact of vibrations is a common cause for concern in many cities throughout the world on account of both the consequences of such vibrations on buildings, especially those in structurally weak conditions, and on people in terms of annoyance. On the other hand, the measured data contain information on the dynamic characteristics of the structures, such as modal parameters (frequencies, damping ratios and mode shapes). Several techniques of experimental modal analysis are nowadays well established and make it possible to extract modal parameters from the measurements of the dynamical response. Books on this topic are by (Bendat & Piersol, 1980; Ewins, 2000; Juang, 1994; Maia & Silva, 1997; Van Overschee & De Moor, 1996). A knowledge of modal parameters is a basic step for updating a finite element model which not only replicates the real response (Friswell & Mottershead, 1995), but also enables to build damage identification procedures based on the variation of the structural response (Morassi & Vestroni, 2009; Vestroni & Capecchi, 1996). Furthermore, periodical repetition of the measurement process over time, together with observation of possible variation of modal parameters, forms the basis for a structural health monitoring procedure (Farrar et al., 2001). These issues are especially important for ancient buildings, marked by complex geometry, heterogeneous materials and in poor conditions, which are often very sensitive to deterioration.
Experimental modal analysis usually deals with frequency response functions (FRF) in the frequency domain or impulse response functions in the time domain and requires that the response to an assigned input is measured. In civil structures, the system should be excited with heavy shakers (De Sortis et al., 2005), which makes these tests expensive and often impracticable, especially in the case of very large structures. The measurement of the ambient vibration response, which is the response to an unknown input due to natural and human actions (for instance wind, microtremors, traffic), makes it possible to overcome the difficulties that often arise when artificial excitation is used. The drawbacks in this kind of measurements are that there is the need to deal with signals with small amplitude and, furthermore, the hypothesis that the spectrum of the forcing function is approximately flat in the frequency band where the modes are to be estimated, which can not be fully experimentally proved, must be accepted. Of the several ambient vibration modal identification techniques, three will be described in this chapter: peak picking from the power spectral densities (PP) (Bendat

& Piersol, 1980), singular value decomposition (SVD) (Brincker et al., 2001) and stochastic subspace identification (SSI) (Peeters, 2000; Van Overschee & De Moor, 1993; 1994; 1996). The mentioned techniques have been successfully used for the modal identification of numerous civil structures, such as bridges (Ren et al., 2004) or tall buildings (Brownjohn, 2003), but less frequently applied to historical structures and monuments (Gentile & Saisi, 2007; Pau & Vestroni, 2008; 2010). This chapter aims to describe their application to selected cases of historical masonry structures in Italy.

Of late, some of the most important monuments in Rome have been investigated because of the proximity of these structures to a new underground line that is at present under construction. These tests include the recording of the ambient vibration response. The Colosseum, the Basilica of Maxentius and the Trajan Column are some of the investigated monuments. The availability of such data enables a dynamic characterization and identification of modal parameters of the structures, which presents a challenging task in such large and geometrically complex monuments, built with heterogeneous materials. Parts of the results of these experimental tests are reported in the works by (Pau & Vestroni, 2008; 2010). Here, the case of the Trajan Column will be discussed in detail together with another application to a railway masonry bridge of the 19th century. For each of these cases, a comparison between experimental and numerical modal parameters is discussed, in the perspective of the evaluation and updating of the finite element models according to the measured behavior. This comparison may enable the identification of the possible causes of discrepancies between predicted and measured properties. In particular, the information obtained may relate to the current state of a structure: lower natural frequencies than those predicted by the finite element model may indicate deterioration in the stiffness of the structure and anomalous mode shapes may point to the independent motion of structural parts due to major cracks. In many cases, notwithstanding the severe simplifications, mainly regarding the material behavior introduced in the numerical modeling, the comparison between numerical and experimental frequencies and mode shapes provides sufficient agreement, after an adjustment of the mechanical characteristics to tune the two models. This adjustment has shown to have a significant mechanical meaning indicating the effective presence of cracks and discontinuities (Pau & Vestroni, 2010).

2. Ambient vibration modal identification techniques

Very often, when dealing with large engineering structures such as building or bridges, it is impractical to measure the response to an *ad hoc* and controlled artificial excitation for different reasons, such as costs concern or even the unwanted possibility of activating nonlinear phenomena.

Reasonable estimates of modal properties can be obtained from an output-only analysis of the ambient vibration response to the natural dynamic environment. This excitation, which is random in its nature, is due to various human and artificial sources, such as traffic, wind and microtremors. When dealing with output-only analysis of the vibration response, it is fundamental to cope with signals with small amplitude and contaminated by noise. Although the input is unknown, which prevents from measuring the proper FRF, a hypothesis that the spectrum of the forcing function is flat in the frequency band where the modes are to be estimated must be made, which can only be partially proved from experiments. This paragraph describes three techniques of modal identification, which are important for different reasons. The peak picking from the power spectral densities is a frequency domain based technique and is important for historical reasons, since it was one of the first output-only modal identification techniques to be presented in the late '70s (Bendat & Piersol, 1980), and its simplicity. The singular value decomposition is an extension of the peak picking (Brincker et al., 2001). With respect to the peak picking, it enables to deal better with close frequencies and damped modes. Its advantage over other recent techniques consists mainly in its preserving the user's understanding of the data he is dealing with through a frequency approach. In the early '90s, the stochastic subspace identification, which is a time domain technique, was described in research papers (Van Overschee & De Moor, 1993; 1994) and in the fundamental book by (Van Overschee & De Moor, 1996). Today, the SSI is one of the most widespread techniques for output-only modal identification and is implemented not only in commercial softwares for data analysis (Artemis) but also in Matlab routines and freely available software (http://homes.esat.kuleuven.be/ smc/sysid/software/).

2.1 Peak picking

This method is very often used for its simplicity in analysing the ambient vibration response, when the input is unknown (Bendat & Piersol, 1980). The ambient vibration response of a structure cannot be predicted by deterministic models, within reasonable error. Each experiment produces a random time-history that represents only one physical realization of what might occur. In general, the response $x(t)$ of the structure to ambient excitation is recorded for a very long time, even for hours, which enables to cut the random process $x(t)$ into a collection of subregistrations $x_k(t)$ which describe the phenomenon. The Fourier Transforms of the kth subregistrations of two random processes $x_k(t)$ and $y_k(t)$ are respectively:

$$X_k(f,T) = \int_0^T x_k(t) \exp^{-i2\pi ft} \, dt \tag{1}$$

$$Y_k(f,T) = \int_0^T y_k(t) \exp^{-i2\pi ft} \, dt. \tag{2}$$

The auto (or power) spectral density (PSD) and cross-spectral density (CSD) and related coherence function between the two random processes are respectively:

$$S_{xx}(f) = \lim_{T\to\infty} \frac{1}{T} E[\mid X_k(f,T) \mid^2] \tag{3}$$

$$S_{xy}(f) = \lim_{T\to\infty} \frac{1}{T} E[X_k^*(f,T) Y_k(f,T)] \tag{4}$$

$$\gamma_{xy}(f) = \frac{\mid S_{xy}(f) \mid^2}{S_{xx}(f) S_{yy}(f)} \tag{5}$$

where the symbol $E[.]$ indicates an averaging operation over the index k and the asterisk denotes complex conjugate.

Let us now assume that $x(t)$ is the input and $y(t)$ is the output. The auto-spectral and cross-spectral density functions satisfy the important formulae:

$$S_{yy}(f) = |H_{xy}(f)|^2 S_{xx}(f) \quad S_{xy}(f) = H_{xy}(f) S_{xx}(f) \tag{6}$$

where $H_{xy}(f)$ is the frequency response function. The simple peak picking method is based on the fact that the autospectrum (6_1), at any response point, reaches a maximum either when the excitation spectrum peaks or the frequency response function peaks. To distinguish between peaks that are due to vibration modes as opposed to those in the input spectrum, a couple of criteria can be used. The former concerns the fact that in a lightly damped structure, two points must oscillate in-phase or out-of-phase. Then, the cross spectrum (6_2) between the two responses provides this information, which can be used to distinguish whether the peaks are due to vibration modes or not. The second criterion uses the coherence function (5), which tends to peak at the natural frequencies, as the signal-to-noise ratio is maximised at these frequencies.

2.2 Singular value decomposition

The second method referred to also relies only on the response to ambient excitations (output only). The method is based on the singular value decomposition of the response spectral matrix (Brincker et al., 2001), exploiting the relationship:

$$\mathbf{S}yy\,(\omega) = \mathbf{H}^*\,(\omega)\,\mathbf{S}xx\,(\omega)\,\mathbf{H}^T\,(\omega) \tag{7}$$

where $\mathbf{S}xx(\omega)$ ($r \times r$, r number of inputs) and $\mathbf{S}yy(\omega)$ ($m \times m$, m number of measured responses) are the input and output power spectral density matrices, respectively, and $\mathbf{H}(\omega)$ is the frequency response function matrix ($m \times r$). Supposing the inputs at the different points are completely uncorrelated and white noise, $\mathbf{S}xx$ is a constant diagonal matrix, independent of ω. Thus:

$$\mathbf{S}yy\,(\omega) = S\,\mathbf{H}\,(\omega)\,\mathbf{H}^T\,(\omega) \tag{8}$$

whose term jk can be written, by omitting the constant S, as:

$$Syy_{jk}\,(\omega) = \sum_{r=1}^{r}\left(\sum_{p=1}^{n}\frac{\phi_{jp}\phi_{rp}}{\bar{\lambda}_p^2-\omega^2}\right)\left(\sum_{q=1}^{n}\frac{\phi_{kq}\phi_{rq}}{\lambda_q^2-\omega^2}\right). \tag{9}$$

In the neighbourhood of the ith resonance, the previous equation can be approximated by:

$$Syy_{jk}\,(\omega) \cong \sum_{r=1}^{r}\frac{\phi_{ji}\phi_{ri}}{\bar{\lambda}_i^2-\omega^2}\frac{\phi_{ki}\phi_{ri}}{\lambda_i^2-\omega^2} = \frac{\phi_{ji}\phi_{ki}}{\left(\bar{\lambda}_i^2-\omega^2\right)\left(\lambda_i^2-\omega^2\right)}\sum_{r=1}^{r}\phi_{ri}^2. \tag{10}$$

By ignoring the constant $\sum_{r=1}^{r}\phi_{ri}^2$, Syy can thus be expressed as the product of the three matrices:

$$\mathbf{S}yy\,(\omega) = \mathbf{\Phi}\mathbf{\Lambda}_i\mathbf{\Phi}^T \tag{11}$$

which represents a singular value decomposition of the matrix $\mathbf{S}yy$, where:

$$\mathbf{\Lambda}_i = \begin{bmatrix} \frac{1}{\left(\lambda_i^2-\omega^2\right)\left(\bar{\lambda}_i^2-\omega^2\right)} & 0... & 0 \\ 0 & 0... & 0 \\ \vdots & \vdots & \vdots \\ 0 & 0... & 0 \end{bmatrix}. \tag{12}$$

This is valid in the neighbourhood of every natural frequency of the system, that hence emerges as a peak of the first singular value. The first column of the matrix $\mathbf{\Phi}$ contains the first singular vector, which, in the neighborhood of the ith resonance, coincides with the ith eigenvector. This occurs at each resonance, when the prevailing contribution is given by the related mode. This procedure has recently had great diffusion mainly in *in situ* experimental tests and has also been implemented in commercial codes.

2.3 Stochastic subspace identification

The stochastic subspace identification belongs to the wide class of time domain methods. The continuous-time dynamics of a discrete or a discretized (in space) mechanical system in the state-space can be written as:

$$\dot{\mathbf{x}}(t) = \mathbf{A}_c\mathbf{x}(t) + \mathbf{B}_c\mathbf{f}(t) \tag{13}$$

which is a representation deriving from the control theory (Juang, 1994). In this relationship, $\mathbf{x}(t) = [\mathbf{u}(t)^T\ \dot{\mathbf{u}}(t)^T] \in \mathbb{R}^{2n}$ is the state vector of the process. This vector contains the $2n$ states of the system, where $\mathbf{u}(t)$ and $\dot{\mathbf{u}}(t)$ are respectively the displacement and velocity vectors and n is the number of degrees-of-freedom. $\mathbf{A}_c \in \mathbb{R}^{2n\times 2n}$ is the continuous-time state matrix, which is related to the classical matrices of mass $\mathbf{M}$, damping $\mathbf{C}_d$ and stiffness $\mathbf{K}$ by:

$$\mathbf{A}_c = \begin{bmatrix} \mathbf{0} & \mathbf{I} \\ -\mathbf{M}^{-1}\mathbf{K} & -\mathbf{M}^{-1}\mathbf{C}_d \end{bmatrix}, \tag{14}$$

$\mathbf{f}(t) \in \mathbb{R}^n$ is the load vector and $\mathbf{B}_c \in \mathbb{R}^{2n\times n}$ is the system control influence coefficient matrix:

$$\mathbf{B}_c = \begin{bmatrix} \mathbf{0} \\ \mathbf{M}^{-1} \end{bmatrix}. \tag{15}$$

In a vibration experiment, only a subset l of the responses at the n degrees-of-freedom can be measured. The vector of the measured outputs $\mathbf{y}(t) \in \mathbb{R}^l$ is written as: $\mathbf{y}(t) = \mathbf{C}_a\ddot{\mathbf{u}}(t) + \mathbf{C}_v\dot{\mathbf{u}}(t) + \mathbf{C}_u\mathbf{u}(t)$, where $\mathbf{C}_a$, $\mathbf{C}_v$ and $\mathbf{C}_u$ are output location matrices for accelerations, velocities and displacements respectively, which are matrices of zeros and ones made up to select the measured degrees of freedom. The vector $\mathbf{y}(t)$ can be written as:

$$\mathbf{y}(t) = \mathbf{C}_c\mathbf{x}(t) + \mathbf{D}_c\mathbf{f}(t) \tag{16}$$

where $\mathbf{C}_c \in \mathbb{R}^{l\times 2n}$ is the output matrix and $\mathbf{D}_c \in \mathbb{R}^{l\times n}$ is the direct transmission matrix:

$$\mathbf{C}_c = [\mathbf{C}_u - \mathbf{C}_a\mathbf{M}^{-1}\mathbf{K} \quad \mathbf{C}_v - \mathbf{C}_a\mathbf{M}^{-1}\mathbf{C}_u] \quad \text{and} \quad \mathbf{D}_c = \mathbf{C}_a\mathbf{M}^{-1}. \tag{17}$$

Then, in conclusion, the continuous-time state-space model can be written as:

$$\begin{cases} \dot{\mathbf{x}}(t) = \mathbf{A}_c\mathbf{x}(t) + \mathbf{B}_c\mathbf{f}(t) \\ \mathbf{y}(t) = \mathbf{C}_c\mathbf{x}(t) + \mathbf{D}_c\mathbf{f}(t) \end{cases}. \tag{18}$$

It can be shown that the eigenvalues $\mathbf{\Lambda}_c$ and eigenvectors $\mathbf{\Psi}$ of the continuous state-space matrix $\mathbf{A}_c$ which solve the eigenvalue problem $\mathbf{A}_c\mathbf{\Psi} = \mathbf{\Psi}\mathbf{\Lambda}_c$ contain the eigenvalues $\mathbf{\Lambda}$ and eigenvectors $\mathbf{\Theta}$ of the original second-order system:

$$\mathbf{\Lambda}_c = \begin{pmatrix} \mathbf{\Lambda} & \mathbf{0} \\ \mathbf{0} & \mathbf{\Lambda}^* \end{pmatrix}, \quad \mathbf{\Psi} = \begin{pmatrix} \mathbf{\Theta} & \mathbf{\Theta}^* \\ \mathbf{\Theta}\mathbf{\Lambda} & \mathbf{\Theta}^*\mathbf{\Lambda}^* \end{pmatrix}. \tag{19}$$

In practice, experimental data are discrete. Therefore, the model of equation (18) has to be converted to discrete time, in order to fit the models to measurements. The continuous-time equations are discretized and solved at all the discrete time instants $t_k = k\Delta t, k \in \mathbb{N}$, where Δt is the sampling period. Let us suppose to focus the analysis on time-invariant state-space models. These deterministic-stochastic systems, excited both by deterministic and random actions, are described by the following set of difference equations:

$$\begin{cases} \mathbf{x}_{k+1} = \mathbf{A}\mathbf{x}_k + \mathbf{B}\mathbf{f}_k + \mathbf{w}_k \\ \mathbf{y}_k = \mathbf{C}\mathbf{x}_k + \mathbf{D}\mathbf{f}_k + \mathbf{v}_k \end{cases} \tag{20}$$

The vector $\mathbf{x}_k \in \mathbb{R}^{2n}$ is defined as the state vector of the process at the discrete time instant k. This vector contains the numerical values of the $2n$ states of the system. When dealing with mechanical systems, the state vector is $\mathbf{x}_k = [\mathbf{u}_k^T \ \dot{\mathbf{u}}_k^T] \in \mathbb{R}^{2m}$, $\mathbf{f}_k \in \mathbb{R}^n$ and $\mathbf{y}_k \in \mathbb{R}^l$ are respectively the experimental measurements at time instant k of the n inputs and l outputs. $\mathbf{w}_k \in \mathbb{R}^{2n}$ and $\mathbf{v}_k \in \mathbb{R}^l$ are respectively process and measurement noise vectors, which are unmeasurable quantities. The former is due to model inaccuracies, the latter due to measurement inaccuracies. **A** is the discrete state matrix, **B** is the discrete input matrix, **C** is the discrete output matrix and **D** is the direct transmission matrix. They are related to their continuous-time counterparts by the relationships:

$$\begin{array}{ll} \mathbf{A} = e^{\mathbf{A}_c \Delta t} & \mathbf{B} = \left(\int_0^{\Delta t} e^{\mathbf{A}_c \tau} d\tau \right) \mathbf{B}_c = (\mathbf{A} - \mathbf{I}) \mathbf{A}_c^{-1} \mathbf{B}_c \\ \mathbf{C} = \mathbf{C}_c & \mathbf{D} = \mathbf{D}_c. \end{array} \tag{21}$$

These well-established relationships can be found in the literature (Juang, 1994). The hypothesis:

$$\mathrm{E}\left[\begin{pmatrix} \mathbf{w}_p \\ \mathbf{v}_p \end{pmatrix} \begin{pmatrix} \mathbf{w}_q^T & \mathbf{v}_q^T \end{pmatrix} \right] = \begin{pmatrix} \mathbf{Q} & \mathbf{S} \\ \mathbf{S}^T & \mathbf{R} \end{pmatrix} \delta_{pq} \geqq 0 \tag{22}$$

is further added, where E[.] indicates the expected value and δ_{pq} is the Kronecker delta. The matrices $\mathbf{Q} \in \mathbb{R}^{2n \times 2n}$, $\mathbf{S} \in \mathbb{R}^{2n \times l}$ and $\mathbf{R} \in \mathbb{R}^{l \times l}$ are the covariance matrices of the noise terms $\mathbf{w}_k$ and $\mathbf{v}_k$, which are supposed to be independent of each other and both with zero mean. It must be remarked that in output-only modal identification, the input sequence $\mathbf{f}_k$ is unmeasured and only the response $\mathbf{y}_k$ is known. Hence, it is impossible to distinguish the input term $\mathbf{f}_k$ from the noise terms $\mathbf{w}_k$ and $\mathbf{v}_k$ in equation (20). This results in a purely stochastic system:

$$\begin{cases} \mathbf{x}_{k+1} = \mathbf{A}\mathbf{x}_k + \mathbf{w}_k \\ \mathbf{y}_k = \mathbf{C}\mathbf{x}_k + \mathbf{v}_k \end{cases} . \tag{23}$$

In equation (23), the white noise assumption on the terms $\mathbf{w}_k$ and $\mathbf{v}_k$ cannot be omitted. If the input contains some dominant frequency components, they will not be separated from the eigenfrequencies of the system. The stochastic subspace identification then moves from equations (23) to estimate the state-space matrices **A** and **C** from the measured output $\mathbf{y}_k$, with $k = 1, 2, \ldots, N$ and $N \longrightarrow \infty$. The estimate of state-space matrices can be performed by different algorithms. In the applications, the procedure described in the work by (Van Overschee & De Moor, 1994) is used. In short, this procedure is based on selected theorems of linear algebra illustrated in (Van Overschee & De Moor, 1994; 1996), which demonstrate that the state space matrices can be calculated from the knowledge of the block Hankel matrix. This matrix is obtained by casting the finite dimensional output vector $\mathbf{y}_k$ into the columns of a semi infinite $2i \times j$ matrix:

$$U_{0|2i-1} = \left(\begin{array}{ccccc} y_0 & y_1 & y_2 & \cdots & y_{j-1} \\ y_1 & y_2 & y_3 & \cdots & y_j \\ \cdots & \cdots & \cdots & \cdots & \cdots \\ y_{i-1} & y_i & y_{i+1} & \cdots & y_{i+j-2} \\ \hline y_i & y_{i+1} & y_{i+2} & \cdots & y_{i+j-1} \\ y_{i+1} & y_{i+2} & y_{i+3} & \cdots & y_{i+j} \\ \cdots & \cdots & \cdots & \cdots & \cdots \\ y_{2i-1} & y_{2i} & y_{2i+1} & \cdots & u_{2i+j-2} \end{array} \right) \tag{24}$$

where the horizontal line divides *past* inputs from *future* inputs. Once the matrix $\mathbf{A}$ is known, the natural frequencies and mode shapes can be evaluated. In fact, as shown in (Peeters, 2000), the eigenvalues $\mathbf{\Lambda}_d$ and eigenvectors of the discrete state space matrix are related to their continuous counterparts by the relationships:

$$\mathbf{A} = e^{\mathbf{A}_c \Delta t} = e^{\mathbf{\Psi}\mathbf{\Lambda}_c\mathbf{\Psi}^{-1}\Delta t} = \mathbf{\Psi} e^{\mathbf{\Lambda}_c \Delta t}\mathbf{\Psi}^{-1} = \mathbf{\Psi}\mathbf{\Lambda}_d\mathbf{\Psi}^{-1}. \tag{25}$$

That is, the eigenvectors are the same for both systems, while the discrete eigenvalues μ_i are related to the continuous eigenvalues λ_i by:

$$\lambda_i = \frac{\ln(\mu_i)}{\Delta t}. \tag{26}$$

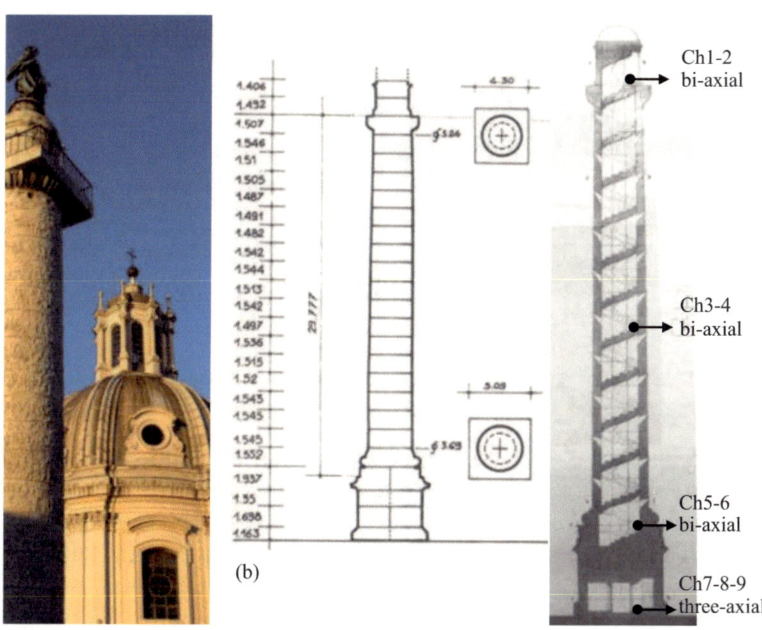

Fig. 1. A view of the Trajan column (a), its survey (b) and accelerometer setup (c).

3. Applications

3.1 The Trajan Column

The Trajan Column is a honorary monument that, in 113 A.D., was dedicated to Trajan the emperor to celebrate his triumph over the Dacians, the inhabitants of the present Romania. Over the surface of the column, a helical bas-relief depicts the story of Trajan's victory. The monument consists of a marble column about 30m tall, with a circular section having an external diameter of 3.55m, placed over a square-section pedestal 6.23m high (Figures 1 a-b). It represents a peculiarity in archaeological heritage because of its slenderness. The column is formed by nineteen cylindrical elements, dug-out to form an internal helical staircase going to the top level. The helical geometry is perturbed by tiny windows along the external surface. The response of this structure was measured by one three-axial set of accelerometers at the base and three biaxial horizontal sets at the upper levels. The measurement points with their related channels are reported in Figure 1c. The recordings were performed at a sampling frequency of $f_s = 300Hz$ for a duration of about 2 hours.

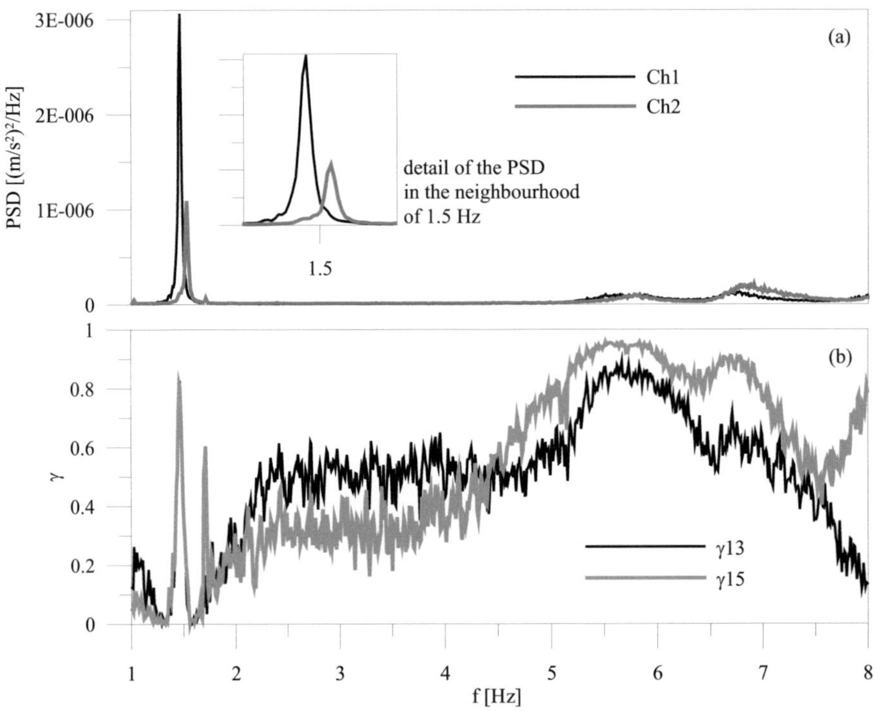

Fig. 2. Power Spectral Densities of accelerations measured on the top of the column (a) and coherence function among channel 1 and channels 3 and 5(b).

As a first step, the power spectral densities of the accelerations are observed. Figure 2a reports the PSDs of the two measurement points on the top of the column, in the frequency band where natural frequencies are expected. Two peaks in the neighborhood of 1.5 Hz emerge quite clearly (see details in Figure 2), while two other peaks appear in the range 5-8 Hz, but with strong damping. In such an unclear situation, the observation of the coherence (Figure 2 b) may be of some help. The coherence peaks are at the same frequencies as those observed

in the PSDs, suggesting that all these four peaks may in fact be representative of natural frequencies, as reported in Table 1.
Analogous results can be obtained from the singular value decomposition. Figure 3 reports the first singular value of the spectral matrix as a function of frequency, showing the two close peaks related to the first and second natural frequencies and the other peaks related to the third and fourth frequencies. The identified frequencies coincide with those detected with the peak picking, as reported in Table 1.

	f_1	f_2	f_3	f_4
PP, SVD	1.46	1.53	5.83	6.83
SSI	1.45	1.52	5.69	6.56
FE	3.14	3.32	15.68	17.52

Table 1. Experimental and numerical natural frequencies [Hz] of the Trajan column

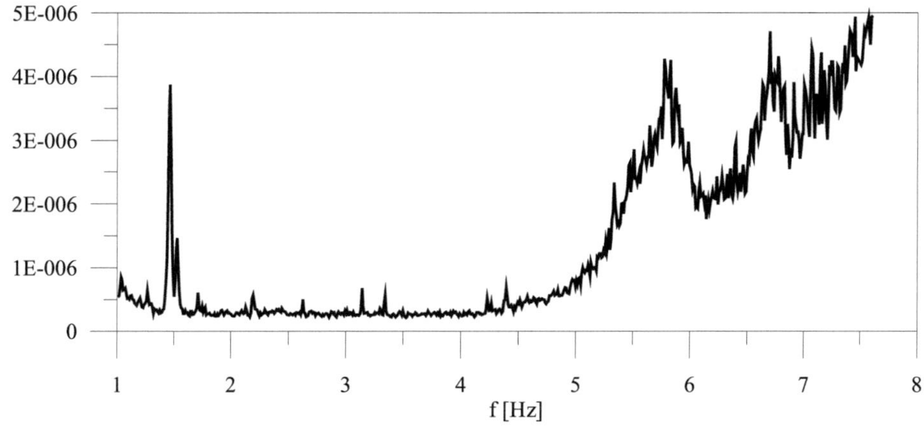

Fig. 3. First singular value of the spectral matrix as a function of frequency.

As a final step, the data are analyzed following the stochastic subspace decomposition. In this case, the evaluation of the model order is fundamental. A good model for modal analysis applications can be obtained by constructing stabilization diagrams, that is, by evaluating a set of models with different orders (Peeters, 2000). A criterion to state when an eigenvalue is stable must be defined; for instance, eigenvalues do not have to change more than 1% when the model order is increased. When an eigenvalue satisfies this stability criterion, its value is determined. Figure 4 shows the eigenvalue stabilization when increasing the model order and enables to define the natural frequencies that are reported in Table 1. The difficulties in the interpretation of the third and fourth frequencies, and related mode shapes, remain, in fact these frequencies stabilize for higher model order than the first and second. These difficulties, which concern in fact all the employed methods, are not surprising. In fact, the third and fourth frequencies are close to 6 Hz, which is the cutoff frequency of the ground, as was observed in other experimental tests on the Colosseum and Basilica of Maxentius (Pau & Vestroni, 2008; 2010). The ground attenuates frequencies which are smaller than 6 Hz and guarantees a white-noise spectrum in the frequency band 0-6 Hz. Therefore, for frequencies higher than 6 Hz, the hypotheses on the input, on which the present modal identification methods are based, are not satisfied.

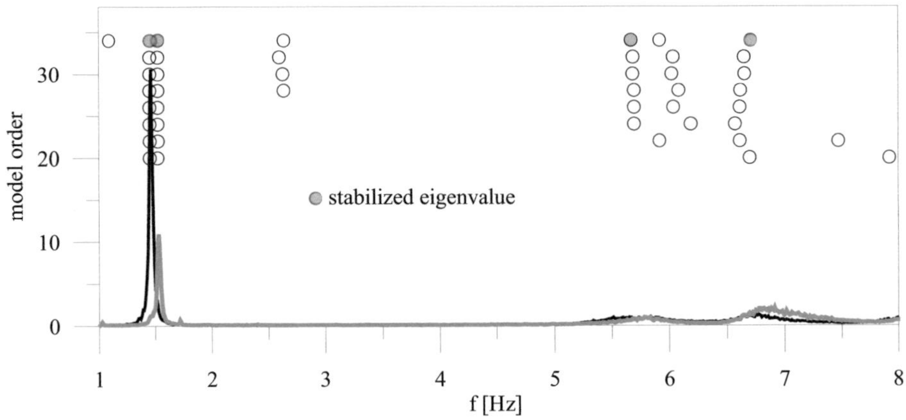

Fig. 4. Eigenvalue stabilization diagram for the Trajan column.

A comparison between the experimental mode shapes is now performed. The modal assurance criterion (MAC), which is a scalar product between the two mode shape vectors under consideration, normalized to the product of the moduli, is a measure of the agreement between two mode shapes. A comparison shows that the differences between the three techniques are very small for the first two modes (MAC$\simeq$ 0.99), but increase for the third and fourth modes (MAC $\simeq$ 0.8), which are identified with great difficulties in all the cases because of the strong damping. However, the results obtained by SVD and SSI agree each other better than those obtained by PP.

As regards the shapes of the modes, the mode shape pairs 1-3 and 2-4 strongly resemble those of a cantilever beam, as shown in Figure 5. For the sake of brevity, this Figure shows only the mode shapes determined by SSI method. Furthermore, the first two modes are nearly contained respectively into the two planes parallel to the base, while the third and fourth mode shapes are contained in planes which are not coincident with the measurement planes. This is also evident from Figure 2, as the peaks related to the first and second frequencies are present only in one of the two spectra, while the peaks related to the third and fourth are present in both the spectra. This experimental result was verified by a laboratory experiment on an axisymmetric clamped cylinder, a pipe with vertical axis, which was tested both in its nominally perfect and perturbed configuration. Figure 6 reports the projection onto the horizontal plane of the vertical planes containing the mode pairs 1-2, 3-4 and 5-6. Different colors relate to different test conditions. The tests show that even in nominally perfect conditions, the planes containing the mode shape pairs corresponding to the clamped beam can be different for each pair, especially for higher modes. Furthermore, each pair is contained in planes which only slightly deviate from orthogonality, consistent with the orthogonality of modes. These results can be ascribed to imperfections in geometry, which cause a deviation from perfect axisymmetry.

In conclusion, a comparison with the results provided by a numerical (FE) model is performed. The column is simply represented as a cantilever beam with varying section. In this model, the Young's modulus E and mass density ρ come from literature values determined by static tests on cores bored into the solid material. The natural frequencies obtained are reported in Table 1. These values are much higher than the experimental ones, and the reason is that the material parameters of the solid material are not representative of the behavior of the assembled system, where the interactions among the blocks have considerable

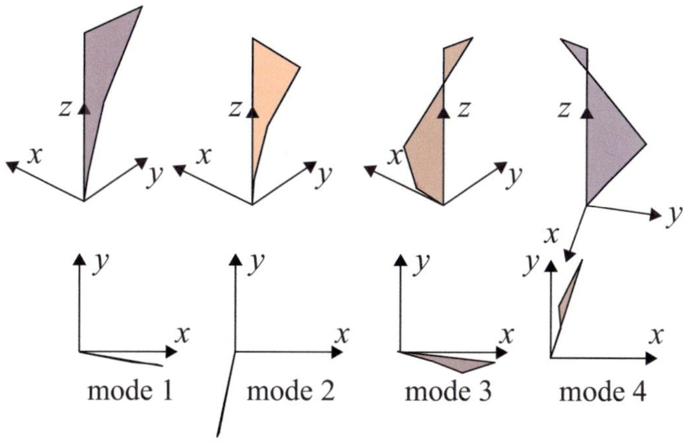

Fig. 5. Experimental mode shapes.

influence. A similar result was found by the authors in the analysis of the response of the Colosseum (Pau & Vestroni, 2008), where a reduction in the elastic modulus based on measurements of the wave propagation velocity in structural parts including joints brought the analytical and experimental results into satisfactory agreement. Here also, the reduction of the ratio E/ρ brings numerical and experimental results into satisfactory agreement. As regards the mode shapes, Table 2 shows that, whichever modal analysis method is used, the experimental modes 1 and 2 agree very well with the numerical ones. By contrast, for the pair 3-4 the mode shapes obtained by the SSI method have better quality.

	1	2	3	4
PP–FE	1.00	1.00	0.47	0.36
SVD–FE	0.98	0.95	0.62	0.55
SSI–FE	0.98	0.98	0.84	0.79

Table 2. MAC between experimental and numerical modes for the Trajan column

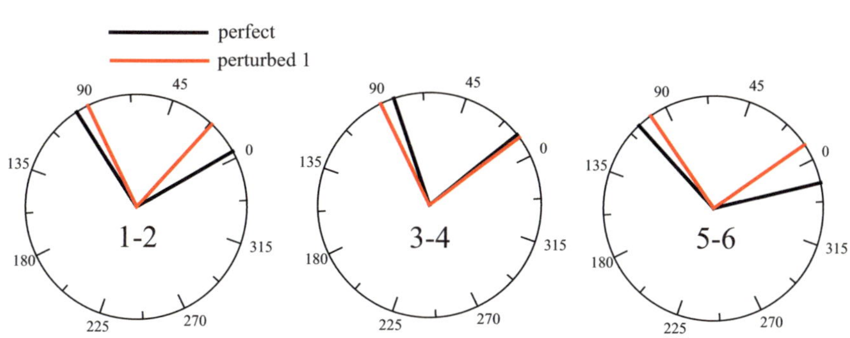

Fig. 6. Top view of the experimental mode shapes of a clamped pipe.

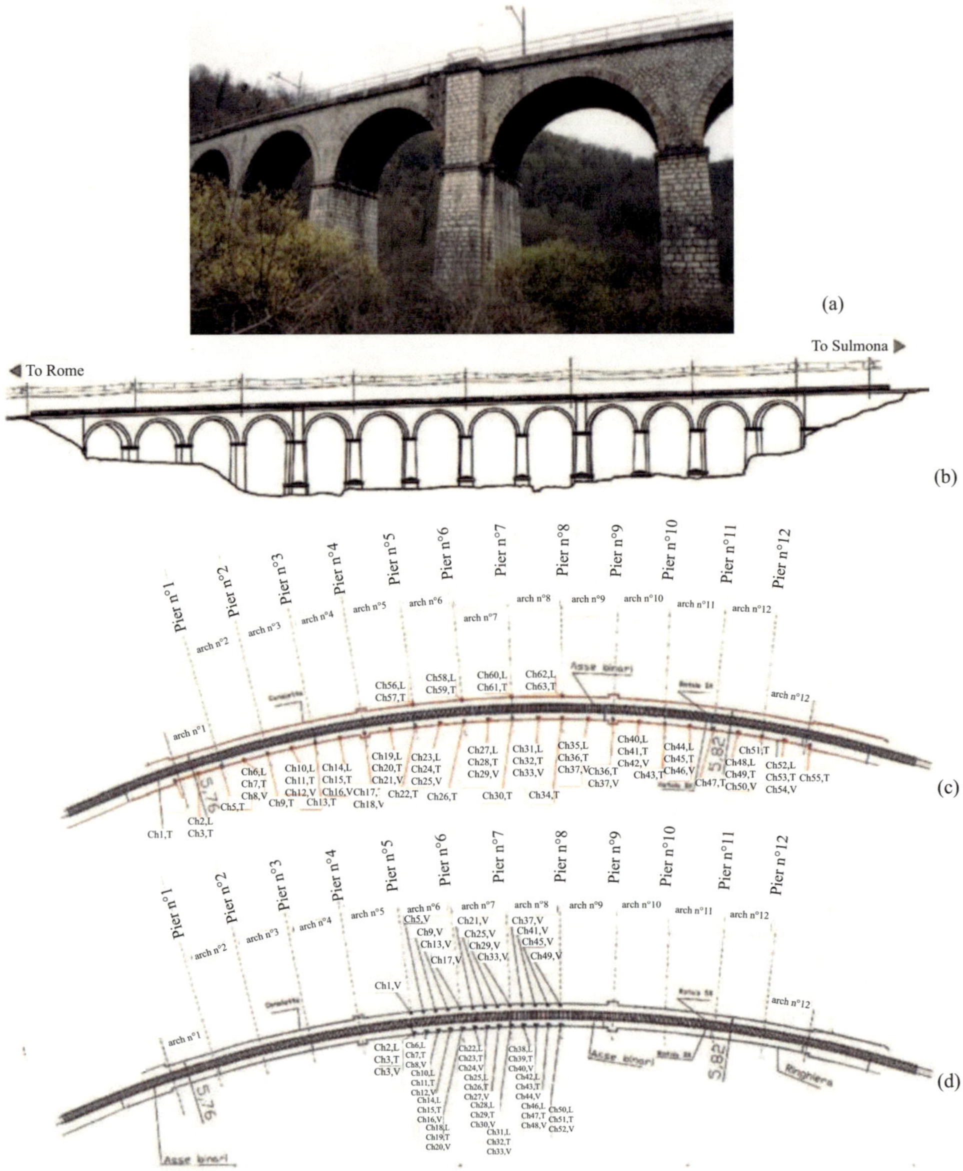

Fig. 7. Picture of the Vallone Scarpa bridge (a), front view (b) and plans (c,d) of the two accelerometer setups.

3.2 The Vallone Scarpa bridge

The Vallone Scarpa bridge was built at the end of the nineteenth century and is located along the Roma-Sulmona railway line, which crosses the central Italian region of Abruzzo. It is a masonry arch viaduct with thirteen bays, each with a span of 10 m. The piers are about 9 m in height (Figure 7). The plan has a radius of curvature of 400 m; the slope of the line is 2.7 %. The ambient vibrations of the bridge were recorded using two different arrangements of accelerometers (Figure 7c,d) at a sampling frequency of $f_s = 120Hz$. The measurement directions were: in the plan of the deck, tangent to the bridge axis (L, longitudinal) and related orthogonal line (T, transverse), and the vertical direction (V) along the viaduct axis. In the first setup (Figure 7c), transverse sensors were placed on top of each pier, together with triaxial sets (T, L, V) located at the middle of each span. In the second setup (Figure 7d), triaxial sets of accelerometers (L,T,V) were placed on the deck edges of the three central bays.

To start with, the PSDs are examined. Figure 8(a) shows the PSDs measured at channels 19-21 of the first arrangement of sensors, placed on the fifth bay of the bridge. This figure immediately points out that, in this frequency range, the vertical and longitudinal components of the modes are much smaller than the transverse ones. These PSDs, at a glance, also enable to detect some peaks that are representative of the first natural frequencies of the structure. However, the identification of their values is very difficult because of the closeness of frequencies and strong damping, as is often the case with masonry structures. For instance, two very close peaks are present in the neighborhood of 3.9 Hz. The coherence function, which is shown in Figure 8(b), also shows peaks in correspondence with the peaks of PSDs, but does not help in resolving the close resonances.

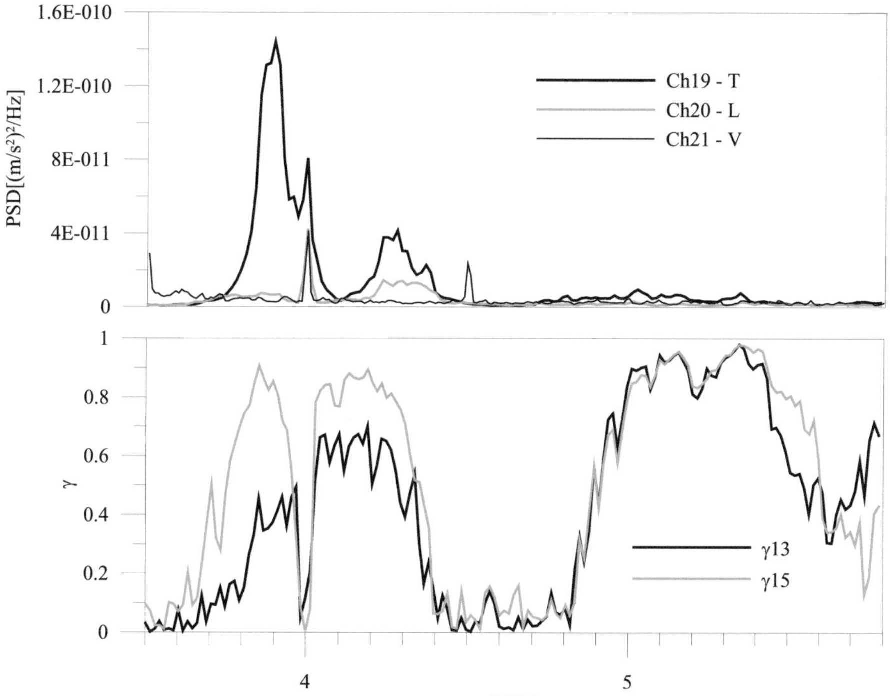

Fig. 8. PSDs (a) and coherence (b) of the accelerations measured on the fifth bay of the Vallone Scarpa bridge.

The technique of singular value decomposition points out analogous difficulties, as can be seen from Figure 9, that reports the first singular value of the spectral matrix for both setups as a function of frequency. An advantage of this technique compared to peak picking is the easier and faster determination of mode shapes, which enables to choose the peaks representative of natural frequencies.

Using the SVD with the first arrangement of sensors, the first five frequencies listed in Table 3 were determined. The stabilization diagram shown in Figure 10 furnishes frequencies,

obtained using SSI, similar to those obtained from SVD and PP (Table 3). Figure 10 also shows that a high model order is necessary to detect natural frequencies, which is computationally very much demanding, especially when using such a large number of accelerometers.
The mode shapes are bending modes of a beam over elastic supports in the plane of the deck, as shown in Figure 11, which depicts the first five modes of the structure. As for frequencies, SVD and SSI provide similar mode shapes. The main component of the mode shape is also in the transverse direction, as shown in Figure 11, which reports a comparison between the longitudinal and transverse components of each mode.

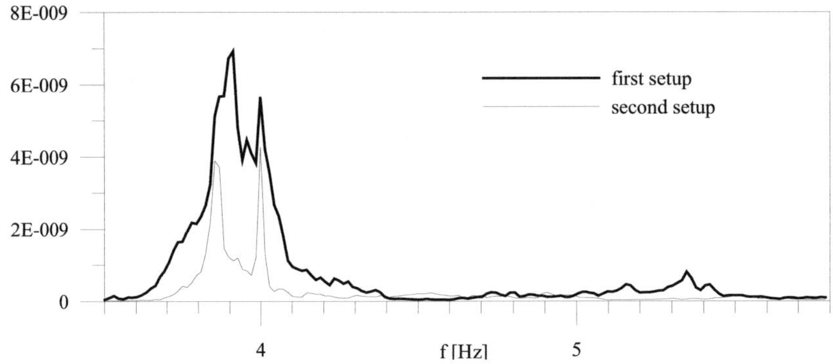

Fig. 9. First singular value of the spectral matrix as a function of frequency.

	f_1	f_2	f_3	f_4	f_5
PP, SVD	3.87	3.88	4.24	4.73	5.49
SSI	3.82	3.98	4.57	4.74	5.33
FE	2.41	2.71	2.92	3.44	3.76

Table 3. Experimental and numerical natural frequencies [Hz] of the Vallone Scarpa bridge

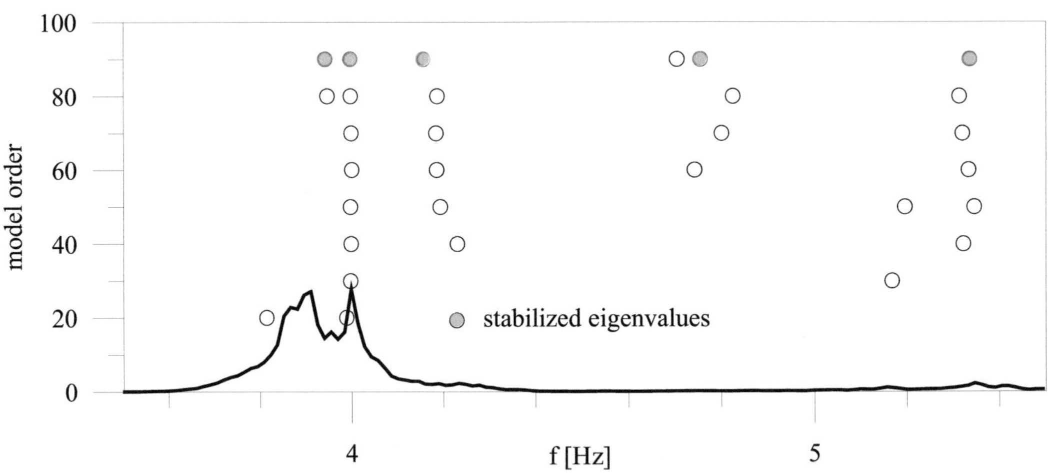

Fig. 10. Eigenvalue stabilization diagram for the Vallone Scarpa bridge.

The measurements performed with the second arrangement of sensors provided similar results, but showed in addition that the identified modes presented a slight rotational component, shown by the opposite sign of the modal displacements measured on the right and left edges of the deck. This is shown in Figure 12, only for modes 3 and 4 for brevity. Furthermore, in the second arrangement of sensors there was a slight variation of frequencies, with an inversion between the order of modes 1 and 2. According to these measurements, the first mode presents one node, while the second one does not have any. This phenomenon is related to the closeness between the two frequencies of the arch in the horizontal plane and to the possible slight variation of the mechanical parameters between a measurement set and the other. A complete explanation of the phenomenon would require a repetition of the measurements and a verification of their robustness with regard to the ambient conditions.

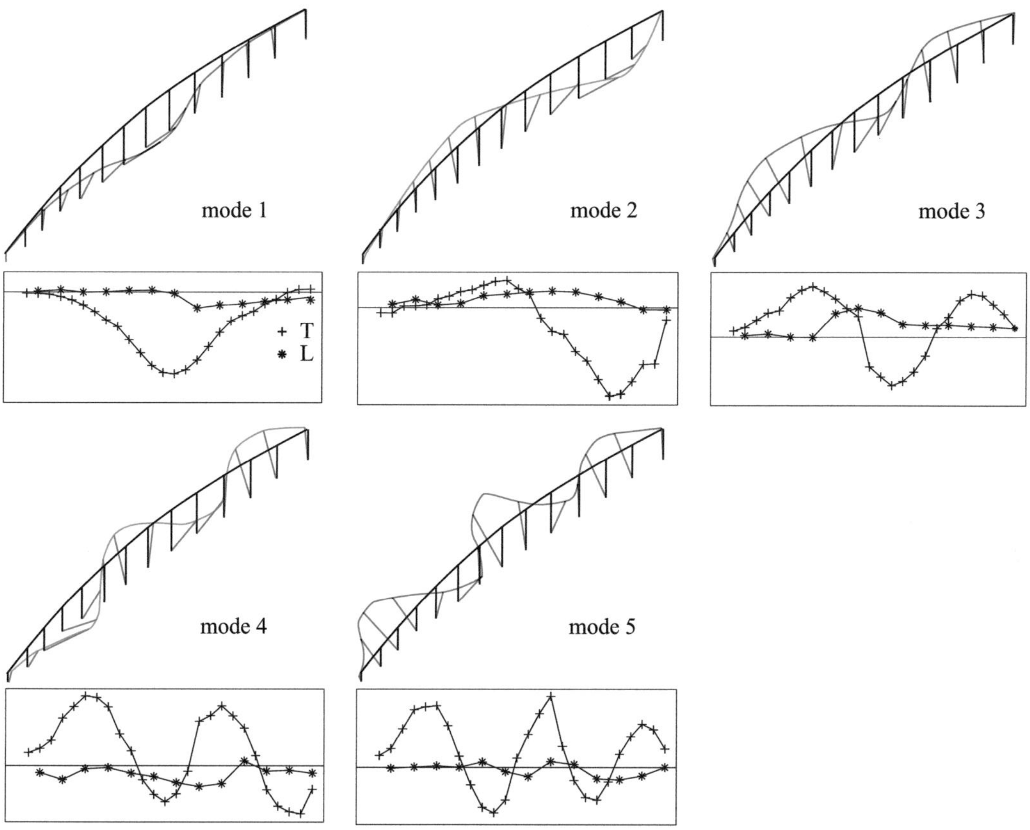

Fig. 11. Mode shapes of the Vallone Scarpa bridge and comparison between longitudinal (crosses) and transverse components (asterisks).

A finite element model of the bridge was also built to perform a modal analysis. The natural frequencies obtained are reported in Table 3 and are lower than those experimentally detected. The observed differences may be considerably reduced with a magnification of the mechanical parameters. In fact, the required updating is opposite in sign and smaller than that needed for the Trajan column. This depends on the fact that in a brick masonry the mechanical behavior of a specimen is much more representative of the behavior of the continuum than in the case of a dry block masonry. Analogous results were observed by the authors in (Pau & Vestroni, 2010). A comparison between experimental and numerical mode shapes, reported in Figure 13, presents a good agreement.

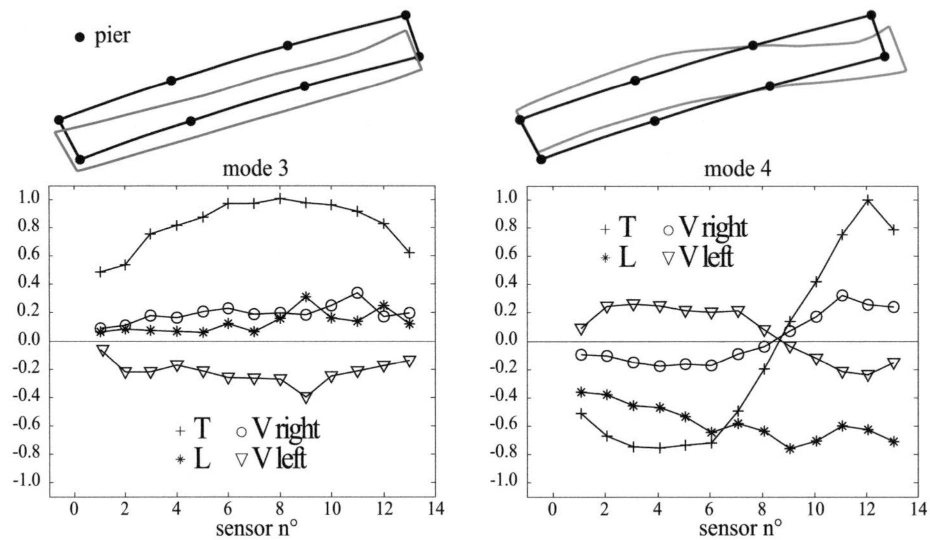

Fig. 12. Comparison between longitudinal, transverse and vertical components of experimental modes 3 and 4.

4. Conclusions

The ambient vibration response of two masonry structures, the Trajan column and the Vallone Scarpa bridge, has been analyzed using three widespread techniques, namely, the peak picking, singular value decomposition and stochastic subspace identification. The two structures are very different in masonry typology, with large blocks connected by clamps and pins for the Trajan column and mixed brick masonry for the Vallone Scarpa bridge. Notwithstanding the low level of excitation, the analysis has shown that the first frequencies of the structures are quite easily detectable. However, when higher frequencies are sought, difficulties may arise due to the lack in the verification of the hypothesis requiring that the input is white noise. Among the three different techniques considered, the SSI is the most demanding from a computational point of view and in general provides a better quality for mode shapes. As regards the comparison with finite element models, it has been shown that after an updating of the material properties of the finite element models, an agreement between experimental and numerical frequencies can be obtained. When dealing with dry masonry structures, this updating can imply a strong reduction of the Young's modulus since the material parameters of the solid material are not representative of the assembly, where the interactions between the blocks have considerable influence. In other types of masonry, such as the brick type, the behavior of a specimen is more representative of the continuum from which it is extracted and the required updating is more limited.

5. Acknowledgements

The authors wish to thank the RFI, Rete Ferroviaria Italiana company, which provided funds for the tests conducted on the Vallone Scarpa Bridge and in particular Dr. Giacomo Kajon, from the Experimental Institute of Italian National Railways, who was responsible for the experimental activities. The experimental tests on the Trajan Column were performed by the personnel of the Laboratory of the Department of Structural and Geotechnical Engineering of La Sapienza University of Rome with the consent of the Roman Soprintendenza Archeologica.

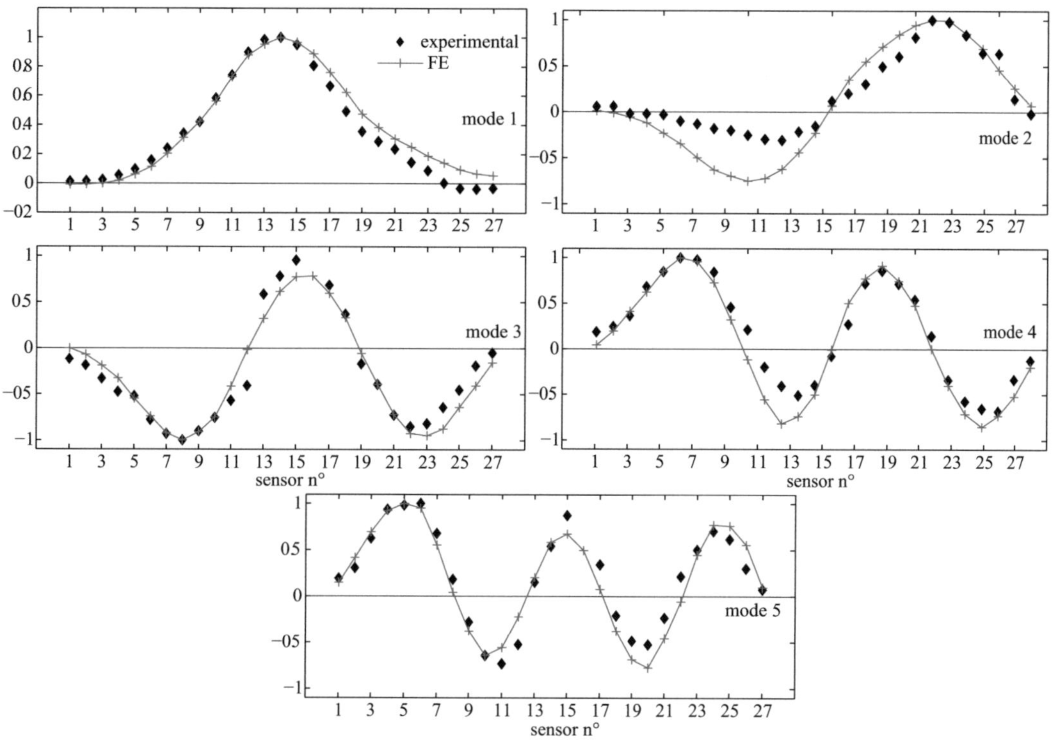

Fig. 13. Comparison between numerical and experimental transverse components of the mode shapes of the Vallone Scarpa Bridge.

6. References

Bendat, J. S. & Piersol, A. G. (1980). *Engineering applications of correlation and spectral analysis*, Wiley, New York.

Brincker, R., Zhang, L. & Andersen, P. (2001). Modal identification of output-only systems using frequency domain decomposition, *Smart Materials and Structures* Vol. 10: 441–445.

Brownjohn, J. M. W. (2003). Ambient vibration studies for system identification of tall buildings, *Earthquake engineering and structural dynamics* Vol. 31(No. 1): 71–95.

De Sortis, A., Antonacci, E. & Vestroni, F. (2005). Dynamic identification of a masonry building using forced vibration tests, *Engineering Structures* Vol. 0(No. 27): 155–165.

Ewins, D. J. (2000). *Modal testing: Theory and practice*, Research Studies Press Ltd., Hertfordshire, U.K.

Farrar, C. R., Doebling, S. W. & Nix, D. A. (2001). Vibration-based structural damage identification, *Philosophical Transactions of the Royal Society of London A* Vol. 359: 131–149.

Friswell, M. I. & Mottershead, J. E. (1995). *Finite element model updating in structural dynamics*, Kluwer Academic, Dordrecht, The Netherlands.

Gentile, C. & Saisi, A. (2007). Ambient vibration testing of historic masonry towers for structural identification and damage assessment, *Construction and building materials* Vol. 21: 1311–1321.

Juang, J. N. (1994). *Applied system identification*, Prentice-Hall, Englewood Cliffs, N.J.

Maia, N. M. M. & Silva, J. M. M. e. (1997). *Theoretical and experimental modal analysis*, Research Studies Press Ltd., Hertfordshire, U.K.

Morassi, A. & Vestroni, F. (2009). *Dynamic Methods for Damage Detection in Structures*, Springer.

Pau, A. & Vestroni, F. (2008). Vibration analysis and dynamic charachterization of the colosseum, *Structural Control and Health Monitoring* Vol. 15: 1105–1121.

Pau, A. & Vestroni, F. (2010). Dynamic characterization of the basilica of maxentius in rome, *Proceedings of the Int. Conf. on Noise and Vibration Engineering ISMA*, Leuven, Belgium.

Peeters, B. (2000). *System identification an damage detection in civil engineering*, PhD thesis, Katholieke Universitaet Leuven.

Ren, W. X., Zhao, T. & Harik, I. E. (2004). Experimental and analytical modal analysis of steel arch bridge, *Journal of Structural Engineering* Vol. 7(No. 130): 1022–1031.

Van Overschee, P. & De Moor, B. (1993). Subspace algorithms for the stochastic identification problem, *Automatica* Vol. 29(No. 3): 649–660.

Van Overschee, P. & De Moor, B. (1994). N4sid: Subspace algorithms for the identification of combined deterministic-stochastic systems, *Automatica* Vol. 30(No. 1): 75–93.

Van Overschee, P. & De Moor, B. (1996). *Subspace Identification for Linear Systems, Theory, Implementation, Applications*, Kluwer Academic Publishers, Boston.

Vestroni, F. & Capecchi, D. (1996). Damage evaluation in cracked vibrating beams using experimental frequencies and finite element models, *Journal of Vibration and Control* Vol. 2: 269–286.

12

Vibration Analysis of Long Span Joist Floors Submitted to Human Rhythmic Activities

José Guilherme Santos da Silva, Sebastião Arthur Lopes de Andrade,
Pedro Colmar Gonçalves da Silva Vellasco,
Luciano Rodrigues Ornelas de Lima and Rogério Rosa de Almeida
State University of Rio de Janeiro (UERJ)
Rio de Janeiro/RJ,
Brazil

1. Introduction

In the last years, building structures are more and more becoming the modern landmarks of urban areas. Designers seem to continuously move the safety border, in order to increase slenderness and lightness of their structural systems. However, more and more steel and composite floors (steel-concrete) are carried out as light weight structures with low frequencies and low damping. These facts have generated very slender composite floors, sensitive to dynamic excitation, and consequently changed the serviceability and ultimate limit states associated to their design.

The increasing incidence of building vibration problems due to human rhythmic activities led to a specific design criterion for rhythmic excitations to be addressed in structural design (Allen et al. 1985); (Almeida, 2008); (Almeida et al., 2008); (Bachmann & Ammann, 1987); (Faisca, 2003); (Ji & Ellis, 1994); (Langer, 2009); (Murray et al., 2003); (Silva et al., 2008). This was the main motivation for the development of a design methodology centred on the structural system dynamical response submitted to dynamic loads due to human activities.

This paper investigated the dynamic behaviour of composite floors (steel-concrete) subjected to the human rhythmic activities. The dynamic loads were obtained through experimental tests conducted with individuals carrying out rhythmic and non-rhythmic activities such as stimulated and non-stimulated jumping and aerobics (Faisca, 2003).

The description of the loads generated by human activities is not a simple task. The individual characteristics in which each individual perform the same activity and the existence of external excitation are relevant factors when the dynamic action is defined. Numerous investigations were made aiming to establish parameters to describe such dynamic loads (Allen et al. 1985); (Bachmann & Ammann, 1987); (Faisca, 2003); (Murray et al., 2003).

The present investigation considered the dynamic loads, based on results achieved through a long series of experimental tests made with individuals carrying out rhythmic and non-rhythmic activities. This investigation described these dynamic loads, generated by human activities, such as jumps with and without stimulation, aerobics, soccer, rock concert audiences and dancing (Faisca, 2003).

The load modelling was able to simulate human activities like aerobic gymnastics, dancing and free jumps. In this paper, the Hanning function was used to represent the human dynamic actions since it was verified that this mathematical representation was very similar to the signal force obtained through experimental tests (Faisca, 2003). Based on the experimental results, human load functions due to rhythmic and non-rhythmic activities were proposed.

The computational model, developed for the composite floors dynamic analysis, adopted the usual mesh refinement techniques present in finite element method simulations implemented in the Ansys program (ANSYS, 2003). In the present computational model, the floor steel joists were represented by three-dimensional beam elements, considering flexural and torsion effects, while the composite slab was represented by shell finite elements.

The investigated structural model was associated to a floor composed by steel joists and a concrete slab. The structural system was a typical floor used as a restaurant with an adjacent dancing area (Almeida, 2008); (Almeida et al., 2008); (Murray et al., 2003); (Silva et al., 2008). The composite (steel-concrete) floor system consisted of long span (14m) joists supported by concrete block walls.

The floor effective weight was estimated to be equal to 3.6kPa, including 0.6kPa for people dancing and dining. The joists effective composite moment of inertia was selected based on its required strength, i.e., $1.1x10^6$ mm^4. This structural system geometry was based on a typical example described in literature (Almeida, 2008); (Almeida et al., 2008); (Murray et al., 2003); (Silva et al., 2008).

The parametric study considered correlations between analytical and numerical results found in literature. The peak acceleration values were compared to the limits proposed by design codes and recommendations (ISO 2631-2, 1989); (Murray et al., 2003), based on human comfort criteria. The results indicated that the limits suggested by the design recommendations were not satisfied. This fact indicated that these rhythmic activities could generate peak accelerations that surpass design criteria limits developed for ensuring human comfort.

2. Human-induced dynamic loads

Floor vibrations induced by human rhythmic activities like: walking, running, jumping or even aerobics consist on a very complex problem. This is due to the fact that the dynamical excitation characteristics generated during these activities are directly related to the individual body adversities and to the specific way in which each human being executes a certain rhythmic task. All these aspects do not contribute for an easy mathematical or physical characterization of this phenomenon.

Human beings have always analysed the most apparent distinctions of the various activities they perform. However the fundamental mechanical analysis of these tasks was not possible before a significant development of the mechanical science. Initially the human motion received an incipient attention from researchers like Borelli in 1679 (Lehmkuhl & Smith, 1985) and the Weber brothers in 1836 (Lehmkuhl & Smith, 1985). The first pioneer on this field was Otto Fischer, a German mathematician that in 1895 made the first study containing a comprehensive evaluation of the forces involved in human motion.

In order to determine the dynamical behaviour of floor structural systems subjected to excitations from human activities, various studies have tried to evaluate the magnitude of these rhythmic loads. The following stage of this research line was the development of a loading platform by Elftman (Lehmkuhl & Smith, 1985), that enable the determination of the ground reactions to the foot forces associated to the human walk motion. The typical force platform is made by an approximate 1m^2 steel plate supported by four small columns at the plate midsides. Load cells were installed at each of the columns to detect the magnitude of the load variation at these points. With these results in hand it was possible to determine the magnitude and direction of the forces transmitted to the supporting surface, denominated ground reaction forces.

Rainer also contributed in this investigation developing more sophisticated load platforms that recorded the ground reaction forces coming from the foot forces associated to the human motion (Rainer et al., 1987). Ebrahimpur developed a 14.2m length x 2m wide platform designed to record the actions from a single individual, or groups of two or four individual walk motion (Ebrahimpur, 1996).

Another load model used to represent the walk motion forces is expressed as a function of tests that recorded the heel impact over the floor. This load type, considered as the main excitation source during the human walk motion, produces a transient response, i.e., when the system is excited by an instantaneous force application. Its graphical representation was presented by Ohmart (Ohmart, 1968) in experiments denominated heel drop tests, where the individual drops its heel over the floor after elevating it to a height corresponding to its weight.

The heel drop test was also made by Murray and Hendrick in different building types (Murray & Hendrick, 1977). A 0.84kN impact force was measured by a seismograph in nine church ceremonial rooms, three slabs located at a shopping mall highest floor, two balcony slabs of a hotel and one slab located at a commercial building second floor. With these results in hand, the structural dynamic responses, in terms of the force amplitudes, frequencies and damping, associated to the investigated structural systems, could be determined.

Murray (Murray, 1975) classified the human vibration perception in four categories, i.e.: the vibration is not noticed by the occupants; the vibration is noticed but do not disturb the occupants; the vibration it is noticed and disturb the occupants; the vibration can compromise the security of the occupants. These categories were established based on 100 heel drop tests performed on composite floors made of steel beams and concrete slabs.

Allen et al. (Allen et al., 1985) proposed minimum values for the natural frequencies of structures evaluated according to the type of occupation and their main characteristics. These values were based on the dynamical load values produced by human rhythmic activities like dancing and aerobics and on the limit acceleration values associated to those activities.

A significant contribution to this field was made in Brazil by Alves (Alves, 1997) and Faisca (Faisca, 2003) based on experiments made with a group of volunteers acting on a concrete platform. These tests enabled the development of approximated descriptions of the loads induced by human activities such as: jumps, aerobics, soccer and rock show audience responses. These tests were executed over two concrete platforms, one rigid and the other flexible, both of them over movable supports. The experimental results analysis, allied to an analytical model, led to the development of load functions associated to synchronous and asynchronous activities that could be used in structural designs intended for stadiums and other related structures.

3. Loads generated by human activities

The description of the loads generated by human activities is not a simple task. The individual characteristics in which each individual perform the same activity and the existence of external excitation are relevant factors when the dynamic action is defined. Numerous investigations were made aiming to establish parameters to describe such loads (Allen et al. 1985); (Bachmann & Ammann, 1987); (Faisca, 2003); (Murray et al., 2003).

Several investigations described the loads generated by human activities as a Fourier series, which consider a static part due to the individual weight and another part due to the dynamic load. The dynamic analysis is performed equating one of the activity harmonics to the floor fundamental frequency, leading to resonance (Almeida, 2008); (Bachmann & Ammann, 1987); (Langer, 2009); (Murray et al., 2003); (Silva et al., 2008).

The present investigation considered the dynamic loads, based on results achieved through a long series of experimental tests made with individuals carrying out rhythmic and non-rhythmic activities (Faisca, 2003). These dynamic loads, generated by human activities, were described such as jumps with and without stimulation, aerobics, soccer, rock concert audiences and dancing.

The load modelling was able to simulate human activities like aerobic gymnastics, dancing and free jumps. In this paper, the Hanning function was used to represent the human dynamic actions since it was verified that this mathematical representation was very similar to the signal force obtained through experimental tests (Faisca, 2003).

The mathematical representation of the human dynamic loading is described by Equation (1). This expression requires some parameters like the activity period, T, contact period with the structure, T_c, period without contact with the model, T_s, impact coefficient, K_p, and phase coefficient, CD, see Fig. 1 and Table 1.

$$F(t) = CD\left\{K_p P\left[0.5 - 0.5\cos\left(\frac{2\pi}{T_c}t\right)\right]\right\}, \text{ for } t \le T_c$$
$$F(t) = 0, \text{ for } T_c < t \le T \qquad (1)$$

Where:

- F(t) : dynamic loading, in (N);
- t : time, in (s);
- T : activity period (s);
- T_c : activity contact period (s);
- P : weight of the individual (N);
- K_p : impact coefficient;
- CD : phase coefficient.

Figure 1 illustrates the phase coefficient variation, CD, for some human activities, initially, considering a few number of individuals and later extrapolating for a larger number of people (Faisca, 2003). Figure 2 presents an example of dynamic action related to human rhythmic activities using the following parameters: T = 0.53s, T_c = 0.43s, T_s = 0.10, K_p = 2.78 and CD = 1.0, see Table 1.

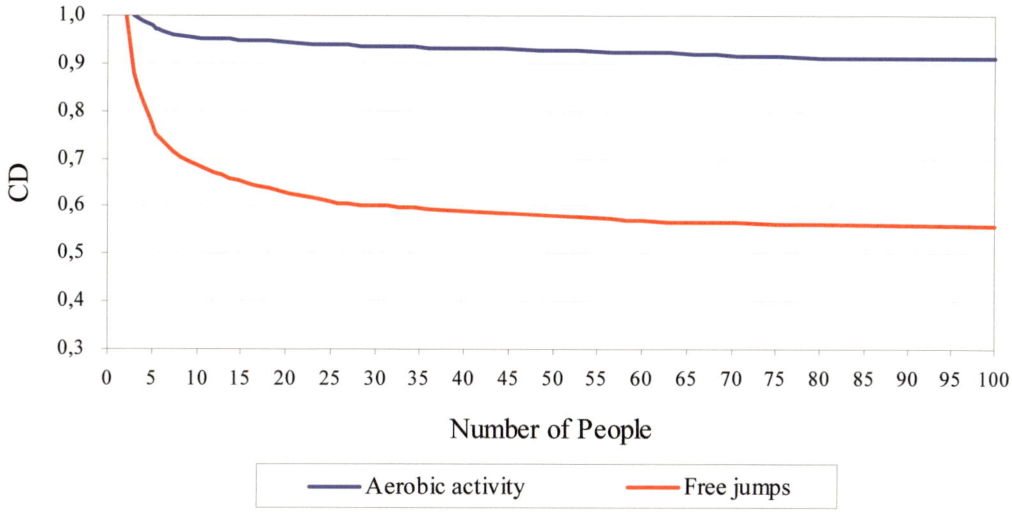

Fig. 1. Phase coefficients for the studied activities (Faisca, 2003)

Activity	T (s)	T_c (s)	K_p
Aerobics	0.44 ± 0.09	0.34 ± 0.09	2.78 ± 0.60
Free jumps	0.44 ± 0.15	0.32 ± 0.09	3.17 ± 0.58

Table 1. Parameters used for human rhythmic activities representation (Faisca, 2003)

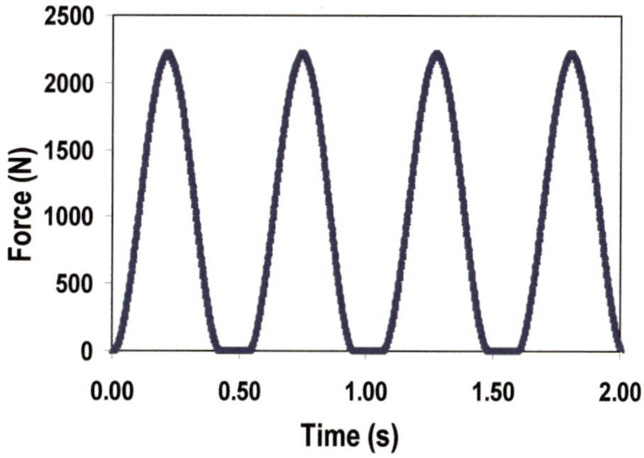

Fig. 2. Dynamic loads induced by dancing associated to the following parameters: T=0.53s, T_c=0.43s, T_s=0.10, K_p=2.78 and CD=1.0

4. Investigated structural model

The investigated structural model was associated to a floor composed by steel joists and a concrete slab, as presented in Figs. 3 to 6. The structural system was a typical floor used as a restaurant with an adjacent dancing area (Almeida, 2008); (Almeida et al., 2008); (Murray et al., 2003); (Silva et al., 2008).

The composite floor system consisted of long span (14m) joists supported by concrete block walls, see Figs. 3 to 6. The floor effective weight was estimated to be equal to 3.6kPa, including 0.6kPa for people dancing and dining. The joists effective composite moment of inertia was selected based on its required strength, i.e., 1.1×10^6 mm^4. This structural system geometry was based on a typical example described in literature (Almeida, 2008); (Almeida et al., 2008); (Murray et al., 2003); (Silva et al., 2008).

The adopted steel sections were made with a 300MPa yield stress steel grade. A 2.05×10^5 MPa Young's modulus was used for the steel joists. The concrete slab had a 30MPa specified compression strength and a 2.4×10^4 MPa Young's Modulus. The structural model geometrical characteristics are illustrated in Table 2.

Main Span	Bottom Chords	Top Chords	Vertical Members	Diagonals
14.0m	⅃L2x(1 ½" x 1/8")	⅂Γ2x(2" x 1/8")	L (½ "x 1/8")	L (½ "x 1/8")

Table 2. Structural model geometric properties

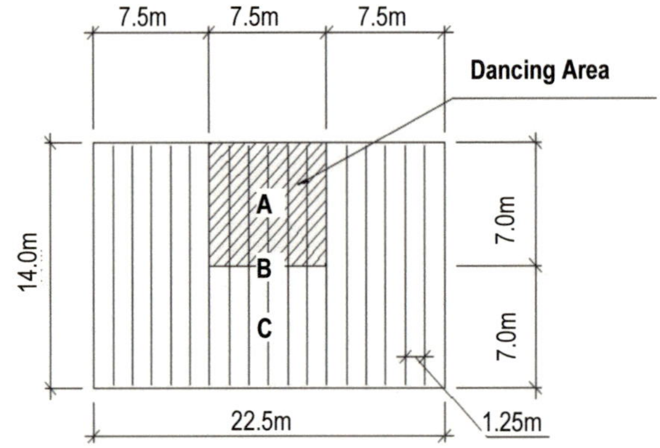

Fig. 3. Dancing floor layout (dimensions in m)

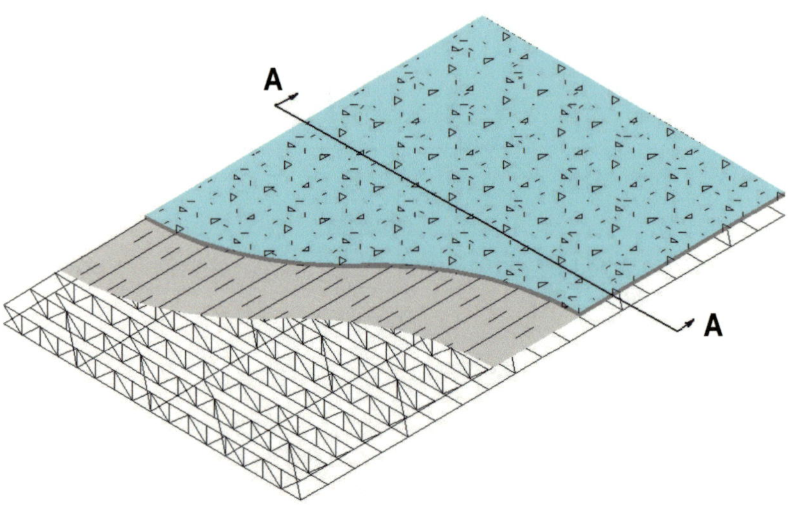

Fig. 4. Structural model three-dimensional view

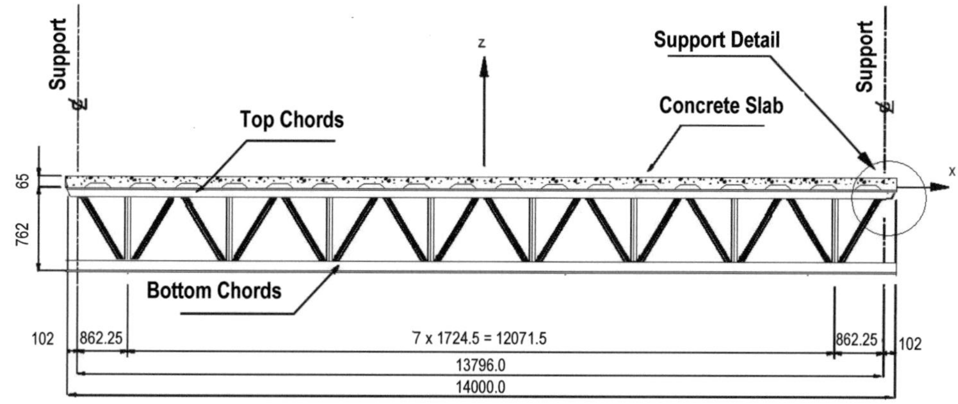

Fig. 5. Composite floor cross section - Section AA (dimensions in mm)

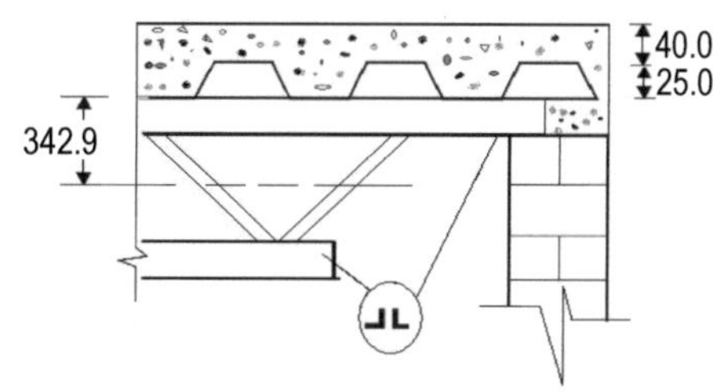

Fig 6. Support details (dimensions in mm)

The human-induced dynamic action was applied to the dancing area, see Figs. 3 and 7. The composite floor dynamical response, in terms of peak accelerations values, were obtained on the nodes A, B and C, to verify the influence of the dynamical loads on the adjacent slab floor, see Figs. 3 and 7. In the current investigation, the human rhythmic dynamic loads were applied to the structural model corresponding to the effect of 1, 3, 6, 9 and 12 individuals practicing aerobics or couples dancing.

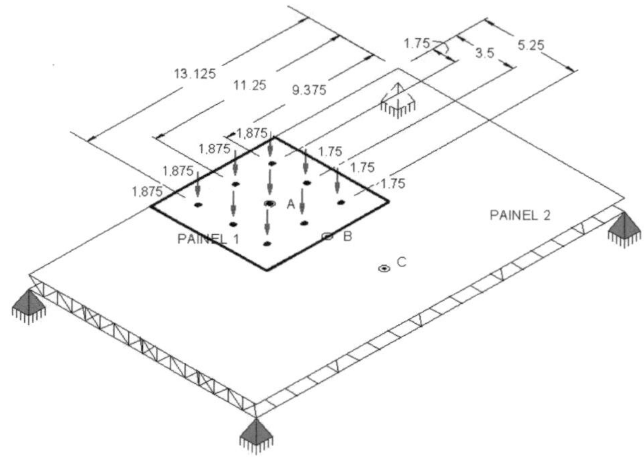

Fig. 7. Load distribution associated to nine individuals acting on the floor (dimensions in m)

The live load considered in this analysis corresponds to one individual for each $4.0m^2$ (0.25 person/m^2), (Bachmann & Ammann, 1987). The load distribution was considered symmetrically centred on the slab panel, as depicted in Fig. 7. The present investigation also assumed that an individual person weight was equal to 800N (0.8kN) (Bachmann & Ammann, 1987) and that the adopted damping ratio was equal to, ξ=3% (ξ = 0.03) in all studied cases (Almeida, 2008); (Almeida et al., 2008); (Murray et al., 2003); (Silva et al., 2008).

5. Finite element modelling

The proposed computational model, developed for the composite floors dynamic analysis, adopted the usual mesh refinement techniques present in finite element method simulations implemented in the ANSYS program (ANSYS, 2003). In the present computational model, the floor steel joists were represented by three-dimensional beam elements (BEAM44), with tension, compression, bending and torsion capabilities (ANSYS, 2003). The composite slab was represented by shell finite elements (SHELL63) (ANSYS, 2003), as illustrated in Fig. 8. In this investigation, it was considered that both materials (steel and concrete) presented total interaction and have an elastic behaviour. The finite element model has 11673 nodes, 5267 three-dimensional beam elements (BEAM44), 6912 shell elements (SHELL63) and 62568 degrees of freedom. The developed computational model is illustrated in Fig. 8.

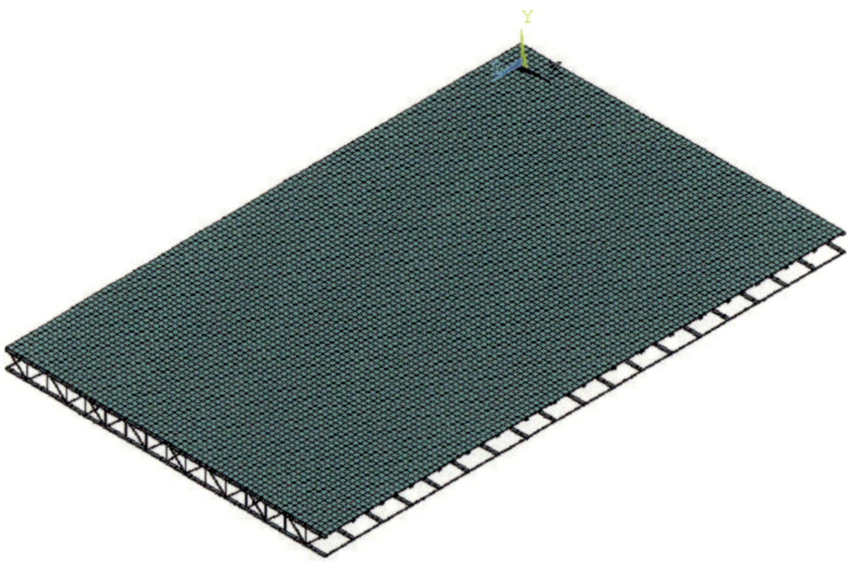

Fig. 8. Composite floor (joists and concrete slab) finite element model

6. Natural frequencies and vibration modes

The composite (steel-concrete) floor natural frequencies were determined with the aid of the numerical simulations, as illustrated in Table 3. The structural system vibration modes were illustrated in Fig. 9.

It can be clearly noticed from Table 3 results, that there is a very good agreement between the structural model fundamental frequency value calculated using finite element simulations and the AISC recommendation (Murray et al., 2003).

Such fact validates the numeric model here presented, as well as the results and conclusions obtained throughout this work. It must be emphasized that the structural model presented vibration modes with a predominance of flexural effects, as illustrated in Fig. 9.

Natural Frequencies (Hz)						AISC*	Error
f_{01}	f_{02}	f_{03}	f_{04}	f_{05}	f_{06}	f_{01}	%
5.70	5.91	6.13	6.42	6.56	7.89	5.80	2.0

*(Murray et al., 2003)

Table 3. Composite floor (steel-concrete) natural frequencies

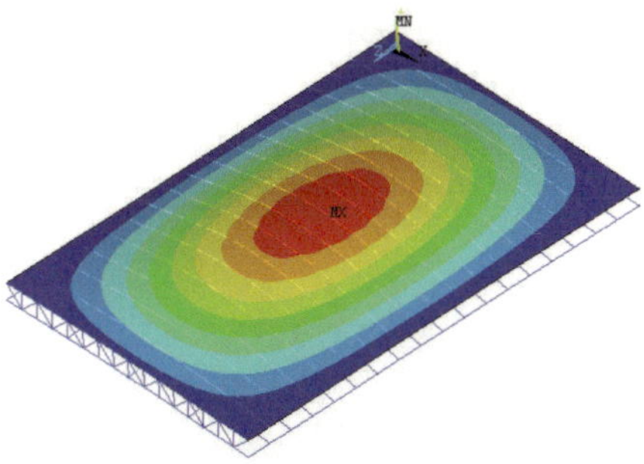

a) Vibration mode associated to the first natural frequency: f_{01}=5.80 Hz.

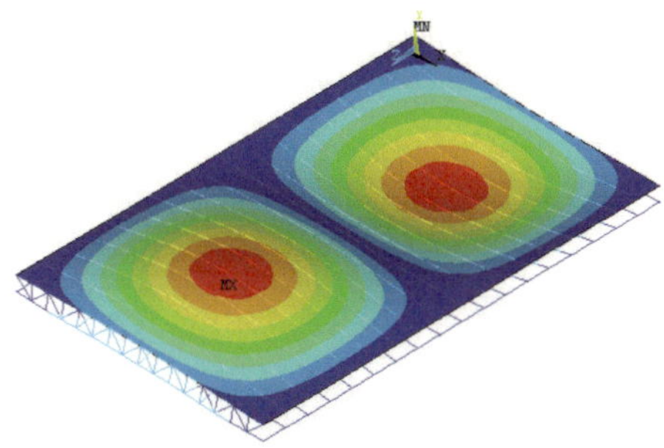

b) Vibration mode associated to the second natural frequency: f_{02}=5.91 Hz.

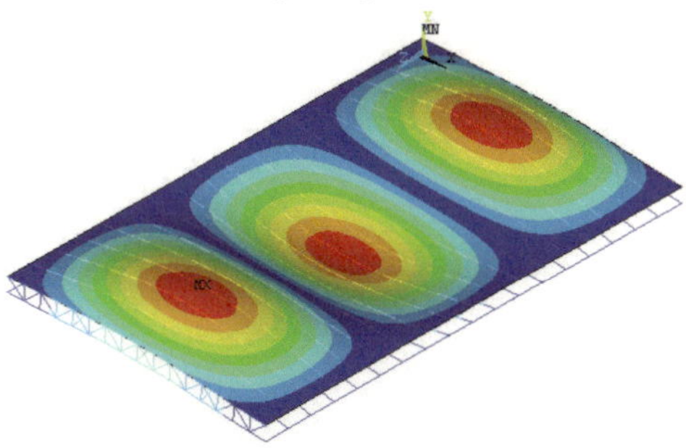

c) Vibration mode associated to the third natural frequency: f_{03}=6.13 Hz.

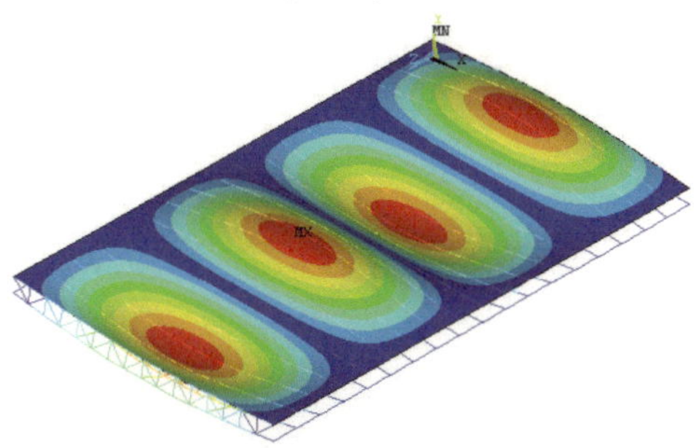

d) Vibration mode associated to the fourth natural frequency: f_{04}=6.42 Hz.

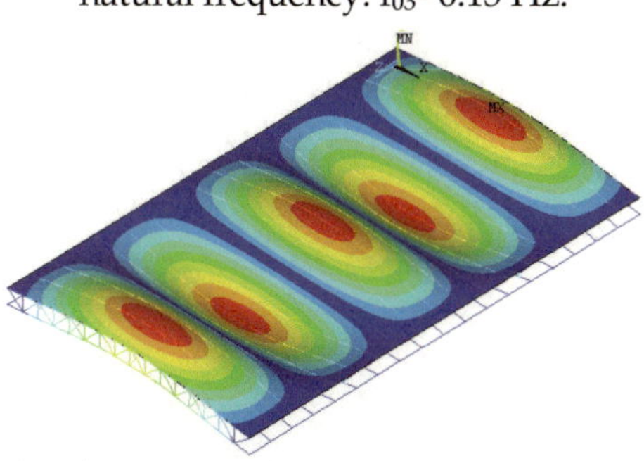

e) Vibration mode associated to the fifth natural frequency: f_{05}=6.56 Hz.

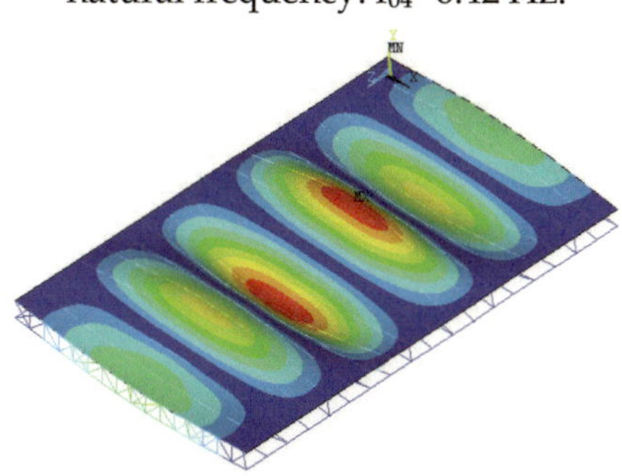

f) Vibration mode associated to the sixth natural frequency: f_{06}=7.89 Hz.

Fig. 9. Composite floor (steel-concrete) vibration modes

7. Time domain analysis

For practical purposes, a linear time-domain analysis was also performed throughout this study. This section presents the evaluation of the structural systems vibrations levels when submitted to dynamic excitations coming from human rhythmic activities (aerobics and dancing).

The composite floor (steel-concrete) dynamic responses were determined through an analysis of its displacements and accelerations. The results of the dynamic analysis were obtained from an extensive numerical analysis, based on the finite element method using the ANSYS program (ANSYS, 2003).

Figures 10 and 11present the vertical displacement and acceleration, respectively, versus time graphs for the analysed composite floor (steel-concrete) at point A (see Figs. 3 and 7), when only one individual is acting on the structural model (aerobics).

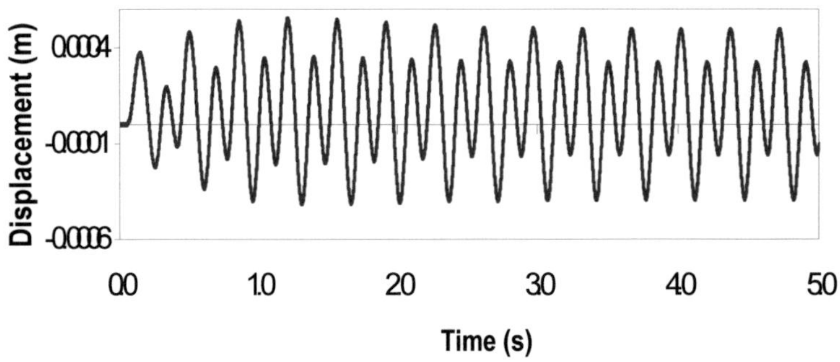

Fig. 10. Composite floor displacement response due to one individual practicing aerobics at Point A (see Figs. 3 and 7): T_c=0.25s, T_s=0.10s, K_p=2.78 and CD=1.

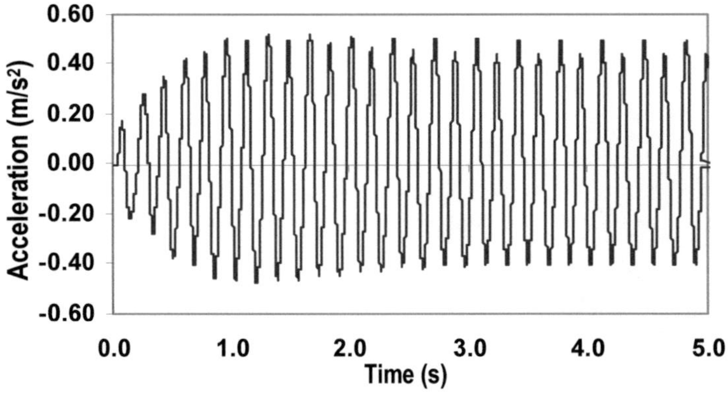

Fig. 11. Composite floor acceleration response due to one individual practicing aerobics at Point A (see Figs. 3 and 7): T_c=0.25s, T_s=0.10s, K_p=2.78 and CD=1.

Figures 10 and 11 show that, the vertical displacement and acceleration, at point A (see Figs. 3 and 7) of the structural model, gradually increase with time until the beginning of the composite floor steady state response, which occurred at the time of approximately 2.0 seconds. From this point (t = 2.0s) onwards, the maximum displacement and acceleration values were, respectively, equal to 0.051 cm and 0.55 m/s^2.
It must be emphasized that considering only one individual acting on the floor (aerobics) the calculated peak acceleration value (a_p = 0.55m/s^2), was higher than limits proposed by design recommendations (a_{lim} = 5%g = 0.50 m/s^2), violating the human comfort criteria (ISO 2631-2, 1989); (Murray et al., 2003).

8. Peak accelerations

The peak acceleration analysis was focused in dancing activities and considered a contact period carefully chosen to simulate dancing activities on the composite floor. The adopted parameters were: T_c, equal to 0.43s (T_c = 0.43s) and the period without contact to the structure, T_s, of 0.10s (T_s = 0.10s). Based on the experimental results (Faisca, 2003), the composite floors dynamic behaviour was evaluated keeping the impact coefficient value, K_p, equal to 2.78 (K_p = 2.78). Tables 4 and 5 depict the peak accelerations, a_p, corresponding to nodes A, B and C, see Figs. 3 and 7, when 1, 3, 6, 9 and 12 dynamical loads, simulating individual dancing, see Table 4, and couples dancing, see Table 5, were applied to the composite floor.

Nodes (see Fig. 3)	Number of individuals - a_p (m/s^2)					ISO 2631-2 and AISC* (m/s^2)
	1	3	6	9	12	
A	0.12	0.24	0.41	0.52	0.69	0.50
B	0.11	0.28	0.53	0.72	0.88	
C	0.07	0.17	0.31	0.42	0.54	
*(ISO 2631-2, 1989); (Murray et al., 2003)						

Table 4. Structural model peak accelerations corresponding to individuals dancing: T_c=0.43s; T_s=0.10s; K_p=2.78.

Nodes (See Fig. 3)	Number of couples - a_p (m/s^2)					ISO 2631-2 and AISC* (m/s^2)
	1	3	6	9	12	
A	0.23	0.47	0.83	1.05	1.36	0.50
B	0.23	0.57	1.04	1.45	1.76	
C	0.13	0.34	0.62	0.83	1.09	
*(ISO 2631-2, 1989); (Murray et al., 2003)						

Table 5. Structural model peak accelerations corresponding to couples dancing: T_c=0.43s; T_s=0.10s; K_p=2.78

It can be verified that the obtained peak acceleration values are proportional to an increase of the number of considered individuals, Tables 4 and 5. These values tend to decrease when the dynamical response obtained on the node C (see Figs. 3 and 7) was compared to the response of nodes A and B (see Figs. 3 and 7), as presented in Tables 4 and 5.

Based on the results presented in Table 4, it was possible to verify that dancing activities on the structural model, represented by Equation (1), led to peak accelerations higher than 0.50 m/s^2 (5%g) (ISO 2631-2, 1989); (Murray et al., 2003), when the composite floors was submitted to six individuals dancing, violating the human comfort criteria. The situation becomes even more significant when nine and twelve individuals were considered in the analysis, see Table 4.

On the other hand, when couples dancing were considered, the human comfort criterion was violated starting from cases associated with only three couples. It must be emphasized that in this situation the peak accelerations presented higher values when compared to individual dancing.

Observing the results illustrated in Tables 4 and 5, it was also possible to verify that the analyzed composite floor presented peak accelerations higher than 5.0% g (ISO 2631-2, 1989); (Murray et al., 2003) and the human comfort criteria was not satisfied even when an adjacent area, where no dancing actions are present (Node C, see Figs. 3 and 7), was investigated, Tables 4 and 5.

9. Final remarks

This paper investigated the dynamic behaviour of composite floors (steel-concrete) when subjected to the human rhythmic activities corresponding to aerobics and dancing effects. The dynamic loads were obtained through experimental tests conducted with individuals carrying out rhythmic and non-rhythmic activities such as stimulated and non-stimulated jumping and aerobics.

The proposed analysis methodology adopted the usual mesh refinement techniques present in the finite element method (FEM). Based on the experimental results (Faisca, 2003), human load functions due to rhythmic and non-rhythmic activities were proposed. The investigated structural system was a typical floor used as a restaurant with an adjacent dancing area. The composite floor system consisted of long span (14m) joists supported by concrete block walls.

The parametric analysis considered correlations between analytical and numerical results found in literature. The results, in terms of maximum accelerations, were compared to the limits proposed by design recommendations, focusing on human comfort considerations.

The results obtained throughout this study indicated that the limits recommended by design standards (ISO 2631-2, 1989); (Murray et al., 2003) were not satisfied for the investigated structural model when subjected dancing load actions. Such fact shows that these rhythmic activities may generate peak accelerations that violated design criteria related to human comfort.

The present investigation also indicated that these dynamic loads can even generated considerable perturbations on adjacent areas, where there is no human rhythmic activity of such kind present. Despite this fact there was still a surpassing of the associated human comfort criteria.

10. Acknowledgements

The authors gratefully acknowledge the support for this work provided by the Brazilian Science Foundations CAPES, CNPq and FAPERJ.

11. References

Almeida, R. R. de. (2008). Análise de vibrações em sistemas estruturais para pisos mistos com joists submetidos a atividades humanas rítmicas (Vibration analysis of composite joist floors subjected to human rhythmic activities), MSc Dissertation (in Portuguese), Civil Engineering Post-Graduate Programme, PGECIV, State University of Rio de Janeiro, UERJ, Rio de Janeiro, Brazil, pp. 1-154.

Almeida, R. R. de; Silva, J.G.S. da; Vellasco, P.C.G. da S.; Andrade, S.A.L. de; Lima, L.R.O de. (2008). Dynamical behaviour of long span joist floors submitted to human rhythmic activities. Proceedings of the 5th European Conference on Steel and Composite Structures, Eurosteel 2008, Graz, Austria, v. A. p. 735-740.

Allen, D.E.; Rainer, J.H.; Pernica, G. (1985). Vibration criteria for assembly occupancies, *Canadian Journal of Civil engineering*, Vol. 12, pp. 617-623.

Alves, N.K.C. (1997). Cargas dinâmicas devido a pessoas em movimento (Dynamic loading due to human movements), MSc Dissertation (in Portuguese), COPPE/UFRJ, Rio de Janeiro, RJ, Brazil, pp. 1-113.

ANSYS, Swanson Analysis Systems (2003). Inc. P.O. Box 65, Johnson Road, Houston, PA. 15342-0065, Version 10.0. Basic analysis procedures. Second edition.

Bachmann, H. & Ammann, W. (1987). Vibrations in structures induced by man and machines. Structural Engineering Document 3e, International Association for Bridges and Structural Engineering.

Ebrahimpur, A.; Haman, A.; Sack, R.L.; Patten, W.N. (1996). Measuring and modelling dynamic loads imposed by moving crowds, *Journal of Structural Engineering*, Vol. 122 (12), pp. 1468-1473.

Faisca, R. G. (2003). Caracterização de cargas dinâmicas geradas por atividades humanas (Characterization of Dynamic Loads due to Human Activities), PhD Thesis (in Portuguese), COPPE/UFRJ, Rio de Janeiro, RJ, Brazil, pp. 1-240.

International Standard Organization - ISO 2631-2 (1989). Evaluation of Human Exposure to Whole-Body Vibration, Part 2: Human Exposure to Continuous and Shock-Induced Vibrations in Buildings (1 to 80Hz), International Standard.

Ji, T. & Ellis, B.R. (1994). Floor vibration induced by dance-type loads: theory and verification, *The Structural Engineer*, Vol. 72, Nº 3, pp. 37-50.

Langer, N.A. dos S.; Silva, J.G.S. da; Vellasco, P.C.G. da S.; Lima, L.R.O de; Neves, L.F. da C. (2009). Vibration analysis of composite floors induced by human rhythmic activities. Proceedings of the 12th International Conference on Civil, Structural and Environmental Engineering Computing, CC 2009, Funchal, Ilha da Madeira, Portugal, CD-ROM, pp. 1-14.

Lehmkuhl, L. & Smith, L.K. (1985). Cinesealogia clínica de brunnstrom, ed. Manole, pp. 472-499.

Murray, T.M. (1975). Design to prevent floor vibration, *Engineering Journal*, Vol. 12(3), pp. 82-87.

Murray, T.M.; Hendrick, W.E. (1977). Floor vibrations and cantilevered construction, *Engineering Journal*, AISC.

Murray, T.M.; Allen, D.E.; Ungar, E.E. (2003). Floor vibration due to human activity. Steel Design Guide Series, AISC, Chicago, USA.

Ohmart, R.D. (1968). An approximate method for the response of stiffened plates to aperiodic excitation studies in engineering mechanics, Report n° 30, The University of Kansas, Center for Research in Engineering Science, Lawrence, Kansas.

Rainer, J.H.; Pernica, G.; Allen, D.E. (1987). Dynamic loading and response of footbridges. Structures Section, Institute for Research in Construction, National Research Council of Canada, Ottawa, Ont. Canada KIA OR6, pp 66-71.

Silva, J.G.S. da; Vellasco, P.C.G. da S.; Andrade, S.A.L. de; Lima, L.R.O de; Almeida, R.R. de. (2008). Vibration analysis of long span joist floors when submitted to dynamic loads due to human activities. Proceedings of the 9th International Conference on Computational Structures Technology, CST 2008, Athens, Greece, CD-ROM, pp. 1-11.

13

Progress and Recent Trends in the Torsional Vibration of Internal Combustion Engine

Liang Xingyu, Shu Gequn, Dong Lihui, Wang Bin and Yang Kang
State Key Laboratory of Engines, Tianjin University, 300072
P. R. China

1. Introduction

With modern machinery industry developing, the application of internal combustion engine is getting wider and research direction is towards high-power, high speed and strong loads. So the issue of torsional vibration of the engine is becoming more prominent. All kinds of work conditions of the engine may have great impacts on the shafting, leading to all sorts of torsional vibration and resonance, and many accidents which lead to much detriment have occurred at home and abroad due to torsional vibration.

As the problem of torsional vibration of the engine is becoming more and more prominent, broad research is made both at home and abroad. This article mainly refers to the literatures on torsional vibration issue published in recent years, summarizes on the modeling of torsional vibration, corresponding analysis methods, appropriate measures and torsional vibration control, and points out the problems to be solved in the study and some new research directions.

2. Modeling of engine crankshaft

2.1 Engine crankshaft modeling method

Crankshaft is the main component of internal combustion engine. Shaft vibration is one of the most important factors affecting engine operation safety. Crankshaft modeling is the base of crankshaft torsional vibration analysis, whose accuracy and simple practical applicability will greatly improve the efficiency and credibility of research results.

At present, there are 3 kinds of most basic shaft models used in analyzing torsional vibration: the first type is simple mass - spring model, the second is continuous mass model, and the third is multi-segment concentrated mass model.

2.1.1 Simple mass - spring model

Simple mass - spring model is the earliest mechanics model in the calculation of shaft vibration [1-6], which was also called lumped parameter model in some literatures. It disperses crankshaft onto the disk with concentration of inertia moment, elastic axis without mass, internal damping and external damping, as shown in figure 1. Each disk rotational inertia includes: the rotational inertia of the crank, the equivalent rotational inertia of connecting rod and piston , transmission system, shock absorber, the rotational inertia of the flywheel, etc.

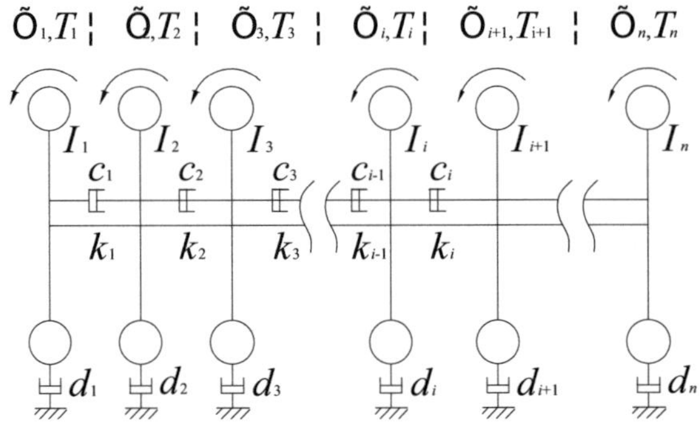

Fig. 1. Simple mass - spring model schematic diagram

This model has certain precision for lower frequency of torsional vibration modal and clear physical concept. It's simple to use and easy to calculate. But since this model is simplified, when precise calculation of the crankshaft is required, its precision is limited. This model is established completely for rigid shaft and rotation parts, so it can not simulate the actual shaft.

2.1.2 Continuous mass model

Continuous mass model is based on continuum theory, regarding shaft as elastomer, established in finite element method. It's also called distributed mass model in some literatures [7-8]. It adopts finite element method in general, dissecting the crankshaft entities directly into finite element calculation model of division. Hence, the mass of the shaft is distributed continuously along the shaft, closer to practice than that of simple mass - spring model. Partial differential equations can be used in this model, which can accurately calculate low frequency and vibration model of the shaft, as well as high frequency and vibration model, solve by numerical method, and also can calculate arbitrary section stress conveniently. But the model is complex and with low speed to calculate, and is easy to cause greater accumulative error. It is more difficult to use this model in system simulation and design. Due to the method of forced vibration calculation, it is hard to realize, thus it's mainly used in the calculation of free vibration.

Recently, two consecutive quality models also have derived from this model: framework model and multi-diameter model.

Framework model is a model, in which, circular cross section straight beam represents main journal and crank pin, and variable cross-section rectangular beam represents crank arm and counterbalance in finite element analysis [9]. For these analyses, circular cross section beam also can represent main journal and crank pin, but the crank arm and counterbalance should be treated as simple rectangular beam. Model schematic diagram is shown in figure 2. In framework model, different structural parts of the crankshaft are substituted by the continuous entities with regular shape, and the original basic shapes of crankshaft are kept. Thus this model has higher precision to analyze the crankshaft vibration.

Multi-diameter model is a model used in elastic wave propagation theory solving torsional vibration of internal combustion engine [10-12]. Assign piston-rod additional mass to two crank arms and simplify a unit crankcase into a group of concentric multi-diameter. Model schematic diagram is shown in figure 3. Because the model has continuous mass distribution, the effect of distribution parameters on shafting vibration characteristics can be considered. It also can adopt different mathematical methods to calculate and compare with simple mass - spring model. This model can have high precision.

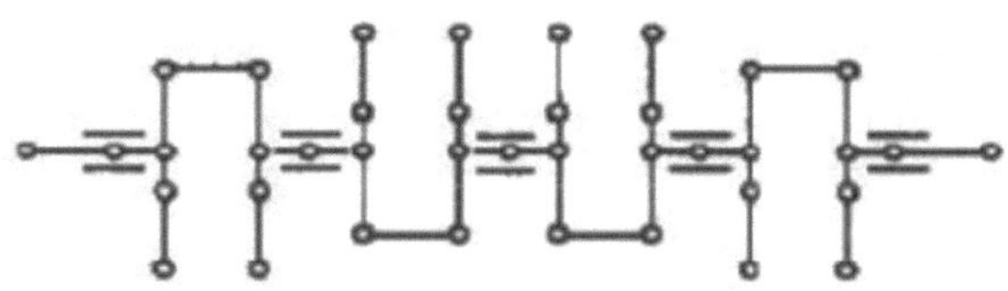

Fig. 2. Framework model schematic diagram

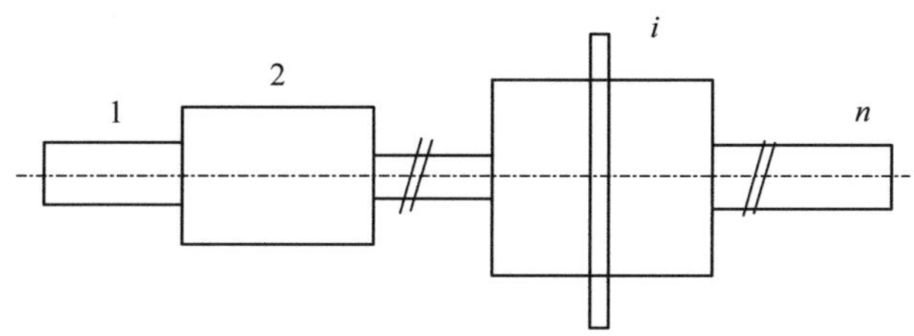

Fig. 3. Multi-diameter model schematic diagram

2.1.3 Multisegment concentrated mass model [13]

This model is similar to the simple mass model in essence. However, it can be separated into dozens to hundreds sections according to the structure characteristics upon analysis demand. It can calculate high order torsional vibration frequencies that can't be determined by simply mass model, and also avoids the large amount of computation that required in the calculation of continuous mass model. Thus it has been widely used.

2.1.4 Soft body dynamics model [14]

In the calculation of flexible multi-body dynamics, flexible body is described as modal flexible body. A flexible body contains a series of modals. In the breakdown steps, each model unit requires obtaining system state variables and calculating the relative amplitude of each characteristic vector, then using linear superposition principle to integrate node deformation of each time step to reflect total deformation of flexible body.

2.1.5 Other axis modeling methods

In recent years, with the further study of shafting vibration, many new modeling methods came up in the engine industry and other related industries.

Continuous beam model was used in the crankshaft load calculation by Li Renxian [15]. The crank and conrod were equalized to the concentrated force acting on a non-equal continuous beam, and all kinds of force were also equalized to simplify. The author analyzed various loads of crankshaft and its changes in an operation cycle comprehensively. This is a simplified model force shown in figure 4. Of course, in order to calculate simply, the author treats both gas load and centrifugal force function load as concentrated loads. If they were expressed as some forms of distributed loads, the calculation might be more accurate. Calculation model may also adopt continuous beam to make it more close to the actual situation of crankshaft.

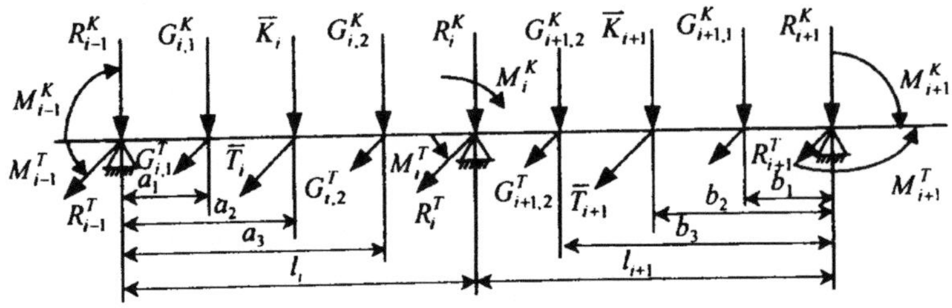

Fig. 4. Continuous beam model schematic diagram

Gu Yujiong[16] used the four-terminal network model in calculating torsional vibration. The author starts with motor control equations and its general solution, equalizing torsional vibration system to the four-terminal network model based on the principle of electromechanical analogy; then adopts mechanical impedance method to obtain frequency equation controlling torsional vibration according to the input impedance of the system and the resonant characteristics, and then obtains the natural frequency, vibration mode and stress distribution, etc. of each order. Four-terminal network model is an accurate low-order model, whose algorithm is convenient and fast. The physical significance is obvious to analyze the mechanical impedance of the system, thus it is a good attempt to model torsional vibration. He Shanghong and Duan Jian [17] also used network method in calculating torsional vibration, and based on dynamic in the process of modeling, which was also a kind of deepening of the method.

Through the analysis of different vibration mathematical models, Xiang Jianhua [18] proposed a graphical modeling method based on system matrix method solving the axis of torsional vibration. The modeling method only requires users providing original torsional vibration mechanics model, and is not restricted by axis branches and modeling scale. In actual implementation, torsional vibration module is divided into two kinds of module unit in this method, which can be used to build various kinds of torsional vibration mechanical model. Model topology relation can be generated through the module traverse and equalisation conversion of the torsional vibration model, and finally the system integration required for solution is integrated.

2.2 Axis modeling research direction

Thanks to the development of modern computer, the precise calculation for shaft can be easily realized by finite element method. So for the continuous mass model, the main development direction is how to make the model have better simulation with material object in computer modeling, thus to peer analysis of the model to material object.

, As the model parameters (especially rigidity parameters) of a simple mass - spring model are obtained by a large number of experience formula and approximate calculations, as its accuracy is hard to ensure, resulting in rather great error between calculated results and actual machine test results. The reason that theoretical calculation result has lager error is often not because of the calculation method of itself, but lies in the accuracy of the model. Since rotational inertia has a relatively accurate analytic calculation, and the torsional vibration damping of the axis is small, to improve the accuracy of the models, focus should be on the amendment of rigidity parameters. Multi-segment concentrated mass model also has similar problem.

The establishment and derivation of a new model should be based on the amendment with material object, adopting all kinds of similar models used in other industries to derive, thus further perfect engine crankshaft torsional vibration model. The influence of damping should not only be considered, the influence of bending-torsional mixture should also be taken into account. Now, many scholars apply model reduction method [19]used in dealing with torsion vibration of steam turbine and generator as well as the method [20]used in identifying parameters of experimental data to the calculation of the engine torsional vibration. This is also the evitable trend of torsional vibration integration.

3 Solving method of torsional vibration of internal combustion engine

3.1 Common method of Torsional vibration

Based on the above-mentioned several shaft models, the common methods and algorithms solving torsional vibration for multi-freedom free vibration calculation include Holzer method, system matrix method and transfer matrix method, etc. The methods for multi-freedom forced vibration calculation include energy method, amplification coefficient method and system matrix method, etc. With the development of computer technology, the traditional manual calculation has been replaced by computers gradually, while some common calculation methods of torsional vibration emerged, such as modal analysis method and finite element method, etc. Various methods are described below.

3.1.1 Holzer method.

Holzer method [1, 2] is always a classic and effective solution in "free- free" system of power machine. The Holzer form method or the Tolle form method derived from its basic principle are often used in engineering. The Holzer method, widely used today, is a numerical calculation method and corresponding calculation program derived from its principle. The advantage of this method is clear physical concept.

From its scientific name, this method can be called "method of sum of torsion moments". The basic idea is: the sum of inertia moment of each lumped mass (disc) should be zero when the shaft doing free vibration without damp, that is

$$\sum I_k \ddot{\varphi}_k = 0$$

Due to the characteristics of simple harmonic oscillator, the relation between the displacement α_k and the acceleration $\ddot{\varphi}_k$ of each inertia I_k is:

$\ddot{\varphi}_k = -p^2 \alpha_k$, namely, $\sum I_k p^2 \alpha_k = 0$. This is the foundation of Holzer method.

This method is effective in estimating low order torsional vibration frequency in initial design stage. This method has simple algorithm and is easy to use, thus is widely used in actual engineering. But its higher-order calculation has lower precision and is time consuming.

3.1.2 System matrix method

System matrix method [1, 2, 21] is a method using each parameter matrices of torsional vibration equation of the axis to solve characteristic root to calculate torsional vibration. All methods which can calculate the eigenvalue and eigenvector of the matrix can be used to calculate the free torsional vibration of multi-mass system.

The basic principle of this method is: for more freedom vibration equation: $\mathbf{I}\ddot{\varphi}+\mathbf{K}\varphi=0$ by assuming the form of solution and input them into equation, the following can be obtained: $\mathbf{KA}=\omega_n{}^2\mathbf{IA}$ let $\lambda=\omega_n^2$ and $\mathbf{H}=\mathbf{I}^{-1}\mathbf{K}$, now the system matrix of $W_M=W_C$ .can be obtained. So free vibration calculation can come down to the question of solving characteristic equations $|\mathbf{D}|=0$.
System matrix method is widely used, not only in free vibration solution, but also in the solution of forced vibration. However, it's generally only suitable for solving low frequencies and its accumulative error would be bigger for calculating high frequencies.

3.1.3 Transfer matrix method

The transfer matrix method [16,22] is a commonly used method for analyzing various vibration problems, which was first introduced by Holzer to analyze crankshaft vibration and calculate the inherent frequency of undamped-free vibration of the shafting.
The basic concept of transfer matrix method is: decomposing the studied system into several two-terminal elements with simple mechanical properties, and building relation between the state vectors of the two terminals of one component by transfer matrix. Then, connect all components one by one, and multiply them together to obtain and solve the transfer matrix. Internal combustion engine shafting, according to its composition configuration characteristics, can be divided into three kinds of components: inertial disks - viscous damper components, elastic elements and even elastomer shaft section components.
The advantage of vibration calculation by transfer matrix method is that the order of transfer matrix will be not affected by the increased unit number, namely, the dimension of matrix will not increase with the increase of the freedom degree of the system, and the calculation method of each order vibration mode is identical. So with simple calculation, convenient programming and less memory for calculation and less time consumption, this method is widely used in the analysis and research of crankshaft vibration. However, when analyzing complex shaft with many freedom degrees by this method, due to the error accumulation of the transfer matrix, the calculation accuracy will decrease, thus the precision of higher frequencies computation is relatively low.

3.1.4 Energy method and amplification coefficient method

Both energy method and amplification coefficient method[2,23] belong to the resonance calculation method of forced torsional vibration, which are basic and the most important calculation methods of torsional vibration before electronic computer popularized. They are still widely used at present. The basic principle of energy method is that the input energy of the exciting moment within a system vibration period is completely consumed by system damping, namely, $W_M=W_C$. Amplification coefficient method was proposed by Tuplin in 1930's for resonance calculation, and then was further developed, becoming a guiding method that the Shipping Standard of British Lloyds Register recommends.

3.1.5 Modal analysis method

The basic thought of modal analysis method [24-25] is to decompose complex multi-freedom system into several sub systems. Firstly, when analyzing, compute the several lower modes of each sub system, then assemble each sub system into an integrated motion differential equation set according to displacement compatibilities or force balance relations between adjacent sub systems to derive comprehensive eigenvalue problem of shrinkage of Freedom

Degree., thus work out the inherent frequency A vibration mode and response of the system. Since modal analysis method reduces the freedom degree of the system, the time consumption and memory for calculation are significantly reduced compared with finite element method. If the sub systems are divided reasonably, its calculation precision is also satisfactory. In addition, modal analysis method can also combine with experimental research [26] to obtain system vibration modal parameters by measuring the transfer function of shaft vibration, e.g., natural frequency, vibration mode, damping, modal inertia and modal rigidity, etc.

3.1.6 Finite element method

Finite element method [1,2,7-10] is a numerical calculation method for solving mathematical physics equation based on variation principle. Its basic thought is to regard complex structure as finite set of discretized units. Each unit is connected into a unity through the common point of the neighboring units, namely, "joints". Take each unit as a continuous component and joint displacement as generalized coordinate. To establish torsional vibration mechanics model of the shafting of internal combustion engine, we need to define which units are selected as well as load positions and sizes, etc. Finite element method is currently accepted as with the highest calculation precision for torsional vibration calculation.

3.1.7 Substructure analysis method of the torsional vibration of systems with branch shafts [27]

In the torsional vibration analysis of shaft systems with branches, the main commonly used methods are transfer matrix method, matrix iterative method and system matrix method, etc [28-31]. But these methods are mainly used to analyze straight string structure or a particular branch. Their calculation efficiency is relatively low for the whole branching structure system. In general, substructure method has already formed systemic theory [32-33], whose basic idea is to divide large and complex structure system into several substructures and calculate the dynamic characteristics information of each subsystem by finite element method, analytical method and experimental method, and then integrate them into the dynamic characteristics of the whole structure system. But substructure method is used less in torsional vibration analysis of shaft systems with branches. Representative method is dynamic substructure matrix method. This method requires working out the compatible relation among all substructures in substructure integrating, which leads to the complicated and tedious modal synthesis process in case the amount of divided substructures is large. Thus this method has certain limit in solving complex shaft systems with branches.

Chu Hua [34] and Z P Mourelatos [35] combined substructure method with transfer matrix method in torsional vibration calculation, which became a new migration substructure method, and the new method was compared with finite element method.

According to the structure of shaft systems with branches, Li Shen and Zhao Shusen [27] put forward a method that divided substructures and integrate step by step based on gear meshing form to obtain the torsional vibration inherent characteristics of the whole system. This method was also used in analyzing the torsional vibration inherent characteristics of the structure of the main transmission system with branches of 650 rolling mill. The comparison with the results calculated by other methods shows the feasibility of substructure graded division and stepwise integration method. This principle has expanded the application scope of the substructure modal synthesis in solving the torsional vibration of shaft systems with branches and effectively solved the problem of complex system, thus provided a good idea for solving the torsional vibration of shaft systems with branches.

3.1.8 New research methods for torsional vibration

In recent years, the number of scholars engaged in research of vibration has continuously increased and new algorithms kept on emerging continuously, such as elastic wave propagation method, eigenvector method and frequency analysis method, etc. According to the theory of torsional elastic wave, Bogacz [36] gave out a method to solve torsional vibration dynamic response by torsion wave method. Shu Gequn and Hao Zhiyong [11,37] also presented a new torsional vibration response calculation method based on the theory of torsional elastic wave, whose basic thought is: the torsional vibration of the shafting is caused by the torsional elastic wave propagation along the shaft; elastic wave propagates along the axis forward and back in traveling wave form; when one traveling wave meets with another after reflection or delay, , both waves will stack into standing wave causing torsional vibration if their phases are appropriate. The method can be used to analyze continuous parameter distribution boundary, transient response and steady-state response of the crankshaft axis with transient boundary conditions and other vibration characteristics. Since it only requires solving linear equations in calculation, its computational complexity is small, thus it is an accurate and fast vibration analysis method. According to electromechanical analog principle, Gu Yujiong [16] put forward a four-terminal network method for analyzing torsional vibration. State vector method, proposed by He Chengbing [38], was widely used in the analysis of torsional vibration. People are also exploring the calculation method of continuous mass model for torsional vibration response of forced vibration. Wang Ke she and Wang Zheng guang[39] used frequency analysis method in the calculation of torsional vibration to combine frequency change with the structural parameters of shafting, which was beneficial to visual analysis. It also worked out analysis mode, resonance frequency and resonant modes, etc.

3.2 Research direction of shafting solving methods

At present, the methods for solving torsional vibration are various and each has its own use. While in general, it shall be developed from the following aspects:

1. Improve the computation efficiency of current methods. For instance, calculation precision is high by finite element method, but its calculation is time consuming and resources occupying, so fewer and dimension-reduced units should be considered in model building when improving this method.
2. Combine the calculation method used in other industries with this direction.
   - 2.1 For example, Shen Tumiao [16] mentioned to apply electrical four-terminal network in the calculation of torsional vibration . This is an example of unified calculation method. In addition, integrating all calculation methods to construct a new method is also one of the research directions.
   - 2.2 Based on the torsional vibration numerical simulation study of the internal combustion engine shaft with the precise time integration method, Lin Sen [40] introduced and deduced the precise time integration based on Duhamel integral, described in detail the calculation characteristics of this method with example and comparison, simulated the torsional vibration of some type of internal combustion engine, compared the results with the calculation results

in literature, and analyzed their similarities and differences briefly, which, to a certain extent, solved the conflict between the accuracy and stability of calculation.

2.3 Along with the development of microcomputer technology, we can use professional software to analyze torsional vibration of internal combustion engine. Tong-Qun Han [41] introduced the functions and characteristics of engine simulation software EXCITE—designer developed by AVL company, analyzed torsional vibration and vibration reduction of the shaft system based on the software targeting at the problem of a car engine flywheel bolt fracture, and put forward correcting measures.

4. Experimental studies on engine crankshaft

4.1 Current torsional vibration measurement methods

Torsional vibration measurement is an important content in the study of crankshaft vibration. Compared with transverse vibration measurement, the extract and analysis of torsional vibration signals are both difficult. There are basically two kinds of torsional vibration measurements: contact measurement and non-contact measurement. The former installs sensor (such as strain gauge, accelerometer, etc) on the shaft, and the measured signal is transmitted to instrument by collector ring or radio signal. Non-contact measurement commonly uses "measuring gear method", which uses shaft encoder, gear, or other repeated structure to measure angular velocity in homogeneity to measure torsional vibration. If designing Doppler test method properly, laser can also be used to measure torsional vibration. The followings are introduction to various methods and analysis of their error sources and applications.

4.1.1 Mechanical measurement

Geiger torsional vibration analyzer is a typical mechanical torsional vibration measuring instrument [2, 23] and was used in torsional vibration study in the earliest stage. This instrument is designed dexterously, whose signal acquisition and signal record are both realized by mechanical devices, simple and practical. It is widely used in the study of torsional vibration. DVL torsional vibration instrument also belongs to this type of torsional vibration instrument. However, the torsional vibration of this method is transmitted to measuring head shelf by belt, the belt elastic vibration will cause distortion. The response bandwidth of mechanical measuring system is very limited, and also because disc springs cannot be too soft, so very low frequency torsional vibration cannot be measured. In addition, the measured signal cannot be analyzed directly by means of modern analytical instrument, thus it has gradually been eliminated.

4.1.2 Contact measurement

Contact measurement [42] is to install sensor (such as inductance, strain gauge, etc.) on crankshaft directly. The measured signals are transmitted to analytical instrument by collector ring or in radio frequency manner. To monitor the dynamic response of shaft or shaft parts (e.g., blade, etc.), the arrangement of strain gauge should eliminate the interference of transverse vibration, and can realize the automatic compensation of influence by temperature. Torsional vibration meters belonging to this measurement method include strain-gauge torsional vibration meter, piezoelectric torsional vibration meter and inductance-type torsional vibration meter, etc. Contact measurement, centered by sensing element, is widely used in the vibration test of internal combustion engine, thanks to its high sensitivity, wide frequency response range and convenience for measured signal record and analysis. But this kind of measuring device system itself has certain rotational inertia, which will inevitably impact on the system under test in measurement. In all kinds of contact measurement, measurement devices, such as sensors, are required being installed on the shaft, which sometimes has to destroy the original shaft structure. This is not allowed in many cases.

4.1.3 Non-contact measurement

The measurement device of non-contact torsional vibration measurement [43,44] is not installed directly on the crankshaft, but collects torsional vibration signals through photoelectric and magnetoelectricity conversion by code disc, gear or other indexing structure on the crankshaft. These kinds of method are based on the principle of "gear testing". When the shaft is rotating, the teeth structure installed on the shaft can induce bell shaped pulse leveling signal sequences on the sensor, whose amplitude and phase might carry the information on axial torsional vibration, which is demodulated by phase detectors into torsional vibration signals. Torsional vibration meters belonging to this measurement methods include TV - 1 torsional vibration meter of British Econocruise Company, VED - 233A torsional vibration meter of American Shaker Company and DTV - 88 torsional vibration meter[45] developed by Shanghai Institute of Electrical Equipment, etc. Non-contact measurement method does not need installing special devices on the shaft, but uses the existing shaft repeated structure, whose measurement preparations is less, and measurement process does not interfere with the normal operation of shaft. It's especially suitable for the long-term monitoring of torsional vibration. At present it has become a major means of torsional vibration measurement.

4.1.4 Laser measurement

Laser Doppler Torsional Vibration Measurement Technique [46-48] is put forward and developed from fluid velocity measurement. When laser beam irradiates on shaft surface, the linear velocity of shaft surface make scattered light produce Doppler frequency shift. The transient angular velocity of shaft represents the transient value of frequency shift volume of the instantaneous axis. Torsional vibration is obtained by removing dc component. 2523 torsional vibration meter launched out by the Denmark B&K Company was a typical representative of Doppler laser torsional vibration meter. In 1994, Ge Weijing [49] and others from Tianjin University applied laser Doppler velocimetry on the torsional vibration measurement of internal combustion engine shaft. Only a smooth section on the surface of the shaft is required, and measuring point is easy to be set up. This method can realize absolute measurement and measurement datum is not required to be specially established. However, since the transverse vibration of the shaft and the form and position errors of the shaft section directly affect measurement precision, it is rather difficult to improve its accuracy.

4.1.5 The latest torsional vibration measurement method

Wang Ting and Cheng Peng [50] introduced a kind of digital measurement system of crankshaft using PC computer. The measurement system consists of angle encoder, self-made count plate, PCL724 digital input/output card and PC computer, etc. installed on the tested crankshaft. Angle encoder is crankshaft angle sensor with high precision, and is connected with crankshaft by flange. The grating disc fixed in the angle encoder has two reticules, e.g., outer and inner rings. The outer ring is a uniform reticule, which can produce CDM signals, and the inner ring is a TRIG reticule for judging tdc signal. The light emitting components in the angle encoder are two infrared light emitting diodes, and there are two infrared light receptors respectively corresponded with CDM reticule and TRIG reticule. When angle encoder operates together with crankshaft, a TRIG signal will be outputted in each rotation, and a series of CDM square-wave pulse signal will be outputted. Thus, the crankshaft torsional vibration can be directly reflected on the time width of the CDM square-wave pulses outputted by angle encoder. Count each CDM pulse width with frequency division by high count circuit board. The counted data is inputted into PC by parallel data I/O card PCL - 724. Then crankshaft torsional angle can be obtained after program processing. The measurement system measures torsional angle directly, thus it's with convenient measurement, high precision and simple process.

4.2 Research direction of torsional vibration measurement of internal combustion engine

The focus and future development of torsional vibration measurement is to improve its accuracy and real-time performance to realize the torsional vibration monitoring of internal combustion engine in operation, especially the monitoring of severe torsional vibration caused by emergencies, such as the severe torsional vibration by transient large torque incentive resulted from cylinder flameout. At the same time, eliminating the interference of lateral vibration and establishing reliable measurement datum are still the problem requiring to be solved. Finally, the problem of system calibration should also be solved.

5. The latest research direction of torsional vibration of internal combustion engine

The traditional research methods of torsional vibration can not meet the needs on the precise study. In recent years, many scholars have continuously broadened research field and scope to further explore the various problems of shaft torsional vibration, making the research on torsional vibration closing to ideal level unceasingly. Some main research directions of torsional vibration in resent years are introduced as follows.

5.1 Nonlinear research

With the further research on torsional vibration of the shaft, many nonlinear vibration problems are met [51-55]. At the same time, crank shaft is a complex nonlinear system, thus it often needs to consider all sorts of complex nonlinear factors to construct a model that can reflect actual system. However, current relevant studies are mostly on single degree-of-freedom nonlinear vibration problems that considers single factor, which obviously cannot meet the need on the accurate calculation of crankshaft vibration. Therefore, it is necessary to further consider the crankshaft nonlinear vibration problems with multiple nonlinear factors, multi-degree-of-freedom and even continuous mass distribution. In the current calculation models, equivalent moment of inertia (constant) is usually adopted to consider the inertia of piston and connecting rod. But in fact, the inertia of engine crank module is:

$$I = I_0\left[1 - \varepsilon\cos(2\varphi)\right]$$

Where ε is variable inertia coefficient, a value below 1. We can see from the formula, the moment of inertia of crank component is a variable related with rotation angle. Sheng Gang [56] researched on the solution methods of some simplified models of single cylinder engine and established the equation of motion of crankshaft vibration of multi-cylinder engine under the condition of considering variable inertia. At the same time, in literature [1], the problems of variables caused by machining error and assembling error were also considered. In literature [57], forward and inverse Fourier transformation was applied to numerically solve nonlinear torsional vibration system. While Lin Ruilin [58] took the diesel engine shaft with a third-order rigidity component as research object, deduced the calculation formula and numerical calculation formula iterative procedure for solving periodic response of nonlinear torsional vibration by incremental harmonic balance method (IHB). This method is used to solve linear and nonlinear torsional vibration response of diesel engine shaft. Compared with the existing methods, it is more effective to solve strong nonlinear vibration response. What's more, it has virtues of less operation time and accurate calculated result.

The discussion about nonlinear components is mainly concentrated on the non-linear shock absorber, coupling and other components. Literature[59] of as early as 1987 analyzed the nonlinear problems of diesel engine shaft with piecewise linear components (cylindrical spring-loaded buffers), and calculated vibration response of shaft by step-by-step integration method. Gong Xiansheng[55] introduced theoretical and experimental research on the calculation method of steady state vibration response of marine propulsion shafting with hysteretic nonlinear coupling subjected to eccentric mass exciting force action. Farshidianfar [60] solved nonlinear problems of driving shaft by substructure modeling, and compared the results with the results of whole structure modeling.

The research on nonlinear torsional vibration has made many important achievements [50]. But so far, the majority of nonlinear torsional vibration problems are still analyzed by some approximate methods or by ignoring nonlinear factors, in most cases, the results obtained have greater errors compared with actual results. Therefore, there are still many problems waiting to be solved in further exploring the nonlinear problem, mainly including:

1. The modeling, system parameters identification method and test of complex nonlinear torsion vibration problems;
2. Accurate solving methods for multi-degree-of-freedom strong nonlinear torsional vibration problems;
3. Self-excited vibration of complex nonlinear torsional vibration system;
4. Decoupling, numerical calculation and optimization methods of complex nonlinear structure.

5.2 Coupling vibration analysis

Torsional vibration of shaft has huge harm on the system, so people paid attention to and researched on it at very early period. However, many phenomenon produced by vibration in practice need to take longitudinal/bending/tortional vibration together into account. The bending vibration caused by unbalanced mass has certain weight on the torsional direction and can couple to the torsional vibration; on the other hand, torsion also has certain weight on horizontal and vertical directions, thus couples to the bending vibration. In recent years,

significant progress has been made in the aspects of theoretical calculation method and testing technology of longitudinal/bending/tortional coupling vibration.
Li Bozhong [62] discussed about the axial vibration problems caused by torsional vibration and established a relatively simple analysis model for this kind of model. In the following two literatures[63,64] of same series, the longitudinal twist coupling vibration was tested and further analyzed and the coupled vibration model was established, and the model calculation was compared and analyzed with actual measurement. In paper [65], the author put forward a kind of spring - mass model with non-linear rigidness being used in calculating torsional - vertical coupled vibration of engine shaft. It explained the doubled-frequency problem of the torsional - longitudinal coupling, and also revealed the presence of quadruple frequency and octuple frequency in the longitudinal - torsional coupled response. It is more reasonable than just simply giving an assumption doubled-frequency excitation torque in the right of the motion equation. Zhang Yong and Jiang Zikang[66,67,68] adopted distributed mass model in analyzing bending - torsional coupling vibration of shaft, which divided the actual unit shaft system structure into several sections with equal diameter according to orders in simplifing, treated each segment as continuous mass, and listed the vibration differential equation of each segment, then united them to solve. Finally, some results of the analysis for bending - torsional coupling vibration of shaft by numerical method were given. In literature [69], system matrix model was established for longitudinal twisting coupling vibration of shaft, whose general rule of coupling vibration was studied based on the calculation and analysis of the practical examples of longitudinal twisting coupling vibration of shaft. In this paper, the test equipment used for measuring coupling vibration is only eddy current sensor for non-contact measuring the condition of axis vertical vibration. In contrast, multi-dimensional measurement is relatively rare.
Okamura[70] and Shen Hongbin[71] used the longitudinal / bending / tortional vibration test device in all research processes of shaft vibration. This kind of measuring device can acquire three-dimensional vibration signals simultaneously. As shown in figure 5, an electromagnetic sensor (measuring torsional vibration signal) and three acceleration sensors (of which, one measuring longitudinal vibration signals and the other two measuring bending vibration signals) are mounted on its shell. It shows that testing technology has also been developed from single parameter measurement to multi-parameter measurement method.

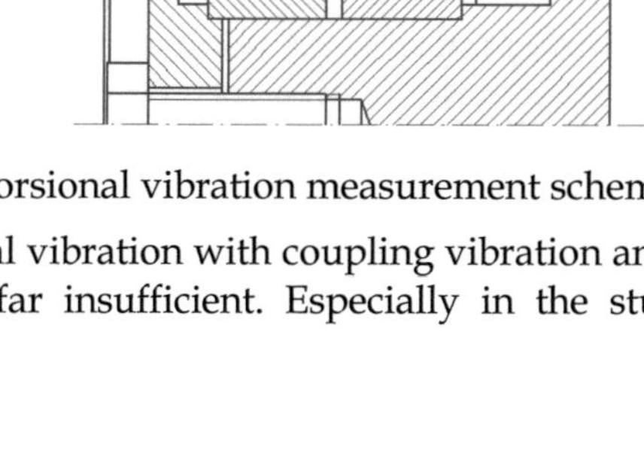

Fig. 5. Three-dimensional torsional vibration measurement schematic diagram

For the research of torsional vibration with coupling vibration and transversal vibration, the current research level is far insufficient. Especially in the study of theoretical models, traditional method can not unify the physical model and mathematical model of coupling vibration simultaneously. In view of this, the future research on coupling vibration can roughly focus on the following aspects:

1. Mechanism research on coupling vibration and relay relation of mutual excitation;
2. Model unification and solve using universal algorithm;
3. Precise treatment of measuring equipment, long distance measurement and the implementation of long-term test, etc.

5.3 The analysis of torsional vibration response based on multi-body dynamics of soft body crank shaft

The forces on engine are very complex. Traditional analysis method is to calculate rotation inertia and reciprocating inertia produced by each force based on the motion analysis of each component, then combine them with the maximum combustion pressure of gas to solve the force on the main body and excitation force of shaft vibration. This is a very complicated process[73]. By using mechanical system simulation software ADAMS, by establishing crankshaft multi-body dynamics model including pistons, connecting rod, crankshaft and flywheel, we can not only calculate the motion law and the force among each component, but can also further analyze balance and vibration. Due to the interaction between inertial load and transverse bending deformation of shaft and the coupling behavior with lubrication problem, the bearing load problem based on rigid body dynamics becomes complicated, and there exist errors in calculation precision. If transforming engine crankshaft into flexible body, the tiny deformation can guarantee the completely accurate dynamic equation to deformation generalized coordinates first-order items. In order to sufficiently study the effect of crankshaft flexible body on the calculation results of dynamics, based on the finite element analysis of crankshaft system and by establishing rigid-flexible coupling multi-body dynamics system model with multiple degrees, Liang Xingyu and Shu Gequn[72] analyzed the torsional vibration response of crankshaft system that constitutes main flexible body, and obtained the time history response of system dynamics, and then made assessment on the power quality and safety of the system. Then it measured the torsional vibration of the crankshaft free end of an inline four cylinder diesel engine with a newly developed test device. Through calculation and comparison between sub-harmonic analysis of test results, both reflected higher equality, and explained the correctness of rigid-flexible coupling multi-body dynamics system model.

5.4 The method of compensating divisional error in shafts torsional vibration measurement and program implementation [74]

Now non-contact measurement method are generally used for torsional vibration measurement, namely, by using repeated structure in the shaft, pulses are produced in non-contact sensor, and the interval dimension reflects the transient angular velocity dimension of the shaft. The shaft torsional vibration information can be obtained by processing interval data of the pulses. When measuring torsional vibration by this method, the indexing error of the shaft repeated structure directly influences the precision of measurement results. If indexing error is very great, the measurement results will have serious distortion. In the measurement of torsional vibration by non-contact measurement method in practice, selected sensors mainly include photoelectric encoder, hall sensor and photoelectric sensor. The three kinds of sensors in practical measurement have their advantages and disadvantages. Photoelectric encoder has large indexing number and small indexing error, so its measurement is more accurate. But its installation is inconvenient and it requires using shaft coupling to connect with measured shaft. If encoder shaft is eccentric with measured shaft after installation, transmission eccentric error will be introduced[75]. Hall sensor requires gearing disc with equal division to measure. Since the teeth number of equal division disc is usually less than the indexing number of encoder, and gear disc has certain indexing error in processing, its accuracy is lower than that of encoder. But its installation is convenient. It can be directly installed on the measured shaft, or can directly measure by gear disc on the shaft without additional modification on the shaft. The use of photoelectric sensor is of the most convenience. It needs only uniformly pasting a certain number of reflective strips on the component with circular surface of the shaft. If the shaft is very thick, those reflective strips can be directly pasted on the shaft. However, currently, the reflective strips can only be manually pasted. Great degree error will definitely occur leading to the distortion of measurement results. So in the actual torsional vibration test, if the indexing error of selected repeated structure can not be ignored, such as selected manually pasted turntable of reflective strips, test results should be dealt with to compensate for the effects of indexing error, then correct torsional vibration information can be calculated. Guo Wei-dong [74] described the compensation principle of indexing error in detail and listed the compensation program of indexing error compiled based on LabVIEW. So we could find out from test results that the data curves after the compensation of indexing error became smooth, and the effect that indexing error on measured results is obviously reduced and test result is more accurate.

6. Control technologies in torsional vibration of internal combustion engine

For the internal combustion engine with reciprocating motion, due to the property of periodical work, the torque on the shaft is a periodic compound harmonic torque, and then forms excitation source. When the frequency of the excitation source is equal to the inherent vibration frequency, resonance phenomenon will occur, and torsional vibration will be subjected to huge dynamic amplification effect, then the torsional stress on the shaft greatly increases, leading to various accidents on the shaft, and even fracture. These are the causes and consequences of torsional vibration.

To avoid the destructive accident of torsional vibration of internal combustion engine, it's not only required to conduct detailed calculation of torsional vibration in design phase, torsional vibration measurement is also required timely after manufacturing completion. This can not only check and modify the theoretical calculation results, but also detect and so as to solve the torsional vibration problems promptly.

Based on the above analysis, main vibration control technology includes two parts: study on the avoidance of vibration and on shock absorber.

6.1 Study on the avoidance of vibration

If great torsional vibration does exist on internal combustion engine according to the calculation of and actual test on torsion vibration, proper measures shall be taken to avoid or remove it.

There are a lot of preventive measures for avoiding torsional vibration [61], classified roughly into the following two methods.

6.1.1 Frequency adjustment method [2, 76]

According to torsional vibration characteristics, when the frequency of excitation torsional vibration is equal to some inherent frequency n_w of torsional vibration system, extremely severe dynamic amplification phenomenon will occur, namely resonance phenomenon, thus the possibility of $w = n_w$ shall be avoided, i.e., avoidance of the most severe conditions of dynamic amplification means the possibility of avoidance of all consequences caused by excessive torsional vibration. The basic concept of this method is that let w actively avoid n_w. The main measures of this kind of method include: inertia adjustment method and flexibility adjustment method, etc. By adjustment, let the natural vibration frequency of the system itself avoid excitation frequency. Reduce vibration stress to be within the instantaneous allowable stress range, thus avoid the damage on engine by bigger torsional vibration. This method is one of the most widely applied measures in torsional vibration prevention measures, not only because of it being a simple and feasible measure, but also because of it being effective and reliable when meeting the requirement of frequency modulation. But its disadvantage is small scale of frequency modulation, which restricts its practical application.

6.1.2 Vibration energy deducing method [23]

Incentive torque is the power source causing torsional vibration. Since the input system energy of incentive torque is the source of maintaining torsional vibration., if the vibration energy of input system can be reduced, the magnitude order of torsional vibration can also be reduced immediately. One way is to change the firing sequence of internal combustion engine. When the dangerous torsional vibration is deputy critical rotation speed within machine speed range, this method might be used to reduce the dangerous torsional vibration and reduce the risk degree. The second method is to change crank arrangement. Deliberately choosing unequal interval firing in multi-cylinder engine and appropriately choosing crank angle to change crank arrangement can let some simple harmonic torsional vibration in any main-subsidiary critical speed counteract mutually to avoid dangerous torsional vibration. The third method is to choose the best relative position between crank and power output device, make the disturbance torque between them counteract mutually, which can reduce the torsional vibration of the crankshaft.

6.1.3 Impedance coordination method

Considering the complexity of solving above problems by the conventional dynamics method, energy wave theory can be used to solve this problem. According to energy wave theory and by coordinating the impedance of various component loops, resonance can be avoided to realize the target of reducing vibration intensity. Impedance coordination method can modify the inferior design in design phase, or design directly correct transmission shaft system, to ensure the shaft working with sound dynamic characteristics without resonance and reducing dynamic load.

6.2 Study on shock absorber

As is known to all, engine installed on shock absorber can greatly reduce the vibration transmitted to the foundation. Likewise, torsional vibration can also be eliminated before it reaches the foundation. If vibration reducing device is installed on the front head of the crankshaft of the engine, then shock absorber will absorb the torsional vibration of rotating shaft generated by engine. It shows the important role of shock absorber in internal combustion engine system. The technical requirements on shock absorber are very high, mainly including: elastic material strength should be reliable in use and storage, the fixation with metal should be firm, rigid fluctuation range in installation stage should be small, and technical characteristics do not change with time.

Now, main shock absorbers include the following kinds: dynamic shock absorber, damping shock absorber and dynamic-damping shock absorber.

6.2.1 Dynamic shock absorber [2, 23, 77]

This kind of shock absorber is connected with crankshaft by spring or short shaft. By the dynamic effect of shock absorber at resonance, an inertia moment with the size and frequency same with excitation torque, but direction opposite to excitation torque is produced at the vibration reduction location to achieve the purpose of vibration reduction. This kind of shock absorber doesn't consume the energy of the shaft. They can be divided into two types: one type is constant fm dynamic shock absorber, namely undamped elastic shock absorber, shown in the schematic of figure 6, and the other type is variable fm dynamic shock absorber, such as undamped tilting shock absorber, drawing as shown in figure 7.

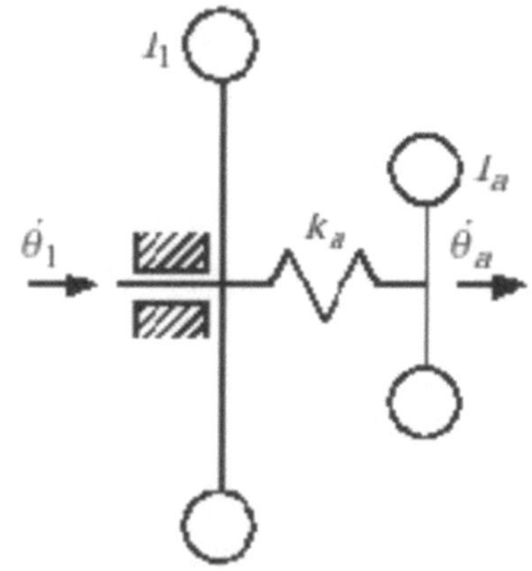

Fig. 6. Undamped elastic shock absorber schematic diagram

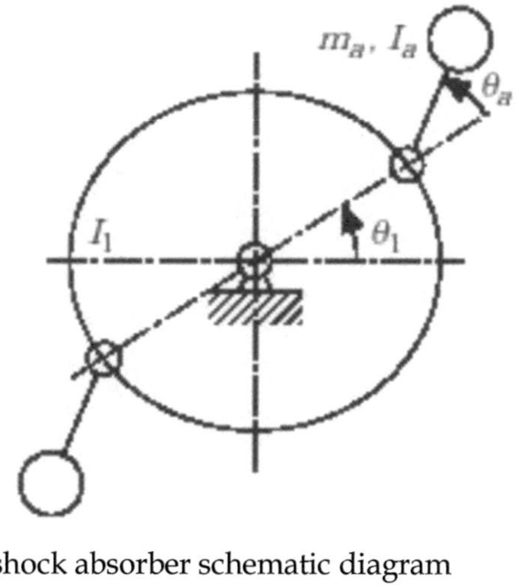

Fig. 7. Undamped tilting shock absorber schematic diagram

6.2.2 Damping shock absorber

Damping shock absorber achieves the purpose of vibration reduction by damping consuming excitation energy (shown in schematic diagram 8).The main type is silicone oil damper[78,79], whose shell is fixed to the crankshaft, high viscosity silicone oil is filled between ring and shell. When the shaft is under torsional vibration, the shell and the crankshaft vibrate together, and the ring moves relatively with the shell due to inertia effect. Silicone oil absorbs vibration energy by friction damper, thus reduce vibration for the torsional vibration system. This kind of shock absorber is widely used with simple structure, good effect of vibration isolation and reliable and durable performance.

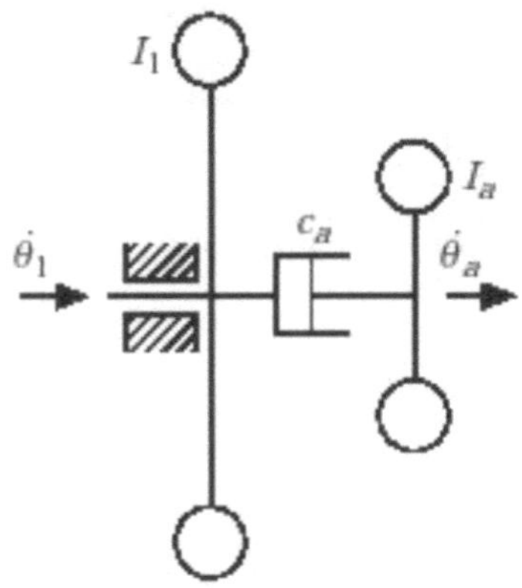

Fig. 8. Damping shock absorber schematic diagram

6.2.3 dynamic-damping shock absorber

Dynamic damping shock absorber features both of the above effects, such as rubber flexible shock absorber [80], rubber silicone oil shock absorber, silicone oil spring shock absorber [81], etc., shown in schematic diagram 9. Theoretically, the effect of dynamic damping shock absorber is the best, since it can not only produce dynamic effect by elastic, but also consume excitation energy by damping. But the elastic elements, such as springs and rubber, etc., that connect the shock absorber and crankshaft often work under great amplitude and high stress, thus the process is relatively complex and the cost is higher.

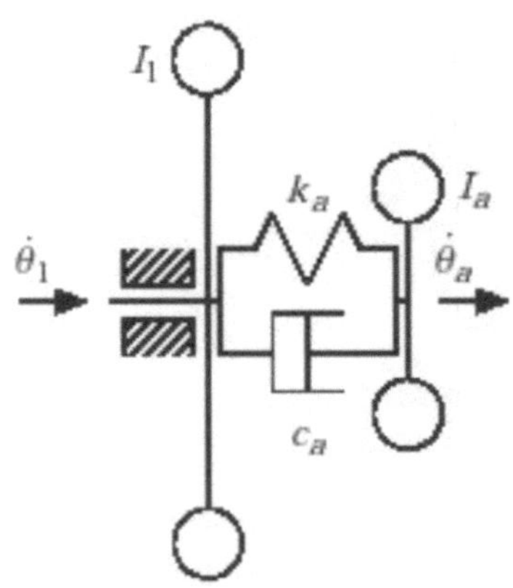

Fig. 9. Dynamic-damping shock absorber schematic diagram

6.2.4 Study on new shock absorber

Along with the deepening of the research on shock absorber, many new shock absorbers have appeared. Several typical kinds are listed as below:

Yan Jiabin [82] proposed an elastic metal shock absorber, which fixed two disks connected with elastic materials at the ends of the crankshaft. disks are tightened in the way that one disk rotates in the opposite direction of the other disk (see figure 10). If loosen both disks simultaneously, they will complete torsional vibration with low amplitude, till stop. At this moment, one section of the shaft rotates in one direction, and another section of the shaft rotates in the other direction. In this case, one end face of the crankshaft will produce displacement. Due to the effect of vibration absorption, the vibration will proceed with reduced amplitude but constant speed, which depends on the internal friction or delayed quantity of elastic material. There are three kinds of elastic shock structure of shock absorber: welding metal elastic elements, combination elastic metal components and welding-combination elastic metal elements. Welding-combination elastic metal elements consist of driving and inertia members that connect each other with elastic material. This kind of shock absorber is suitable for application with simple structure and convenient maintenance.

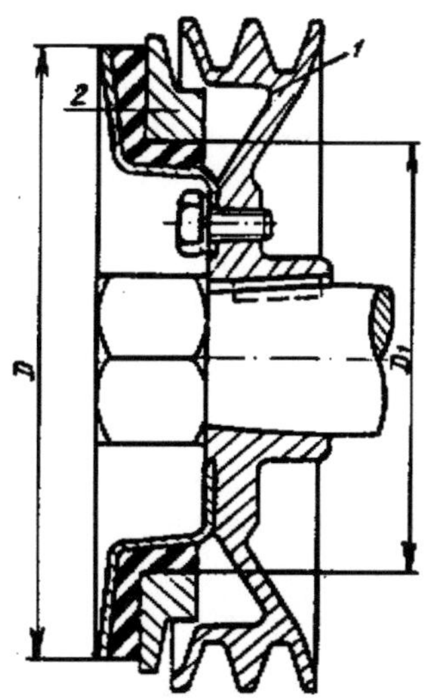

Fig. 10. Monolayer thin-type elastic metal shock absorber

Huo Quanzhong [83] and Hao Zhiyong[84] introduced the research on driving control shock absorber. Figure 2 is the diagram of the shock absorber. The shock absorber itself is similar to a dc motor, whose stator and the shell of the shock absorber compose as a whole entity, and rotor is connected with the shell by radial leaf spring, forming a dynamic shock absorber. The shell of the shock absorber is fixed on the main vibration body. According to the conditions of main vibration body, the regulation apparatus produces control signals with fixed size, phase and frequency, which, by power amplifying, make armature generate control torque (namely, electromagnetic torque). Active torsional vibration shock absorber is feasible both in theory and practice. What's more, its damping effect is better than that of dynamic shock absorber.

Liu Shengtian [85] proposed a double-mass flywheel torsional vibration shock absorber, which was a new type of torsional vibration shock absorber occurred in the middle of

1980's. The double-mass flywheel torsional vibration shock absorber at early period was to remove torsional vibration absorber from the clutch driven plate, place it among engine flywheels, thus double-mass flywheel torsional vibration shock absorber was formed. The basic structure of double-mass flywheel torsional vibration shock absorber has three major parts, i.e. the first mass, the second mass and the shock absorber between the two masses. Relative rotation can exist between the first and the second masses, which are connected with each other by shock absorber.

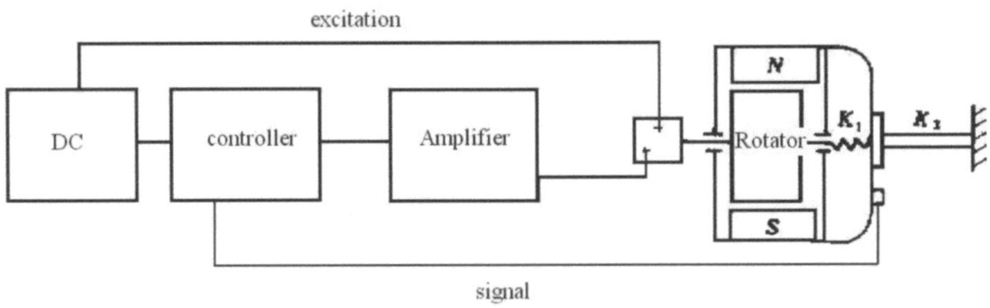

Fig. 11. Active control shock absorber functional diagram

Double mass flywheel torsional vibration shock absorber can very effectively control torsional vibration and noise of automobile power-transmission system. Compared with the traditional clutch disc torsional vibration shock absorber, its effect of damping and isolation of vibration is not only better within the common engine speed range, it can also realize effective control over idle noise. After developing the double-mass flywheel torsional vibration shock absorber, the author also introduced hydraulic pressure into shock absorber and developed hydraulic double-mass flywheel shock absorber [86], which is the latest structure style in the family of double-mass flywheel torsional vibration shock absorber. It lets the technology of car powertrain and noise control of torsional vibration step further into the direction of excellent performance and simple structure.

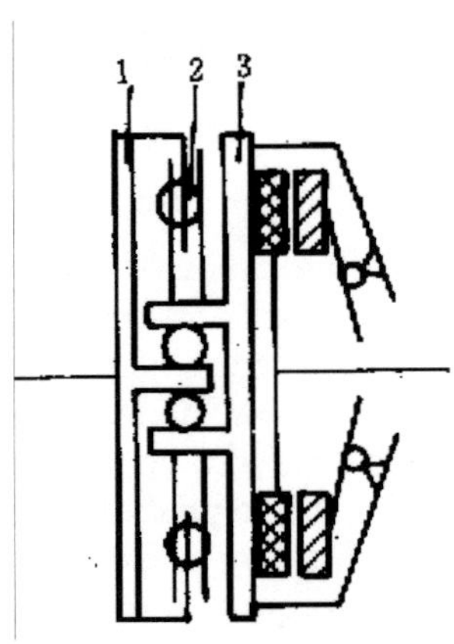

Fig. 12. Double-mass flywheel torsional vibration shock absorber

M Hosek, H Elmal [87] introduced the design process of a kind of FM tilting shock absorber, which was developed based on centrifugal tilting shock absorber, and was named as centrifugal delay type resonator by the author. Based on the study of centrifugal tilting shock absorber, the author installed a sliding globule between the end of pendulum and the rotary table, thus, when the rotation of the shaft fluctuates, the pendulum will delay duet to the effect of damper, while the sliding globule will coordinate actively with the changes. By this shock absorber, minor disturbance can be quickly completely eliminated; and broadband disturbance, especially the disturbance that obviously increases speed, also can be completely eliminated.

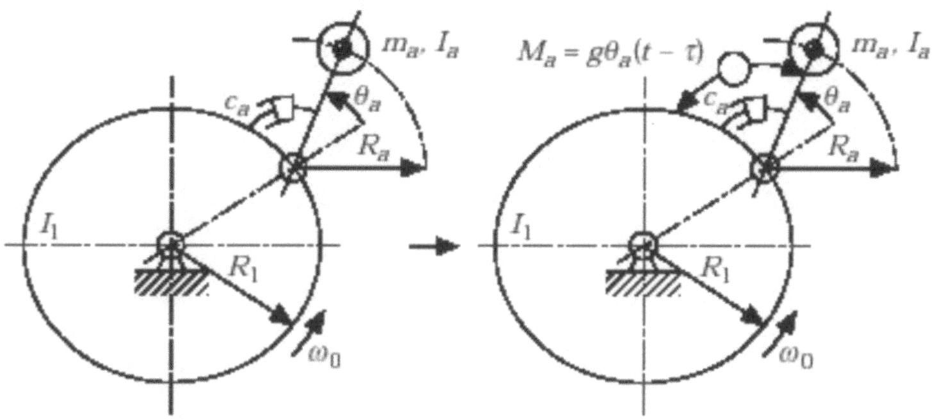

Fig. 13. Centrifugal delay resonator

Shu Gequn [88,89] presented a research approach of coupling shock absorber. Since torsion vibration is the most dangerous vibration mode in shaft vibration, torsional vibration shock absorber is the main damping device, and for coupling damping, bending shock absorber or lateral shock absorber will be installed on the basis of torsional vibration shock absorber. Through the experimental research, the author concluded that, compared with single torsion vibration damper, after installing bending vibration shock absorber, due to the effective damping act on shaft bending vibration, twist/bending shock absorber can control engine vibration and noise effectively. Normally, the parameters setting of bending vibration shock absorber depends on the bending vibration model of crankshaft, but due to its effect on torsion vibration reduction, the design of coupling shock absorber should consider its damping effect on torsional vibration and bending vibration of the shaft. Shu Gequn [90] investigated the effect of bending shock absorber on the performance of twist/bending shock absorber by theoretical analysis.

Through the above analysis, we can see that torsional vibration absorber is being developed towards the aspects of broadband, high efficiency, being timely, multi-function, etc. So research on torsional vibration shock absorber still has considerable prospect, worthy more efforts from scholars. The following aspects can be studied and explored.

1. Study on active control of torsional vibration of internal combustion engine;
2. Study on shock absorber with coupling between torsional vibration with longitudinal vibration and transverse vibration, etc;
3. Study on integrating torsion vibration absorber with clutch or other components of internal combustion engine;
4. Finite element optimization design of torsional vibration shock absorber [91].

7. Utilization of torsional vibration of internal combustion engine crankshaft

Restricted by various factors, we can only decrease the degree of torsional vibration, while the occurrence of torsion vibration is inevitable. Torsional vibration is directly related to the various incentive factors of internal combustion engine, such as the combustion sequence of various cylinders, the change of crankshaft inertia and the sudden change of loads, etc. So how to use torsional vibration signals to monitor the changes of these quantities is the main purpose of utilizing torsional vibration. The utilization of torsional vibration is mainly embodied in identifying faults by torsional vibration signals[92, 93].

7.1 The progress of torsional vibration utilization

The diagnosis of diesel engine faults by the change of torsional vibration parameters of the shaft is a new fault diagnosis technology. Torsional vibration signals of diesel engine shaft often have strong repeatability and regularity, and fault diagnosis by torsional vibration signals is used to diagnose cylinder flameout. Diagnosing cylinder flameout fault by torsional vibration signal of diesel engine has been developed in recent years. The work process fault of diesel engine cylinder directly affect the changes of torsional vibration characteristics, and such changes of torsional vibration parameters also reflect directly the work state of cylinders. Torsional vibration signals of diesel engine shaft can be used as the basis for fault diagnosis. Ying Qiguang explored this issue at the beginning of 1990's of the 20th century, who thought that diagnosis of the technical condition and fault of diesel engine by the response characteristics of the frequency, amplitude (and phase) and damping of shaft torsional vibration is a new and promising fault diagnosis technology. The author judged cylinder flameout fault by the comparison of amplitude size between normal torsional vibration and in the circumstance of cylinder flameout. This method is convenient and intuitive. However, it requires normally torsional vibration amplitude figure for comparison under the same conditions, so its application is limited [94]. In the application of fault diagnosis by torsional vibration, Lin Dayuan and Shu Gequn studied on the sensitivity of various torsional vibration modals and frequency response characteristics on crack by torsional vibration modal experiment, who recommended modal damping, damping attenuation factor, frequency response function modal and self-spectral modal as the optimum evaluation factors for the crack fault, and further discussed the change law between crack and the above evaluation factors, thus provided an effective method of intermittent diagnosis for the engine stops[95,96].

7.2 The development direction of torsional vibration utilization

Fault diagnosis by torsional vibration signal of internal combustion engine crankshaft is a new type of fault diagnosis theory. Developing this theory towards new application field is an inevitable trend in internal combustion engine industry. Thus, the development direction of torsional vibration is mainly oriented to broader fault diagnosis fields and continuously make this achievement become more mature and its application become more skilled.

7.3 Sub-conclusions

1. Damping technology becomes more mature.
2. Multi-function shock absorbers are innovated constantly.
3. Application fields of torsional vibration become more and more wide.
4. Modern design theory is used unceasingly in control field.

8. Conclusions

Research on torsional vibration of internal combustion engine will become more and more deepen with the development of science and technology. Corresponding new research methods will appear in modal building, solving, test and control of the shaft model, making research contents more wide, method more scientific, object more specific and application more direct.

9. Acknowledgements

Authors wish to express their sincere appreciation to the financial support of the NSFC(50906060) and State Key Laboratory of Engines.

10. References

Tan Daming. Vibration control of internal combustion engine [M]. Southwest Jiaotong university press, 1993.

Yu Qi. Torsional vibration of internal combustion engine [M].Beijing: National Defense Industry Press, 1985.

W J Hshen. on the Vibration Analysis of Multi-Branch Torsional System[J]. Journal of Sound and Vibration,1999,224(2):209-220.

Nestorides E J. A Handbook on Torsional Vibration[M]. London: Cambridge University Press, 1958.

Wilson W K. Practical So1ution of Torsional Vibration Problem[M]. London: Campman and Hall, 1963.

A Boysal and H Rahnejat. Torsional Vibration Analysis of a Multi-body Single Cylinder Internal Combustion Engine Model[J]. Appl. Math. Modeling,1997,21(8):481-493.

Wang Changmin, Zhu Dejun. Calculation of Torsional Stiffness of Engine Crankthrow by Finite Element Method [J]. Transactions of CSICE, 1991,9 (2) : 177-183.

Qu xiaoming, Gu yanhua, Wang xiaohu. Modeling and calculation of torsional vibration of internal combustion engine [J]. Internal combustion Engines, 2000,4:6-9.

Li Huizhen, Zhang Deping, Crankshaft Torsional Vibration Calculation by Finite Element Method [J]. Transactions of Csice, 1991,9 (2) : 157-162.

K Koser, F Pasin. Continuous Modeling of the Torsional Vibrations of the Drive Shaft of Mechanisms[J]. Journal of Sound and Vibration,1995,188(1):17-24.

Hao Zhiyong, Shu Gequn. An Investigation on the Torsional Elastic Wave Theory for the Calculation of Crankshaft [J]. Torsional Vibration Response in I. C. Engines. Transactions of Csice, 2000,18 (1) : 29-32.

J Q Pan, J Pan, R S Ming, T Lin. Three-Dimensional Response and Intensity of Torsional Vibration in a Stepped Shaft [J]. Journal of Sound and Vibration,2000,236(1):115-128.

Chen Yong, Lu Songyuan, Gu Fang, Wang Yumin. The Calculation and Analysis of Torsional Vibration Characteristic of Turbine Generator Rotor System. [J]. Turbine Technology, 1999,41(5):276-279.

Liang Xingyu, Shu Gequn, Li Donghai, Shen Yinggang. Torsional Vibration Analysis Based on Multi-body Dynamics of Flexible Crankshaft System [J]. Chinese Combustion Engine Engineering, in 2008, the fourth period.

Li Renxian. Calculation of Loads on Crankshafts of Internal Combustion Engines Based on Continuous Beam Model [J]. Journal of Southwest Jiaotong University, 1998,33(2): 164-169.

Gu Yujiong, Yang Kun, He Chengbing, ect. Simulations on Torsional Vibration Characteristics of Turbogenerators Based on Four-Terminal Network Method [J]. China Mechanical Engineering, 1999,10(5):540-542.

He Shanghong, Duan Jian. Network method for dynamic modeling of complex shafting torsional vibration system [J]. The Chinese Journal of Nonferrous Metals, 2002,12(2): 388-392.

Xiang Jianhua, Liao Ridong, Zhang Weizheng. Study and Application of Graphic Modeling Techniques of Shaft Torsional Vibration Based on System Matrix Method [J]. Acta Armamentarii, 2005-03-001

Yu Yinghui, Zhang Baohui. Development And Prospect of Research on Turbine-Generator Shaft Torsional Oscillation [J]. Automation Of Electric Power Systems, 1999,23(10):56-60.

He Chengbing, Gu Yujiong, Yang Kun. Summarization on torsional vibration model and algorithm of turbogenerator unit [J]. Journal of North China Electric Power University, 2003,30(2):56-60.

Liu Hui, Xiang Changle, Zheng Muqiao. Torsional Vibration on the Complex Shaft System of a Vehicular Powertrain [J]. Journal of Beijing Institute of Technology, 2002,22(6):699-703.

Honda Ym, Wakabayashi K, Matsuki K. Torsional Vibration Characteristics of a Diesel Engine Crankshaft with a Torsional Rubber Damper by Using Transition Matrix Method [J]. JSAE Review, 1996,17(1):89.

Li BoZhong, Chen ZhiYan, Ying QiGuang. The torsional vibration of internal combustion engine crankshaft system[M]. Beijing: National Defense Industry Press, 1984.

G D Jenning, R G Harley. Modal Parameters and Turbo-Generator Torsional Behavior[C]. IEEE 3rd AFRICON Conference,1992:517-520.

Wang Guozhi, Modal Analysis Method in Research on Torsional Vibration of Diesel Engine Shafting System [J]. Transactions of Csice, 1986,4(3):249-259.

Shen Tumiao. Dynamic Measurement and Calculation of Twisting Stress of Crankshaft and Flywheel System in Diesel Type TC 387 [J]. Transactions of Csice, 1993, 11(6):243-248.

Li Shen, Zhao Shusen. Study on Substructure Method for Torsional Vibration of a Branch Shafting System [J]. Journal of Vibration and Shock, 2007-10-033.

Qu Zhihao, Chai Shaokuan, Ye Qianyuan. Analysis of Dynamic Characteristics and Chatter Of A 1420 Cold Tandem Rolling Mill [J]. Journal of Vibration and Shock, 2006, 25(4): 25-29.

Xiang Jianhua, Liao Ridong, Zhang Weizheng. Study and Application of Graphic Modeling Techniques of Shaft Torsional Vibration Based on System Matrix Method [J]. Acta Armamentarii, 2005, 26(3): 294-298.

Wei Haijun. Modification of Some Formulas for Calculating Shaft Torsional Vibration [J]. Journal of Vibration and Shock, 2006, 25 (2): 166-168.

Ji Chen, Zeng Fanming. Study on the general model and system matrix method for torsion vibration calculation of complex shafting [J]. Ship & Ocean Engineering, 2006(4): 55-57.

Li Lin, Sheng Jun. Mixed Interface Direct Component Modal Synthesis Method [J]. Chinese Journal of Applied Mechanics, 2005 22 (2): 315-319.

Xiang Shuhong, Qiu Jibao, Wang Dajun. The Resent Progresses on Modal Analysis and Dynamic Sub-Structure Methods [J]. Advances In Mechanics, 2004, 34(3): 289-303.

Chu Hua, She Yinghe. A Substructure Transfer Matrix Method in Torsional Vibration Analysis of Rotor System [J]. Journal of Southeast University, 1989,19(3):87-95.

Z P Mourelatos. An Efficient Crankshaft Dynamic Analysis Using Sub structuring with RITZ VECTORS[J]. Journal of Sound and Vibration, 2000,238(3):495-527.

Bogacz, Szo1c T, Irretier H. An Application of Torsional Wave Analysis to Turbo generator Rotor Shaft Response[J]. Trans. of ASME, Journal of Vibration and Acoustics, 1992, 114(2):149-153.

Shu Gequn, Hao Zhiyong. An Investigation on the Torsional Elastic Wave Theory for the Calculation of Multi-Stepped Shaft Torsional Vibration Response [J]. China Mechanical Engineering, 1999,10(11):1277-1279.

He Chengbing, Gu Yujiong, Yang Kun. The Analysis of the Shaft Torsional Vibrations Response of Turbine-generator Units Based on State-Space Method [C]. Asia-Pacific Vibration Conference, China, 2001.

Wang Ke she, Wang Zheng guang. Frequency analysis method of shafting crankle vibration [J]. Journal of Beijing Institute of Machinery, 2001,16(4):11-18.

Lin Sen, Zhang Hongtian, Geng Ruiguang, et al. Study on numerical emulator of shafts' torsion vibration of internal-combustion engine based upon time step integration method [J]. Journal of Heilongjiang Institute of Technology, 2008, 22(1):1-7.

Tong-Qun Han, Sheng-Hua Yao, Sheng-Jun Wu. Analysis of Torsional Vibration of Crankshaft System of a Vehicle Diesel Engine Based on EXCITE-designer Software. Journal of Hubei Automotive Industries Institute, 2005, 19(4): 5-8.

Yang Yanfu, Measurement and Application of Torsional Vibration in Engine[J]. Ccec Science & Technology,1997,4:1-6.

Huang Diangui. Experiment on the Characteristics of Torsional Vibration of Rotor-to-stator Rub in Turbo Machinery[J]. Tribology International,2000,33:75-79.

Xia Qingjie. 16V240ZJC Diesel Engine Torsional Vibration Measurement and Analysis[J]. Si Ji Science & Technology,1998,4:16-19.

Guo Li, Li Bo. , Assessment of Torsional Vibration Measurement Method[J], Grinder and Grinding. 2000,3:53-56.

N A Halliwell. The Laser Torsional Vibrometer: A Step Forward in Rotating Machinery Diagnostics[J]. Journal of Sound and Vibration, 1996,190(3):399-418.

T J Miless, M Lucast, N A Halliwell. Torsional and Bending Vibration Measurement on Rotors Using Laser Technology[J]. Journal of Sound and Vibration, 1999,226(3): 441-467.

Hua Jianwen, Liu Liren. Study of Measurement by Torsional Vibration Transducer Type MM0071 [J]. Laser & Infrared, 1996,26(3):193-195.

Ge Weijing, Wang Weisheng. Small Internal Combustion Engine, 1994,23(2):53-57.

Wang Ting, Cheng Peng. Development of Computer Testing System of Crankshaft Torsional Vibration [J]. Natural Science Journal of Jilin University of Technology, 201,21(3):74-77.

B O AL-Bedoor. Modeling the Coupled Torsional and Lateral Vibrations of Unbalanced Rotors[J]. Computer Methods in Applied Mechanics and Engineering, 2001,190: 5999-6008.

S F Asokanthan, P A Meehans. Non-linear Vibration of a Torsional System Driven by a Hook's Joint[J]. Journal of Sound and Vibration, 1999,226(3):441-467.

K Koser, F Pasin. Torsional Vibration of the Drive Shafts of Mechanisms[J]. Journal of Sound and Vibration, 1997,199(4):559-565.

E Brusa, C Delprete, G Genta. Torsional Vibration of Crankshafts Effects of Non Constant Moments of Inertia[J]. Journal of Sound and Vibration, 1997,205(2):135-150.

Gong Xiansheng Xie Zhijiang Tang Yike. Research on Steady Vibration Response of Shafting with Nonlinear Coupling [J]. Chinese Journal of Mechanical Engineering, 2001,37(6):19-23.

Sheng Gang, Chen Zhiyan, Li Liangfeng. Torsional Vibration of the Diesel Crankshaft System with Variable Inertia [J]. Transactions of Csice, 1991,9(2):143-149.

Zhu Menghua. Direct-Inverse Fourier Transformation Technique for Responses of Non-Linear Torsional Vibration of Diesel Shafting [J]. Transactions of Csice, 1992,10(1):47-52

Lin Ruilin, Huang Cihao, Tang Kaiyuan. The Incremental Harmonic Balance Method for CalculatingResponse of Non-Linear Torsional Vibration of a Diesel Shafting [J].Transactions of Csice, 2002,20(2):185-187.

Zhang Zhihua, Tang Mi. The Numerical Computation on Torsional Vibration of Diesel Nonlinear Shafting, Transactions of Csice, 1987,5(4): 353-361.

Farshidianfar A, Ebrahimi M, Bartlett H. Hybrid Modeling and Simulation of the Torsional Vibration of Vehicle Driveline Systems[J]. Journal of Automobile Engineering, 2001.215(4):217-229.

Wen Bangchun, Li Yinong, Han Qingkai, Analytical Approach and Application of Nonlinear Vibration Theory [M]. Shenyang: Northeast University Press, 2001.

Li Bozhong, Song Tianxiang, Song Xigeng. On Coupled Vibrations of Reciprocating Engine Shaft Systems(Part 1) Axial Vibration Caused by Torsional Vibration [J]. Transactions of Csice, 1989,7(1): 1-6.

Li Bozhong, Song Xigeng, Song Tianxiang. On Coupled Vibrations of Reciprocating Engine Shaft Systems(Part 2)- Progressive Torsional-Axial Continued Vibrations [J].Transactions of Csice, 1990,8(4):317-322.

Song Xigeng, Song Tianxiang, Li Bozhong. On Coupled Vibrations of Reciprocating Engine Shaft Systems (Part 3)- Calculation Method of Coupled at Same and Double Frequencies [J]. Transactions of Csice, 1994,12(2):115-120.

Du Hongbing, Chen Zhiyan, Jing Bo. The Mathematical Model for the Coupled Torsional-Axial Vibration of Internal Combustion Engine Shaft System[J]. Chinese Combustion Engine Engineering 1992,13(2):66-74.

Zhang Yong, Jiang Zikang. Numerical Analyses of the Coupling of the Flexural and Torsional Vibrations of Rotary Shaft System[J]. Turbine Technology,1999,41(5): 280-283.

Zhang Yong, Jiang Zikang. Analysis of the coupled flexural-torsional vibrations of rotary shaft systems [J]. Journal of Tsinghua University(Science And Technology), 2000,40(6):80-83.

Zhang Yong, Jiang Zikang. Mathematic model of coupled bending and torsional vibration of shaft systems [J]. Journal of Tsinghua University(Science And Technology), 1998,38(8):114-117.

Zhang Hongtian, Zhang Zhihua, LIU Zhigang. A Study on Coupled Axial and Torsional Vibration of Marine Propulsion Shafting [J]. Shipbuilding of China, 1995,129(2):68-76.

0kamura H et. Simple Modeling and Analysis for Crankshaft Three-Dimensional Vibration[J]. Journal of Vibration and Acoustics, 1995,117(1): 70-86.

Shen Hongbin Shu Gequn. An Investigation into Dynamic Measurement of Longitudinal/Bend/Tortional Vibration in Shaft System of Internal Combustion Engines [J]. Small Internal Combustion Engine, 2000, 29(6):9-10.

Liang Xingyu, Shu Gequn, Li Donghai, Shen Yinggang. Torsional Vibration Analysis Based on Multi-body Dynamics of Flexible Crankshaft System [J]. Chinese Combution Engine Engineering, In 2008, the fourth period.

Natarajan M, Frame E A, Naegeli D, et al. Oxygenates for ad vanced petroleum-based diesel fuels: part1. Screening and selection methodology for the oxygenates[C]. SAE 2001-01-3631.

Guo Wei-dong; Gu Wen-gang; ZhanG Xiao-ling. The Method of Compensating Division Error in Shafts Torsional Vibration Measurement and Implementation with Program [J]. Machine Design & Research, 2007, 23(2): 88-89.

Zhang Xiaoling, Tang Xikuan. Analysis of torsional vibration measuring errors and their correct method [J]. Journal Of Tsinghua University (Science And Technology), 1997, 37(8): 9-12.

Xu Xintian. Measure of the torsional vibration control [J]. Wuhan Shipbullding, 1999, 1:17-18.

C T Lee, S W Shaw. The Non-linear Dynamic Response of Paired Centrifugal Pendulum Vibration Absorbers [J]. Journal of Sound and Vibration, 1997,203(5): 731-743.

Weng Wenhua. Silicone Oil Damper [J]. CCEC Science & Technology, 1997,1:19-23.

Luo Zhouquan. Characters of Shafting Torsional Vibration and Analysis on Damping Measures of 6180 Series Marine Engine [J]. Design and Manufacture of Diesel Engine, 2000, 92 (3):3-7.

Zhou Shengfu, Li Cuncai. Function and Behaviour of Torsional Damper of Crankshaft [J]. Vehicle Engine, 1993,4:23-28.

Chen Xi'en. Study on Nonlinear High-damping Dampers in calculation and measurement of torsional vibration[J], China Ship Survey, 1996, 1:26-32.

Yan Jiabin. Engine Crankshaft Torsional Vibration Damper, World Rubbre Industry, 1998, 25(4):26-30.

Huo Quanzhong, Liang Jie, Hao Zhiyong. Operation Principle and Experimental Study of an Active Torsional Vibration Absorber [J]. Transactions of Csice, 1992,10(4): 329-334.

Hao Zhiyong, Gao Wenzhi. A Study on Active Control and Simulation Test for Torsional Vibration of Turbogenrator Shaft System [J]. Power Engineering, 1999,19(5): 338-341.

Liu Shengtian. Double Mass Flywheel Type Torsional Vibration Damper [J]. Automobile Technology, 1997,1: 23-27.

Liu Shengtian, Hydraulic dual-mass flywheel torsional vibration damper, [J]. Motor Transport,1998,9: 17-19.

M Hosek, H Elmal, N Olgac. A Tunable Torsional Vibration Absorber: the Centrifugal Delayed Resonator[J]. Journal of Sound and Vibration, 1997,205(2):151-165.

Shu Gequn, Hao Zhiyong. Experiments on the Crankshaft Torsional/Bending Vibration Damper in Internal Combustion Engine [J]. Journal of Tianjin University, 1997,30(6): 806-811.

Shu Gequn, Lü Xingcai, QIin De, Su Yanling. Experimental Study on Hybrid Damper of HighSpeed Diesel Engine Crankshaft [J]. Transactions of Csice, 2000,20(6): 537-540.

Shu Gequn. Effect of Bending Vibration Damper on Properties of Torsional [J]. Transactions of Csice, 1998,16(3): 348-353.

Zhao Qian, Hao Zhiyong. Optimal Design of Torsional Damper of Diesel Engine Crankshaft System [J]. Transactions of the Chinese Society of Agricultural Machinery, 2000,31 (5): 91-93.

Song Jingbo. Internal combustion Engines, 1993, (3):38-40.

Ying Qiguang, Bao Defu and Others. A New Fault-diagnosing Technique of Diesel Engine based on Shafting Vibration [J]. Ship Engineering, 1995, (4):33-35.

Ying Qiguang, Li Bozhong. A New Effective Method to Diagnose the Troubles in the Technical Condition of the Diesel Engine [J]. Journal of Shanghai Maritime University, 1990, (3):20-30.

Lin Dayuan, Wang Xu, Song Xueren. Diagnosis of Crankshaft Cracks via Measurement of Crankshaft—Flywheel System Torsional Vibration Mode [J]. Transactions of Csice, 1989, 7(3):215-222.

Shu Gequn, Lü Xingcai. Calculation Method for Continuous Distribution Model of Torsional Vibration of a Crankshaft [J]. Transactions of The Chinese Society of Agricultural Machinery, 2004,35 (4) 36-39.

14

A Plane Vibration Model for Natural Vibration Analysis of Soft Mounted Electrical Machines

Ulrich Werner
Siemens AG, Industry Drive Technologies, Large Drives, Products Development
Germany

1. Introduction

Large electrical machines, which operate at high speeds, are often designed with flexible shafts and sleeve bearings, because of the high circumferential speed of the shaft journals. Especially for industrial applications, the foundations of this kind of machines are often designed as soft foundations (Fig. 1), because of plant specific requirements. Therefore often a significant influence of the soft foundation on the vibrations exists (Gasch et al., 1984; Bonello & Brennan, 2001). Additionally to the mechanical parameters - such as e.g. mass, mechanical stiffness and damping - an electromagnetic field in the electrical machine exists, which causes an electromagnetic coupling between rotor and stator and also influences the natural vibrations (Schuisky, 1972; Belmans et al., 1987; Seinsch, 1992; Arkkio et al., 2000; Holopainen, 2004; Werner, 2006). The aim of the chapter is to show a plane vibration model for natural vibration analysis, of soft mounted electrical machines, with flexible shafts and sleeve bearings, especially considering the influence of a soft foundation and the electromagnetic field. Based on a simplified plane vibration model, the mathematical correlations between the rotor and the stator movement, the sleeve bearings, the electromagnetic field and the foundation, are shown. For visualization, the natural vibrations of a soft mounted 2-pole induction motor (rated power: 2 MW) are analyzed exemplary, especially focusing on the influence of the foundation, the oil film stiffness and damping and of the electromagnetic field.

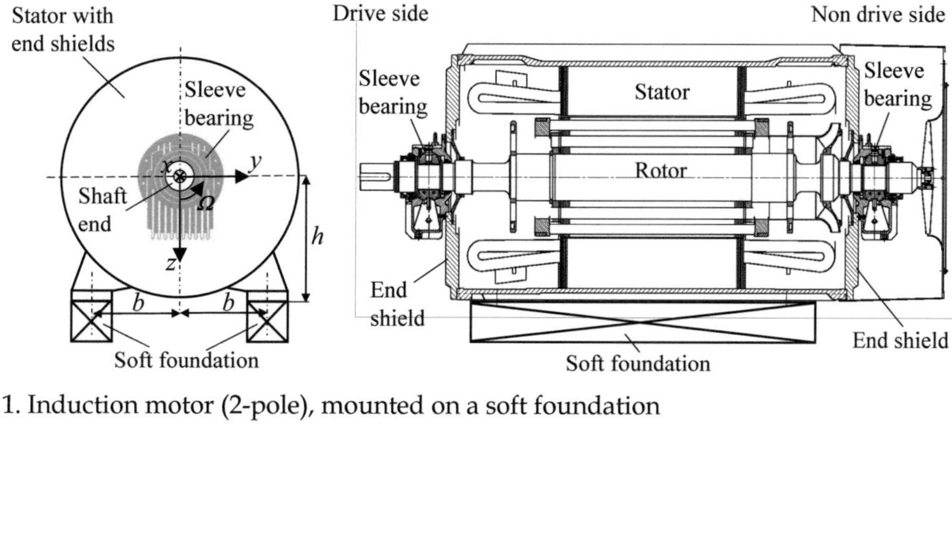

Fig. 1. Induction motor (2-pole), mounted on a soft foundation

2. Vibration model

The vibration model is a simplified plane model (Fig. 2), describing the natural vibrations in the transversal plane (plane y, z) of a soft mounted electrical machine. Therefore no natural vibrations regarding the translation in the x-axis, the rotation at the y-axis and the rotation at the z-axis are considered. The plane model is based on the general models in (Werner, 2008; Werner, 2010), but especially focusing here on the natural vibration analysis.

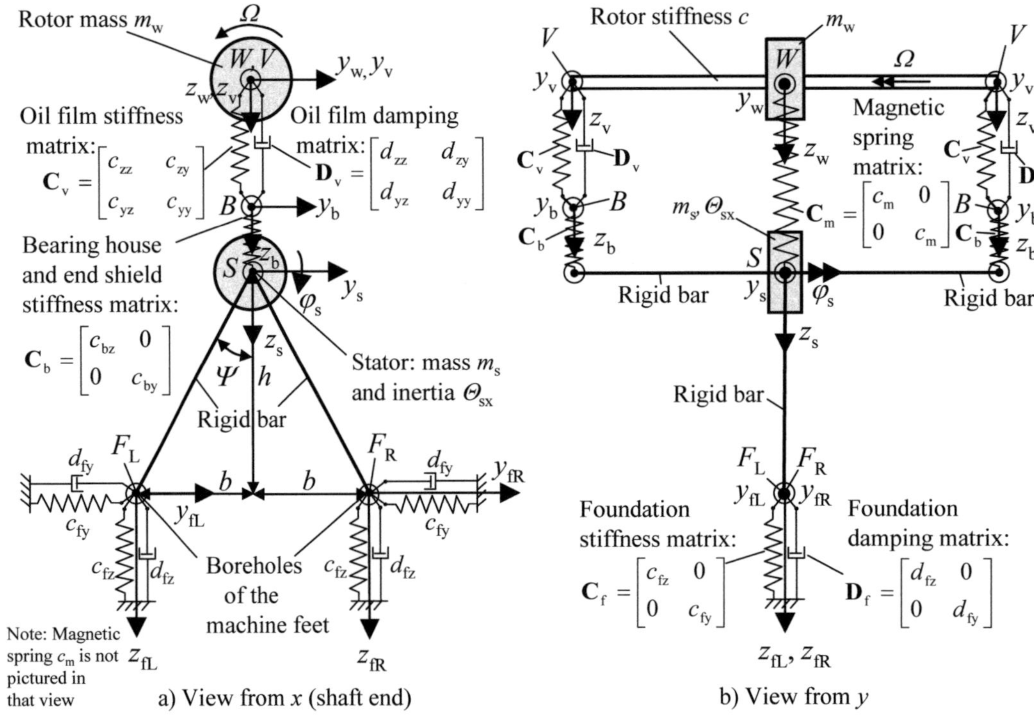

Fig. 2. Vibration model of a soft mounted electrical machine

The model consists of two masses, rotor mass m_w, concentrated at the shaft - rotating with angular frequency Ω - and stator mass m_s, which has the inertia Θ_{sx} and is concentrated at the centre of gravity S. The moments of inertia of the rotor are not considered and therefore no gyroscopic effects. Shaft journal centre point V describes the movement of the shaft journal in the sleeve bearing. Point B is positioned at the axial centre of the sleeve bearing shell and describes the movement of the bearing housing. The rotor mass is mechanically linked to the stator mass by the stiffness of rotor c and the oil film stiffness matrix $\mathbf{C}_v$ and the oil film damping matrix $\mathbf{D}_v$ of the sleeve bearings, which contain the oil film stiffness coefficients (c_{yy}, c_{yz}, c_{zy}, c_{zz}) and the oil film damping coefficients (d_{yy}, d_{yz}, d_{zy}, d_{zz}) (Fig. 3). The cross-coupling coefficients - stiffness cross-coupling coefficients c_{yz}, c_{zy} and damping cross-coupling coefficients d_{yz}, d_{zy} - cause a coupling between vertical and horizontal movement and the vertical oil film force F_z and the horizontal oil film forces F_y (Tondl, 1965; Glienicke, 1966; Lund & Thomsen, 1978; Lund & Thomsen, 1987; Gasch et al. 2002; Vance et al., 2010), which is mathematically described in (1).

$$\begin{bmatrix} F_z \\ F_y \end{bmatrix} = \begin{bmatrix} c_{zz} & c_{zy} \\ c_{yz} & c_{yy} \end{bmatrix} \cdot \begin{bmatrix} z_v - z_b \\ y_v - y_b \end{bmatrix} + \begin{bmatrix} d_{zz} & d_{zy} \\ d_{yz} & d_{yy} \end{bmatrix} \cdot \begin{bmatrix} \dot{z}_v - \dot{z}_b \\ \dot{y}_v - \dot{y}_b \end{bmatrix} \tag{1}$$

For cylindrical shell bearings the cross-coupling stiffness coefficients are usually not equal ($c_{zy} \neq c_{yz}$). This leads to an asymmetric oil film stiffness matrix C_v, which is the reason that vibration instability may occur (Tondl, 1965; Glienicke, 1966; Lund & Thomsen, 1978; Lund & Thomsen, 1987; Gasch et al. 2002; Vance et al., 2010). In this model it is assumed that the drive side and the non drive side values are the same, and the bearing housing and end shield stiffness matrix C_b is also assumed to be same for the drive side and non drive side. The stiffness and damping values of the oil film are calculated by solving the Reynolds-differential equation, using the radial bearing forces, which are caused by the rotor weight and static magnetic pull. The stiffness and damping values of the oil film are assumed to be linear regarding the displacements of the shaft journals relative to the bearing housings.

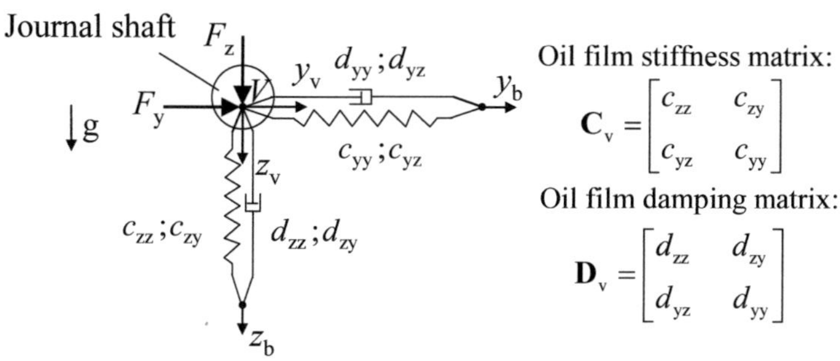

Fig. 3. Oil film forces

Damping of the bearing housings and the end shields are not considered because of the usually low damping ratio. For electrical machines, an additional magnetic stiffness matrix C_m between the rotor and the stator exists, which describes the electromagnetic coupling between the rotor and stator. The magnetic spring constant c_m has a negative reaction. This means that a radial movement between the rotor and stator creates an electromagnetic force that tries to magnetize the movement (Schuisky, 1972; Belmans et al., 1987; Seinsch, 1992; Arkkio et al., 2000; Holopainen, 2004; Werner, 2006). Here the magnetic spring coefficient c_m is defined to be positive, which acts in the direction of the magnetic forces. Electromagnetic field damping effects, e.g. by the rotor cage of an induction motor, are not considered in this paper. The stator structure is assumed to be rigid when compared to the soft foundation. The foundation stiffness matrix C_f and the foundation damping matrix D_f connect the stator feet, F_L (left side) and F_R (right side), to the ground. The foundation stiffness and damping on the right side is assumed to be the same as on the left side. The stiffness values c_{fy} and c_{fz} and the damping values d_{fy} and d_{fz} are the values for each machine side. The coordinate systems for V (z_v; y_v) and B (z_b; y_b) have the same point of origin, as well as the coordinate systems for the stator mass m_s (z_s; y_s) and for the rotor mass m_w (z_w; y_w). They are only shown with an offset to show the connections through the various spring and damping elements.

3. Natural vibrations

To calculate the natural vibrations, it is necessary to derive the homogenous differential equation, which is assumed to be linear.

3.1 Derivation of the homogenous differential equation system

The homogenous differential equation system can be derived by separating the vibration system into four single systems - (a) rotor mass system, (b) journal system, (c) bearing house system and (d) stator mass system - (Fig. 4).

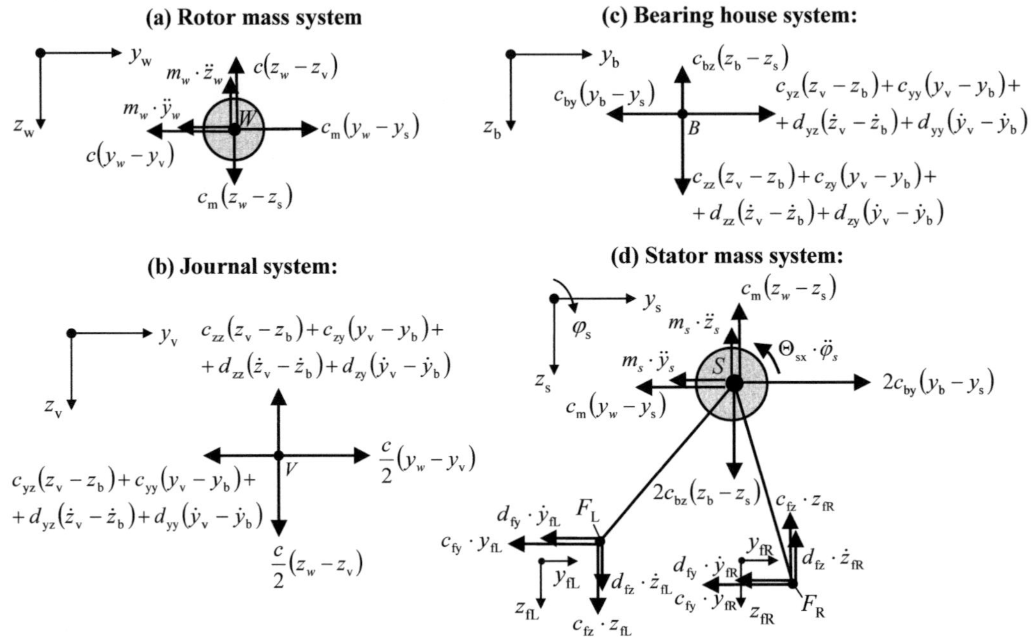

Fig. 4. Vibration system, split into four single systems

The equilibrium of forces and moments for each single system (Fig. 4) leads to following equations for each single system:

- Rotor mass system (Fig. 4a):

$$\uparrow: \quad m_w \cdot \ddot{z}_w + c \cdot (z_w - z_v) - c_m \cdot (z_w - z_s) = 0 \tag{2}$$

$$\rightarrow: \quad m_w \cdot \ddot{y}_w + c \cdot (y_w - y_v) - c_m \cdot (y_w - y_s) = 0 \tag{3}$$

- Journal system (Fig. 4b):

$$\uparrow: \quad c_{zz}(z_v - z_b) + c_{zy}(y_v - y_b) + d_{zz}(\dot{z}_v - \dot{z}_b) + d_{zy}(\dot{y}_v - \dot{y}_b) - \frac{c}{2}(z_w - z_v) = 0 \tag{4}$$

$$\rightarrow: \quad c_{yz}(z_v - z_b) + c_{yy}(y_v - y_b) + d_{yz}(\dot{z}_v - \dot{z}_b) + d_{yy}(\dot{y}_v - \dot{y}_b) - \frac{c}{2}(y_w - y_v) = 0 \tag{5}$$

- Bearing house system (Fig. 4c):

$$\uparrow: \quad c_{zz}(z_v - z_b) + c_{zy}(y_v - y_b) + d_{zz}(\dot{z}_v - \dot{z}_b) + d_{zy}(\dot{y}_v - \dot{y}_b) - c_{bz} \cdot (z_b - z_s) = 0 \tag{6}$$

$$\rightarrow: \quad c_{yz}(z_v - z_b) + c_{yy}(y_v - y_b) + d_{yz}(\dot{z}_v - \dot{z}_b) + d_{yy}(\dot{y}_v - \dot{y}_b) - c_{by} \cdot (y_b - y_s) = 0 \tag{7}$$

- Stator mass system (Fig. 4d):

$$\uparrow: \quad m_s \cdot \ddot{z}_s + c_m \cdot (z_w - z_s) - 2c_{bz} \cdot (z_b - z_s) + c_{fz} \cdot z_{fR} + d_{fz} \cdot \dot{z}_{fR} - c_{fz} \cdot z_{fL} - d_{fz} \cdot \dot{z}_{fL} = 0 \tag{8}$$

$$\rightarrow: \quad m_s \cdot \ddot{y}_s + c_m \cdot (y_w - y_s) - 2c_{by} \cdot (y_b - y_s) + c_{fy} \cdot y_{fR} + d_{fy} \cdot \dot{y}_{fR} + c_{fy} \cdot y_{fL} + d_{fy} \cdot \dot{y}_{fL} = 0 \tag{9}$$

$$S): \quad \Theta_{sx} \cdot \ddot{\varphi}_s + b \cdot (c_{fz} z_{fR} + d_{fz} \dot{z}_{fR} + c_{fz} z_{fL} + d_{fz} \dot{z}_{fL}) - h \cdot (c_{fy} y_{fR} + d_{fy} \dot{y}_{fR} + c_{fy} y_{fL} + d_{fy} \dot{y}_{fL}) = 0 \tag{10}$$

The equations (2)-(10) lead to a linear homogenous differential equation system (11) with 13 degrees of freedom (DOF = 13), with the mass matrix $\mathbf{M}_o$, the damping matrix $\mathbf{D}_o$ and the stiffness matrix $\mathbf{C}_o$, which have the form 13x13.

$$\mathbf{M}_o \cdot \ddot{\mathbf{q}}_o + \mathbf{D}_o \cdot \dot{\mathbf{q}}_o + \mathbf{C}_o \cdot \mathbf{q}_o = \mathbf{0} \tag{11}$$

The coordinate vector $\mathbf{q}_o$ is a vector with 13 rows described by:

$$\mathbf{q}_o = (z_s; z_w; y_s; y_w; \varphi_s; z_v; z_b; y_v; y_b; z_{fR}; z_{fL}; y_{fR}; y_{fL})^T \tag{12}$$

The linear homogenous differential equation system can be reduced into a system of 9 DOF, by considering the cinematic constraints between the stator mass and the machine feet.

3.2 Kinematic constraints between stator mass and machine feet

The kinematic constraints are derived for translation of the stator mass and for angular displacement of the stator mass and for the superposition of both.

3.2.1 Kinematic constraints for translation of the stator mass

If the stator mass centre S makes only a translation (z_s, y_s) without angular displacement ($\varphi_s = 0$) the kinematic constraints between stator mass centre S and the machine feet F_L and F_R can be described as follows:

$$z_{fL} = z_{fR} = z_s \; ; \; y_{fL} = y_{fR} = y_s \tag{13}$$

3.2.2 Kinematic constraints for angular displacement of the stator mass

If the stator mass centre S only makes an angular displacement (φ_s) without translation ($z_s = y_s = 0$) the kinematic constraints between the angular displacement (φ_s) of the stator mass centre S and the translation of the machine feet F_L and F_R are shown in Fig. 5.

The displacements of the machine feet on the left side of the machine can be described as follows:

$$z_{fL} = -u_{fL} \cdot \sin\beta = -2 \cdot l \cdot \sin\left(\frac{\varphi_s}{2}\right) \cdot \sin\beta \tag{14}$$

$$y_{fL} = -u_{fL} \cdot \cos\beta = -2 \cdot l \cdot \sin\left(\frac{\varphi_s}{2}\right) \cdot \cos\beta \tag{15}$$

The angle β is described by:

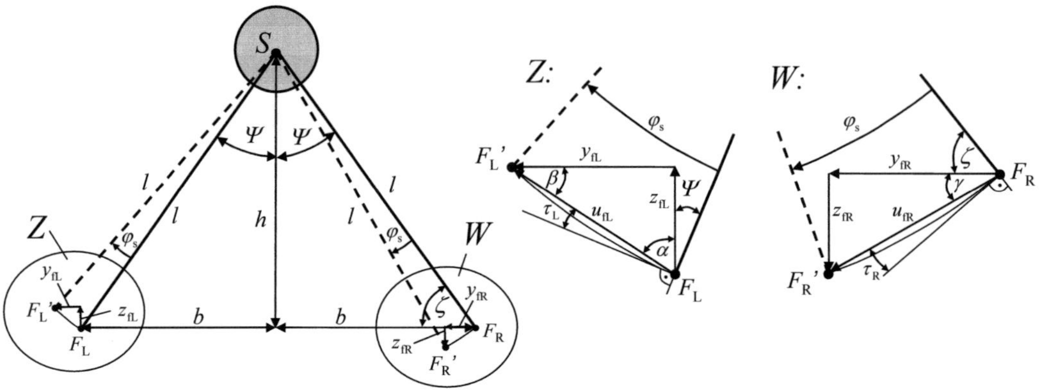

Fig. 5. Angular displacement φ_s of the stator mass centre S

$$\beta = 90° - \alpha = \tau_L + \Psi = \frac{\varphi_s}{2} + \Psi \tag{16}$$

The displacements of the machine feet on the right side of the machine can be described as follows:

$$z_{fR} = u_{fR} \cdot \sin\gamma = 2 \cdot l \cdot \sin\left(\frac{\varphi_s}{2}\right) \cdot \sin\gamma \tag{17}$$

$$y_{fR} = -u_{fR} \cdot \cos\gamma = -2 \cdot l \cdot \sin\left(\frac{\varphi_s}{2}\right) \cdot \cos\gamma \tag{18}$$

The angle γ is described by:

$$\gamma = 90° - (\zeta + \tau_R) = 90° - \left(\zeta + \frac{\varphi_s}{2}\right) \tag{19}$$

For small angular displacements φ_s of the stator mass centre S ($\varphi_s << \Psi$ and $\varphi_s << \zeta$) following linearizations can be deduced:

$$\sin\left(\frac{\varphi_s}{2}\right) \to \frac{\varphi_s}{2} \tag{20}$$

$$\beta = \frac{\varphi_s}{2} + \Psi \to \beta \approx \Psi \tag{21}$$

$$\gamma = 90° - \left(\zeta + \frac{\varphi_s}{2}\right) \to \gamma \approx 90° - \zeta \tag{22}$$

With these linearizations the displacements of the machine feet on the left side and on the right side can be described as follows:

$$z_{\mathrm{fL}} = -l \cdot \varphi_{\mathrm{s}} \cdot \sin \Psi = -b \cdot \varphi_{\mathrm{s}} \tag{23}$$

$$y_{\mathrm{fL}} = -l \cdot \varphi_{\mathrm{s}} \cdot \cos \Psi = -h \cdot \varphi_{\mathrm{s}} \tag{24}$$

$$z_{\mathrm{fR}} = l \cdot \varphi_{\mathrm{s}} \cdot \sin(90° - \zeta) = b \cdot \varphi_{\mathrm{s}} \tag{25}$$

$$y_{\mathrm{fR}} = -l \cdot \varphi_{\mathrm{s}} \cdot \cos(90° - \zeta) = -h \cdot \varphi_{\mathrm{s}} \tag{26}$$

3.2.3 Kinematic constraints for superposition of translation and angular displacement

For superposition of the translation and angular displacement of the stator mass centre *S* following kinematic constraints can be derived:

$$z_{\mathrm{fL}} = z_{\mathrm{s}} - b \cdot \varphi_{\mathrm{s}} \tag{27}$$

$$y_{\mathrm{fL}} = y_{\mathrm{s}} - h \cdot \varphi_{\mathrm{s}} \tag{28}$$

$$z_{\mathrm{fR}} = z_{\mathrm{s}} + b \cdot \varphi_{\mathrm{s}} \tag{29}$$

$$y_{\mathrm{fR}} = y_{\mathrm{s}} - h \cdot \varphi_{\mathrm{s}} \tag{30}$$

Therefore, it is possible to describe the translations of the machine feet (z_{fL}; y_{fL}; z_{fR}; y_{fR}) by the movement of the stator mass (z_{s}, y_{s}, φ_{s}).

3.3 Reduced homogenous differential equation system

With the kinematic constraints (27)-(30) the differential equation system (11) - with 13 DOF - can be reduced to a differential equation system of 9 DOF. By deriving the reduced differential equation system, it is necessary to consider, that the negative vertical displacement of the machine foot F_{L}, related to the coordinate system in Fig. 4 is considered in the direction of the vertical forces in F_{L}. Therefore the displacement z_{fL} has to be described negative $z_{\mathrm{fL}} \rightarrow -z_{\mathrm{fL}}$, as well as the velocity $\dot{z}_{\mathrm{fL}} \rightarrow -\dot{z}_{\mathrm{fL}}$. With this boundary condition and with the kinematic constraints (27)-(30) the equations for the stator system (8)-(10) become:

$$\uparrow: \quad m_{\mathrm{s}} \cdot \ddot{z}_{\mathrm{s}} + c_{\mathrm{m}} \cdot (z_{\mathrm{w}} - z_{\mathrm{s}}) - 2c_{\mathrm{bz}} \cdot (z_{\mathrm{b}} - z_{\mathrm{s}}) + 2c_{\mathrm{fz}} \cdot z_{\mathrm{s}} + 2d_{\mathrm{fz}} \cdot \dot{z}_{\mathrm{s}} = 0 \tag{31}$$

$$\rightarrow: \quad m_{\mathrm{s}} \cdot \ddot{y}_{\mathrm{s}} + c_{\mathrm{m}} \cdot (y_{\mathrm{w}} - y_{\mathrm{s}}) - 2c_{\mathrm{by}} \cdot (y_{\mathrm{b}} - y_{\mathrm{s}}) + 2c_{\mathrm{fy}} \cdot (y_{\mathrm{s}} - h \cdot \varphi_{\mathrm{s}}) + 2d_{\mathrm{fy}} \cdot (\dot{y}_{\mathrm{s}} - h \cdot \dot{\varphi}_{\mathrm{s}}) = 0 \tag{32}$$

$$S\circlearrowright: \quad \Theta_{\mathrm{sx}} \cdot \ddot{\varphi}_{\mathrm{s}} - 2d_{\mathrm{fy}} h \cdot \dot{y}_{\mathrm{s}} + 2(d_{\mathrm{fy}} h^2 + d_{\mathrm{fz}} b^2) \cdot \dot{\varphi}_{\mathrm{s}} - 2c_{\mathrm{fy}} h \cdot y_{\mathrm{s}} + 2(c_{\mathrm{fy}} h^2 + c_{\mathrm{fz}} b^2) \cdot \varphi_{\mathrm{s}} = 0 \tag{33}$$

Therefore, it is now possible to derive the reduced homogenous differential equation system, which only has 9 DOF:

$$\mathbf{M} \cdot \ddot{\mathbf{q}} + \mathbf{D} \cdot \dot{\mathbf{q}} + \mathbf{C} \cdot \mathbf{q} = \mathbf{0} \tag{34}$$

The mass matrix $\mathbf{M}$ and coordinate vector $\mathbf{q}$ are described by:

$$\mathbf{M} = \begin{pmatrix} m_s & 0 & 0 & 0 & 0 & 0 & 0 & 0 & 0 \\ 0 & m_w & 0 & 0 & 0 & 0 & 0 & 0 & 0 \\ 0 & 0 & m_s & 0 & 0 & 0 & 0 & 0 & 0 \\ 0 & 0 & 0 & m_w & 0 & 0 & 0 & 0 & 0 \\ 0 & 0 & 0 & 0 & \Theta_{sx} & 0 & 0 & 0 & 0 \\ 0 & 0 & 0 & 0 & 0 & 0 & 0 & 0 & 0 \\ 0 & 0 & 0 & 0 & 0 & 0 & 0 & 0 & 0 \\ 0 & 0 & 0 & 0 & 0 & 0 & 0 & 0 & 0 \\ 0 & 0 & 0 & 0 & 0 & 0 & 0 & 0 & 0 \end{pmatrix} \; ; \; \mathbf{q} = \begin{pmatrix} z_s \\ z_w \\ y_s \\ y_w \\ \varphi_s \\ z_v \\ z_b \\ y_v \\ y_b \end{pmatrix} \tag{35}$$

The damping matrix **D** is described by:

$$\mathbf{D} = \begin{pmatrix} 2d_{fz} & 0 & 0 & 0 & 0 & 0 & 0 & 0 & 0 \\ 0 & 0 & 0 & 0 & 0 & 0 & 0 & 0 & 0 \\ 0 & 0 & 2d_{fy} & 0 & -2d_{fy}h & 0 & 0 & 0 & 0 \\ 0 & 0 & 0 & 0 & 0 & 0 & 0 & 0 & 0 \\ 0 & 0 & -2d_{fy}h & 0 & 2\left(d_{fy}h^2 + d_{fz}b^2\right) & 0 & 0 & 0 & 0 \\ 0 & 0 & 0 & 0 & 0 & 2d_{zz} & -2d_{zz} & 2d_{zy} & -2d_{zy} \\ 0 & 0 & 0 & 0 & 0 & -2d_{zz} & 2d_{zz} & -2d_{zy} & 2d_{zy} \\ 0 & 0 & 0 & 0 & 0 & 2d_{yz} & -2d_{yz} & 2d_{yy} & -2d_{yy} \\ 0 & 0 & 0 & 0 & 0 & -2d_{yz} & 2d_{yz} & -2d_{yy} & 2d_{yy} \end{pmatrix} \tag{36}$$

The stiffness matrix **C** is described by:

$$\mathbf{C} = \begin{pmatrix} 2(c_{fz}+c_{bz})-c_m & c_m & 0 & 0 & 0 & 0 & -2c_{bz} & 0 & 0 \\ c_m & c-c_m & 0 & 0 & 0 & -c & 0 & 0 & 0 \\ 0 & 0 & 2(c_{fy}+c_{by})-c_m & c_m & -2c_{fy}h & 0 & 0 & 0 & -2c_{by} \\ 0 & 0 & c_m & c-c_m & 0 & 0 & 0 & -c & 0 \\ 0 & 0 & -2c_{fy}h & 0 & 2\left(c_{fy}h^2 + c_{fz}b^2\right) & 0 & 0 & 0 & 0 \\ 0 & -c & 0 & 0 & 0 & 2c_{zz}+c & -2c_{zz} & 2c_{zy} & -2c_{zy} \\ -2c_{bz} & 0 & 0 & 0 & 0 & -2c_{zz} & 2(c_{zz}+c_{bz}) & -2c_{zy} & 2c_{zy} \\ 0 & 0 & 0 & -c & 0 & 2c_{yz} & -2c_{yz} & 2c_{yy}+c & -2c_{yy} \\ 0 & 0 & -2c_{by} & 0 & 0 & -2c_{yz} & 2c_{yz} & -2c_{yy} & 2(c_{yy}+c_{by}) \end{pmatrix} \tag{37}$$

3.4 Solution of the reduced homogenous differential equation system

The natural vibrations can be derived by solving the homogeneous differential equation (34). Therefore usually a complex ansatz is used. So the homogeneous differential equation is described complex, with the vector $\underline{\mathbf{q}}$ as a complex vector (underlined = complex value), the mass matrix **M**, the damping matrix **D** and the stiffness matrix **C**.

$$\mathbf{M}\cdot\underline{\ddot{\mathbf{q}}} + \mathbf{D}\cdot\underline{\dot{\mathbf{q}}} + \mathbf{C}\cdot\underline{\mathbf{q}} = \mathbf{0} \quad \text{with: } \underline{\mathbf{q}} = (\underline{z}_s;\underline{z}_w;\underline{y}_s;\underline{y}_w;\underline{\varphi}_s;\underline{z}_v;\underline{z}_b;\underline{y}_v;\underline{y}_b)^T \tag{38}$$

The complex ansatz – with the complex eigenvalue $\underline{\lambda}$ and the complex eigenvectors $\underline{\hat{\mathbf{q}}}$ –

$$\underline{\mathbf{q}} = \underline{\hat{\mathbf{q}}} \cdot e^{\underline{\lambda} \cdot t} \quad \text{with:} \quad \underline{\hat{\mathbf{q}}} = (\underline{\hat{z}}_{\mathrm{s}}; \underline{\hat{z}}_{\mathrm{w}}; \underline{\hat{y}}_{\mathrm{s}}; \underline{\hat{y}}_{\mathrm{w}}; \underline{\hat{\varphi}}_{\mathrm{s}}; \underline{\hat{z}}_{\mathrm{v}}; \underline{\hat{z}}_{\mathrm{b}}; \underline{\hat{y}}_{\mathrm{v}}; \underline{\hat{y}}_{\mathrm{b}})^T \tag{39}$$

leads to the eigenvalue equation:

$$[\mathbf{C} + \underline{\lambda} \cdot \mathbf{D} + \underline{\lambda}^2 \cdot \mathbf{M}] \cdot \underline{\hat{\mathbf{q}}} = \mathbf{0} \tag{40}$$

To get the complex eigenvalues $\underline{\lambda}$, it is necessary to solve the determination equation:

$$\det[\mathbf{C} + \underline{\lambda} \cdot \mathbf{D} + \underline{\lambda}^2 \cdot \mathbf{M}] = 0 \tag{41}$$

This leads to a characteristic polynomial of 12th grade:

$$\sum_{n=0}^{12} A_{\mathrm{n}} \cdot \underline{\lambda}^n = 0 \tag{42}$$

With a numerical solution of this polynomial, n complex eigenvalues $\underline{\lambda}_{\mathrm{n}}$ – with the real parts α_{n}, which describe the decay of each natural vibration and the imaginary parts ω_{n}, which describe the corresponding natural angular frequencies – can be calculated. The eigenvalues occur mostly conjugated complex (j: imaginary unit $\rightarrow j^2 = -1$):

$$\underline{\lambda}_{\mathrm{n}} = \alpha_{\mathrm{n}} \pm j \cdot \omega_{\mathrm{n}} \tag{43}$$

With the complex eigenvalues $\underline{\lambda}_{\mathrm{n}}$ the complex eigenvectors $\underline{\hat{\mathbf{q}}}_{\mathrm{n}}$ can be calculated. Therefore the natural vibrations can be described by:

$$\underline{\mathbf{q}} = \sum_{n=1}^{12} \underline{\hat{\mathbf{q}}}_{\mathrm{n}} \cdot \underline{k}_{\mathrm{n}} \cdot e^{\underline{\lambda}_{\mathrm{n}} \cdot t} \tag{44}$$

The factors $\underline{k}_{\mathrm{n}}$ can be used, to adapt the natural vibrations to the starting conditions. Using the calculated real part α_{n} and the imaginary part ω_{n} of each complex eigenvalue λ_{n} the modal damping D_{n} of each natural vibration mode can be derived (Kellenberger, 1987).

$$D_{\mathrm{n}} = \frac{-\alpha_{\mathrm{n}}}{\sqrt{\alpha_{\mathrm{n}}^2 + \omega_{\mathrm{n}}^2}} \tag{45}$$

3.5 Stability of the vibration system

If the oil film stiffness matrix $\mathbf{C}_{\mathrm{v}}$ of the sleeve bearings is non symmetric ($c_{zy} \neq c_{yz}$) – for e.g. sleeve bearings with cylindrical shell the cross-coupling coefficients of the stiffness matrix are mostly unequal ($c_{zy} \neq c_{yz}$) – also the system stiffness matrix $\mathbf{C}$ (37) gets non symmetric. This may lead to instabilities of the vibration system (Gasch et al., 2002), which occur if the real part of one or more complex eigenvalues gets positive, leading to negative modal damping values (45). The oil film stiffness and damping coefficients are a function of the rotary angular frequency Ω of the rotor.

$$c_{\mathrm{ij}} = c_{\mathrm{ij}}(\Omega)\,; \;\; d_{\mathrm{ij}} = d_{\mathrm{ij}}(\Omega) \quad \text{with} \;\; i, j = z, y \tag{46}$$

To find the limit of stability of the vibration system, the rotary angular frequency Ω has to be increased, until the real part of one or more complex eigenvalues becomes zero. Then the limit of stability is reached at the rotary angular frequency $\Omega = \Omega_{\text{limit}}$. At the limit of stability the natural angular frequency of the critical mode becomes ω_{limit} and no damping exists ($\alpha_{\text{limit}} = 0$). So the critical complex eigenvalue at the limit of stability becomes:

$$\underline{\lambda}_{\text{limit}} = \pm j \cdot \omega_{\text{limit}} \quad \text{with } \alpha_{\text{limit}} = 0 \tag{47}$$

With this complex eigenvalue the complex eigenvector can be calculated. So the undamped natural vibration at the limit of stability can be described by:

$$\underline{\mathbf{q}}_{\text{limit}} = \underline{\hat{\mathbf{q}}}^{+}_{\text{limit}} \cdot \underline{k}^{+}_{\text{limit}} \cdot e^{j\omega_{\text{limit}} \cdot t} + \underline{\hat{\mathbf{q}}}^{-}_{\text{limit}} \cdot \underline{k}^{-}_{\text{limit}} \cdot e^{-j\omega_{\text{limit}} \cdot t} \tag{48}$$

At the limit of stability, that means at the rotary angular frequency of Ω_{limit}, which represents the rotor speed n_{limit} (= $\Omega_{\text{limit}}/2\pi$), the undamped mode (with $\alpha_{\text{limit}} = 0$) oscillates with the natural angular frequency of ω_{limit}, as a self exciting vibration.

4. Example

In this chapter the natural frequencies of a 2-pole induction motor (Fig. 1), mounted on a rigid foundation and also mounted on a soft steel frame foundation, is analyzed.

4.1 Data of motor, sleeve bearing and foundation

The machine data, sleeve bearing data and foundation data are shown in Table 1. First the stiffness data of the foundation are chosen arbitrarily. The damping ratio D_f of the steel frame foundation is assumed to be 0.02, which is common for a welded steel frame.

Machine data		**Sleeve bearing data**	
Rated power	P_N = 2000 kW	Type of bearing	Side flange bearing
Number of pole pairs	p = 1	Bearing shell	Cylindrical
Rated voltage	U_N = 6000 V	Lubricant viscosity grade	ISO VG 32
Rated frequency	f_N = 50 Hz	Nominal bore diameter	d_b = 110 mm
Rated torque	M_N = 6.4 kNm	Bearing width	b_b = 81.4 mm
Rated speed	n_N = 2990 r/min	Ambient temperature	T_{amb} = 25°C
Mass of the stator	m_s = 7200 kg	Lubricant supply temp.	T_{in} = 40°C
Mass of the rotor	m_w = 1900 kg	Mean relative bearing clearance (DIN 31698)	Ψ_m = 1.6 ‰
Moment of inertia of the stator	Θ_{sx} = 1550 kgm²		
Height of the centre of gravity	h = 560 mm	**Foundation data**	
Distance between feet	$2b$ = 1060 mm		
Rotor stiffness	c = 155.7 kN/mm	Vertical foundation stiffness at each motor side	c_{fz} = 133 kN/mm
Magnetic spring constant	c_m = 7.15 kN/mm	Horizontal foundation stiffness at each motor side	c_{fy} = 100 kN/mm
Vertical stiffness of bearing house and end shield	c_{bz} = 570 kN/mm	Damping ratio of the steel frame foundation	D_f = 0.02
Horizontal stiffness of bearing house and end shield	c_{by} = 480 kN/mm		

Table 1. Data of induction motor, sleeve bearings and foundation

4.2 Oil film stiffness and damping coefficients

The oil film stiffness and damping coefficients of the sleeve bearings are calculated for each rotor speed in steady state operation, using the program SBCALC from RENK AG.

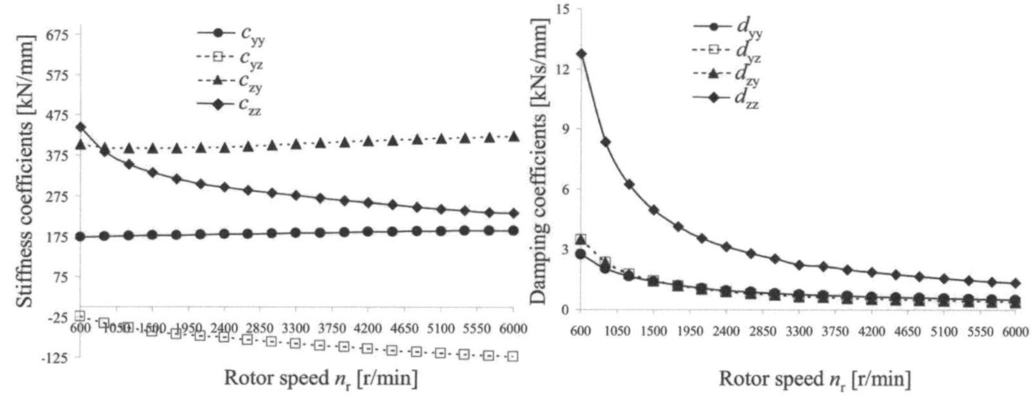

Fig. 6. Oil film stiffness and damping coefficients for different rotor speeds

4.3 Used FE-Program

To calculate the natural vibrations and to picture the mode shapes the finite element program MADYN is used. A simplified finite element model is used (Fig. 7), which is based on the model in Fig. 2. The degrees of freedom of the nodes are chosen in such a way, that only movements in the transversal plane (y-z plane) occur.

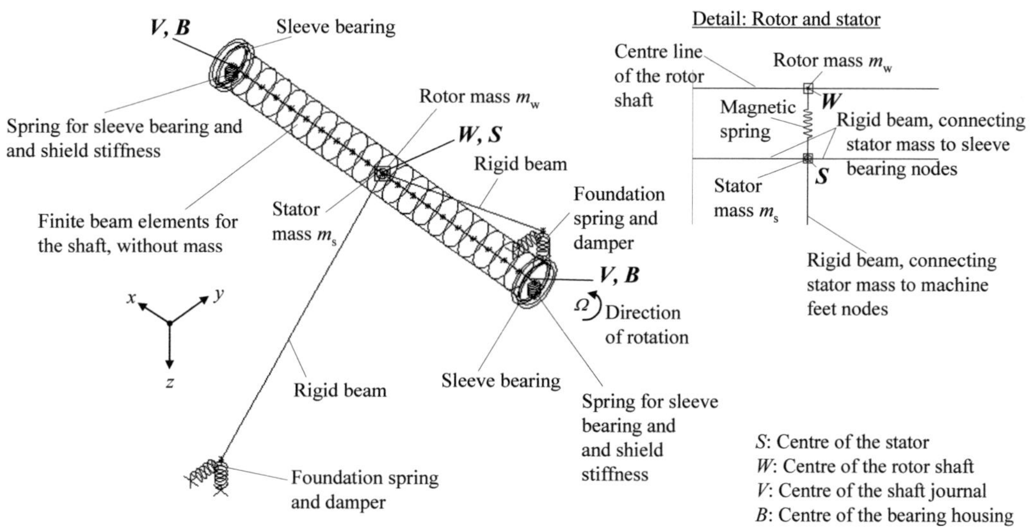

Fig. 7. Finite element model

Additionally the analytical formulas from chapter 3 could be validated with this finite element model, by comparing the calculated eigenvalues, calculated by the analytical formulas - which were solved by using the mathematic program MATHCAD - with the eigenvalues, calculated with the finite element program MADYN.

4.4 Natural vibrations; motor mounted on a rigid foundation

Before the natural vibrations of the motor, mounted on the soft steel frame foundation, are analyzed the natural vibrations of the motor, mounted on a rigid foundation are calculated. Therefore the foundation stiffness values are assumed to be infinite high ($c_{fz} = c_{yz} \rightarrow \infty$).

4.4.1 Natural vibrations at rated speed

First the natural vibrations at rated speed are calculated. The mode shapes are pictured in Fig. 8. In the 1st mode the rotor mass - shaft centre point *W* - moves on an elliptical orbit, which is run through forwards. The semi-major axis of the orbit is about 34° shifted out of the horizontal axis. The orbit of rotor mass is larger than the orbits of the shaft journals. The orbits of the shaft journals - shaft journal points *V* - have the same orientation as the orbit of the rotor mass and are also run through forwards. The orbits of the bearing housing points *B* are much smaller than the orbits of the shaft journal points *V*, but are also run through forwards. Their semi-major axes are about 28° shifted out of the horizontal axis. Because of the infinite stiffness of the foundation no movement of the stator mass occurs. In the 2nd mode the semi-major axes of all orbits have the nearly the same orientation, shifted about 8° out of the horizontal axis. All orbits are run through forwards. In this mode the largest orbits are the orbits of the shaft journal points *V*. In the 3rd mode the semi-major axes of the

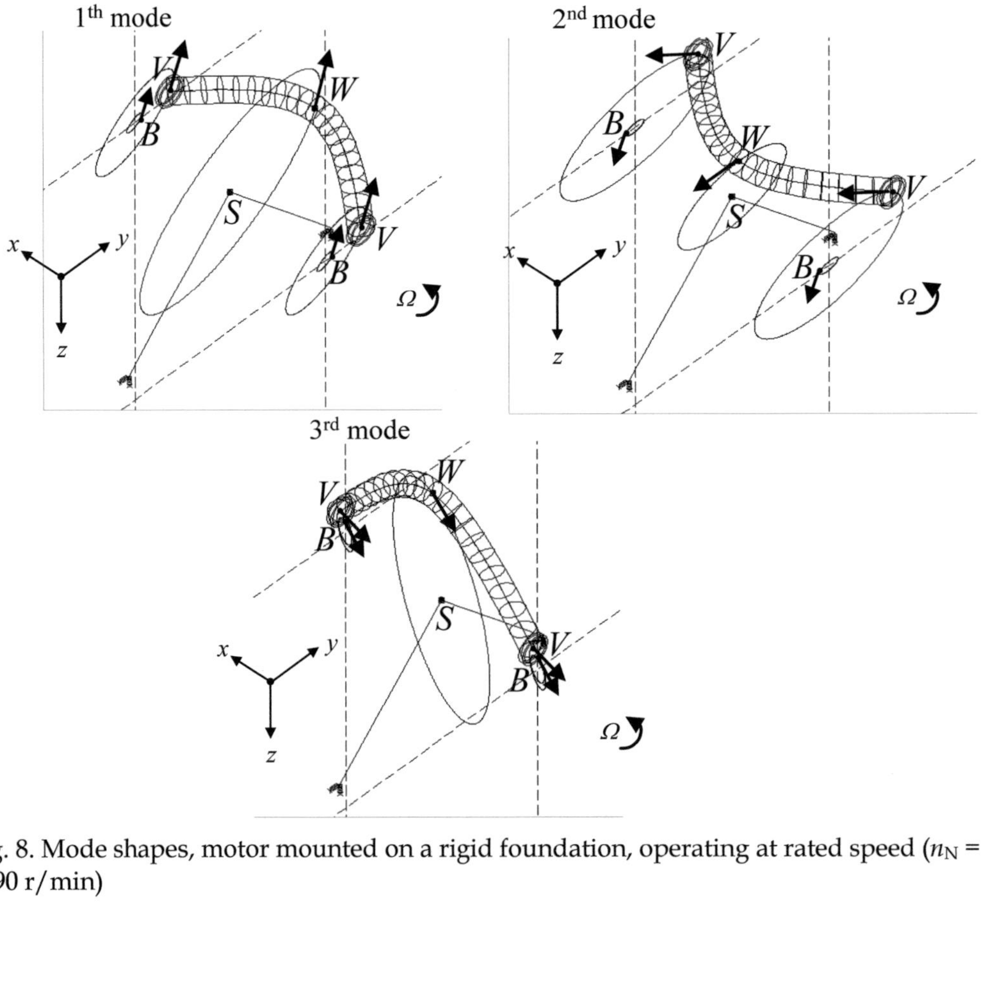

Fig. 8. Mode shapes, motor mounted on a rigid foundation, operating at rated speed (n_N = 2990 r/min)

shaft centre point *W* and of the shaft journal points *V* are shifted about 15° out of the vertical axis. The semi-major axes of the bearing housing points *B* are shifted about 20° out of the vertical axis. All orbits are run through backwards. In this mode the orbit of the shaft centre point *W* is much larger than the orbits of the shaft journal points *V* and the orbits of the bearing housing points *B*, which are nearly equal to each other. This leads to a strong bending of the rotor shaft with only small orbits in the sleeve bearings.
The natural frequencies and the modal damping values are shown in Table 2. Because of the assumption of an infinite high foundation stiffness ($c_{fz} = c_{yz} \rightarrow \infty$) only three natural vibrations occur with three natural frequencies f_1, f_2, f_3 and three modal damping values D_1, D_2,D_3. The modal damping values are here described in percentage.

Modes n	Natural frequency f_n [Hz]	Modal damping D_n[%]
1	33.15	5.31
2	34.62	68.24
3	41.17	3.82

Table 2. Natural frequencies and modal damping, motor mounted on a rigid foundation, operating at rated speed (n_N = 2990 r/min)

4.4.2 Critical speed map

In this chapter the natural frequencies and the modal damping for different rotor speeds are calculated and a critical speed map is derived (Fig. 9).
Fig. 9 shows how the natural frequencies f_n and the modal damping values D_n change with the rotor speed n_r, caused by the changing of the oil film stiffness and damping coefficients. Where the rotary frequency ($\Omega/2\pi$) meets the natural frequencies critical speeds regarding the 1x excitation may occur if the modal damping value is low at this rotor speed. Usually, if the modal damping value D_n is higher than 20% no critical resonance vibrations are expected and the rotor speed is usually not assumed to be a critical speed. Here two critical speeds have to be passed to reach the operating speed. The 1st critical speed occurs at about a rotor speed of 2070 r/min with a modal damping value of about 15%. The 2nd critical speed occurs at a rotor speed of 2475 r/min with a modal damping value of about 3.5%. Fig. 9 shows that a separation margin larger than 15% for the critical speeds to the operating speed (2990 r/min) is given, which is required in many standards and specifications. Fig. 9 shows additionally that limit of stability is reached at a rotor speed of about 3900 r/min. Here the modal damping of mode 1 gets zero.

4.4.3 Stiffness variation map regarding the electromagnetic stiffness

In this chapter the influence of the electromagnetic stiffness between the rotor and the stator on the natural frequencies is analyzed. Therefore the magnetic spring constant c_m, which describes the electromagnetic stiffness between rotor and stator, is variegated by a factor k_{cm}, called magnetic stiffness factor. The rated magnetic spring constant is $c_{m,rated}$ = 7.15 kN/mm (Table 1). The magnetic stiffness factor k_{cm} is variegated in the range of 0…2 and the influence on the natural frequencies and modal damping values are calculated for operation at rated speed (Fig. 10).

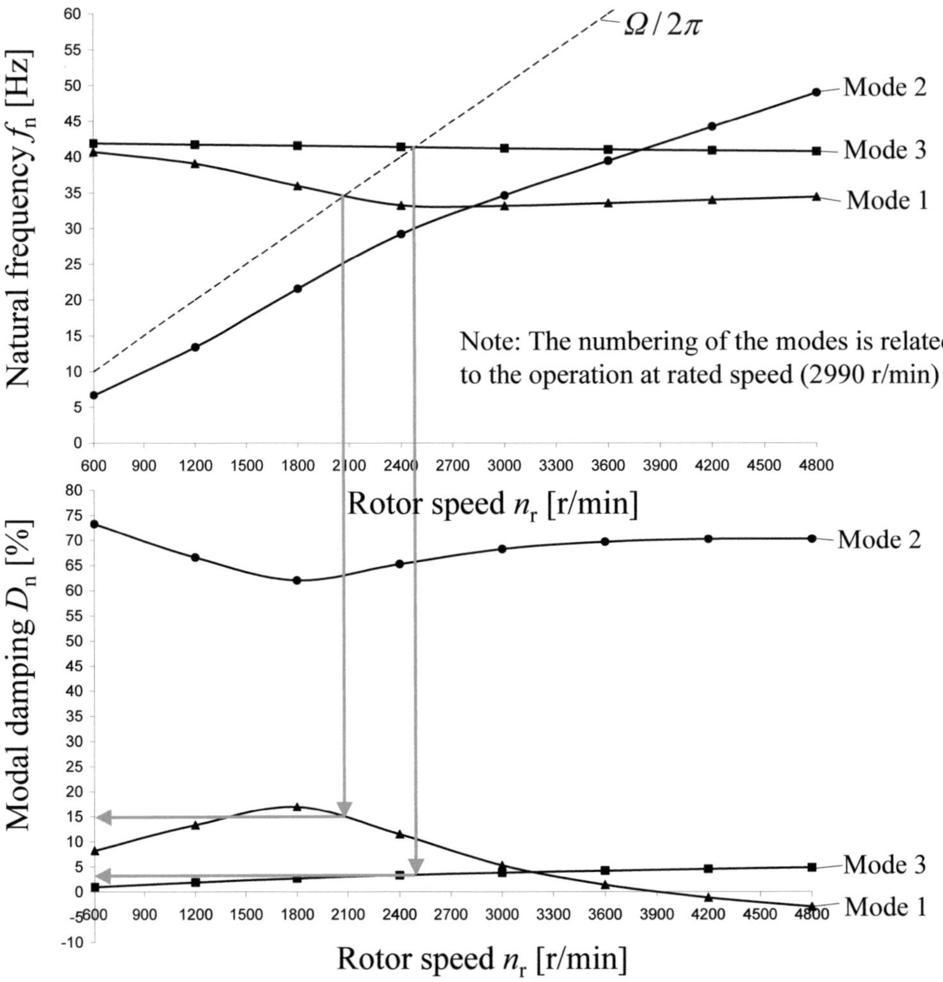

Fig. 9. Critical speed map, motor mounted on a rigid foundation

$$\text{Magnetic spring constant:} \qquad c_{\mathrm{m}} = k_{\mathrm{cm}} \cdot c_{\mathrm{m,rated}} \tag{49}$$

Fig. 10 shows that mode 1 and mode 3 are clearly influenced by the magnetic spring constant. Their natural frequencies and modal damping values change with the magnetic stiffness factor, whereas mode 2 is hardly influenced by the magnetic spring value. The reason is that the orbits of rotor mass are larger than the orbits of the shaft journals for mode 1 and mode 3 (Fig. 8), contrarily to mode 2, where the orbits of the shaft journals are larger. Therefore the influence of the magnetic spring constant, which acts at the rotor mass, is higher for mode 1 and mode 3.

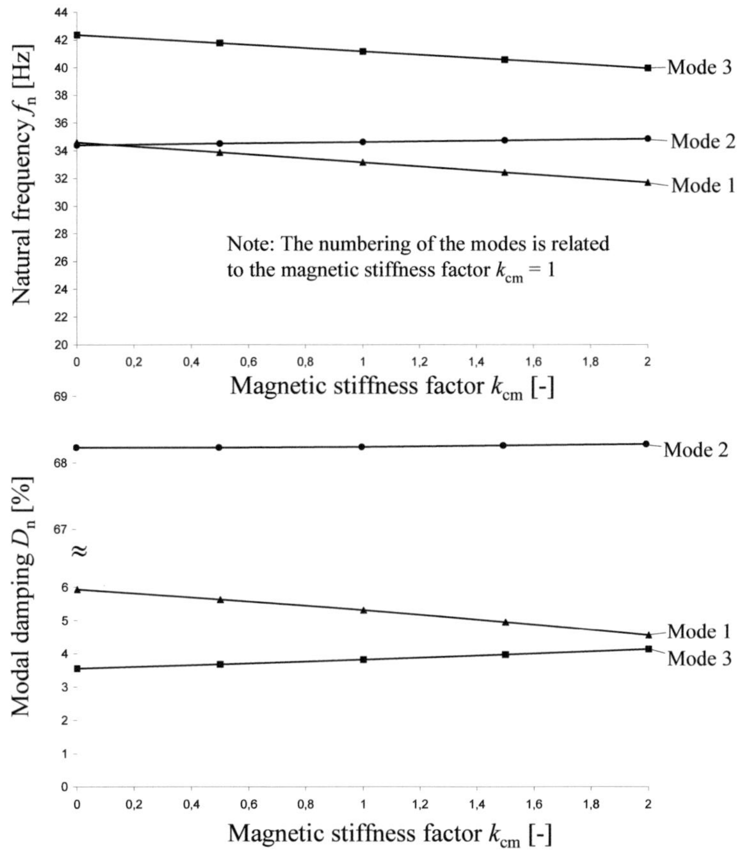

Fig. 10. Stiffness variation map regarding the electromagnetic stiffness, motor mounted on a rigid foundation, operating at rated speed (n_N = 2990 r/min)

4.5 Natural vibrations; motor mounted on a soft steel frame foundation

After the natural vibrations of the rigid mounted motor are analyzed, the natural vibrations of the motor, mounted on a soft steel frame foundation, are now investigated. The foundation data are described in Table 1.

4.5.1 Natural vibrations at rated speed

Again, the natural vibrations at rated speed are calculated first. The natural frequencies are calculated once without considering of the foundation damping (D_f = 0) and once with considering of the foundation damping (D_f = 0.02). The mode shapes without considering foundation damping are pictured in Fig. 11, which can be assumed to be equal to the mode shapes with considering foundation damping.

The first two modes are nearly rigid body modes of the soft mounted machine. Rotor and stator are nearly acting like a one-mass system. The orbits of the rotor and of the stator are nearly straight lines. In the 1st mode the rotor mass and the stator mass oscillate in phase to each other nearly in horizontal direction, while stator mass makes a lateral buckling at the x-axis, in the same direction as its horizontal movement. In the 2nd mode the rotor mass and the stator mass also oscillate in phase to each other, but in vertical direction. Nearly no buckling of

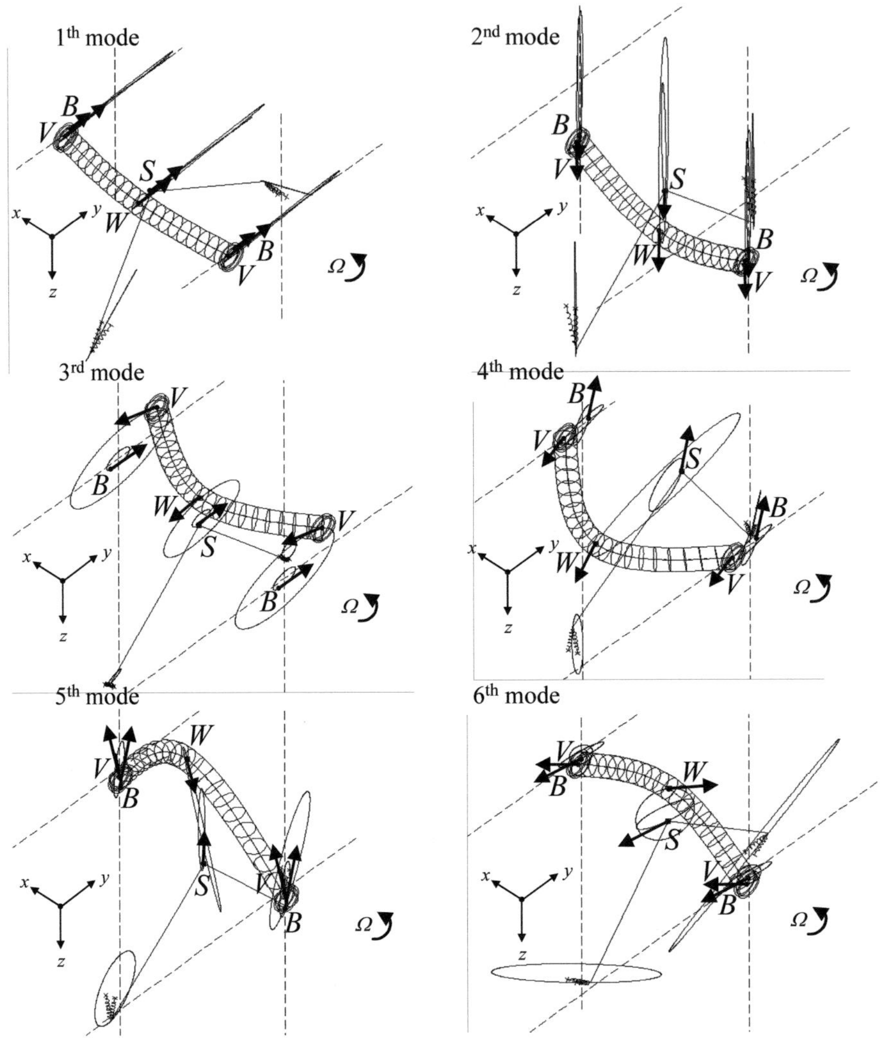

Fig. 11. Mode shapes, motor mounted on a soft steel frame foundation (c_{fz} = 133 kN/mm; c_{fy} = 100 kN/mm), without considering foundation damping (D_f =0), operating at rated speed (n_N = 2990 r/min)

the stator mass occurs. For the higher modes stator and rotor behave like a two- mass system and elliptical orbits of the rotor mass and stator mass occur. In the 3th mode the semi-major axes of orbits of the rotor mass, the bearing housings and the shaft journals are shifted about 12° out of the horizontal axis, whereas the semi-major axis of the orbit of the stator mass is only shifted 5° out of the horizontal axis. All orbits are run through forwards. The rotor mass and the stator mass oscillate out of phase to each other, as well as the shaft journals to the bearing housings. The largest orbits are the orbits of the shaft journals, compared to the other orbits. Because of the large relative orbits between the shaft journals and the bearing housings, the modal damping of this mode is very high, due to the oil film damping of the sleeve bearings. In the 4th mode the semi-major axis of the orbit of the rotor mass is shifted about 14° out of the horizontal axis. The same is valid for the shaft journals, whereas the semi-major axis of the stator orbit is shifted about 47° out of the horizontal axis. The semi-major axes of the orbits of the bearing housings are shifted about 62° out of the horizontal axis. All orbits are still run through forwards. In the 5th mode the semi-major axis of the orbit of the rotor mass is shifted about 12° out of the vertical axis. The other orbits lie nearly in vertical direction. The stator mass and the rotor mass oscillate out of phase to each other. The orbit of the stator mass and the orbits of the bearing housing are run through forwards, while the orbit of the rotor mass and the orbits of the shaft journals are run through backwards. In the 6th mode the semi-major axes of the orbits of the stator mass and of the bearing housings are shifted about 80° out of the vertical axis, while the semi-major axes of the orbits of the rotor mass and of the shaft journals are shifted about 45° out of the vertical axis. All orbits are run through backwards. Additionally the 6th mode shows a strong lateral buckling of the stator mass at the x-axis, which leads to large orbits at the motor feet. Contrarily to the 1st mode the lateral buckling of the stator mass is contrariwise to its horizontal movement, which means that if the stator mass moves to the right the lateral buckling is to the left. To consider the influence of the foundation damping on the natural vibrations, a simplified approach is used. Referring to (Gasch et al., 2002), the damping ratio D_f of the foundation can be described by the damping coefficients d_{fq}, stiffness coefficients c_{fq} of the foundation and the stator mass m_s, as a rough simplification.

$$d_{fq} = D_f \cdot m_s \cdot \sqrt{2 \cdot c_{fq} / m_s} \quad \text{with: } q = z,y \tag{50}$$

The calculated natural frequencies and modal damping of each mode shape with and without considering foundation damping are shown in Table 3. It is shown that considering the foundation damping influences the natural frequencies only marginal, as expected. But the modal damping values of some modes are strongly influenced by the foundation damping. The modal damping values of the first two modes are strongly influenced by the foundation damping, because the modes are nearly rigid body modes of the motor on the foundation. Also the modal damping of the 6th mode is strongly influenced by the foundation damping, because large orbits of the motor feet occur in this mode shape, compared to the other orbits.

Modes n	Without foundation damping ($D_f = 0$)		With foundation damping ($D_f = 0.02$)	
	Natural frequency f_n [Hz]	Modal damping D_n [%]	Natural frequency f_n [Hz]	Modal damping D_n [%]
1	16.05	-0.11	16.05	0.95
2	25.35	0.51	25.33	1.84
3	35.22	65.75	35.23	65.72
4	37.72	6.97	37.67	7.36
5	48.50	3.39	48.54	4.24
6	52.63	1.0	52.61	4.17

Table 3. Natural frequencies and modal damping, motor mounted on a soft steel frame foundation ($c_{fz} = 133$ kN/mm; $c_{fy} = 100$ kN/mm) with and without considering foundation damping ($D_f = 0.02$ and $D_f = 0$), operating at rated speed ($n_N = 2990$ r/min)

4.5.2 Critical speed map

Again, a critical speed map is derived to show the influence of the rotor speed on the natural frequencies and the modal damping and to derive the critical speeds (Fig. 12).

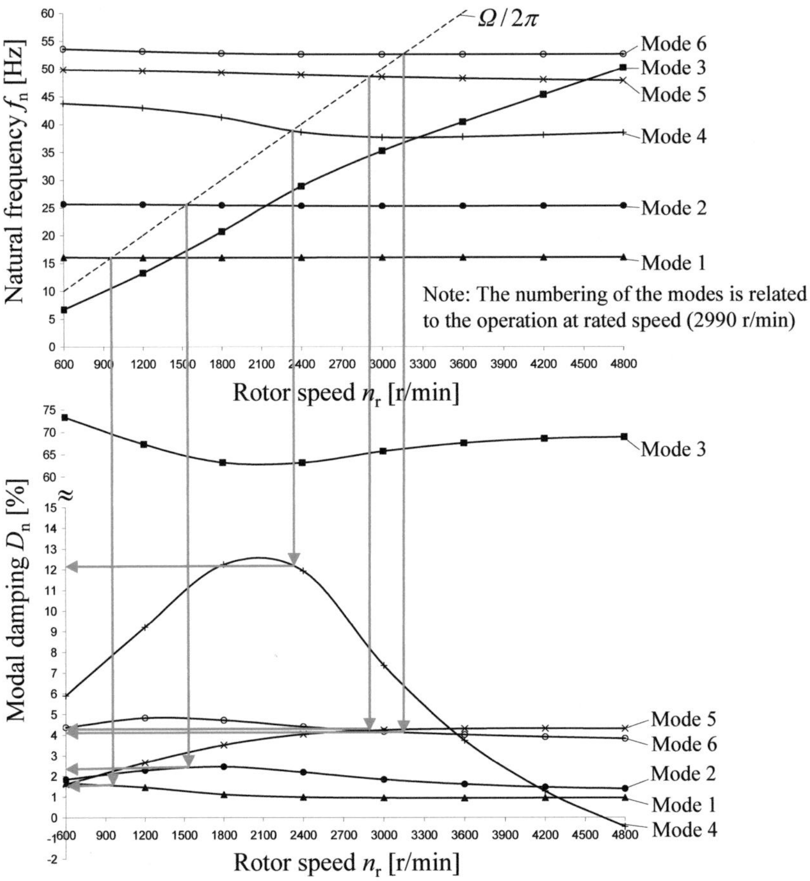

Fig. 12. Critical speed map, motor mounted on a soft steel frame foundation (c_{fz} = 133 kN/mm; c_{fy} = 100 kN/mm; D_f =0.02)

Critical speed	Critical speed [r/min]	Modal damping D_n [%]
1	950	1.6
2	1540	2.3
3	2340	12.2
4	2900	4.3
5	3160	4.2

Table 4. Critical speeds, motor mounted on a soft steel frame foundation (c_{fz} = 133 kN/mm; c_{fy} = 100 kN/mm; D_f =0.02)

Fig. 12 shows that the limit of stability is here reached at about 4650 r/min, because the modal damping of mode 4 gets zero at this rotor speed. For the rigid foundation the limit of stability is already reached at a rotor speed of about 3900 r/min. But contrarily to the rigid mounted motor here four critical speeds have to be passed before the operating speed (2990 r/min) is reached. Additionally a 5th critical speed is close above the operating speed. The critical speeds and the modal damping in the critical speeds are shown in Table 4.

Table 4 shows that two critical speeds (4th and 5th) with low modal damping values are very close to the operating speed (2990 r/min), having less than 5% separation margin to the operating speed. Therefore resonance vibrations problems may occur. The conclusion is that the arbitrarily chosen foundation stiffness values are not suitable for that motor with a operation speed of 2990 r/min. To find adequate foundation stiffness values, a stiffness variation of the foundation is deduced and a stiffness variation map is created (chapter 4.5.4). But preliminarily the influence of the electromagnetic stiffness on the natural frequencies and modal damping values is investigated for the soft mounted motor.

4.5.3 Stiffness variation map regarding the electromagnetic stiffness

In this chapter the influence of the electromagnetic stiffness on the natural frequencies and the modal damping values at rated speed is analyzed again, but now for the soft mounted motor. Again the magnetic stiffness factor k_{cm} is variegated in a range of 0....2 and the influence on the natural frequencies and the modal damping values is analyzed. Fig. 13

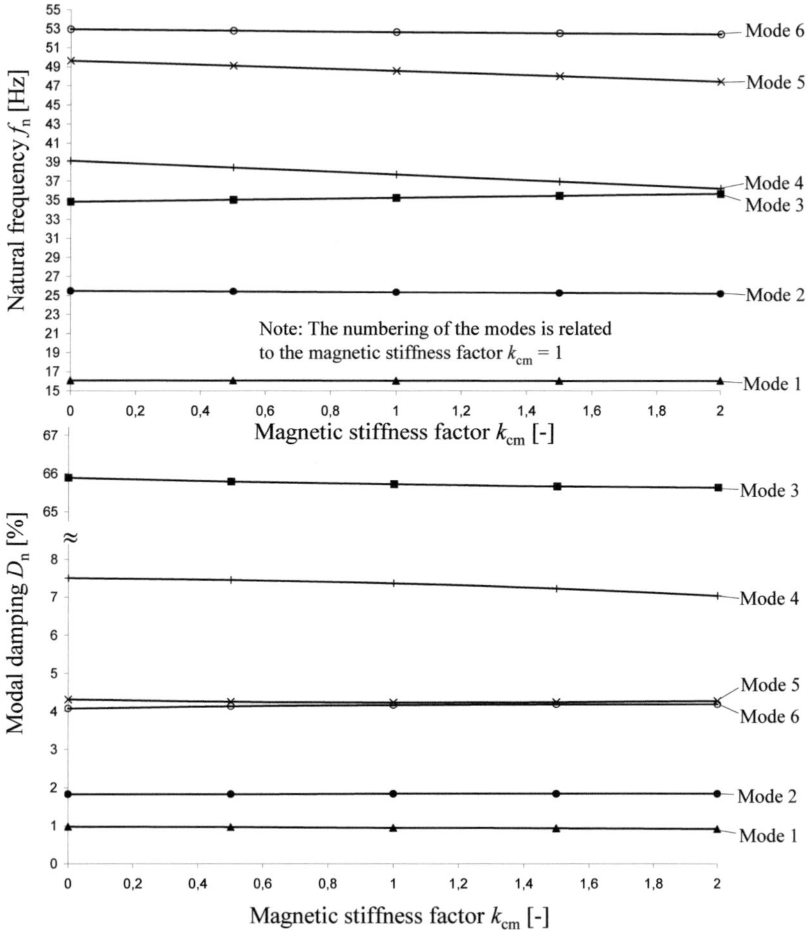

Fig. 13. Stiffness variation map regarding the electromagnetic stiffness, motor mounted on a soft steel frame foundation (c_{fz} = 133 kN/mm; c_{fy} = 100 kN/mm; D_f = 0.02), operating at rated speed (n_N = 2990 r/min)

shows that mainly the natural frequencies of the 4th mode and the 5th mode are influenced by the magnetic spring constant. The natural frequencies of the other modes are hardly influenced by the magnetic spring constant. The reason is that for the 4th mode and the 5th mode the relative orbits between the rotor mass and the stator mass are large, compared to the other orbits. Large orbits of the rotor mass and of the stator mass occur for these two modes and both masses - the rotor mass and the stator mass - vibrate out of phase to each other (Fig. 11), which lead to large relative orbits between these two masses. Therefore, the electromagnetic interaction between these two masses is high and therefore a significant influence of the magnetic spring constant on the natural vibrations occurs for these two modes. In the 1st and 2nd mode the motor is acting like a one-mass system (Fig. 11) and nearly no relative movements between rotor mass and stator mass occur. Therefore the electromagnetic coupling between rotor and stator has nearly no influence on the natural frequencies of the first two modes. The 3th mode is mainly dominated by large relative orbits between the shaft journals and the bearing housings - compared to the other orbits - leading to high modal damping. A relative movement between the rotor mass and the stator occurs, but is not sufficient enough for a clear influence of the electromagnetic coupling. The 6th mode is mainly dominated by large orbits of the motor feet, compared to the other orbits. Again the relative movement of the stator and rotor is not sufficient enough that the electromagnetic coupling influences the natural frequency of this mode clearly. The modal damping values of all modes are only marginally influenced by the magnetic spring constant, only a small influence on the modal damping of the 4th mode is obvious.

4.5.4 Stiffness variation map regarding the foundation stiffness

The foundation stiffness values c_{fz} and c_{yz} are changed by multiplying the rated stiffness values $c_{fz,rated}$ and $c_{fy,rated}$ from Table 1 with a factor, called foundation stiffness factor k_{cf}.

$$\text{Vertical foundation stiffness:} \quad c_{fz} = k_{cf} \cdot c_{fz,rated} \tag{51}$$

$$\text{Horizontal foundation stiffness:} \quad c_{fy} = k_{cf} \cdot c_{fy,rated} \tag{52}$$

Therefore the vertical foundation stiffness c_{fz} and the horizontal foundation stiffness c_{fy} are here changed in equal measure by the foundation stiffness factor k_{cf}. The influence of the foundation stiffness at rated speed on the natural frequencies and on the modal damping is shown in Fig. 14.

It is shown that for a separation margin of 15% between the natural frequencies and the rotary frequency $\Omega/2\pi$ the foundation stiffness factor k_{cf} has to be in a range of 2.5…3.0. If the foundation stiffness factor is smaller than 2.5 the natural frequency of the 5th mode gets into the separation margin. If the foundation stiffness factor is bigger than 3.0 the natural frequency of the 4th mode gets into the separation margin. Both modes - 4th mode and 5th mode - have a modal damping less than 10% in the whole range of the considered foundation stiffness factor (k_{cf} = 0.5…4). Because of the low modal damping values of these two modes, the operation close to the natural frequencies of these both modes suppose to be critical. Therefore the first arbitrarily chosen foundation stiffness values ($c_{fz,rated}$ = 133 kN/mm; $c_{fy,rated}$ = 100 kN/mm) have to be increased by a factor of k_{cf} = 2.5…3.0. With the increased foundation stiffness values the foundation can still be indicated as a soft foundation, because the natural frequencies of the 1st mode and the 2nd mode - the mode shapes are still the same as in Fig. 11 - are still low, lying in a range between 24 Hz and 26 Hz for the 1st mode and between 33 Hz and 35 Hz for the 2nd mode.

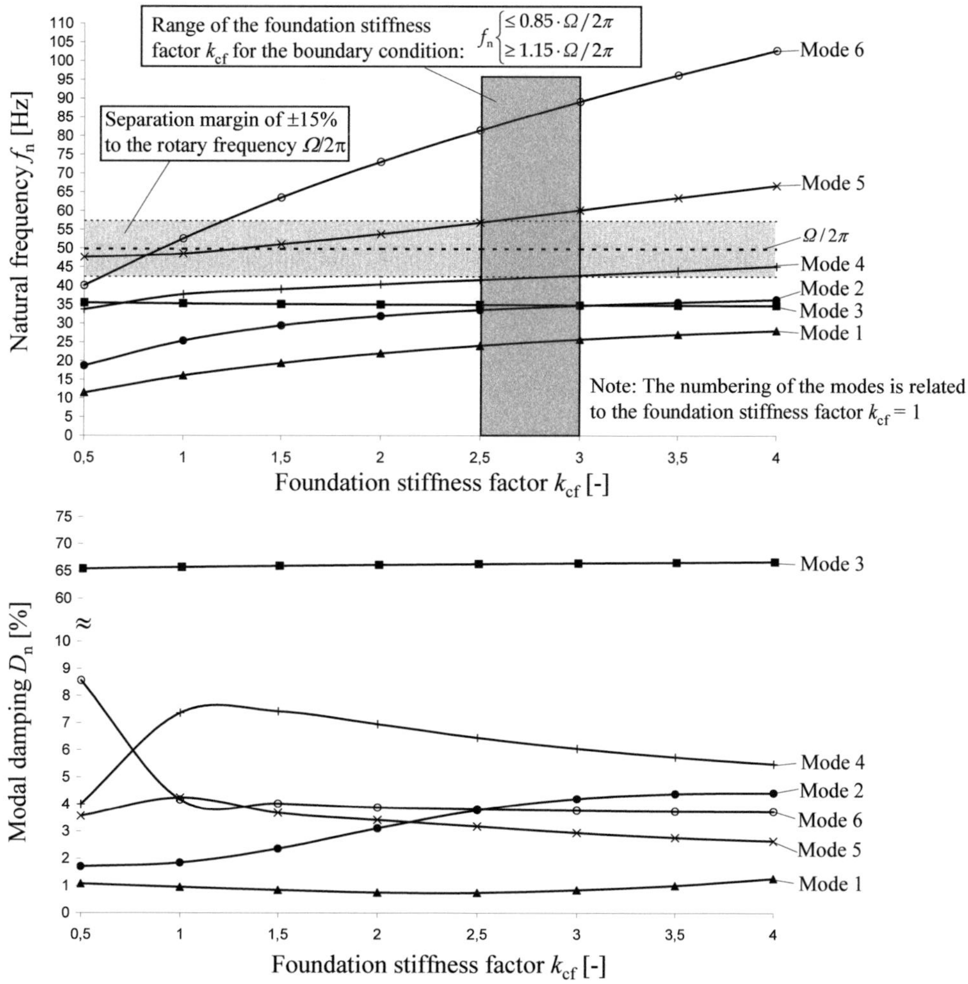

Fig. 14. Stiffness variation map regarding the foundation stiffness, motor mounted on a soft steel frame foundation, operating at rated speed (n_N = 2990 r/min)

5. Conclusion

The aim of this paper is to show a simplified plane vibration model, describing the natural vibrations in the transversal plane of soft mounted electrical machines, with flexible shafts and sleeve bearings. Based on the vibration model, the mathematical correlations between the rotor dynamics and the stator movement, the sleeve bearings, the electromagnetic and the foundation, are derived. For visualization, the natural vibrations of a soft mounted 2-pole induction motor are analyzed exemplary, for a rigid foundation and for a soft steel frame foundation. Additionally the influence of the electromagnetic interaction between rotor and stator on the natural vibrations is analyzed. Finally, the aim is not to replace a detailed three-dimensional finite-element calculation by a simplified plane multibody model, but to show the mathematical correlations based on a simplified model.

6. References

Arkkio, A.; Antila, M.; Pokki, K.; Simon, A., Lantto, E. (2000). Electromagnetic force on a whirling cage rotor. *Proceedings of Electr. Power Appl.*, pp. 353-360, Vol. 147, No. 5

Belmans, R.; Vandenput, A.; Geysen, W. (1987). Calculation of the flux density and the unbalanced magnetic pull in two pole induction machines, pp. 151-161, *Arch. Elektrotech,* Volume 70

Bonello, P.; Brennan, M.J. (2001). Modelling the dynamic behaviour of a supercritcial rotor on a flexible foundation using the mechanical impedance technique, pp. 445-466, *Journal of sound and vibration,* Volume 239, Issue 3

Gasch, R.; Nordmann, R. ; Pfützner, H. (2002). *Rotordynamik,* Springer-Verlag, ISBN 3-540-41240-9, Berlin-Heidelberg

Gasch, R.; Maurer, J.; Sarfeld W. (1984). The influence of the elastic half space on stability and unbalance of a simple rotor-bearing foundation system, *Proceedings of Conference Vibration in Rotating Machinery,* pp. 1-11, C300/84, IMechE, Edinburgh

Glienicke, J. (1966). *Feder- und Dämpfungskonstanten von Gleitlagern für Turbomaschinen und deren Einfluss auf das Schwingungsverhalten eines einfachen Rotors,* Dissertation, Technische Hochschule Karlsruhe, Germany

Holopainen, T. P. (2004) *Electromechanical interaction in rotor dynamics of cage induction motors,* VTT Technical Research Centre of Finland, Ph.D. Thesis, Helsinki University of Technology, Finland

Kellenberger, W. (1987) *Elastisches Wuchten*, Springer-Verlag, ISBN 978-3540171232, Berlin-Heidelberg

Lund, J.; Thomsen, K. (1987). Review of the Concept of Dynamic Coefficients for Fluid Film Journal Bearings, pp. 37-41, *Journal of Tribology, Trans. ASME,* Vol. 109, No. 1

Lund, J.; Thomsen, K. (1978). A calculation method and data for the dynamics of oil lubricated journal bearings in fluid film bearings and rotor bearings system design and optimization, pp. 1-28, *Proceedings of Conference ASME Design and Engineering Conference,* ASME , New York

Schuisky, W. (1972). Magnetic pull in electrical machines due to the eccentricity of the rotor, pp. 391-399, *Electr. Res. Assoc. Trans.* 295

Seinsch, H-O. (1992). *Oberfelderscheinungen in Drehfeldmaschinen,* Teubner-Verlag, ISBN 3-519-06137-6, Stuttgart

Tondl, A. (1965). *Some problems of rotor dynamics*, Chapman & Hall, London

Vance, J.M.; Zeidan, F. J.; Murphy B. (2010). *Machinery Vibration and Rotordynamics,* John Wiley and Sons, ISBN 978-0-471-46213-2, Inc. Hoboken, New Jersey

Werner, U. (2010). Theoretical vibration analysis of soft mounted electrical machines regarding rotor eccentricity based on a multibody model, pp. 43-66, Springer, *Multibody System Dynamics,* Volume 24, No. 1, Berlin/Heidelberg

Werner, U. (2008). A mathematical model for lateral rotor dynamic analysis of soft mounted asynchronous machines. *ZAMM-Journal of Applied Mathematics and Mechanics,* pp. 910-924, Volume 88, No. 11

Werner, U. (2006). *Rotordynamische Analyse von Asynchronmaschinen mit magnetischen Unsymmetrien,* Dissertation, Technical University of Darmstadt, Germany, Shaker-Verlag, ISBN 3-8322-5330-0, Aachen

15

Time-Frequency Analysis for Rotor-Rubbing Diagnosis

Eduardo Rubio and Juan C. Jáuregui
CIATEQ A.C., Centro de Tecnología Avanzada
Mexico

1. Introduction

Predictive maintenance by condition monitoring is used to diagnose machinery health. Early detection of potential failures can be accomplished by periodic monitoring and analysis of vibrations. This can be used to avoid production losses or a catastrophic machinery breakdown. Predictive maintenance can monitor equipments during operation. Predictions are based on a vibration signature generated by a healthy machine. Vibrations are measured periodically and any increment in their reference levels indicates the possibility of a failure.

There are several approaches to analyze the vibrations information for machinery diagnosis. Conventional time-domain methods are based on the overall level measurement, which is a simple technique for which reference charts are available to indicate the acceptable levels of vibrations. Processing algorithms have been developed to extract some extra features in the vibrations signature of the machinery. Among these is the Fast Fourier Transforms (FFT) that offers a frequency-domain representation of a signal where the analyst can identify abnormal operation of the machinery through the peaks of the frequency spectra. Since FFT cannot detect transient signals that occur in non-stationary signals, more complex analysis methods have been developed such as the wavelet transform. These methods can detect mechanical phenomena that are transient in nature, such as a rotor rubbing the casing of a motor in the machine. This approach converts a time-domain signal into a time-frequency representation where frequency components and structured signals can be localized. Fast and efficient computational algorithms to process the information are available for these new techniques.

A number of papers can be found in the literature which report wavelets as a vibration processing technique. Wavelets are multiresolution analysis tools that are helpful in identifying defects in mechanical parts and potential failures in machinery. Multiresolution has been used to extract features of signals to be used in classifications algorithms for automated diagnosis of machine elements such as rolling bearings (Castejón et al., 2010; Xinsheng & Kenneth, 2004). These elements produce clear localized frequencies in the vibration spectrum when defects are developing. However, a more complex phenomena occurs when the rotor rubs a stationary element. The impacts produce vibrations at the fundamental rotational frequency and its harmonics, and additionally yield some high frequency components, that increase as the severity of the impacts increases (Peng et al., 2005).

Rotor dynamics may present light and severe rubbing, and both are characterized by a different induced vibration response. It is known that conditions that cause high vibration

levels are accompanied by significant dynamic nonlinearity (Adams, 2010). The resonance frequency is modified because of the stiffening effect of the rubbing on the rotor (Abuzaid et al., 2009). These systems are strongly nonlinear and techniques have been applied for parameter identification. These techniques have developed models that explain the jump phenomenon typical of partial rub (Choi, 2001; Choi, 2004).
The analysis of rubbing is accomplished with the aid of the Jeffcott rotor model for lateral shaft vibrations. This model states the idealized equations of rotor dynamics (Jeffcott, 1919). Research has been done to extend this model to include the nonlinear behavior of the rotor system for rubbing identification. It has been shown that time-frequency maps can be used to analyze multi-non-linear factors in rotors. They also reveal many complex characteristics that cannot be discovered with FFT spectra (Wang et al., 2004). Other approaches use analytical methods for calculating the nonlinear dynamic response of rotor systems. Second-order differential equations which are linear for non-contact and strongly nonlinear for contact scenarios have been used (Karpenko et al., 2002). Rub-related forces for a rotor touching an obstacle can be modeled by means of a periodic step-function that neglects the transient process (Muszynska, 2005).
In this chapter the phenomenon of rotor rubbing is analyzed by means of a vibrations analysis technique that transforms the time-domain signal into the time-frequency domain. The approach is proposed as a technique to identify rubbing from the time-frequency spectra generated for diagnostic purposes. Nonlinear systems with rotating elements are revised and a nonlinear model which includes terms for the stiffness variation is presented. The analysis of the signal is made through the wavelet transform where it is demonstrated that location and scale of transient phenomena can be identified in the time-frequency maps. The method is proposed as a fast diagnostic technique for rapid on-line identification of severe rubbing, since algorithms can be implemented in modern embedded systems with a very high computational efficiency.

2. Nonlinear rotor system with rubbing elements

Linear models have intrinsic limitations describing physical systems that show large vibration amplitudes. Particularly, they are unable to describe systems with variable stiffness. To reduce the complexity of nonlinear problems, models incorporate simplified assumptions, consistent with the physical situation, that reduce their complexity and allow representing them by linear expressions. Although linearized models capture the essence of the problem and give the main characteristics of the dynamics of the system, they are unable to identify instability and sudden changes. These problems are found in nonlinear systems and the linear vibration theory offers limited tools to explain the complexity of their unpredictable behavior. Therefore, nonlinear vibration theories have been developed for such systems.
The steady state response of the nonlinear vibration solution exhibits strong differences with respect to the linear approach. One of the most powerful models for the analysis of nonlinear mechanical systems is the Duffing equation. Consider the harmonically forced Duffing equation with external excitation:

$$\ddot{x} + c\dot{x} + \alpha x + \beta x^3 = acos(\omega t + \phi) \tag{1}$$

Curves of response amplitude versus exciting frequency are often employed to represent this vibration behavior as shown in Fig. 1. The solid line in this figure shows the response curve for a linear system. The vertical line at ω/ω_n=1 corresponds to the resonance. At this point vibration amplitude increases dramatically and it is limited only by the amount of damping in the system. It is important to ensure that the system operates outside of this frequency to avoid excessive vibration that can result in damage to the mechanical parts. In linear systems amplitude of vibrations grows following a straight line as excitation force increases.

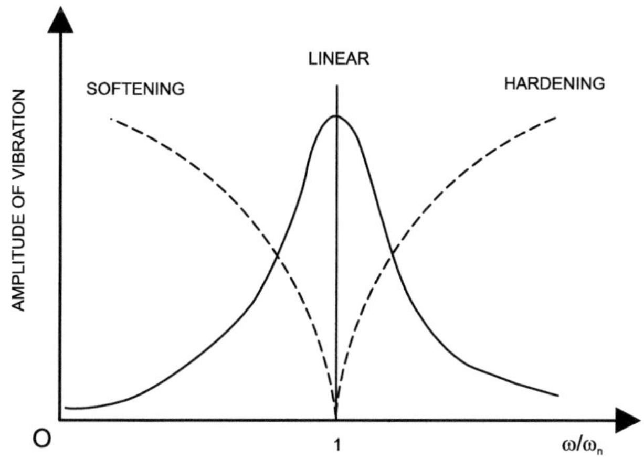

Fig. 1. Resonant frequency dependency in nonlinear systems

In nonlinear systems the motion follows a trend that is dependent upon the amplitude of the vibrations and the initial conditions. The resonance frequency is a function of the excitation force and the response curve does not follow a straight line. When the excitation force increases, the peak amplitude "bends" to the right or left, depending on whether the stiffness of the system hardens or softens. For larger amplitudes, the resonance frequency decreases with amplitude for softening systems and increases with amplitude for hardening systems. The dashed lines in Fig. 1 show this effect.

When the excitation force is such that large vibration amplitudes are present, an additional "jump" phenomenon associated with this bending arises. This is observed in Fig. 2. Jump phenomenon occurs in many mechanical systems. In those systems, if the speed is increased the amplitude will continue increasing up to values above $1.6\omega_n$.

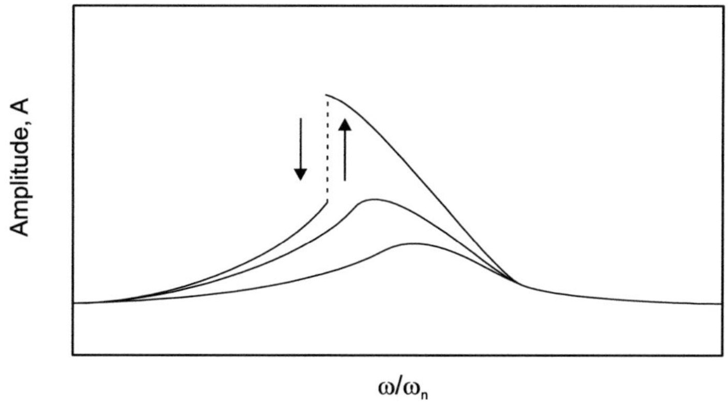

Fig. 2. Jump phenomenon typical of nonlinear systems

When the excitation force imposes low vibration amplitudes, or there is a relative strong damping, the response curve is not very different from the linear case as it can be observed in the two lower traces. However, for large vibration amplitudes the bending effect gets stronger and a "jump" phenomenon near the resonance frequency is observed. This phenomenon may be observed by gradually changing the exciting frequency ω while keeping the other parameters fixed. Starting from a small ω and gradually increasing the frequency, the amplitude of the vibrations will increase and follow a continuous trend. When frequency is near resonance, vibrations are so large that the system suddenly exhibits a jump in amplitude to follow the upper path, as denoted with a dashed line in Fig. 2. When reducing the excitation frequency the system will exhibit a sudden jump from the upper to the lower path. This unusual performance takes place at the point of vertical tangency of the response curve, and it requires a few cycles of vibration to establish the new steady-state conditions.

There is a region of instability in the family of response curves of a nonlinear system where such amplitudes of vibration cannot be established. This is shown in Fig. 3. It is not possible to obtain a particular amplitude in this region by forcing the exciting frequency. Even with small variations the system is unable to restore the stable conditions. Therefore, from the three regions depicted in this figure, only the upper and lower amplitudes of vibration exist. The same applies for a hardening system but with the peaks of amplitude of vibrations bending to the right.

A rotor system with rub impact is complex and behaves in a strong nonlinearity. A complicated vibration phenomenon is observed and the response of the system may be characterized by the jump phenomena at some frequencies. Impacts are associated with stiffening effects; therefore, modeling of rotor rub usually includes the nonlinear term of stiffness.

When the rotor hits a stationary element, it involves several physical phenomena, such as stiffness variation, friction, and thermal effects. This contact produces a behavior that worsens the operation of the machine. Rubbing is a secondary transient phenomenon that arises as a result of strong rotor vibrations. The transient and chaotic behavior of the rotor impacts generate a wide frequency bandwidth in the vibrational response.

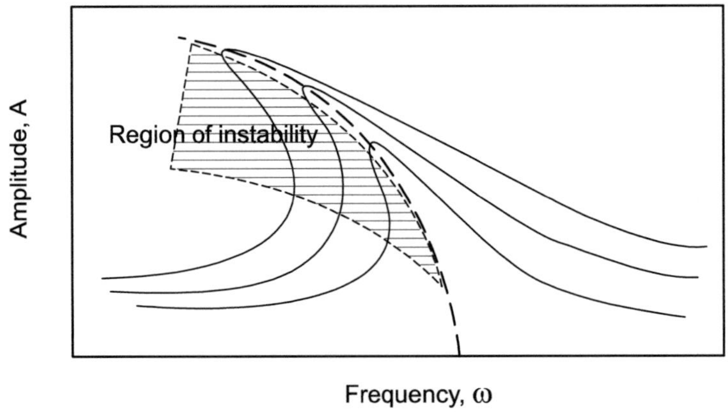

Fig. 3. Region of instability

Dynamics of the rotor rubbing can be studied with the Jeffcott rotor model (Jeffcott, 1919). This model was developed to analyze lateral vibrations of rotors and consists of a centrally mounted disk on a flexible shaft. Rigid bearings support the ends of the shaft as shown in Fig. 4. The model is more representative of real rotor dynamics for the inclusion of a damping force proportional to the velocity of the lateral motion. The purpose of this model was to analyze the effect of unbalance at speeds near the natural frequency, since the vibration amplitude increase considerably in this region.

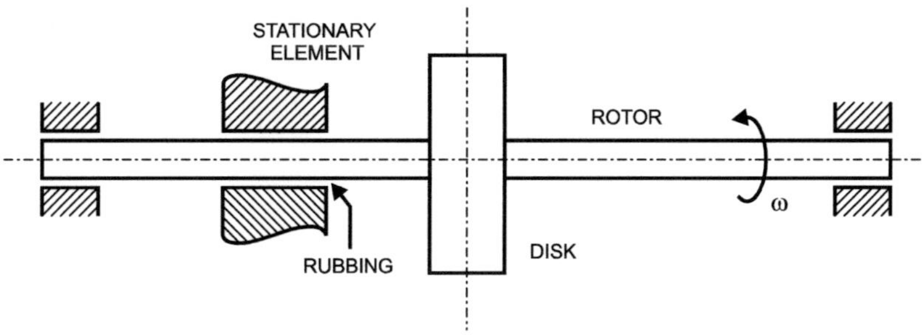

Fig. 4. Diagram of a rotor rubbing with a stationary element

Modifying the Jeffcott´s model, the rubbing phenomenon can be studied. A stationary element can be added to the model to take rubbing into consideration. A diagram of the forces that are involved during the rub-impact phenomenon is shown in Fig. 5.

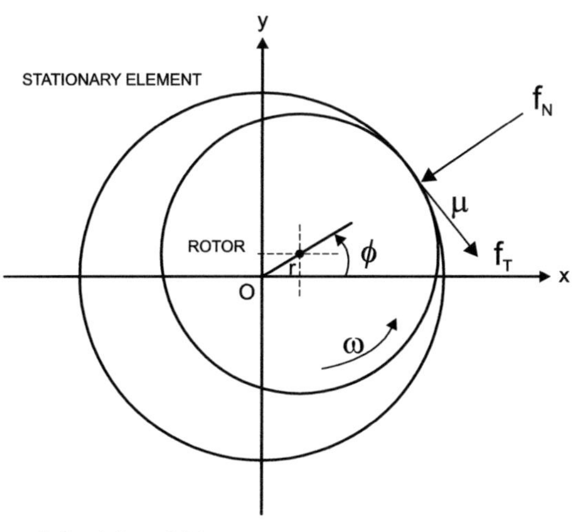

Fig. 5. A Jeffcott rotor model with rubbing

At the contact point, normal and tangential forces are described by the following expressions:

$$f_N = K_R(R - c) \tag{2}$$

$$f_T = \mu f_T \tag{3}$$

Where K_R is the combined stiffness of the shaft and the contact stiffness.
This is valid for

$$R \geq c \tag{4}$$

otherwise

$$f_N = f_T = 0 \tag{5}$$

For a Cartesian coordinate system forces are represented as:

$$f_x = -f_N \cos\phi + f_T \sin\phi \tag{6}$$

$$f_y = -f_N \sin\phi - f_T \cos\phi \tag{7}$$

And the motion equations of the rotor system are described by:

$$M\ddot{x} + C\dot{x} + K(\omega)x = F_x + m\rho\omega^2 \cos(\varphi_0 + \omega t) \tag{8}$$

$$M\ddot{y} + C\dot{y} + K(\omega)y = F_y + MP\omega^2 \sin(\varphi_0 + \omega t) \tag{9}$$

Where K is the stiffness of the system and C is the damping coefficient of the system.
The contact of the rotor with the stationary element creates a coupling of the system that causes a variation in the stiffness because of the non-continuous f_N and the model becomes nonlinear. The rotor rubs the element only a fraction of the circumferential movement and the stiffness value varies with respect to the rotor angular position.
The nonlinear behavior can be related to the stiffness variation. As shown in Fig. 6, the system´s stiffness can be related to the shaft stiffness K_S, and it increases to K_R during contact. This increment can be estimated using the Hertz theory of contact between two elastic bodies placed in mutual contact.

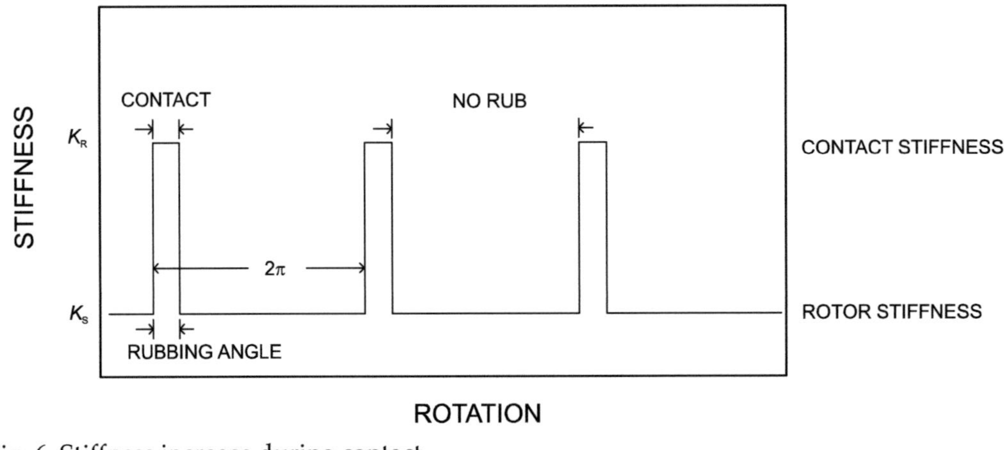

Fig. 6. Stiffness increase during contact

Assuming that the system´s stiffness can be represented as a rectangular function, then the stiffness variation can be approximated as a Taylor series such that

$$K(\omega) = \alpha + \beta^3\omega \tag{10}$$

3. Vibrations analysis with data-domain transformations

The vibrational motion produced by a rotating machine is complicated and may be analyzed by transforming data from the time-domain to the frequency-domain by means of the

Fourier Transform. This transform gives to the operator additional information from the behavior of the machine that a signal in time-domain cannot offer.
Fourier developed a theory in which any periodic function *f(t)*, with period T, can be expressed as an infinite series of sine and cosine functions of the form:

$$f(t) = \frac{a_0}{2} + \sum_{n=1}^{\infty} (a_n cos\, n\omega_T t + b_n\, sin\, \omega_T t) \tag{11}$$

Where ω denotes the fundamental frequency and 2ω, 3ω, etc., its harmonics. This series is known as the Fourier series expansion and a_n and b_n are called the Fourier coefficients. By this way, a periodic waveform can be expanded into individual terms that represent the various frequency components that make up the signal. These frequency components are integer multiples of ω.
The following identity can be used to extend the Fourier series to complex functions:

$$e^{ix} = cos(x) + isin(x) \tag{12}$$

$$f(t) = \sum_{-\infty}^{\infty} c_n e^{int} \tag{13}$$

Where c_n can be obtained by the following integration:

$$c_n = \frac{1}{2\pi} \int_{-\pi}^{\pi} f(t) e^{-int} dt \tag{14}$$

This applies to periodic functions on a 2π interval.
Fourier series can be extended to functions with any period T with angular frequency ω=2π/T. Sine and cosine functions have frequencies that are multiples of ω as in Eq. (11).
For non-periodic functions, with period T, discrete frequencies $n\omega$ separated by Δω=2π/T, and taking the limit as T→∞, $n\Delta\omega$ becomes continuous and the summation can be expressed as an integral. As a result, the continuous Fourier transform for frequency domain is defined as:

$$F(\omega) = \int_{-\infty}^{+\infty} f(t) e^{-i\omega t}\, dt \tag{15}$$

While for time domain the inverse Fourier transform is defined as:

$$f(t) = \frac{1}{2\pi} \int_{-\infty}^{+\infty} F(\omega) e^{i\omega t}\, d\omega \tag{16}$$

And $f=f(t)$ satisfying the condition

$$\int_{-\infty}^{+\infty} |f(t)|^2 dt < \infty \tag{17}$$

Since computers can't work with continuous signals, the Discrete Fourier Transform (DFT) was developed and implemented through the Fast Fourier Transform (FFT). The FFT is a fast algorithm for computing the DFT that requires the size of the input data to be a power

of 2. FFT is a helpful engineering tool to obtain the frequency components from stationary signals. However, non-stationary phenomena can be present in signals obtained from real engineering applications, and are characterized by features that vary with time.
A difficulty that has been observed with FFT is that the complex exponentials used as the basis functions have infinite extent. Therefore, localized information is spread out over the whole spectrum of the signal. A different approach is required for this type of signals. Time-frequency methods are used for their analysis and one of the most used methods is the Short Time Fourier Transform (STFT). This was the first time-frequency technique developed. The solution approach introduces windowed complex sinusoids as the basis functions.
The STFT is a technique that cuts out a signal in short time intervals, which can be assumed to be locally stationary, and performs the conventional Fourier Transform to each interval. In this approach a signal $S_t(\omega)$ is multiplied by a window function $h(\tau)$, centered at τ, to obtain a modified signal that emphasises the signal characteristics around τ:

$$S_t(\omega) = \frac{1}{2\pi}\int_{-\infty}^{\infty} e^{-j\omega t} s(\tau)h(\tau - t)d\tau \tag{18}$$

Frequency distribution will be reflected around τ after applying the Fourier Transform to this window. The spectral density of the energy at time τ can be written as follows:

$$P(t,\omega) = |S_t(\omega)|^2 = \left|\frac{1}{2\pi}\int_{-\infty}^{\infty} e^{-j\omega\tau} s(\tau)h(t - \tau)d\tau\right|^2 \tag{19}$$

As expected, a different spectrum will be obtained for each time. A Spectrogram, which is the time-frequency distribution, can be constructed with the resulting spectra. Resolution in time and frequency depends on the width of the windows $h(\tau)$. Large windows will provide a good resolution in the frequency domain, but poor resolution in time domain. Small windows will provide good resolution in time domain, but poor resolution in frequency domain. The major disadvantage of this approach is that resolution in STFT is fixed for the entire time-frequency map. This means that a single window is used for all the frequency analysis. Therefore, only the signals that are well correlated in the time interval and frequency interval chosen will be localized by the procedure. It may be thought of as a technique to map a time-domain signal into a fixed resolution time-frequency domain.
This drawback can be surpassed with basis functions that are short enough to localize high frequency discontinuities in the signal, while long ones are used to obtain low frequency information. A new transform called wavelet transform achieves this with a single prototype function that is translated and dilated to get the required basis functions.
The wavelet transform is a time-frequency representation technique with flexible time and frequency resolution. Conversely to the STFT where the length of the windows function remains constant during the analysis, in the wavelet approach a function called the mother wavelet is operated by translation and dilation to build a family of window functions of variable length:

$$\psi_{s\tau}(t) = \frac{1}{\sqrt{s}}\psi\left(\frac{t-\tau}{s}\right) \tag{20}$$

Where $\psi(t)$ is the mother wavelet function, s the scale parameter, and τ the time shift or dilation parameter. Based on the mother wavelet function, the wavelet transform is defined as:

$$W_{\psi}x(s,\tau) = \int_{-\infty}^{\infty} x(t)\,\overline{\psi_{s\tau}}(t)dt \tag{21}$$

And $W_{\psi}x(s,\tau)$ are the wavelet coefficients.
The wavelet transform is different from other techniques in that it is a multiresolution signal analysis technique that decomposes a signal in multiple frequency bands. By operating over s and τ the wavelets permit to detect singularities, which makes it an important technique for nonstationary signal analysis.
Due to this characteristic, the wavelet transform is the analysis technique that we found more suitable for the analysis of the rubbing phenomenon.

4. Experimental methodology

An experimental test rig was implemented to get a deeper understanding of the main characteristics of the rubbing phenomenon, and to apply the wavelet analysis technique in the processing and identification of the vibrations produced by the rub-impact of the system. Elements were included to run experiments under controlled conditions. Fig. 7 shows the experimental set-up.

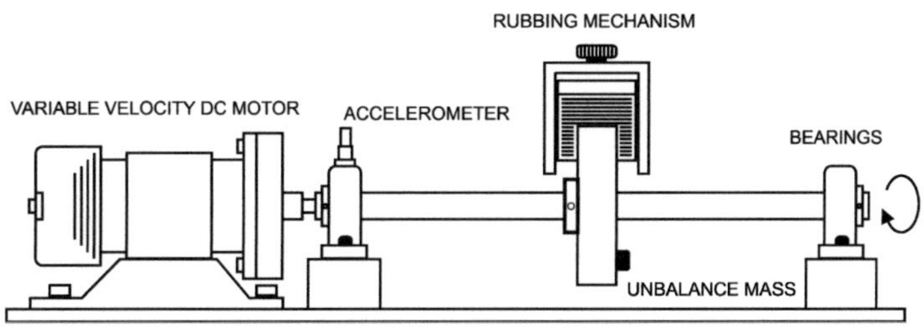

Fig. 7. Test rig for the rubbing experiments

The experimental system is composed of a shaft supported by ball bearings and coupled to an electrical motor with variable rotational speed. The velocity of the motor was controlled with an electronic circuit. A disk was installed in the middle of the shaft, which was drilled to be able to mount bolts of different masses to simulate unbalance forces. An adjustable mechanism was designed in order to simulate the effect of a rotor rubbing a stationary element. The position of the device, acting as the stationary element, was adjusted with a threaded bolt that slides a surface to set the clearance between the rotating disk and the rubbing surface. The shaft and disk were made of steel, and the rubbing device of aluminium alloy. Light and severe rubbing were simulated by controlling the speed of the rotor. Low velocities caused light rubbing while high velocities generated severe impact-like rubbing vibrations. Both types of rubbing were analyzed with the proposed methodology.
An accelerometer was used to measure the vibrations amplitude. Output of the accelerometer was connected to a data acquisition system to convert analog signals to digital data with a sampling rate of 10 kHz. An antialias filter stage was included to get a band limited input signal.
Experimental runs were carried out for fixed and variable rotor velocities. Fixed velocities were tested for values between 350 rpm and 1900 rpm. Continuous variable velocity experiments were also carried out to simulate a rotor system under ramp-up and ramp-down conditions, to verify the preservation of the scale and temporal information with the processing technique used.

Daubechies 4 wavelet transform was implemented to convert the signal from the time-domain to the time-frequency domain. Scaling and wavelet functions for this transform are shown in Fig. 8.

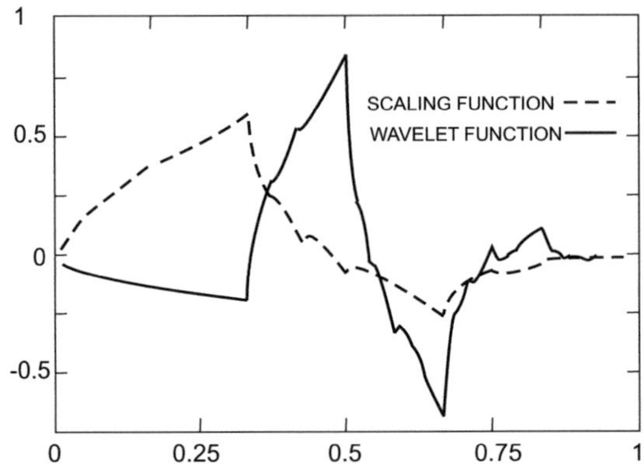

Fig. 8. Daubechies D4 scaling and wavelet functions

Implementation of the continuous wavelet transform is impractical, especially for on-line detection devices for process monitoring purposes. This implementation consumes a significant amount of time and resources. An algorithm for the discrete wavelet transform (DWT) is used to overcome this situation. It is based on a sub-band coding which can be programmed with a high computational efficiency. Subband coding is a multiresolution signal processing technique that decomposes the signal into independent frequency subbands.

With this approach, the DWT applies successive low-pass and high-pass filters to the discrete time-domain signal as shown in Fig. 9. This procedure is known as the Mallat algorithm.

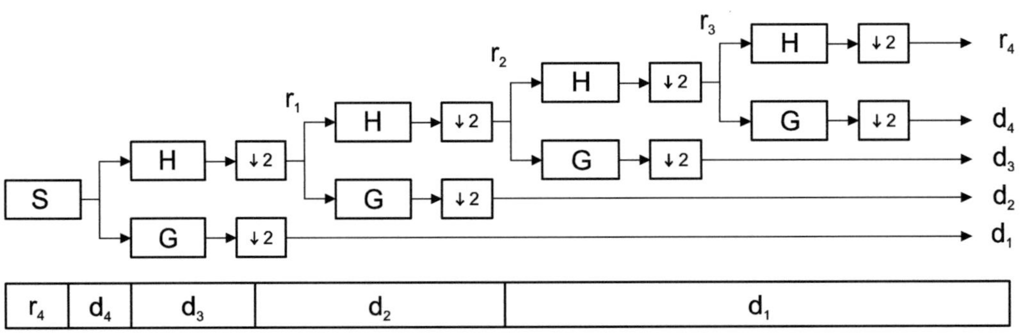

Fig. 9. Algorithm for the sub-band decomposition of the input signal

The algorithm uses a cascade of filters to decompose the signal. Each resolution has its own pair of filters. A low-pass filter is associated with the scaling function, giving the overall picture of the signal or low frequency content, and the high-pass filter is associated with the wavelet function, extracting the high frequency components or details. In Fig. 9 the low-pass filter is denoted by H and the high-pass filter is denoted by G. Each end raw is a level of decomposition. A sub-sampling stage is added to modify the resolution by two at each step of the procedure. As a result of this process, time resolution is good at high frequencies, while frequency resolution is good at low frequencies.

For each transform iteration the scale function to the input data is applied through the low-pass and high-pass filters. If the input array has N cells, after applying the scale function, the down-sampling by two, which follows the filtering, halves the resolution and an array with N/2 values will be obtained. With the low-pass branch, coarse approximations are obtained. The high-pass filtered signal will reflect the fluctuations or details content of the signal. By iterating recursively a signal is decomposed into sequences. The successive sequences are lower resolution versions of the original data.
The implemented form of the Daubechies 4 wavelet transform has a wavelet function with four coefficients and a scale function with four coefficients. The scale function is:

$$A_i = H_0 S_{2i} + H_1 S_{2i+1} + H_2 S_{2i+2} + H_3 S_{2i+3} \tag{22}$$

Where scaling coefficients are defined as

$$H_0 = \frac{1+\sqrt{3}}{4\sqrt{2}} \tag{23}$$

$$H_1 = \frac{3+\sqrt{3}}{4\sqrt{2}} \tag{24}$$

$$H_2 = \frac{3-\sqrt{3}}{4\sqrt{2}} \tag{25}$$

$$H_3 = \frac{1-\sqrt{3}}{4\sqrt{2}} \tag{26}$$

The wavelet function is:

$$C_i = G_0 S_{2i} + G_1 S_{2i+1} + G_2 S_{2i+2} + G_3 S_{2i+3} \tag{27}$$

Where wavelet coefficients are defined as

$$G_0 = H_3 \tag{28}$$

$$G_1 = -H_2 \tag{29}$$

$$G_2 = H_1 \tag{30}$$

$$G_3 = -H_0 \tag{31}$$

Each wavelet and function value is calculated by taking the product of the coefficients with four data values of the input data array. The process is iterated until desired results are reached.

5. Experimental results and discussion

The methodology described in the previous section was applied and experimental runs were carried out with the aid of the test rig to obtain a deeper comprehension of the rubbing phenomenon. Fig. 10 shows results for time and frequency domains for the rotor rubbing and no-rub experimental runs. The upper row corresponds to time-domain signals, while the lower row shows the frequency-domain signals.

With no-rub (upper left), the signal in time-domain is characterized by a uniform trace with a small dispersion of data produced by the low-level noise of the measuring system. However, when rub occurs (upper right), as acceleration is the measurement engineering unit, even for low level rubbing the energy content of the signal is high, and spikes appear at the location of each rub-contact.

The spectral distribution of the signal obtained when the rotor is rubbing shows the wide frequency bandwidth in the vibrational response, produced by the chaotic behavior of the rubbing phenomenon.

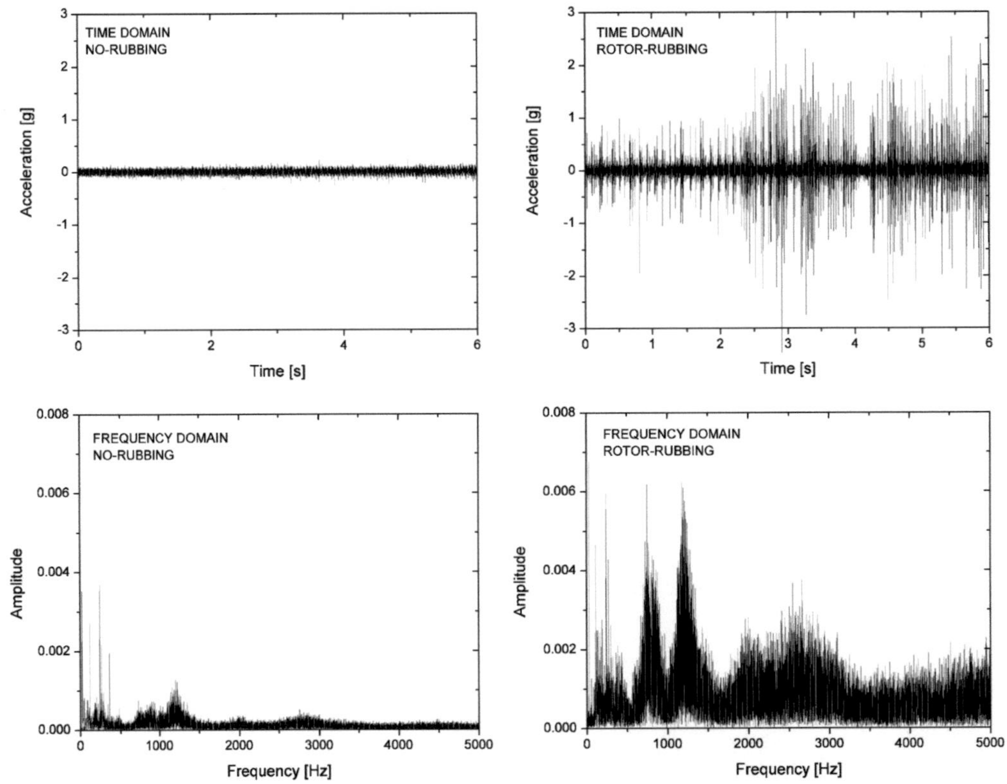

Fig. 10. Spectral distribution of the vibrations for the rotor rubbing and no-rubbing

When rubbing is present, the response is dependent on the angular frequency of the rotor. For low rotor velocities rub generates low vibration amplitudes as shown in the acceleration values in the upper graph of Fig. 11. This can be considered a light-rubbing, but when the rotor velocity is high, the time-domain response of the vibrations produced is quite different and get closer to an impact response characterized by spikes with high acceleration values. This response can be seen in the lower graph of Fig. 11. The amplitudes of vibrations for light rubbing are within ±0.1 g, while for severe rubbing peak values reach ±1 g, about ten times higher.

Processing results of the signals for the rotor with rubbing and without rubbing to obtain the spectral distribution are shown in Fig. 12. The graph localizes the natural frequency of the test rig for both cases. As explained in the introductory section, the nonlinear nature of the system produced a slight different vibrations pattern when rubbing is present. Solid line shows the natural frequency for the rub-free experiments. However, the dotted line corresponds to the experiments with rubbing induced to the rotor, and as expected there is a shift in the natural frequency.

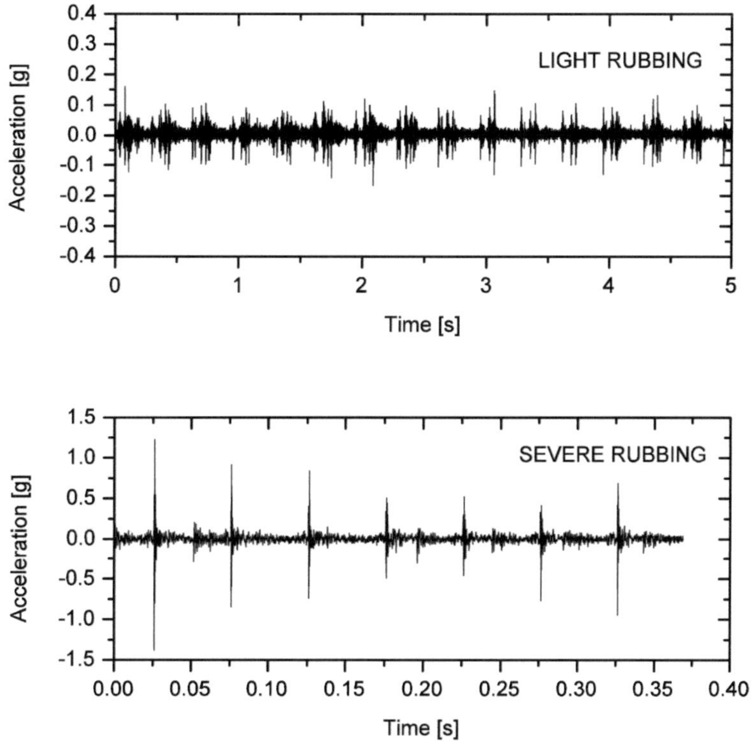

Fig. 11. Vibrations amplitude for light rubbing and severe rubbing

Frequency shift occurs to the right, with the trend to move to the high frequency side of the spectrum, which means that the system is hardening as a result of the stiffness increase produced by the contact of the rotor with the stationary element. The amplitude of the natural frequency also increases as a product of the higher energy content of the rub-impacts.

The signal of the vibrations was processed to transform the data from the time-domain to the time-frequency domain. Wavelet transform Daubechies 4 was used for the transformation and results are shown in Fig. 13.

As stated before, a vector is obtained with this procedure which is the same size as the original vector. Recalling the subband coding, upper half of the vector contains the high frequency content of the information (subband 1). From the remaining data, upper half contains the next subband with mid-frequency content (subband 2), and so on. This way, the low frequency content of the information is coded and located in the lower part of the vector while the high frequency content is coded into the higher indexes of the vector. Indexes represent the transformed values in the resulting vector which amplitude is a function of the correlation between the input signal and the mother wavelet. A higher value for the index means a stronger correlation and therefore a major content of that frequency corresponding to a particular value of scale and translation.

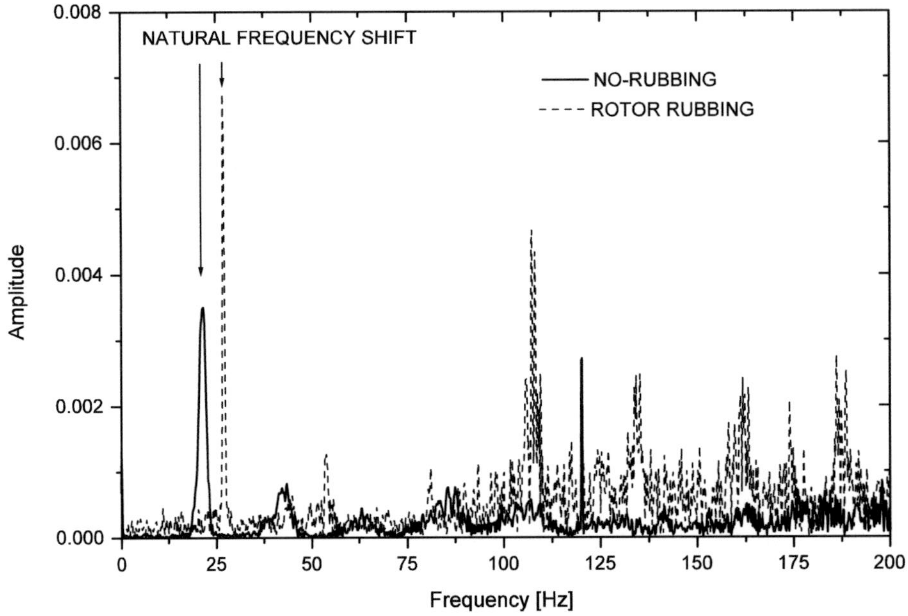

Fig. 12. Resonance frequency dependence observed for a rotor with rubbing

Taking this into consideration, it can be observed that for light rubbing the correlation gets stronger for mid-value indexes, which means that light rubbing is characterized by frequencies that fall in the lower middle of the frequency spectrum.

On the other side, rub-impacts produced by the contact of the rotor with the stationary element at high rotational frequencies, are characterized by spikes with a high frequency content. The wavelet transformation enhances this type of rubbing as can be observed in the upper half of the vector for severe rubbing shown in Fig. 13 (subband 1), although some rubbing information can be found in the next subband. As both light and severe rubbing may be present in a rotor, the sum of the frequency content produced by the phenomena reflects again a wide spectral distribution in the vibrational response.

To test the wavelet approach as a rubbing detection technique, especially for severe rubbing where it is desirable to assess alert signals before a catastrophic failure occurs as it can happens under some rubbing conditions, a vibration signal which presents rub-impacts was chosen. The test data are shown in Fig. 14.

There are two spikes in the graph produced by the rotor rubbing at high velocity rotation. These spikes can be treated as singularities of transient nature whose occurrence cannot be predicted. A technique like wavelets that analyzes a signal by comparison of a basis wavelet that is scaled and translated to extract frequency and location information is ideal for this situation. The procedure enhances these singularities and makes it easier their detection as it is demonstrated next.

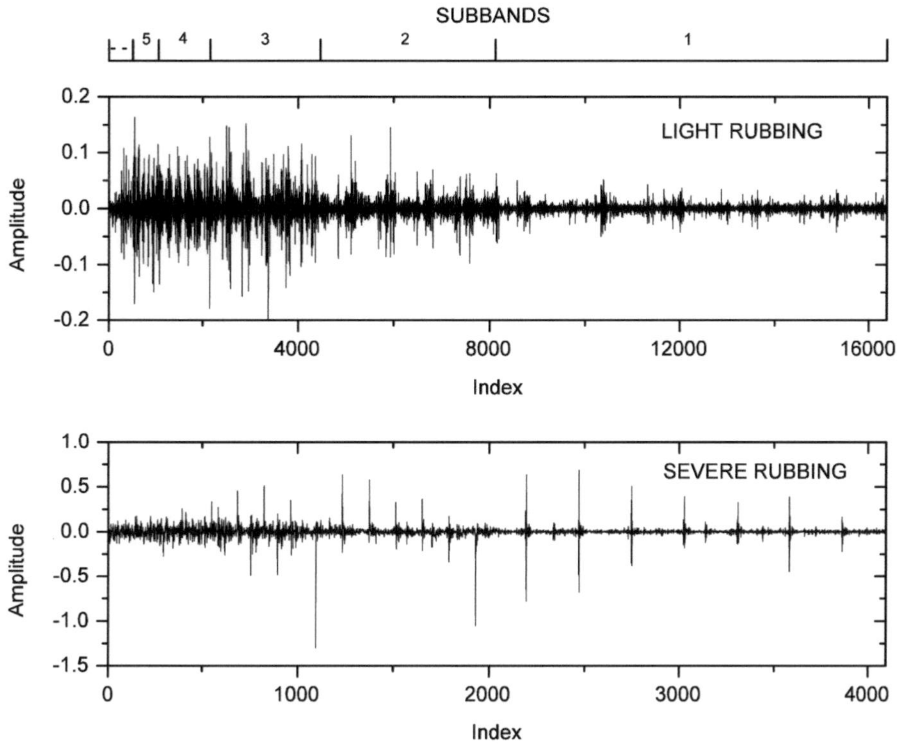

Fig. 13. Rotor rubbing signal transformed to time-frequency domain with wavelets

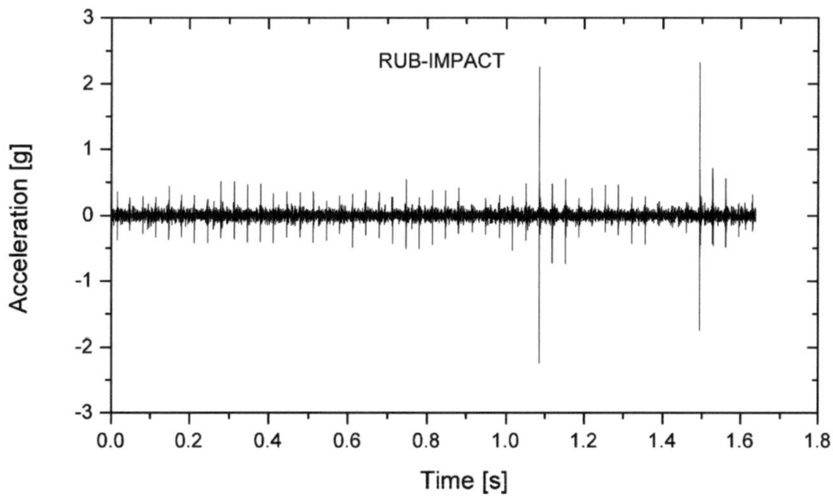

Fig. 14. Time-domain vibrations with rub-impact

Graphs showing the details of the impacts are shown in Fig. 15. The signal is characterized by a sudden excitation that generates a mechanical oscillation that grows to peak amplitude and decays as the impact energy dissipates. Each impact is characterized by only a few cycles that analyzed with the traditional FFT would not have enough energy to obtain a clear spectral definition.

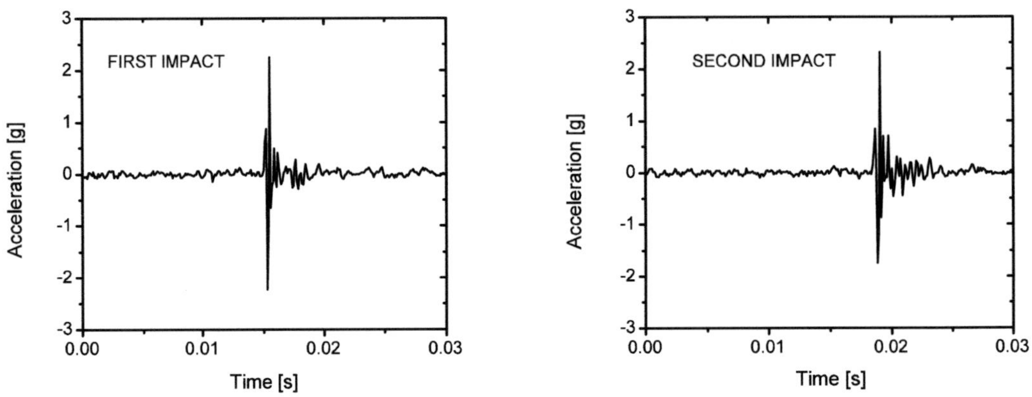

Fig. 15. Details of the transient impacts analyzed

A wavelet decomposition of this signal was made and the main subbands are shown in Fig. 16. Subbands are associated with their corresponding frequency range according to the sampling rate established during the experiments. The graph shows the frequency content between 78 Hz and 5 kHz.

From this graph it is observed that the subband with the higher frequency content encompasses the information of the transient signals. The correlation technique enhances the spikes giving amplitude values higher than the coefficients where impacts are not present. This makes it easier to establish a discrimination criterion and an estimation of their values to determine the presence and severity of the rubbing for diagnostic purposes. Additionally, as the transient signals produced are in the first subband, only the first level of decomposition in the wavelet transformation is necessary reducing the computing time that it takes to make the analysis and optimizing the detection process.

The experimental results of the vibrations presented in the previous discussion were analyzed through one of the approaches that wavelets offer to the vibration analyst. This is a time-frequency representation of the data from which it can be extracted the information of interest to apply the necessary processes and criteria for the rubbing detection. This approach permits the characterization of the signal from which it can be obtained the necessary information for the implementation of the technique for the design of testing equipment with automatic detection and recognition of the rubbing phenomena.

Another type of representation of the information that wavelets offer are the time-frequency maps. These are contour plots where the wavelet coefficient values are plotted against the time and scale parameter, that is, translation and frequency. One axis represents time, the other axis frequency, and the amplitude of the vibrations is color-coded. The contour maps permit to visualize the whole picture of the frequencies present in the signal as well as their occurrence or location in time.

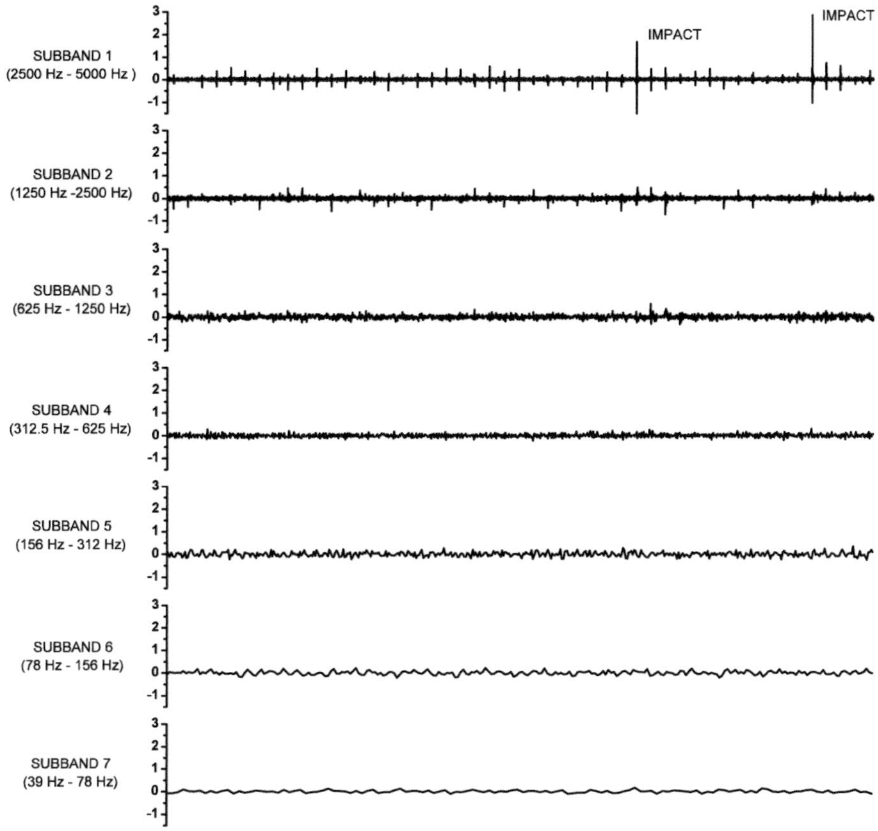

Fig. 16. Subband coding with wavelets of the vibrational response with impact-rubbing

The light rubbing data was processed with commercially available software by means of the Morlet continuous wavelet transform and results are shown in Fig. 17. In this graph, the color coding is red for low amplitude vibrations throughout blue for high amplitude vibrations. It can be observed the intermittent nature of the rubbing and, in concordance with Fig. 13, that the main vibrations are limited to frequencies below 1 250 Hz.

An analogous process was applied to severe rubbing data and results are shown in Fig. 18. The image shows that mid-range frequencies get stronger while high frequencies appear as a result of the increase in the vibrations amplitude as in the lower graph of Fig. 13. Upper spots in the time-frequency map (rub-impacts) appear elongated and lower spots stretched due to the compromise between the time and frequency resolution of the technique as stated in the introduction.

Additionally, an experimental run was carried out varying continuously the rotating conditions to obtain a sweep from a low to a high velocity and then decreasing the velocity until a minimum value. Results are shown in Fig. 19.

It can be seen that as time runs throughout the experiment, velocity increases and higher frequency components appear. These components get their peak value near the middle of the time axis where the maximum velocity is reached, and then begin to fade showing the trend of the higher frequencies to disappear as velocity decreases. This graph confirms the wide spectral bandwidth of the rubbing phenomena.

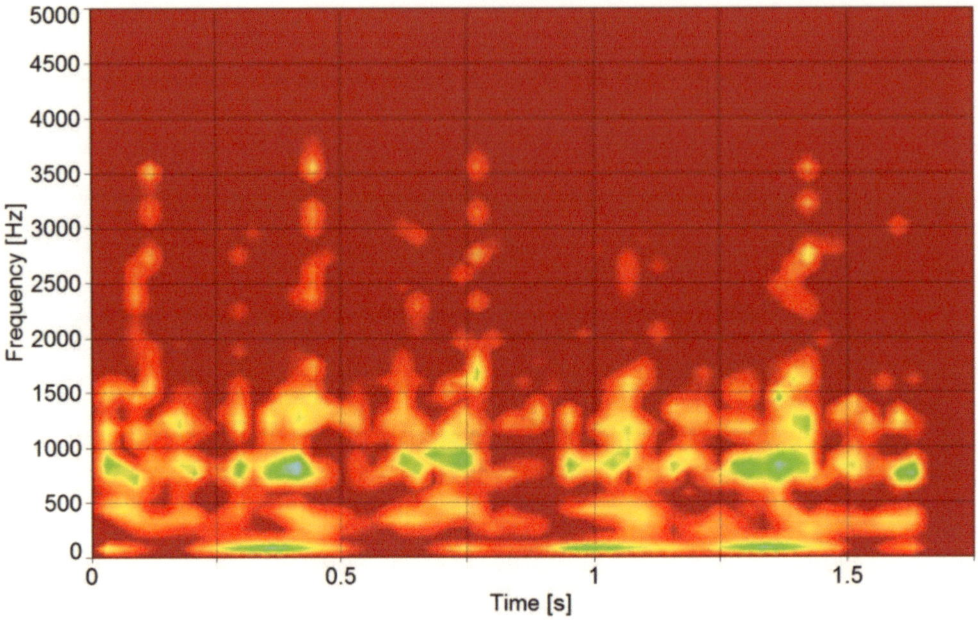

Fig. 17. Time-Frequency map for light rubbing

It is important to notice the evolution of the frequencies as time passes by. There is an unsteady variation of frequencies, and in Fig. 19 it can be seen how they have an unsymmetrical pattern even with speed variations.

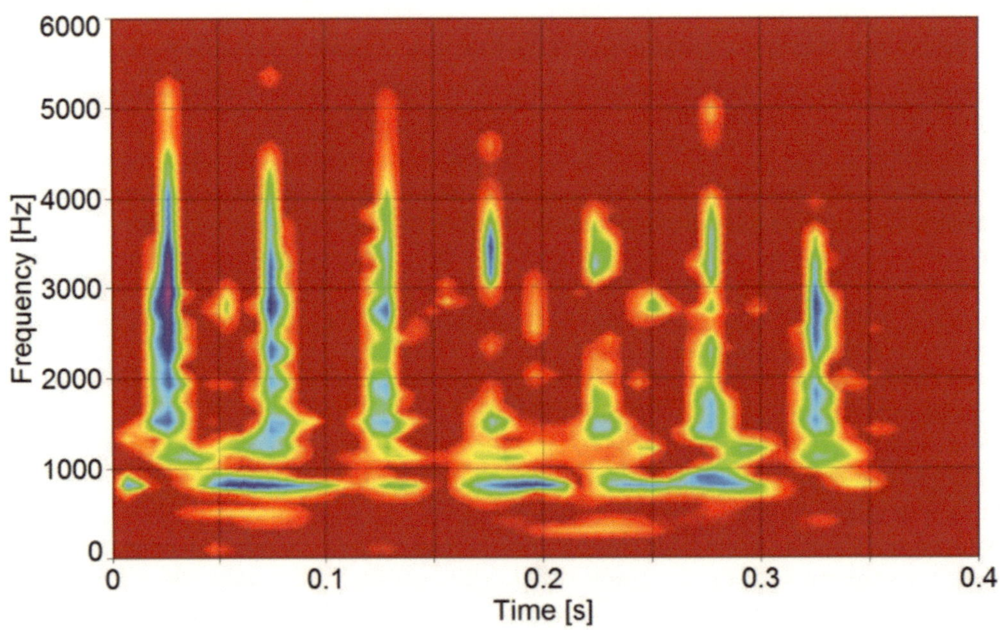

Fig. 18. Time-frequency map for severe rubbing

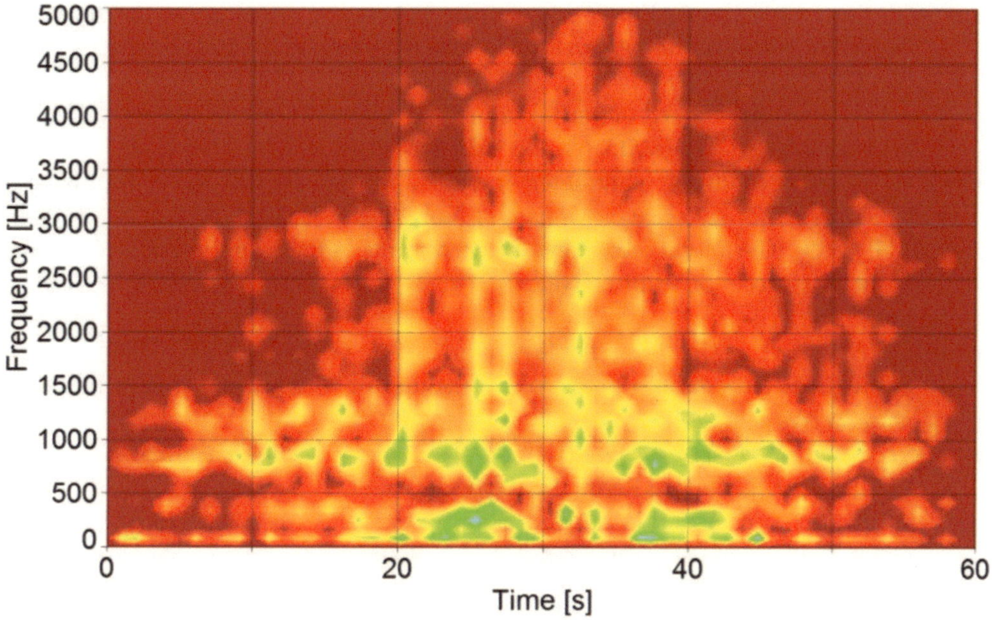

Fig. 19. Time-frequency map for a run-up and run-down rotor velocity sweep

6. Conclusions

Rotor rubbing has been analyzed with a methodology that processes the vibrations signal in such a way that time and scale information is preserved. It was demonstrated that with this approach vibrations of transient nature can be studied through a controlled subband coding scheme and time-frequency spectra. The technique revealed additional information that traditional processing techniques cannot, such as FFT. Experimental results showed that light rubbing presents a vibrational response characterized by a rich frequency content spectrum, and that severe rubbing is more adequately described as an impact-like transient behavior. Accordingly, impacts could be identified and localized with wavelets in the upper-frequency subbands which resulted after the algorithm was applied.

Since rotor-to-stator contact changes the effective stiffness of the coupled elements, a frequency shift was identified that shows the nonlinear response of the system. Time-frequency maps evidenced again the wide spectral response and differences between light and severe rotor rubbing, and location in time of the rub-impacts. The processing algorithm can be implemented with a high computational efficiency for on-line inspection systems for continuous machinery condition monitoring.

7. References

Abuzaid, M.A. ; Eleshaky, M.E. & Zedan, M.G. (2009). Effect of partial rotor-to-stator rub on shaft vibration. *Journal of Mechanical Science and Technology*, Vol. 23, No. 1, 170-182.

Adams, M.L. (2010). *Rotating Machinery Vibration*, 84, CRC Press, Taylor & Francis Group, ISBN 978-1-4398-0717-0, Boca Raton, FL, USA.

Castejón, B; Lara, O. & García-Prada, J.C. (2010). Automated diagnosis of rolling bearings using MRA and neural networks. *Mechanical Systems and Signal Processing*, Vol. 24, No. 1, 289-299.

Choi, Y.-S. (2001). On the contact of partial rotor rub with experimental observations. *KSME International Journal*, Vol. 15, No. 12, 1630-1638.

Choi, Y.-S. (2004). Nonlinear parameter identification of partial rotor rub based on experiment. *KSME International Journal*, Vol. 18, No. 11, 1969-1977.

Jeffcott, H.H. (1919). The lateral vibration of loaded shafts in the neighborhood of a whirling speed-the effects of want of balance. *Philosophical Magazine, Series 6,* Vol. 37, 304-314.

Karpenko, E.V. ; Wiercigroch, M. ; Pavlovskaia, E.E. & Cartmell, M.P. (2002). Piecewise approximate analytical solutios for a Jeffcott rotor with a snubber ring. *International Journal of Machine Sciences*, Vol. 44, No. 3, 475-488.

Muszynska, A. (2005). *Rotordynamics*, 646-648, CRC Press, Taylor & Francis Group, ISBN 978-0-8247-2399-6, Boca Raton, FL, USA.

Peng, Z.K. ; Chu, F.L. & Tse, P.W. (2005). Detection of rubbing-caused impacts for rotor-stator fault diagnosis using reasigned scalogram. *Mechanical Systems and Signal Processing*, Vol. 19, No. 2, 391-409.

Xinsheng, L. & Kenneth, A.L. (2004). Bearing fault diagnosis based on wavelet transform and fuzzy inference. *Mechanical Systems and Signal Processing*, Vol. 18, No. 5, 1077-1095.

Wang, F. ; Xu, Q. & Li, S. (2004). Vibration analysis of cracked rotor sliding bearing systems with rotor-stator rubbing by harmonic wavelet transform. *Journal of Sound and Vibration*, Vol. 271, No. 3-5, 507-518.

16

Analysis of Vibrations and Noise to Determine the Condition of Gear Units

Aleš Belšak[1] and Jurij Prezelj[2]
[1]*University of Maribor, Faculty of Mechanical Enginnering*
[2]*University of Ljubljana, Faculty of Mechanical Enginnering*
Slovenia

1. Introduction

The main goal of maintenance is to maintain the characteristics of a technical system at the most favourable or still acceptable level. Maintenance costs can be reduced, operation can become more reliable and the frequency and complexity of damages can be reduced. To evaluate the condition of a technical system, it is necessary to collect precise data, and the to analyse, compare and process this data properly.
Gear units are the most frequent machine parts or couplings. They consist of a housing, toothed wheels, bearings and a lubricating system and are of various types and sizes. Durable damages in gear units are often a resulta of the following factors: geometrical deviations or unbalanced component parts, material fatigue, which is a result of engagement of a gear pair, or damages caused to roller bearings.
To monitor the condition of mechanical systems, methods for measuring vibrations and noise are usually used; data about a gear unit can be acquired in this way. Afterwards certain tools are used to analyse the data. Features that denote the presence of damages and faults are identified in this way.

2. Noise source identification

A visualization method of complex noise sources on the basis of an acoustic camera is presented. The method is based upon a new digital signal processing algorithm. This algorithm makes it possible to visualize all types of different complex noise sources from their far area. It is possible to observe monopole, dipole or quadropole noise sources, which occur simultaneously. In addition to this, reflections from hard walls, and refraction and scattering of sound waving can be observed.
The difference between the acoustic camera operation and the acoustic ray reconstruction method is great. Signals from all microphones, located along the ring or the cross of the acoustic camera, are processed in a complex way, by means of the acoustic camera algorithm. On the basis of this algorithm, delays are appropriately corrected in time domain - in relation to time, i.e. to the path length of sound waving from the elementary source to an individual microphone located in the camera - and not by means of phases as in frequency domain as in relation to the sound ray reconstruction method.

Sound waving travels along paths ri of various lengths from the elementary acoustic source $V(x_j)$ to an individual microphone on the ring of an acoustic camera (Fig. 1). Paths travelled by sound waving $|r_i|$ are of different lengths and, consequently, signal delays Δi of the same sound waving, produced at the elementary sound source $V(x_j)$, are different as well.

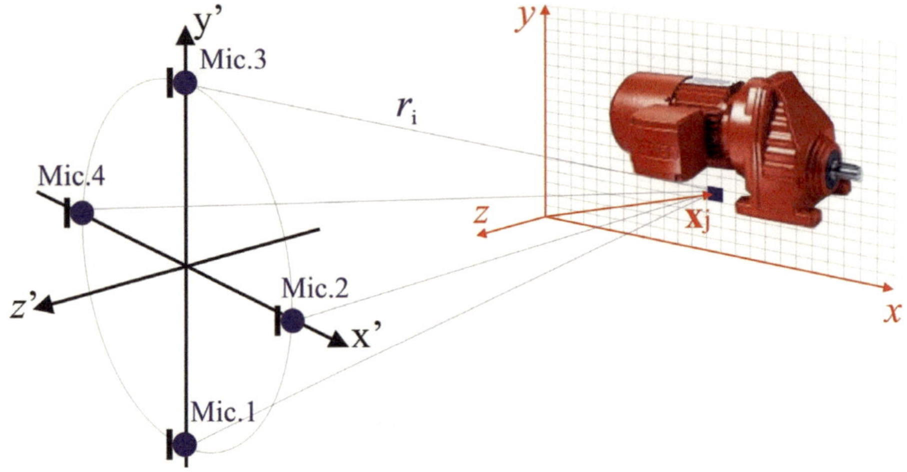

Fig. 1. Path length of an elementary source to individual microphones on an acoustic camera

Figure 2 swows an electrical signal of four microphones. The sound path from the elementary source to microphone 2 is the shortest, and the signal of microphone 2 is the fastest. The second fastest is the signal of microphone 1, the third and the fourth are signals of microphones 3 and 4. Acoustic holography calculation is based upon these delays in time.

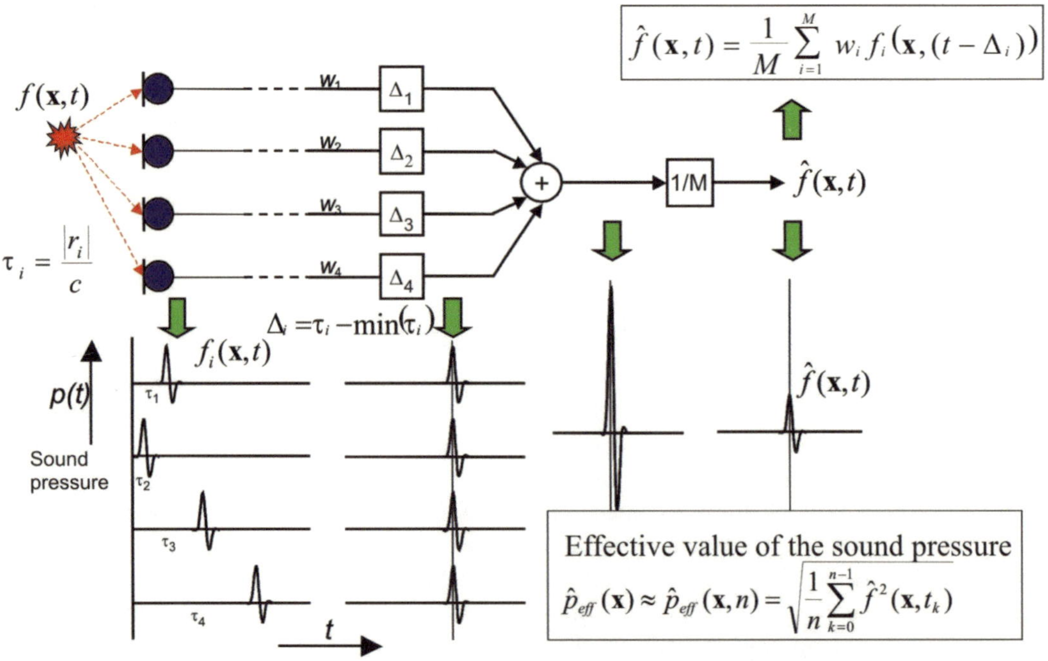

Fig. 2. Acoustic holography calculation method in relation to acoustic camera

Heinz Interference Transformation algorithm, which represents the basis for an acoustic camera, creates a pseudo inverse acoustic field with interference integrals by approximating the original acoustic source in the best way possible (by moving it forward and backward simultaneously). Time-negative reconstruction in a time positive way is realized, using this algorithm. The result is a surface of equivalent acoustic pressure at the point of greatest emission.

If we assume that sound waving from each elementary source reaches each microphone on the ring of the acoustic camera, signals arriving from different microphones can be shifted and integrated in time. For each elementary source, a new signal $f(x_j,t)$ is obtained, using the following equation:

$$\hat{f}(\mathbf{x}_j,t)=\frac{1}{M}\sum_{i=1}^{M}w_i f_i(\mathbf{x}_j,(t-\Delta_i)) \tag{1}$$

Afterwards the effective value of the sound pressure $f_{eff}(x_j,t)$ can be calculated on the basis of this signal:

$$p_{eff}(\mathbf{x}_j)\approx p_{eff}(\mathbf{x}_j,n)=\sqrt{\frac{1}{n}\sum_{k=0}^{n-1}\hat{f}^2(\mathbf{x}_j,t_k)} \tag{2}$$

Effective sound pressure represents a mean square value of the acoustic pressure, caused by the elementary acoustic signal at the spot of emission. The corresponding point in the acoustic image must be coloured appropriately. Ths depends on the position of the elementary source and on the value of its effective acoustic pressure. Areas with high effective sound pressure are usually coloured red, and areas with lower effective sound pressure are blue, which gradually fades until these areas become white. For each elementary source, the procedure must be repeated in order to obtain the entire acoustic image of the acoustic source. In case of more acoustic sources, it is possible to find out, on the basis of the acoustic image, which acoustic source at the measurement spot is the one that contributes mostly to effective acoustic pressure.

The resolution of place and time of an acoustic image, produced with an acoustic camera, has an impact upon the form of sound signals. An impulse of sound pressure has an ideal form in relation to the algorithm of an acoustic camera, and pure sine-shaped form of acoustic waving is the least favourable sound pressure phenomenon. All real sound pressure phenomena can be placed between these two forms. The sinus function, i.e. the Fourier area, represents the basis for most of the acoustic theory. This includes the theory of image method in a nearby field and the theory of acoustic ray reconstruction method. Pure sine-shaped form is very rare in relation to real sound/noise signals. Consequently, the application possibilities of the acoustic camera algorithm are much wider than the application possibilities of other algorithms developed so far.

The acoustic camera is the only acoustic source visualization method functioning exclusively in time domain; it is not necessary to use the Fourier transform to calculate the acoustic image. This means that the method using the acousting camera is not limited in the same way as are methods using the Fourier transform. Frequency analysis is part in the user system but the algorithm calculates the acoustic image first and only afterwards the Fourier transform.

The measurement system in relation to the acoustic camera of the GFaI with dRec48C192 and 32 phase coordinated microphones is presented in Figure 3. The microphones are located on a ring, in concern to the work in a free acoustic field. For an acoustic camera, prepolarised condensation microphones with linear frequency of 23 kHz (-3 dB) are used. Their response is slowly reduced from 6 dB per decade to 40 kHz. It is required to use microphones with such a high frequency area to achieve adequate resolution of the acoustic image. It is possible to achieve higher resolution in relation to higher sampling frequencies or better phase coordination of microphones.
The resolution of an analogue-digital converter is 21 bits. The highest sampling frequency is 196 kHz per channel. Digitalised signals are stored in this converter temporarily, during measurement. After data transfer to a personal computer, taking a few seconds, it is possible to calculate the sound source acoustic image.

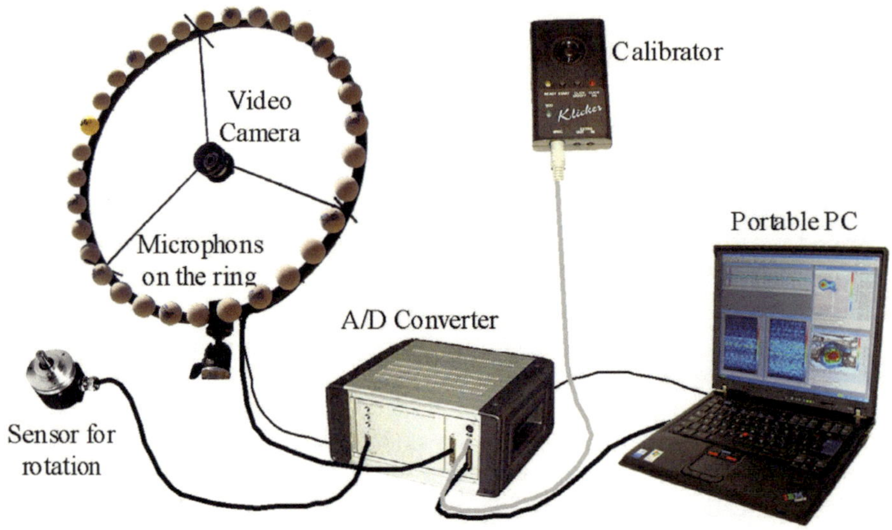

Fig. 3. An acoustic camera system GFaI for visualisation of acoustic sources

3. Adaptive time-frequency identification

A gear unit is a set of elements enabling the transmission of rotating movement. Although it is a complex dynamic model, its movement is usually periodical, and faults and damages represent a disturbing quantity or impulse. Local and time changes in vibration signals denote disturbance, consequently, time-frequency changes can be expected. This idea is based on kinematics and operating characteristics.
The presence of cracks in gear units is the most important factor with a negative impact upon the reliability of operation and quality of operation of gear units. It is usually possible to determine deviations from reference values on the basis of a frequency spectrum. As mentioned, it is impossible to identify changes of a frequency component in time as a gear unit is a complex mechanical system with changeable dynamic reaction; this makes the approach based on time-frequency methods more appropriate.
It often happens that some frequency components in signals are present from time to time only. In such cases classical frequency analysis does not suffice to determine when certain

frequencies appear in a spectrum. If time-frequency analysis is used, it can be determined not only in what way frequency components of non-stationary signals change with time but also their intensity levels.

Frequency analysis is often used in diagnostics, but good results are obtained more or less only in relation to periodical processes without local changes. A presence of a damage or fault leads to changes in dynamic parameters of a mechanical system. This influences the frequency spectrum. Monitoring frequency reaction is one of the most common spectral methods to identify the condition of a gear unit. With classical frequency analysis, time description of vibration is transformed into frequency description, and changes within a signal are averaged within the entire time period. This means that local changes are lost in the average of the entire function of vibrations. As a result, it is very difficult if not impossible to define local changes.

These deficiencies are eliminated with the use of the time-frequency analysis: local changes that deviate from the global periodical oscillation are expressed with the appearance or disappearance of individual frequency components in a spectrogram. A signal is presented simultaneously in time and frequency.

Individual frequency components often appear only from time to time in signals related to technical diagnostics. On the basis of classical frequency analysis of such signals, it is not possible to determine when certain frequencies appear in the spectrum. The aim of time-frequency analysis is to describe in what way frequency components of non-stationary signals change with time and to determine their intensity levels.

Fourier, adaptive and wavelet transforms and Gabor expansion are representatives of various time-frequency algorithms. The basic idea of all linear transforms is to perform comparison with elementary function determined in advance. On the basis of various elementary functions, different signal presentations are acquired.

Qian improved and concluded adaptive transform of a signal to a large extent although many authors had been developing algorithms without interference parts, which make individual transforms less useable as opposed to Cohen's class.

Adaptive transform of a signal x(t) is expressed as follows:

$$x(t) = \sum_{p} B_p \cdot h_p(t) \tag{3}$$

where analysis coefficients are determined by means of the following equations

$$B_p = \langle x, h_p \rangle \tag{4}$$

expressing similarity between the measured signal x(t) and elementary functions $h_p(t)$ of transform. The original signal represents the starting point with parameter values p=0 and $x_0(t)=x(t)$. In the set of desired elementary functions, $h_0(t)$ is searched for that is most similar to $x_0(t)$ in the following sense

$$\left|B_p\right|^2 = \max_{h_p} \left|\langle x_p(t), h_p(t) \rangle\right|^2 \tag{5}$$

for p = 0. The next step includes the calculation of the remaining $x_1(t)$

$$x_{p+1}(t) = x_p(t) - B_p \cdot h_p(t) \tag{6}$$

Without giving up the generalisation idea, hp(t) is to have a unit of energy representation of a signal.

$$\left\|h_p(t)\right\|^2 = 1 \tag{7}$$

The energy in the remaining signal

$$\left\|x_{p+1}(t)\right\|^2 = \left\|x_p(t)\right\|^2 - \left|B_p\right|^2 \tag{8}$$

The equation (6) is repeated in order to find $h_1(t)$ that would suit best $x_1(t)$, etc. One elementary function $h_p(t)$ that suits best $x_p(t)$ is found in each step. The primary purpose of adaptive signal representation is to identify a set of elementary functions $\{h_p(t)\}$, most similar to time-frequency structure of a signal, and at the same time satisfy equations (3) and (4).
If Wigner-Ville distribution is used for both sides of the equation (3), and if equations are organised into two groups, the result is as follows:

$$P_{WV}x(t,\omega) = \sum_p B_p^2 \cdot P_{WV}h_p(t,\omega) + \sum_{p \neq q} B_p \cdot B_q * P_{WV}\left(h_p, h_q\right)(t,\omega) \tag{9}$$

The first group represents elementary signal components and the second one represents cross interference terms.
A new time-dependent adaptive spectrum can be defined in the following way:

$$P_{ADT}(t,\omega) = \sum_p \left|B_p\right|^2 \cdot P_{WV}h_p(t,\omega) \tag{10}$$

As an adaptive spectrum based on representations, it is called an adaptive spectrogram. As opposed to Wigner-Ville distribution, it contains no interferences and no cross terms, and it also satisfies the condition of energy conservation.

$$\left\|x(t)\right\|^2 = \frac{1}{2 \cdot \pi} \cdot \iint P_{ADT}(t,\omega) \cdot dt \cdot d\omega \tag{11}$$

The basic issue related to linear presentations is the selection of elementary functions. When it comes to Gabor expansion, a set of elementary functions comprises time-shifted and frequency modulated prototype window function w(t). In relation to wavelets, elementary functions are acquired by scaling and shifting a mother wavelet $\psi(t)$. In these two examples, structures of elementary functions are determined in advance. Elementary functions related to adaptive representation are rather demanding.
As adaptive transform permits arbitrary elementary functions, it is, generally speaking, independent from the choice of elementary functions $h_p(t)$.
Elementary functions, used for adaptive representation of a signal with equation (3), are very general but this is not always so in practice. It is desirable that elementary functions are localised in regard to time and frequency in order to emphasize time dependence of a signal. Also it must be possible to use the presented algorithm in a relatively simple way. In relation to adaptive representation, a Gauss type signal with its very favourable features is considered a basic choice.

The calculation of an adaptive spectrogram begins in a wide time range of a measured signal. Then the range must be decreased, depending on what the goals are. Fourier integral is among the elementary operations of searching for a suitable elementary function, and so the described calculation process is very effective. The accuracy of approximation depends primarily on the size of time-frequency interval. With narrower intervals, the representation is more accurate, but the calculation time is longer. This means that it is necessary to find a compromise between the accuracy of approximation and its efficiency.

4. Practical example

The measurements were performed in the test plant of the Laboratory for Pumps, Compressors and Technical Acoustics of the Faculty of Mechanical Engineering, University of Ljubljana. The room in which the tests were carried out was not specially adjusted for performing acoustic measurements as the noice level was between 36 and 42 dB(A). This level can be achieved also in an industrial environment that is located adequately far away from intense noice sources.

A single stage gear unit with a helical gear unit with straight teeth integrated into it was used.

Two pairs of spur gear-units, built in a single stage gear-unit, were used for noise measurements. One of the pairs had a crack and the other one was without it. Thus, tests were carried out, using faultless and faulty gear units.

The aim of the measurements was to determine the presence of individual changes in a gear unit. The measured signal of a faultless gear unit and the signal of a faulty gear unit were compared to determine the gear unit condition.

Measurements were carried out under operating conditions normally associated with the relevant type of a gear unit. A ground gear pair used was a standard gear pair, with the teeth quality 6, but it had a crack in the tooth root of a pinion. It is presented in Fig. 4.

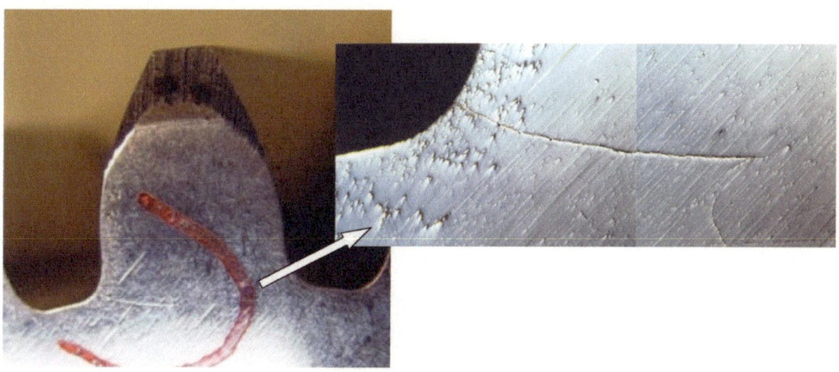

Fig. 4. A pinion with a crack in the tooth root

Adaptive time frequency transform was used to determine the presence of a crack in the tooth root, whereby the LabVIEW software tools, including the author's own software modules, were used [9].

The length of the signal of measured values was 1 s; on an average, the signal was composed of 192000 measuring points. At the time of measurement, the rotational frequency was 28,5Hz. The number of teeth of the pinion was 18, and of the gear unit 99.

In Fig. 5, the acoustic image with sound level of a gear unit is presented, where the engagement area of a gear pair can be observed as a noice source, wheras in Fig. 6, the adaptive spectrogram of noise source is presented; it is not possible to note any rhythmic pulsation of harmonics, with the exception of typical frequencies, defined on the basis typical frequencies components. Some pulsation sources are indicated (but not expressed) and their stohastics. It is very interesting to monitor the increase or decrease (even complete disappearance) in appropriate frequency components with pulsating frequency.

In Fig. 7, the acoustic image of a gear unit with a crack in the tooth root is presented, where it is possible to note the engagement area of a gear pair as a noice source, and in Fig. 8 and Fig. 9, the adaptive spectrogram of noise source is presented. Rhythmic pulsation of some frequency can be observed. This is typical for meshing frequency 515 Hz. Pulsating is expressed only in relation to the presence of a crack. Pulsation reflects a single engagement of a gear pair with a crack.

To determine the presence of a crack in the tooth root, adaptive transform was used for vibration analysis. In relation to adaptive spectrogram, adaptive representation for signal decomposition, prior to Wigner-Ville distribution, was used.

A fine adaptive time-frequency resolution is characteristic of an adaptive spectrogram due to limited features of elementary functions. Consequently, time-frequency resolution of the transform is adapted to signal characteristics. As an elementary function, it is possible to apply Gauss function (impulse) and linear chirp with Gauss window. If a signal contains linear chirps resulting from a linear change in the rotational frequency of a gear unit, it is possible to use an adaptive spectrogram to determine in what ways a possible frequency modulation is reflected in the time-frequency domain. The transform calculation time increases, along with the larger amount of data and the increased number of cycles required to search for an adequate elementary function.

Fig. 5. An acoustic image with noise source of a gear unit of a faultless gear

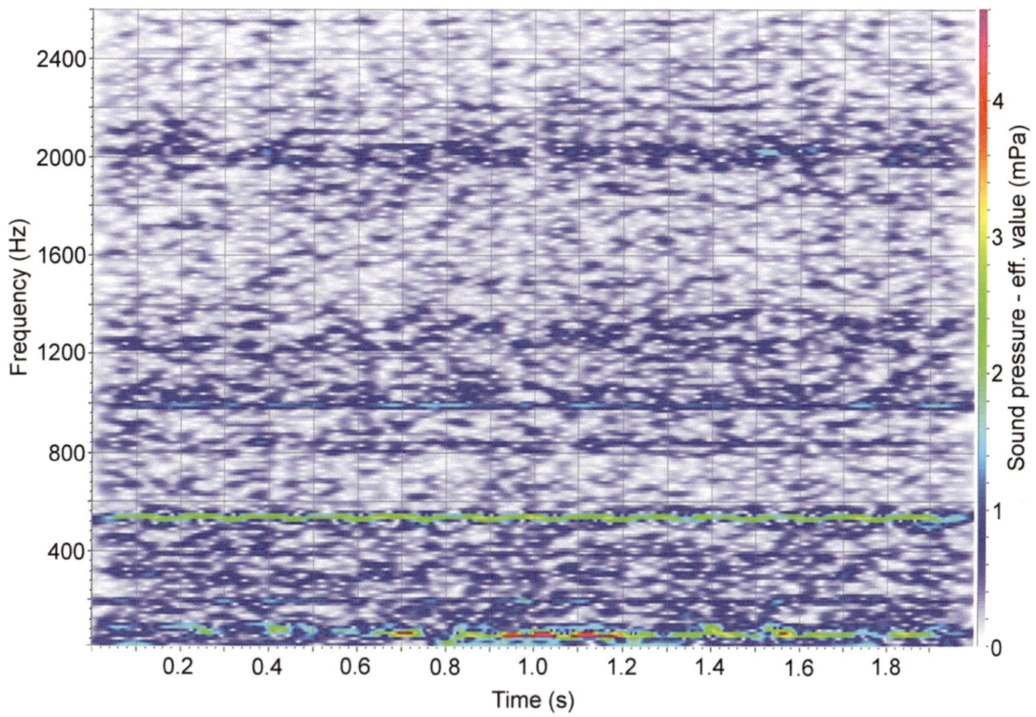

Fig. 6. A adaptive spectrogram of a gear unit of a faultless gear

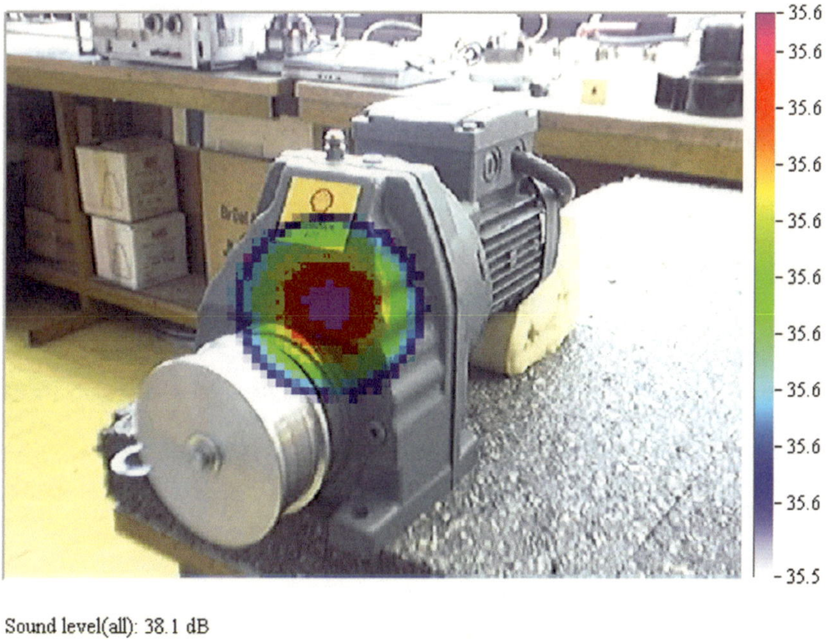

Fig. 7. An acoustic image with noise source of a gear unit with a crack in the tooth root

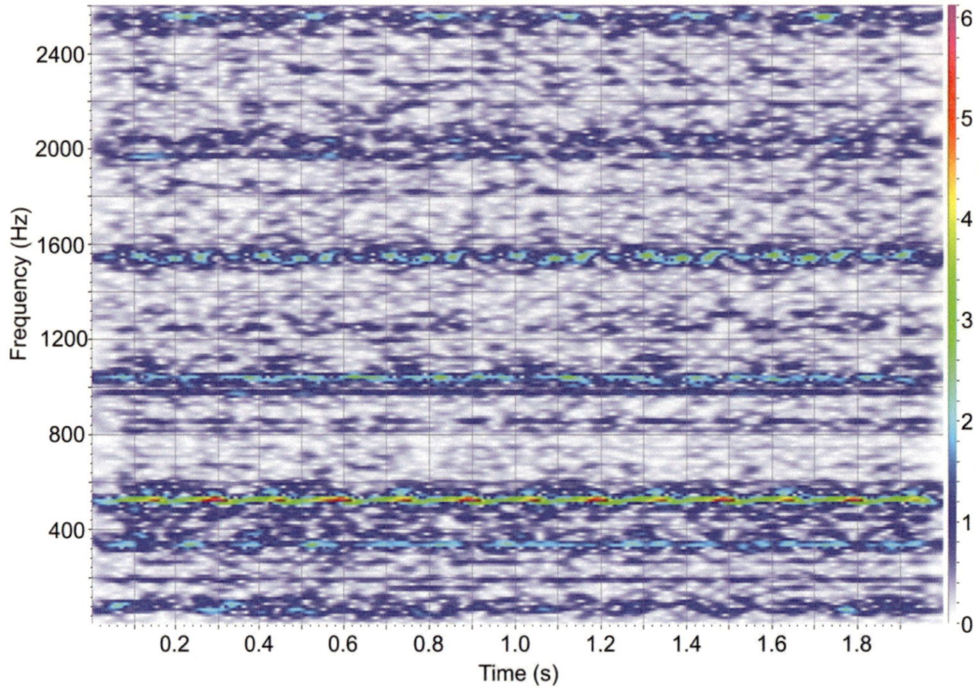

Fig. 8. A adaptive spectrograms of a gear unit with a crack in the tooth root; position 1

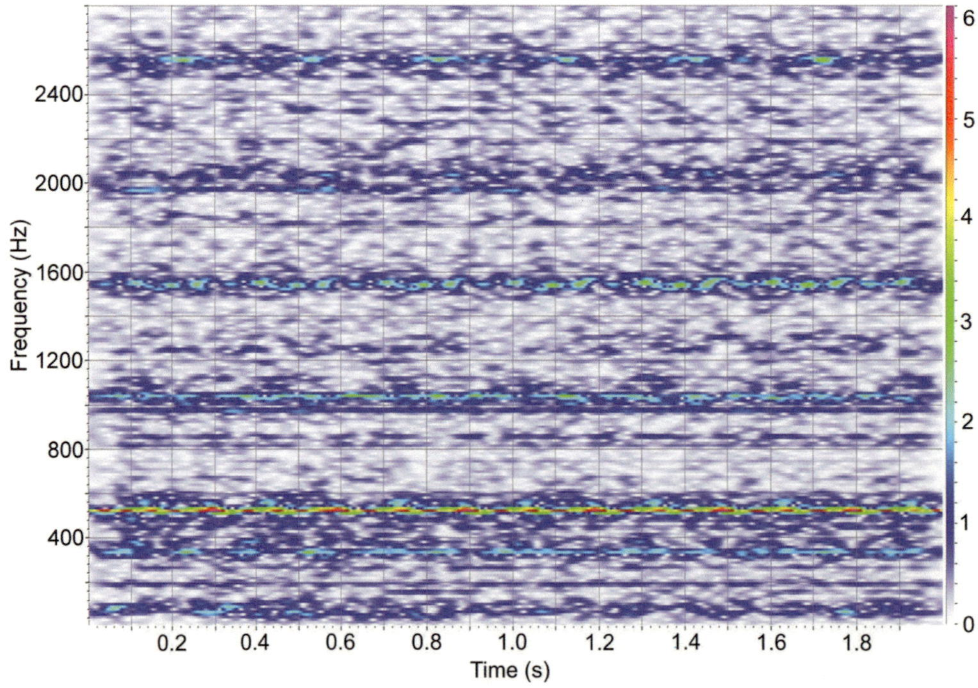

Fig. 9. A adaptive spectrograms of a gear unit with a crack in the tooth root; position 2

The vibration signal of measured values was 1 s long and composed of, on an average, 50000 measuring points. At the time of measurement rotational frequency was 20 Hz. Adaptive spectrograms in relation to Gabor transforms are presented for comparison. The length of the window is 6800 points, which is 15% more that the length of the period of one rotation of a gear pair.

Calculation time required for adaptive spectrogram is at least 10 times longer than the calculation time for the Gabor transform, but the resolution of the adaptive transform is, on an average, two times better.

Fig. 10 shows Gabor spectrogram; no rhythmic pulsation of harmonics can be noted, with the exception of typical frequencies, determined on the basis of power spectrum. When it comes to adaptive spectrogram (Fig. 11), with a higher level of energy accumulation in the origins, it is possible to note some pulsation sources but they are not very expressed. It is very interesting to monitor how appropriate frequency components with rotational frequency of 20 Hz increase or descrease or even completely disappear. This phenomenon is typical of the 3rd harmonic, 1530 Hz is expressed, only in relation to the presence of a crack. The phenomenon is much more expressed in relation to the adaptive spectrogram (Fig. 13) than in relation to the Gabor spectrogram (Fig. 12).

The spectrogram evaluation can be based on an average spectrogram, which represents an amplitude spectrum of a Fourier or adaptive transform of a measured signal, and on observing pulsating frequencies of individual frequency components.

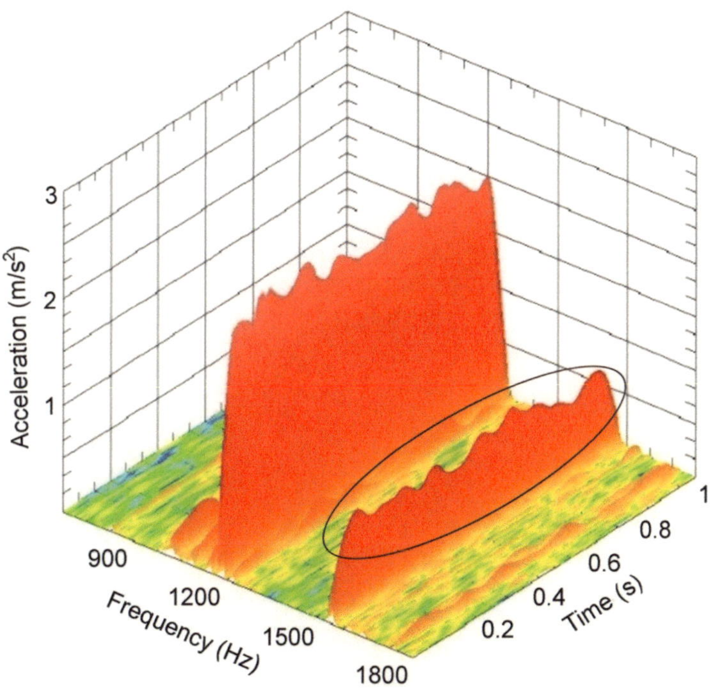

Fig. 10. Gabor's spectrogram of a faultless gear unit

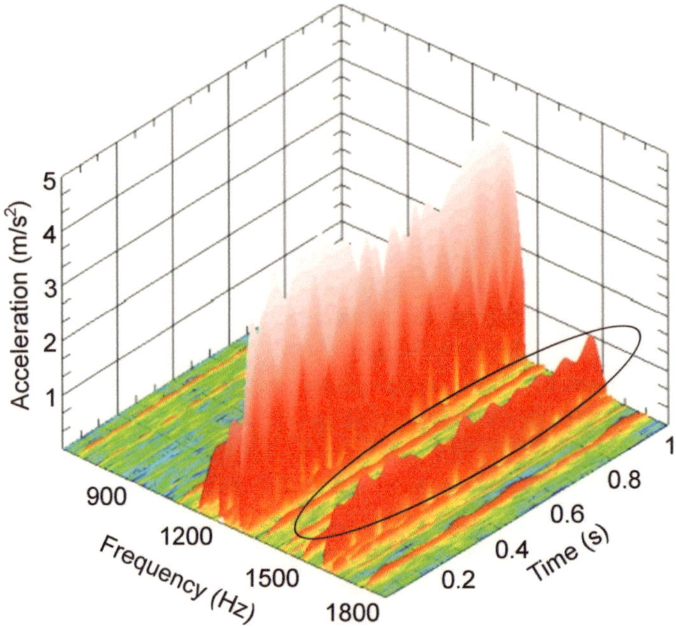

Fig. 11. Adaptive spectrogram of a faultless gear unit

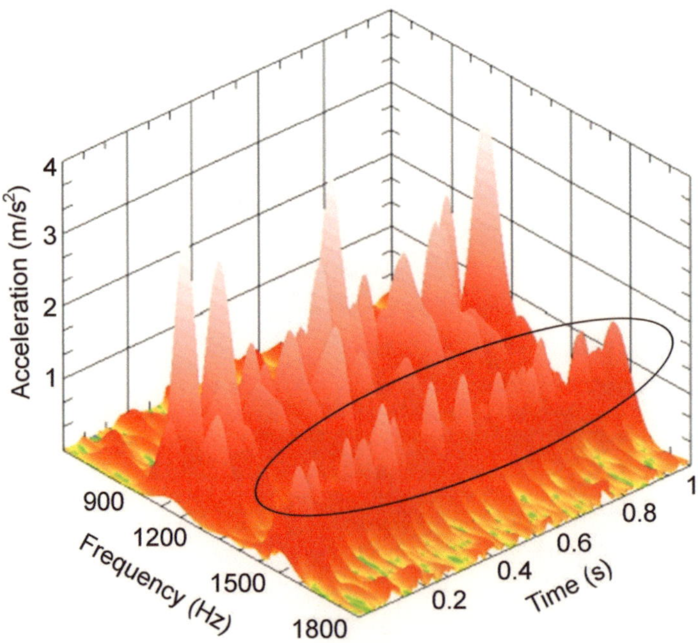

Fig. 12. Gabor's spectrogram of a gear unit with a pinion with a crack

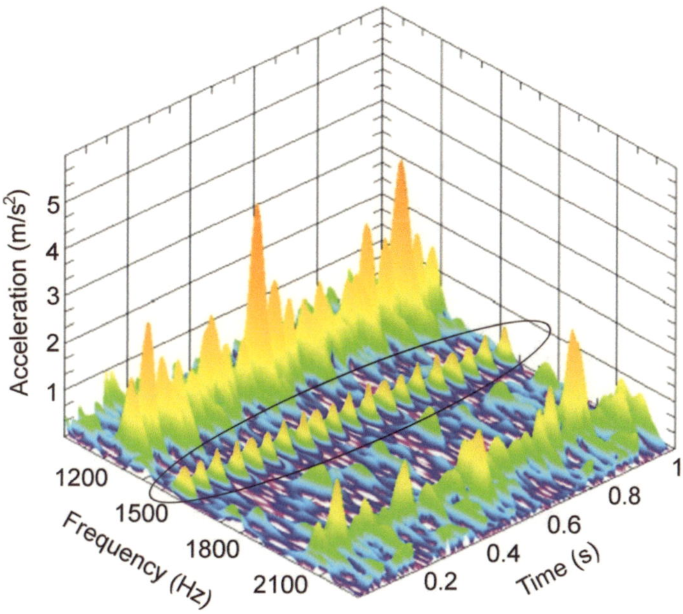

Fig. 13. Adaptive spectrogram of a gear unit with a pinion with a crack

5. Conclusion

The resolution in time and in place achieved with the use of an acoustic camera with its specific algorithm, which functions in time domain, and of specifically located microphones for acoustic source visualization is better than with any other acoustic system.

Industrial gear units were used for noise analysis, the purpose of which was to identify faults. The use of the presented methods can improve both, the safety of operation and the reliability of monitoring operational capabilities.

The reliability of monitoring life cycle of a gear unit is improved with the use of appropriate spectrogram samples and the achievement of a clear presentation of the pulsation of individual frequency components, which, along with the average spectrum, for a criterion for evaluating the condition of a gear unit.

When it comes to life cycle design, it is necessary to use an adequate method or criterion to monitor the actual condition of a device and particularly of its vital component parts, which can have a considerable impact upont the operational capability. If faults and damages are detected in time, it is possible to control the reliability of operation to a great extent. The prediction of the remaining life cycle of a gear unit is improved with the use of reliable fault identification methods.

In this contribution, fault identification in industrial gear units is based on vibration analysis; it increases the safety of operation and, consequently, of monitoring operational capabilities.

The life cycle of a gear unit can be monitored more reliably with the use of appropriate spectrogram samples and a clear presentation of the pulsation of individual frequency components that, in addition to the average spectrum, represent a criterion for evaluating the condition of a gear unit. Adaptive time-frequency representation is clearer, without increased dissemination of signal energy into the surroundings, and it enbles reliable fault identification.

6. References

Christensen, J.J. and Hald, J. (2004). *Beamforming*, B&K technical reviev 1,

Hald, J. (2005). *Combined NAH and Beamforming Using the Same Array*, B&K technical reviev 1,

Heinz, G. (2004). *Locating Noise Sources, A Comparison Between Different Noise Localization Techniques*, GFaI Report 001-06-21,

Fellner, W. (2004). *Die Akustische kamera,* Eine revolutionäre Lösung zum Orten Schallemissionen, Newsletter für professionelle schall und schwingungmesstechnik, Nr.7/2004, Wien

Suresh, S. (1998). *Fatigue of Materials*, Cambridge University Press

Buch, A. (1988). *Fatigue Strength Calculation*, Trans Tech Publications

Stephens, R.I. ; Fatemi, A. ; Stephens R.R ; Fuchs H.O. (2001). *Metal Fatigue in Engineering,* John Wiley & Sons Inc., New York

Belšak, A. (2006). *Časovno-frekvenčna analiza stanja zobniških gonil*, doctoral thesis, University of Maribor, Faculty of Mechanical engineering

Belšak, A. (2004). Razvoj sistema za odkrivanje napak zobniškega gonila, master thesis, University of Maribor, Faculty of Mechanical engineering

Qian S., Chen D. (1996). *Joint Time-Frequency Analysis*, Prentice Hall

Fladrin, P. (1999). *Time-Frequency/Time-Scale Analysis*, Academic Press

Mertins, A. (1999). *Signal Analysis*, John Wiley & Sons Inc., New York

Bendar, J.S.; Piersol, A.G. (2000). *Random Data*, John Willey & Sons

Rohatgi, V.K. ; Saleh, A.K. (2001). An Intruduction to Propability and Statistics, John Willey & Sons

Belšak, A. ; Flašker, J. Detecting cracks in the tooth root of gears, *Engineering Failure Analysis*, Vol. 14(8), pp. 1466-1475

Robert, J. ; Marks, II. (1991). *Introduction to Shannon Sampling and Interpolation Theory*, Springer Verlag, New York

Carmona, R. ; Hwang, W.L. ; Torresani B. (1998). *Practical Time-Frequency Analysis*, Academic Press, San Diego

Feichtinger, H. ; Stroms T. (1998). *Gabor Analysis and Algorithms: Theory and Applications*, Birkhasuer

JefWu, C. F. ; Hamada, M. (2000)., *Experiments*, John Willey & Sons

17

Methodology for Vibration Signal Processing of an On-load Tap Changer

Edwin Rivas Trujillo[1],
Juan C. Burgos Diaz[2] and Juan C. García-Prada[3]
[1]*Universidad Distrital Francisco José de Caldas-Facultad de ingeniería*
Carrera 7 No 40-53 Bogotá
[2]*E.P.S Universidad Carlos III de Madrid -Departamento de Ingeniería Eléctrica*
C/ Butarque 15, 28911Madrid,
[3]*E.P.S Universidad Carlos III de Madrid -Departamento de Ingeniería Mecánica*
C/ Butarque 15, 28911Madrid,
[1]*Colombia*
[2,3]*España*

1. Introduction

A great percentages of transformer failures occur in the on-load tap changer (OLTC) (R. Jongen & P. Morshuis et al.,2007) Fig 1. The OLTC is the only moving part of a power transformer and its elements suffer from great mechanical and electrical stress.
The condition of some of the elements of the OLTC (diverter switch, resistors, contacts, etc.) can be evaluated through periodic inspections. However, these inspections require having the transformer out of service for a relatively long period. In addition, it is expensive and requires a skilled staff.
Vibration is a suitable parameter to monitor the mechanical condition of an OLTC (P. Kang,et al.,2000). Vibrations from a working OLTC are often readily available, but obtaining them from a defective OLTC is not as easy.
Previous studies (P. Kang et al.,2001) show that OLTC failures can be diagnosed by using vibration envelope analysis. Wavelet analysis is an effective technique for extracting the main characteristics of the vibration signal, over the whole spectrum, without requiring a dominant frequency band in the vibration signal.
The purpose of this paper is to present a methodology implemented to find the OLTC diagnostic indicators (Number of vibration bursts, Vibration burst amplitude, Time between vibration bursts, Main frequency bands in the burst, energy of the vibration bursts). To obtain these indicators pre-processing and processing of the vibration signal is needed.
In the pre-processing stage the signal is synchronized, normalized and then Hilbert transform is applied to obtain the envelope. In the signal processing stage a technique in time-frequency domain, Discrete Wavelet Transform, is used and then a threshold based on

preserved energy is applied in order to determine the characteristic bursts of the vibration signal both the OLTC in good condition and with faults.

2. OLTC

An OLTC (Fig. 1) modify the transformer voltage ratio in response to voltage variations in the electrical system, in order to maintain transformer output voltage. The OLTC changes the tapping connection of the transformer winding without disconnecting it out of service.
An OLTC comprises the following elements:

- *Tap Selector*
- *Diverter Switch*

An OLTC always switches between two consecutive taps. The tap selector pre-selects the next tap to work on and does not switch load current. The tap selector is inserted in the transformer tank or attached to it, and it is refrigerated by the same oil refrigerating the windings.
The tap selector is driven by a motor drive mechanism synchronized with the diverter switch. The tap selector and the mechanical energy accumulator of the diverter switch are the first elements to move when a tap change is ordered. The tap selector moves only to the tap preceding the current tap or to the tap next the current tap.
The diverter switch is the element transferring the load current, without load interruption, from the current tap to the pre-selected tap.

3. Vibration signal measurements

A test program was carried out to get vibration signals with different types of failures. As it is no easy to record in field vibration signals from a given OLTC with different kind of failures, failures were simulated physically in the laboratory.
An OLTC (Fig. 1) type D I 1200 150/110 12 23 3 W, from an 80 MVA single-phase autotransformer 250.000/150.000 ±10x1517/17000 V was used. The OLTC is rated at 1200 A and has 12 taps, 23 regulation positions, an inverter type pre-selector and 3 intermediate positions when the inverter is in operation.
The OLTC was not mounted on the transformer tank and was not immersed in oil. So pattern vibration changes slightly because no viscous lubrication is provided. Nevertheless the experiment is valid in the essence.
The vibration measurements were obtained using an accelerometer, which measures tank vibrations during the tap-change operation. A piezoelectric accelerometer was used with a sensitivity of 1.02±0.02pC/ms^2 and a frequency range of 1 to 12 kHz. The data acquisition card had PCMCIA architecture, 12-bit resolution, 50KS/s per channel and variable input range of ±0.05V to ±10V. The measuring system is shown in Fig. 2.
Vibration signal was recorded from the OLTC in good working condition. Fig. 3 shows the vibration signal during a tap change in which the tap selector is operative. Six bursts can be distinguished in the vibration signal.

- Burst 1: Starting of the motor drive.
- Burst 2: Activation of the Geneva drive. The tap selector contact attached to the previous tap (n-1) starts moving.
- Burst 3: Moving contacts of the tap selector reach the final tap (n+1).

- Burst 4: The movement of the Geneva drive ends.
- Burst 5: Group of bursts corresponding to the diverter switch's activation.
- Burst 6: The motor drive stops.

Fig. 1. OLTC used in the experimental measurements

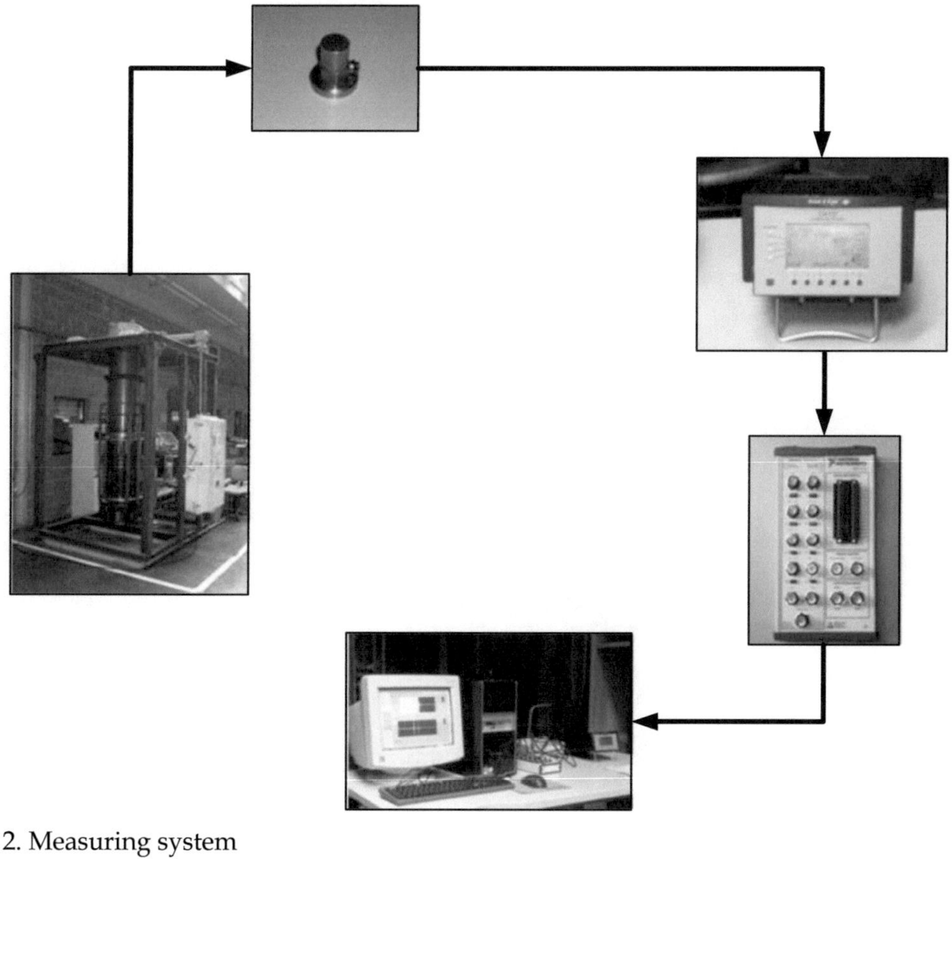

Fig. 2. Measuring system

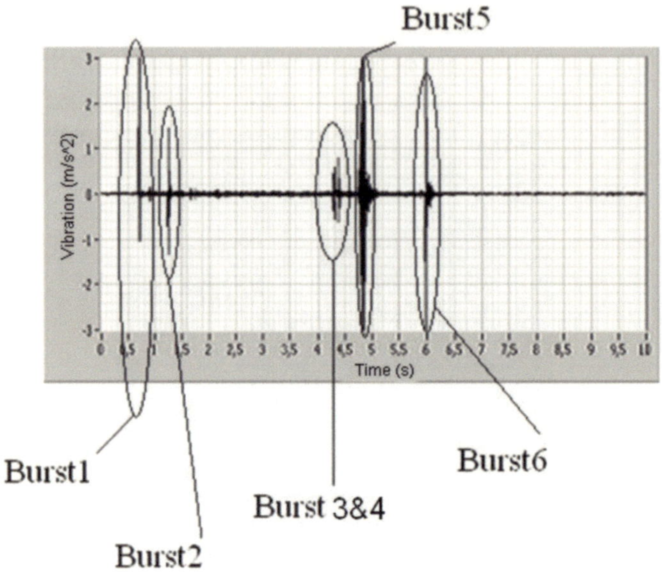

Fig. 3. Vibration signal during a tap change

4. Test programme

Some of the failures simulated in the experiment were: Broken contact bar (Fig. 4) and tap selector contacts damaged by arcing (Fig. 5), worn selector contacts, etc.

Fig. 4. Broken contact bar

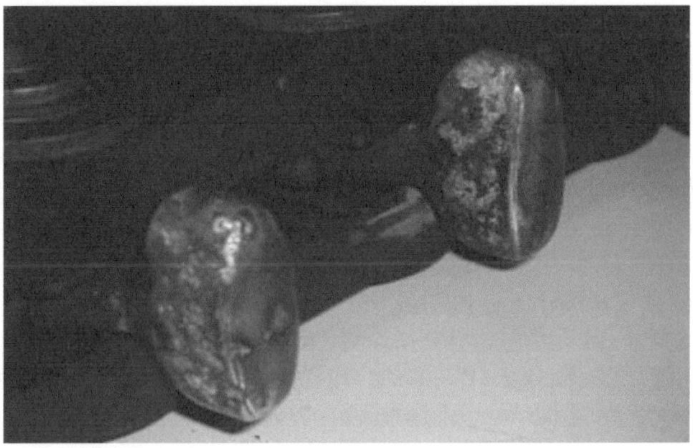

Fig. 5. Tap selector fixed contacts damaged by arcing

5. Vibration signal treatment

Vibration signal analysis is used to detect on load tap changer failures. According to (P. Kang & D. Birtwhistle, 1998) the main parameters for diagnosis are: Number of vibration bursts, vibration burst amplitude, time between vibration bursts.
When two consecutive changes are made from the same initial tap to the same final tap, the obtained records are not identical. Large differences can be observed in amplitudes and in time between vibration bursts. Consequently, some kind of signal processing must be used when using vibration monitoring to detect OLTC faults.
The OLTC vibration signals are subjected to the treatment process shown in Fig. 6. The purpose and meaning of each treatment is summarised below.

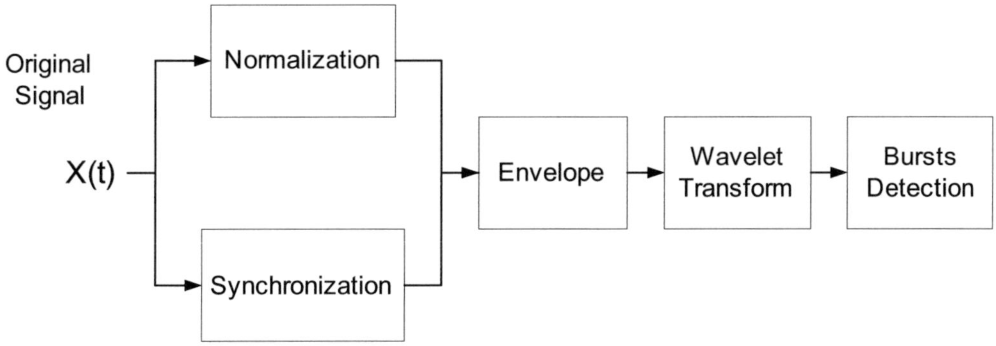

Fig. 6. Process for obtaining the wavelet envelope spectrum

A. Normalization and synchronization of signals

In first place vibration signals are synchronised, so that they start at the same time. Then, signals are normalised using their RMS value.
As in our study the sampling rate is 50kS/s, in a 10s record there are 500 000 samples (Fig.2). By eliminating the samples with no information will the signals are reduced to 267,001-269,001 samples.

B. Envelope analysis

The main information in the vibration signal is in the signal envelope. So, once the vibration signals are normalised and synchronised, the Hilbert transform is used to obtain the envelope (A. I. Zayed., 1998).

The analytic signal corresponding to a real signal $x(t)$ is defined as $x(t)+j\overline{x}(t)$, where the real part of the analytical signal is the original signal and the imaginary part, $\overline{x}(t)$ is the Hilbert transform of $x(t)$. The magnitude of this complex analytical signal forms the signal envelope and is always a positive function.

As mentioned the vibration signal is not exactly the same when repeated a tap change. Even some spurious bursts can appear. To solve this problem the vibration signal is converted into energy with the same number of intervals for each input signal. The average energy signal is computed, obtaining signals containing bursts with a similar number of samples.

For the correct detection of bursts present in the envelope of the vibration signal is divided into six time intervals, thus reducing the signal to 640 samples (Fig. 7).

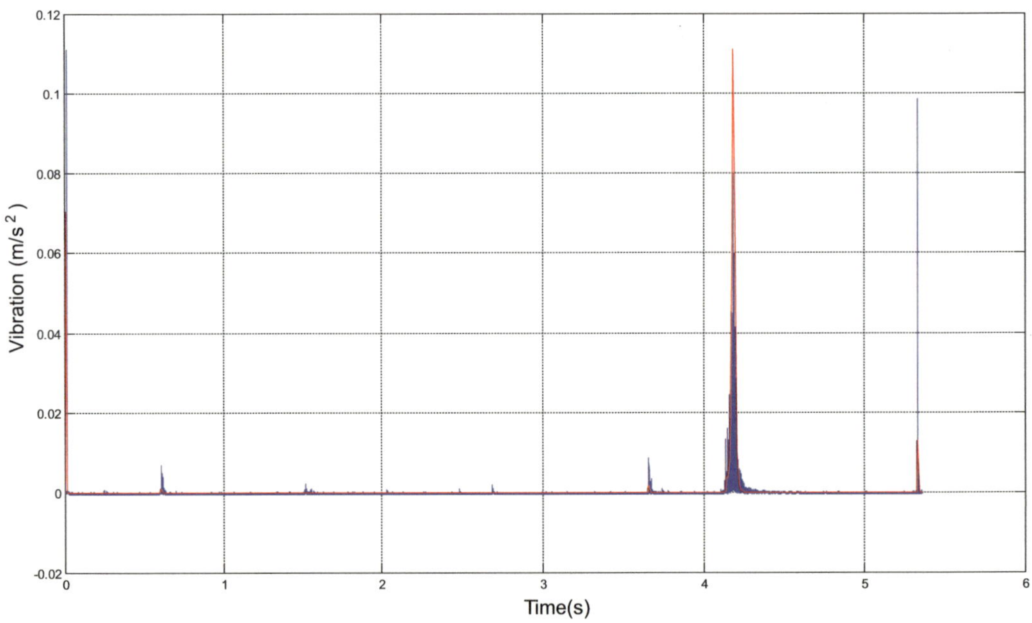

Fig. 7. Average energy of the envelope in regular time intervals

C. Discrete wavelet transform

Envelope analysis based on Hilbert transform is used as Wavelet decomposition pre-processing.

Discrete wavelet transform (DWT) is used to decompose a signal, through two types of filters (low pass and high pass filter). The number of times the signal is filtered is determined by the applied decomposition level. In this way, a set of coefficients (approximation and detail coefficients) are obtained at each decomposition level.

Discrete wavelet transform (DWT) of a signal f (t) is defined by (S. Mallat, 1988):

$$DWTf(t) = \sum_{-\infty}^{j} \sum d_j[n] + \sum_{n \in Z} a_j[n] \psi_{jn} , \; j \in Z \tag{1}$$

$a_j[n]$: Approximation coefficients
$d_j[n]$: Detail coefficients
ψ_{jn} : scaling function.
These coefficients represent part of the original signal, for certain frequency bands and time intervals.
For the implementation of discrete wavelet transform to the envelope of the vibration signal in a first instance we must select the mother wavelet and then decompose the signal. The signal is reconstructed using Inverse Wavelet Transform (IDWT).

Mother wavelet selection

In order to have an indicator relating the correlation between the processed signal and the selected mother wavelet Parseval's theorem is used. This indicator is the sum of the squared spectrum coefficients of the Fourier transform refer to frequency domain.

$$\frac{1}{N} \sum | f[t] |^2 = \sum | b_k |^2 \tag{2}$$

Where N is the sampling period, and b_k are the coefficients of the Fourier transform Spectrum.
Then, we use (1) and (2) to apply Parseval's theorem to the DWT application and we obtain (3)

$$\frac{1}{N} \sum | f[t] |^2 = \sum_{-\infty}^{j} \sum | d_j[n] |^2 + \sum_{n \in Z} | a_j[n] |^2 \tag{3}$$

Parseval's theorem relates the energy of the signal $f(t)$ to the energy in each component and the wavelet coefficients.
Then we use (3) as a comparison index for evaluation of the different mother wavelets that can be used. Table 1 shows the value of the aforementioned index with different mother wavelets when using the original signal and when using the envelope.

Wavelet	Type	Index (Signal)	Index (Envelope)
Daubechies	db9	3500.103468	251.8379024
Symelet	sym8	3497.532913	250.7456702
Coiflet	coif5	3502.697888	253.4763404
Biortogonal	bior 3.1	498043.4396	2420.976428
Biortogonal	bior 3.3	20573.9404	639.0276636

Table 1. The value of the aforementioned index with different mother wavelets

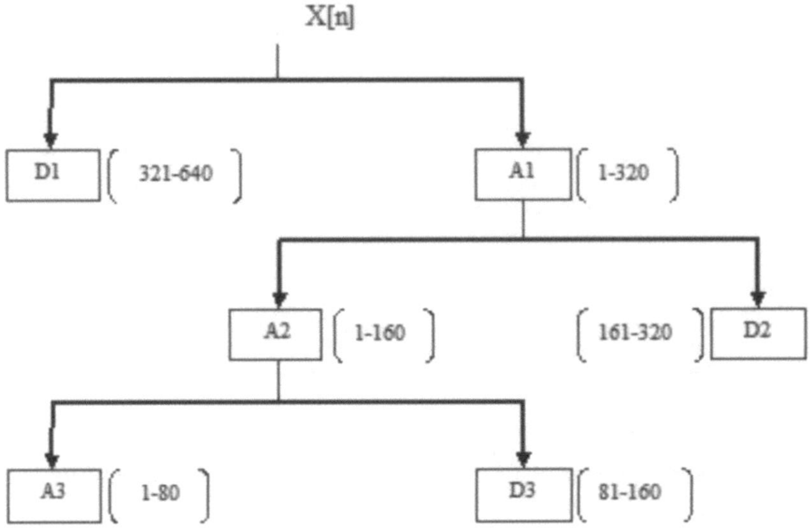

Fig. 8. Wavelet decomposition

From Table 1, it is evident that the best index when evaluating the envelope is the one obtained with the Biorthogonal Wavelet 3.1. Nevertheless, this mother wavelet is not square integrable (E. Rivas, et al.,2009), so the Fourier fast transform cannot be performed since it does not converge. Therefore, a pseudofrequency cannot be associated allowing us to relate one scale in the Wavelet domain with one frequency in the Fourier domain.

Then Bior 3.3 mother wavelet was chosen since this wavelet presents the greatest correlation index when compared to other mother wavelets and it allows characterizing the vibration signal.

Wavelet decomposition at level 3 (Fig. 8) is applied to the envelope of the energy signal (Fig. 7). Signal reconstruction from the coefficients of third level has a correlation of 94% with the original signal.

D. Bursts detection

Once discrete Wavelet Transform is performed, bursts are detected (Fig. 10). A smooth filtering based on threshold is used, to reduce noise and eliminate spurious bursts. To distinguish residual noise in the signal from the true burst, the energy present in the bursts are computed as the energy of residual noise is very low. The indicators used to evaluate the suitability threshold were:

- Preserved energy in the signal (Ep): Measures the proportion of energy preserved in the envelope signal after thresholding.

$$Ep = \frac{\sum_{i=1}^{N} [env_{th}[i]]^2}{\sum_{i=1}^{N} [env\ [i]]^2} x\ 100\% \tag{4}$$

Where env= envelope and $env_{th=}$ smoothed envelope.

- Mean Square Noise (MSE): This indicator is proportional to the distortion in the filtered signal, thus its value should be the lowest possible.

$$MSE = \frac{1}{N}\sum_{i=1}^{N}\left[env[i] - env_{th}[i]\right]^2 \tag{5}$$

- Signal Noise Ratio (SNR): This indicator measures the noise proportion referring to the signal.

$$SNR(db) = 10\log\frac{\sum_{i=1}^{N}\left[env[i]\right]^2}{\sum_{i=1}^{N}\left[env[i] - env_{th}[i]\right]^2} \tag{6}$$

The criterion to choose the threshold value is based on capturing the greatest percentage of energy with a low MSE and a suitable SNR. Fig. 9 shows Ep and MSE values as a function of threshold value.

From this figure it is deduced that the appropriate threshold value is within the range from 0.8% to 1.3% of the envelope RMS value.

In our case a threshold of 1% was used. The average values (72 vibration records) of process indicators were, Ep= 97.59%, MSE=1.28E-08 with SNR= 36. 55db. Fig. 9 shows the reconstructed vibration signal after applying the threshold.

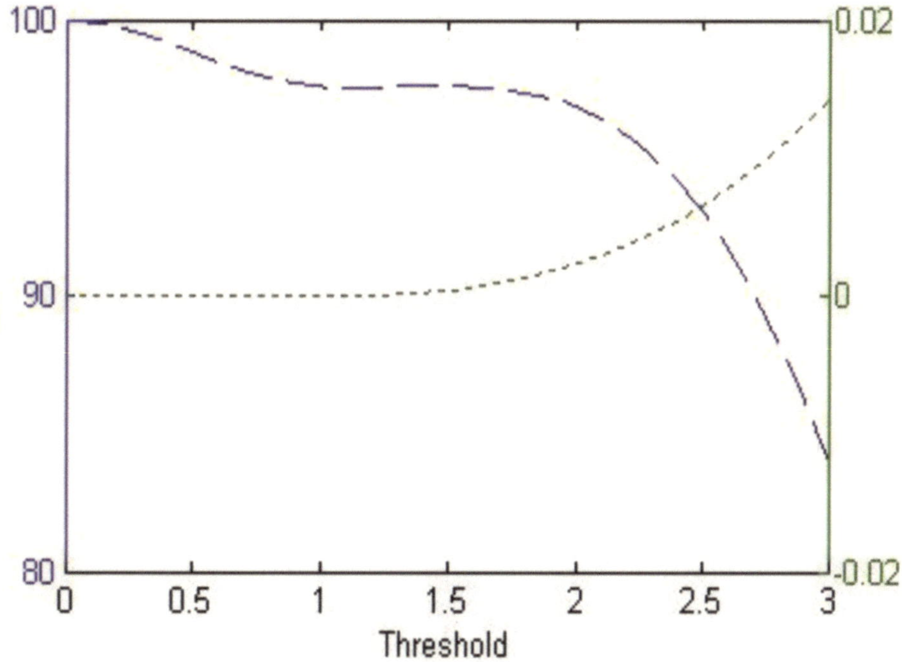

Fig. 9. Preserved energy versus threshold value (discontinuous line) and Mean Square Error (dotted line) vs. threshold

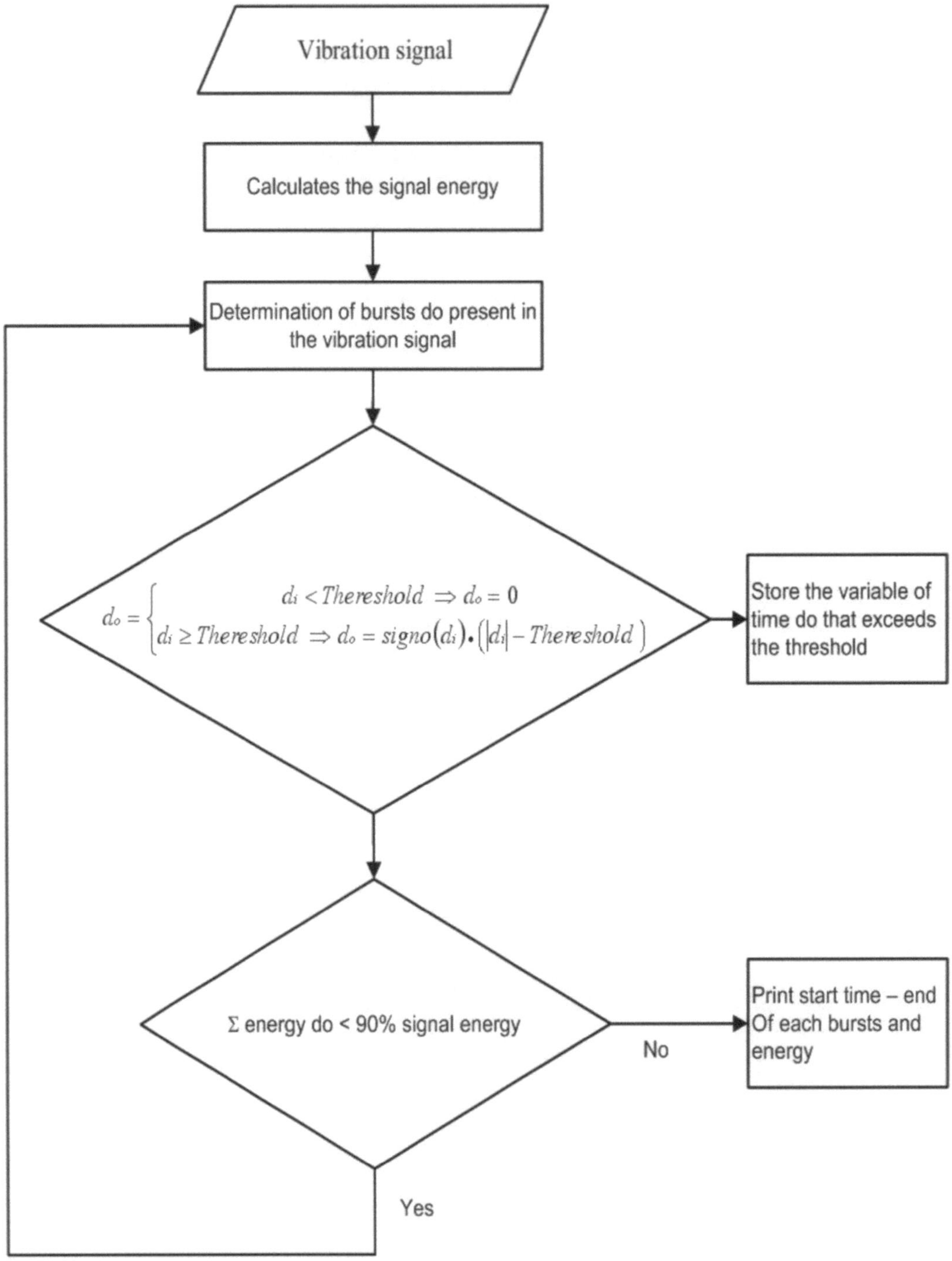

Fig. 10. Bursts detection

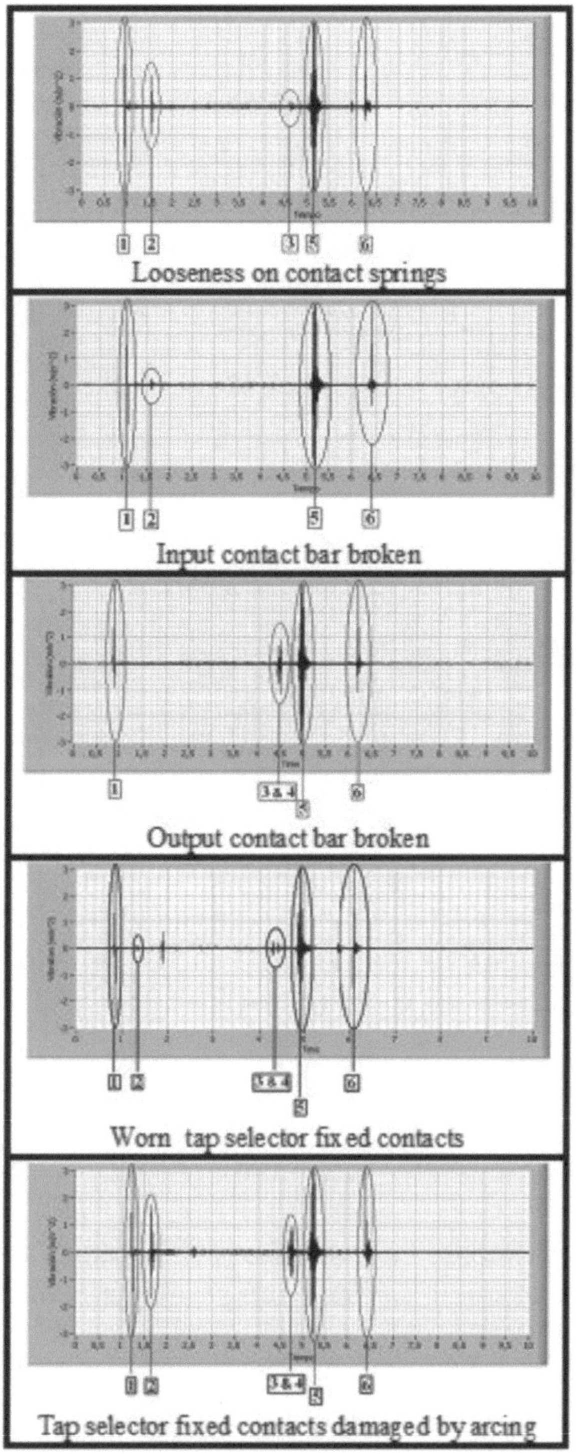

Table 2. Vibration signals for the simulated failures

CONDITION		t2-t1	t3-t2	t4-t3	t5-t4	t6-t5
GOOD	Average time (s)	0.61	3.08	0.12	0.59	1.18
	Standard deviation (s)	0.08	0.01	0.01	0.07	0.22
	Standard deviation/Average time (%)	12.7	0.30	9.33	13.16	18.7
LOOSENESS ON CONTACT SPRINGS	Average time (s)	0.62	3.09			1.22
	Standard deviation (s)	0.01	0.01			0.01
	Standard deviation/Average time (%)	1.61	0.32			1.09
INPUT CONTACT BAR BROKEN	Average time (s)	0.57				1.29
	Standard deviation (s)	0.02				0.01
	Standard deviation/Average time (%)	3.5				0.77
OUTPUT CONTACT BAR BROKEN	Average time (s)				0.55	1.31
	Standard deviation (s)				0.04	0.06
	Standard deviation/Average time (%)				7.23	4.56
WORN CONTACTS	Average time (s)	1.08	2.60	0.14	0.67	1.30
	Standard deviation (s)	0.15	0.34	0.01	0.03	0.01
	Standard deviation/Average time (%)	13.8	13.1	7.14	4.47	0.77
TAP SELECTOR CONTACTS DAMAGED BY ARCING	Average time (s)	0.43			0.39	1.19
	Standard deviation (s)	0.01			0.05	0.01
	Standard deviation/Average time (%)	2.32			12.82	0.84

Table 3. Statistical Study

6. Results

Table 2 shows the vibration signature for some of the simulated failures. The study was carried out on 120 vibration signals, corresponding to 6 different conditions (Table 2) of the OLTC (20 signals for each condition).

Some diagnostic indicators were extracted starting from the vibration signal (frequency spectrum, wavelet coefficient comparison, etc). The best indicator to classify vibration signals was *time between vibration bursts*. As vibration signals are not repetitive a statistical study was carried out to prove the suitability of this indicator. Table 3 shows the results of the statistical study. As it is deduced from Table 3, time between vibration bursts differs significantly when failures do exist and when failures are absent. In certain types of failures some impulses (highlighted in grey in table 3) even disappear from the vibration signal.

7. Conclusion

It is of great importance to obtain indicators to assess the condition of high voltage equipment. Vibration signal can be used to assess the condition of a transformer OLTC.
Main diagnostic indicators can be obtained by using Wavelet Transform to signal vibration.
The use of a mother wavelet with the highest correlation with the vibration signals allows for a higher preservation of the original signal energy (and even with respect to the smoothed signal). But this is not a guarantee for smoothing the obtained signal or spurious burst elimination.
In order to automatically obtain the bursts in the vibration envelope it is necessary to obtain wavelet coefficients of the signal. The proper election of those coefficients depends on the required information and the mother wavelet used and is obtained as a trade-off between the correlation of the envelope of the energy signal and the reconstructed signal.
After analyzing the ability of different diagnostic indicators (number of bursts, vibration burst amplitudes, burst energy, time between vibration bursts, frequency band of bursts) to properly classify different kind of OLTC failures, time between vibration bursts (and eventually burst appearance or disappearance) are proven the most suitable indicators.

8. Acknowledgments

This study has been funded by the Spanish Science and Technology Ministry (MCYT) under project no. DPI2009-07093.

9. References

L.A.L. de Almeida, M. Fontana, F.A. Wegelin, and L. Ferreira "A New Approach for Condition Assessment of On-Load Tap-Changers Discrete Wavelet Transform", Instrumentation and Measurement Technology Conference, 2005. Proceedings of the IEEE. Volume 1, 16-19 May 2005, pp. 653 - 656.

Simas, E.F., de Almeida, L.A.L., de Lima, A.C., "Vibration Monitoring of On-Load Tap Changers Using a Genetic Algorithm" Instrumentation and Measurement Technology Conference, 2005. Proceedings of the IEEE. Volume 3, 16-19, May 2005, pp. 2288 - 2293.

M. Foata, R. Beauchemin, and C. Rajotte, "On-line testing of on-load tap changers with a portable acoustic system," IEEE 9th International Conference on Transmission and Distribution Construction, Operation and Live-Line Maintenance Proceedings, pp. 293 – 298, Oct. 2001.

R. Jongen, P. Morshuis et al. "A statistical approach to processing power transformer failure data" 19th Int Conf on Electricity Distribution CIRED 2007 paper 546. Vienna May 2007.

a. P. Kang, D. Birtwhistle, J. Daly, and D. McCulloch, "Non-invasive on-line condition monitoring of on-load tap changers," in Proceedings of IEEE Power Engineering Society Winter Meeting, Singapore, 2000.

b. P. Kang and D. Birtwhistle "Analysis Of Vibration Signals For Condition Monitoring Of Power Switching Equipment Using Wavelet Transform" (Proc of the IEEE-SP International Symposium on Time Frequency and time scale analysis 1998, Oct 1998, pp. 6-9)

S. Mallat, A Wavelet Tour of Signal processing. Academic Press, 1988.

E. Rivas, J. C. Burgos and J. C. Garcia-Prada. (2009, Condition assessment of power OLTC by vibration analysis using wavelet transform. IEEE Transactions on Power Delivery Vol 24 No 2, pp. 687-694. Abril 2009.

A. I. Zayed., Hilbert transform associated with the fractional fourier transform. Signal Processing Letters, IEEE 5(8), pp. 206-208. 1998.

18

Analysis of Microparts Dynamics Fed Along on an Asymmetric Fabricated Surface with Horizontal and Symmetric Vibrations

Atsushi Mitani[1] and Shinichi Hirai[2]
[1]*Department of Design, Sapporo City University*
[2]*Department of Robotics, Ritsumeikan University*
Japan

1. Introduction

Devices to feed along microparts, such as ceramic chip capacitors and resistors, have become more common, due to their use in sorting, inspecting, and shipping mass produced microparts. In microparts feeding, to feed along microparts in one direction, the driving force applied to each micropart must vary according to the direction of motion of the micropart. Especially, the motion of microparts smaller than submillimeter can be affected by not only inertia but also adhesion which is caused by electrostatic, van der Waal's, intermolecular forces, and surface tension.

Now we have developed a novel microparts feeder applied an asymmetric fabricated surface, for example, sawtoothed surface, as a feeder table (Figure 1) (Mitani, 2006). The asymmetric fabricated surface can feed along microparts in one direction using horizontal and symmetric vibrations because contact between a micropart and the asymmetric fabricated surface varies according to the direction of motion. In order to formulate the dynamics of micropart, we need to analyse driving force and adhesion according to these contacts.

In this chapter, we developed micropart dynamics considering the effect of contact. Sawtoothed silicon wafers with various pitch were applied for feeder table, and also 0603 (size, 0.6 x 0.3 x 0.3 mm: weight, 0.3 mg) capacitors were applied for microparts.

First, we analysed contact between a micropart and a sawtoothed surface. Each surface profile model was approximated by a linear polynomial based on measurements using a microscopy system.

Secondly, we derived dynamics including the effect of adhesion. We analysed contact between both approximated models, because both inertia caused by feeder vibrations and adhesion vary according to contact. Supposing that adhesion occurs in the direction vertical to the tangent and also adhesion acts when the distance between two surface models is smaller than an adhesion limit, we obtained an adhesion model. Then the adhesion model parameters were identified using the results of friction angle measurement experiments of microparts.

Finally, we assessed the dynamics derived above. We firstly conducted feeding experiments of microparts using various pitch of sawtoothed surface with the same driving and environmental conditions. Using these experimental results, we verified driving condition and feeding velocity at each sawtooth pitch, and also we assessed an appropriate driving condition and a feeder surface. Feeding simulations were then executed using dynamics derived with the same parameters as the feeding experiments. These simulation results were compared with experimental results in order to evaluate the derived dynamics.

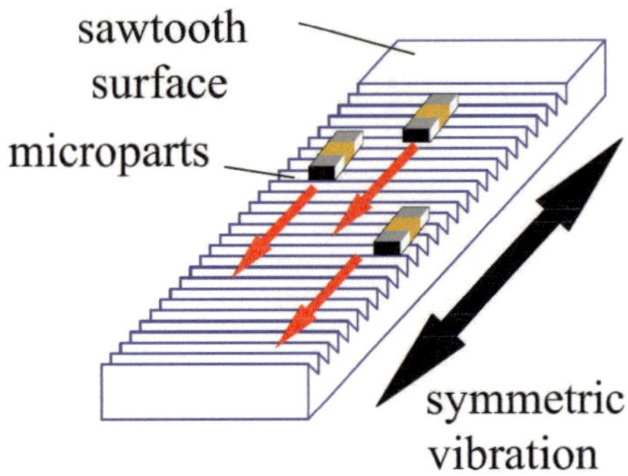

Fig. 1. Diagram of microparts feeding using a sawtoothed surface with symmetric vibrations

2. Related works

Partsfeeder is a key device in factory automation. The most popular feeders are vibratory bowl feeders (Maul, 1997), which use revolving vibrators to move parts along a helical track on the edge of a bowl. Linear feeders as well as an inclined mechanism and oblique vibration for unidirectional feeding (Wolfsteiner, 1999), have also been developed. In all of these systems, the aspect ratio of the horizontal/vertical vibrations must be adjusted to prevent parts from jumping. In our system, however, this adjustment is not necessary because only horizontal vibration is used.

A parts feeding that employs non-sinusoidal vibrations (Reznik, 2001) has been developed. The part moves to its target position and orientation or is tracked during its trajectory by using the difference between the static and sliding friction. Our system realizes unidirectional feeding by symmetric vibration of a sawtoothed surface, which yields different contact forces in the positive and negative directions.

Designing have been tested by simulation (Berkowitz, 1997 & Christiansen, 1996). The focus was mainly on the drive systems such as the structure and actuator, the movement of fed parts was generally neglected. In contrast, the movement of the microparts are considered in the present study.

Attempts have been made to improve the drive efficiency by feedback control systems (Doi, 2001) and nonlinear resonance systems (Konishi, 1997). Our system depends only upon contact between the feeder surface and the micropart. So the driving system is simple and uses an open loop system for feeding.

Micro-electro-mechanical systems (MEMS) technology has been used to mount on a planar board arrays of micro-sized air nozzles which, by turning on or off their air flow, have been used to control the direction of moving microparts (Fukuta, 2004 & Arai, 2002).

It is possible to perform manipulation with ciliary systems (Ebefors, 2000) and vector fields (Oyobe, 2001) without sensors. In this case, there are many actuator arrays on a vibratory plate. Actuator arrays enable control of contact between the vibratory plate and micropart in order to accomplish the target manipulation. However, these studies did not mention the dynamics of the micropart, especially the effects of adhesion forces on its motion. Other various feeding systems using electric-field (Fuhr, 1999), magnetic (Komori, 2005), bimorph piezoelectric actuators (Ting, 2005), and inchworm systems (Codourey, 1995) have been developed. These studies, however, have also not investigated the contact between the feeder surface and the micropart.

3. Measurement tool

For inspection of both the sawtooth surface and the micropart surface profiles, we used the AZ-100 multi-purpose zoom microscopy system (Nikon Instech Co., Ltd.) (Figure 2), which includes a mono zoom optical system that enables on-axis observation and documentation and built-in optics of up to 8 times magnification. In combination with an objective lens of 5 times magnification, we could take pictures at up to 40 times magnification. This microscope also has an automatic stage driven by a stepping motor to control focus height at a resolution of 0.54 µm.
A digital camera is attached to the top of the microscopy system, and captured pictures were forwarded to a computer via USB interface, and saved as bitmap files. The resolution of forwarded pictures taken at 40 times magnification was 0.276 µm/pixel. We used the DynamicEye Real focus image synthesizing software (Mitani Corp.) to analyse these surface profiles. This software can synthesize a three dimensional (3D) model from these pictures according to focus height. Sections of the 3D model are analysed to obtain a surface profile model.

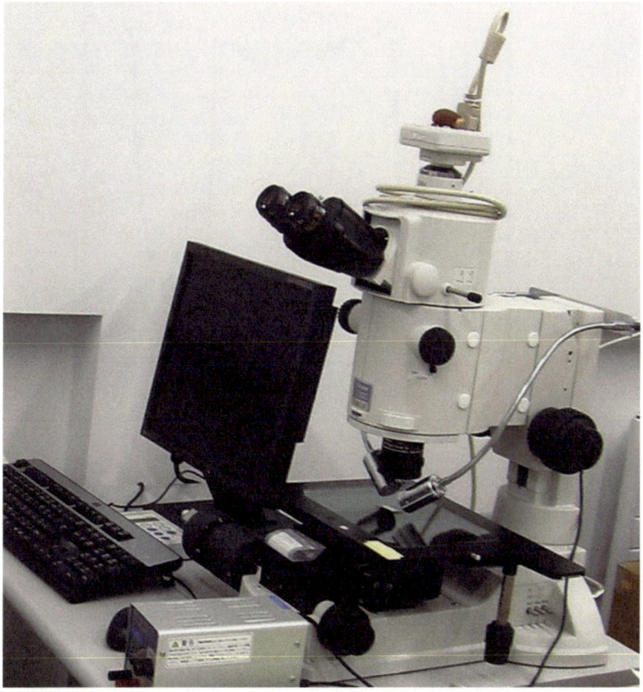

Fig. 2. AZ-100 multi-purpose zoom microscope (Nikon Instech Co., Ltd.)

4. Analysis of micropart surface

4.1 Detail of micropart

We applied a 0603 ceramic chip capacitor, electronic parts used in various mobile devices, as a micropart. As shown in Figure 3, a capacitor consists of a conductor and electrodes with convexities on each end surface. We obtained representative contours along a capacitor using a Form Talysurf S5C sensing-pin surface measurement tool (Taylor Hobson Corp.) (Figure 4). Electrodes contact the feeder because they protrude 10 μm higher than the conductor. To obtain minute profile models of these electrodes, we used the microscopy system mentioned above.

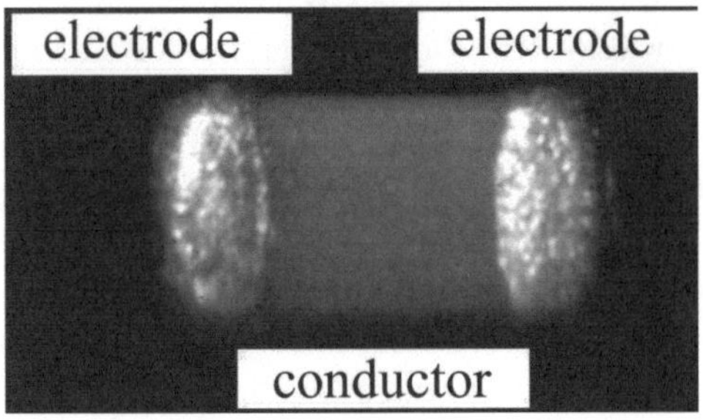

Fig. 3. Ceramic chip capacitor 0603 (size, 0.6 x 0.3 x 0.3 mm: weight, 0.3 mg)

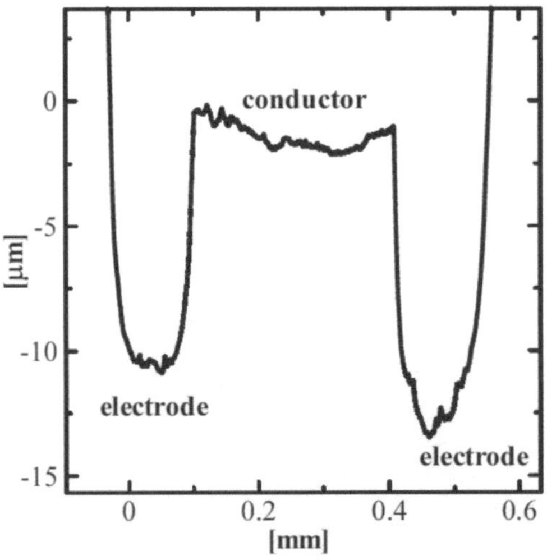

Fig. 4. Section of 0603 capacitor

4.2 Surface model of micropart based on measurements

Using the automatic capture mode of the DynamicEye Real software, we obtained a 3D model of an electrode shown in Figure 5 synthesized from 256 successive pictures with an interval of 0.54 μm in focus height at 40 times magnification. Analysing this 3D model, a numerical model of surface profile at any section was obtained. Figure 5 shows a profile model with a section along the y axis, and also Figure 6 shows a section along the x axis in Figure 4.

Next, we considered a convexity model on the surface of electrode. We assumed that only some higher convexities can contact the sawtoothed feeder surface. We then selected five convexities numbered from #1 to #5 in Figure 6 and 7. Let us approximate each convexity with a second order polynomial as follows:

$$y_p = b' x_p^{\ 2} + c_p, \tag{1}$$

where, c_p is constant, y_p is defined along the vertical line, and also x_p is defined along the horizontal line. Figure 8 shows the profile model of the convexity #1 and its approximation function. Assuming that each approximation function could be rotated around the horizontal line, and be transformed to its minimum value at the position $(x_p, y_p) = (0,0)$ without loss of generality, equation (1) can be rewritten as:

$$y_p = b x_p^{\ 2}, \tag{2}$$

where, $b \equiv -b'$. Averaging five transformed approximation functions, the coefficient was formulated as $b = 0.186$ (Figure 9).

Finally, the surface profile model of electrode convexity was defined by a hyperboloid of revolution of equation (2) around the y_p axis. When considering the z_p axis perpendicular to the $x_p - y_p$ plane with passing the position $(x_p, y_p) = (0,0)$, the trajectory of a point (x_p, y_p) on equation (2) was represented as a circle with a radius r as follows:

$$r = x_p = \sqrt{\frac{y_p}{b}}. \tag{3}$$

Then, the convexity surface model was formulated as:

$$x_p^{\ 2} + z_p^{\ 2} = y_p. \tag{4}$$

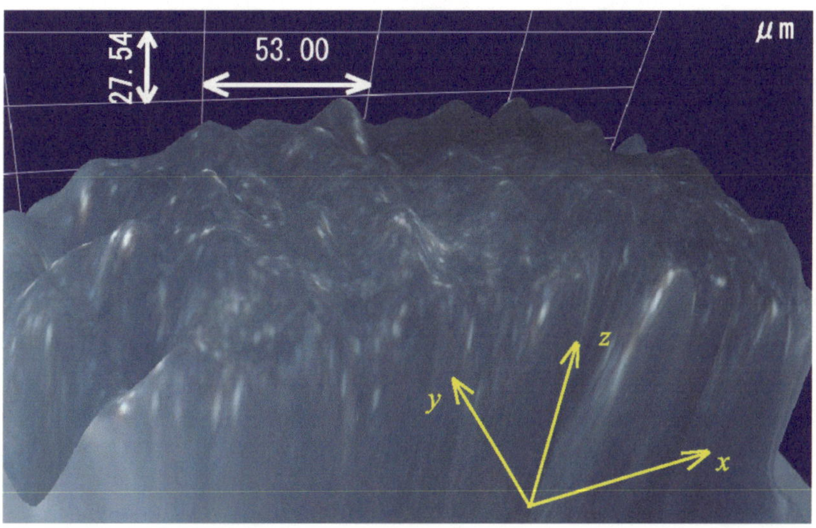

Fig. 5. Synthesized model of capacitor electrode

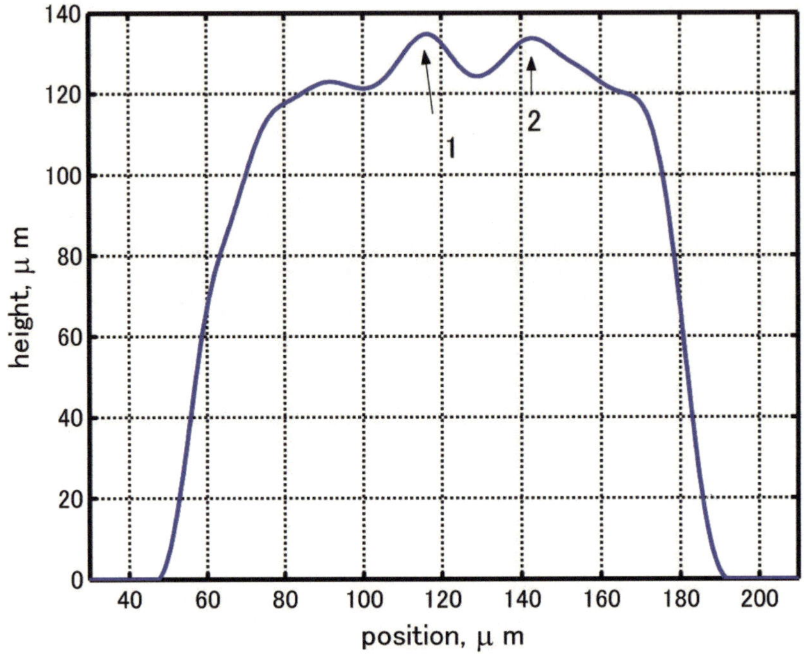

Fig. 6. Profile model along the y axis

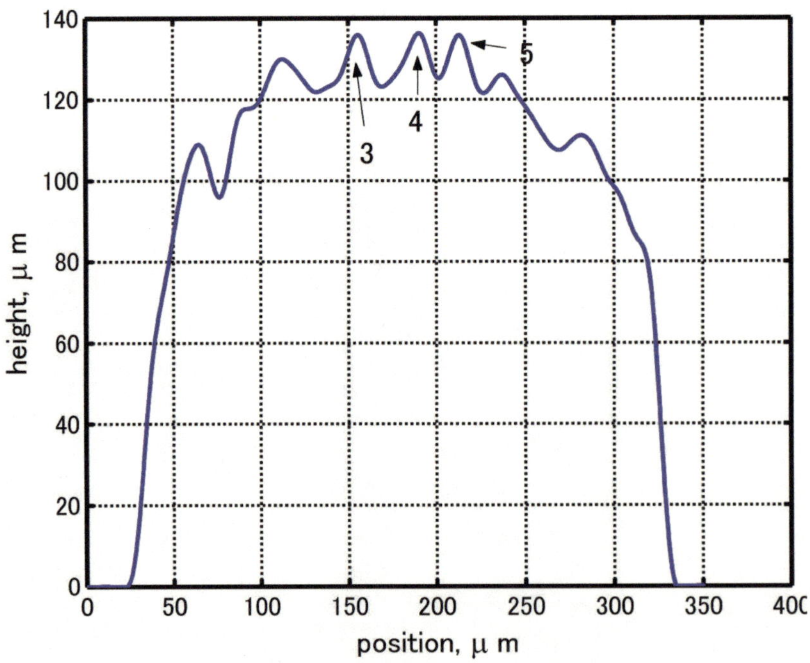

Fig. 7. Profile model along the x axis

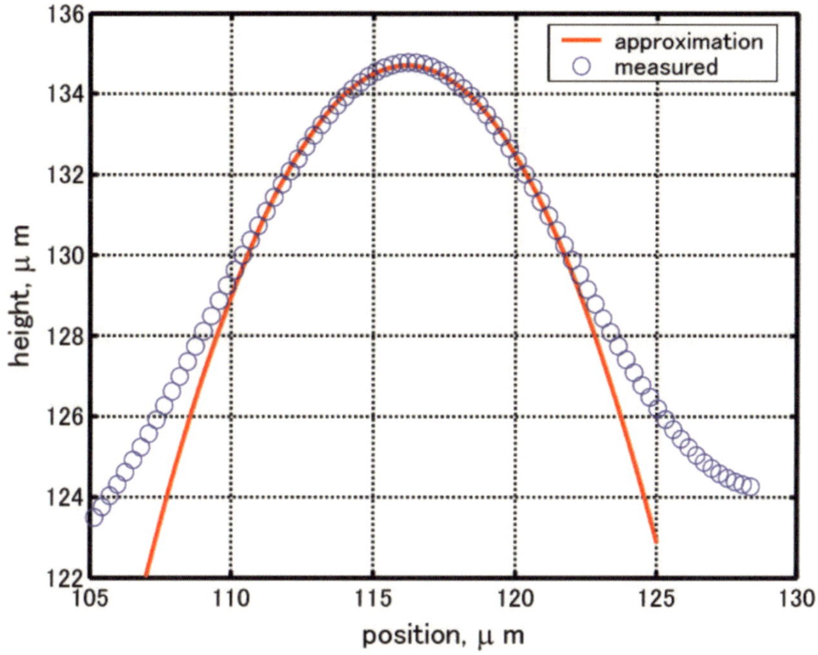

Fig. 8. Profile model of convexity #1 and its approximation

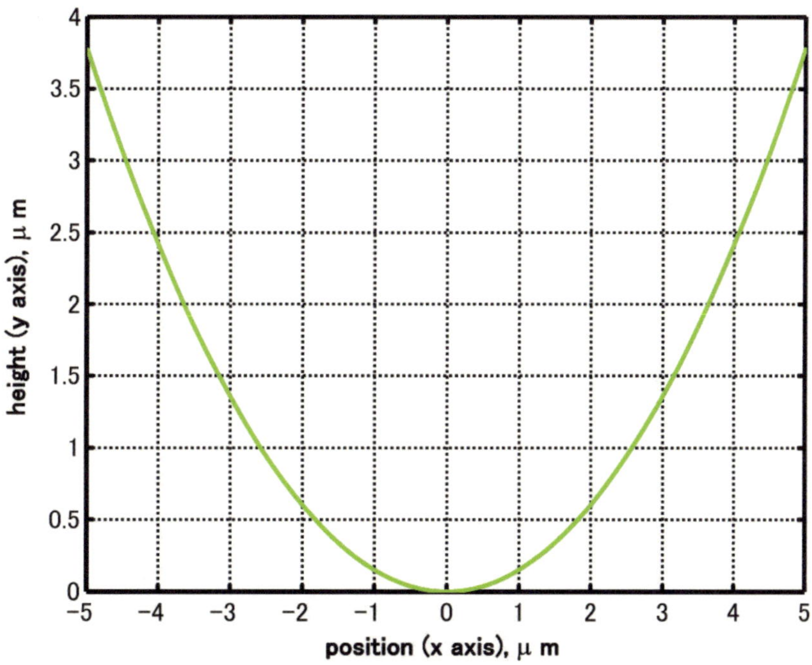

Fig. 9. Convexity model based on measurements: averaged model of five convexities

5. Analysis of sawtoothed feeder surface model

In this study, sawtoothed silicon wafers were applied for feeder surfaces. These surfaces were fabricated by a dicing saw (Disco Corp.), a high-precision cutter-groover using a bevelled blade to cut sawteeth in silicon wafers. Inspecting a sawtoothed silicon wafer using the microscopy system, we obtained a synthesized model (Figure 10) and its contour model (Figure 11). Then we found that these sawtooted surfaces were not perfectly sawtooth shape, but were rounded at the top of sawteeth because of cracks by fabricating errors. So these sawtoothed surfaces were needed to derive surface profile models based on measurements same as Section 4.

Analysing Figure 9 with the DynamicEye Real software, we obtained a numerical model of the top of sawtooth representing with the circle symbol in Figure 12. Defining the feeder coordinate $O-xy$ with the origin O at the maximum value, x axis along the horizontal line, and y axis along the vertical line, this numerical model was approximated with four order polynomials as follows:

$$y = f_s(x) = a_4x^4 + a_3x^3 + a_2x^2 + a_1x + a_0. \tag{5}$$

An approximation function was drawn with a red continuous line in Figure 11 when each coefficient was defined as Table 1. Interpolating other part of sawtooth with straight lines, we obtained surface profile model of sawtoothed surfaces (Figure 13). In this figure, p shows the sawtooth pitch, and θ shows the angle of elevation. In addition, the incline angle of the line HJ was the same as the angle of elevation θ, the line KL was along the y_s axis, and the curve JK was represented by equation (5).

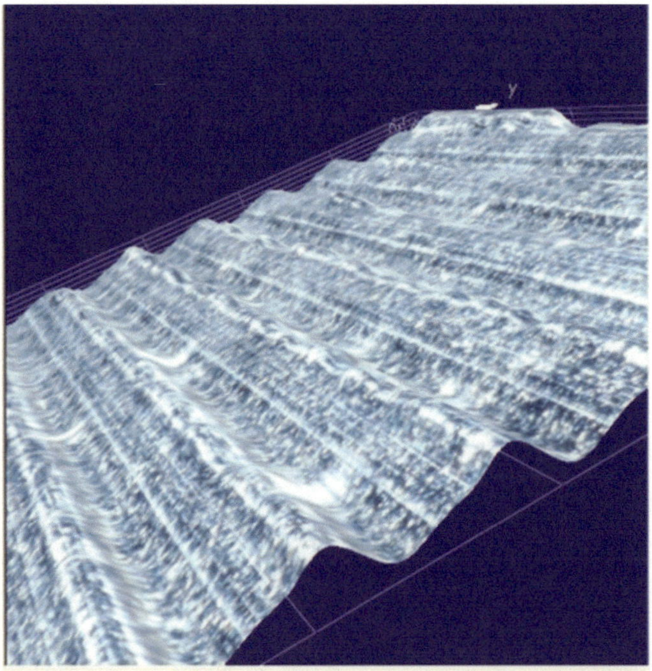

Fig. 10. Synthesized model of sawtoothed surface (p = 0.1 mm and θ=20 deg)

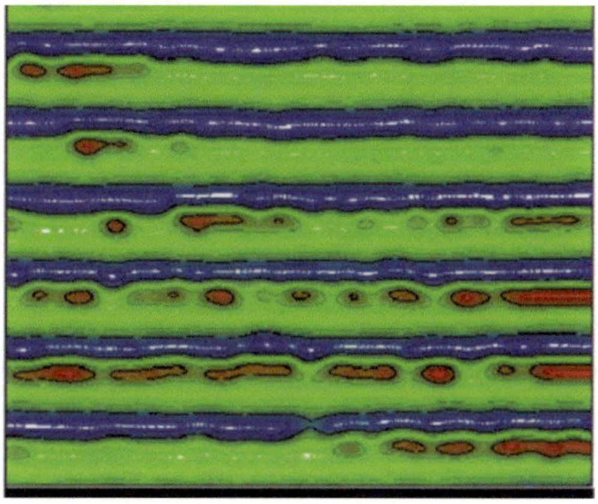

Fig. 11. Contour model

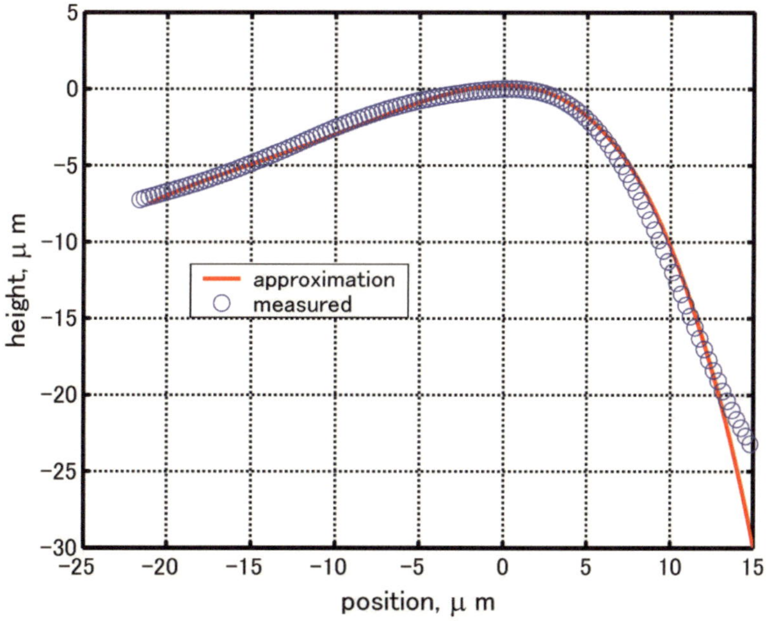

Fig. 12. Measured sawtooth profile and its approximation

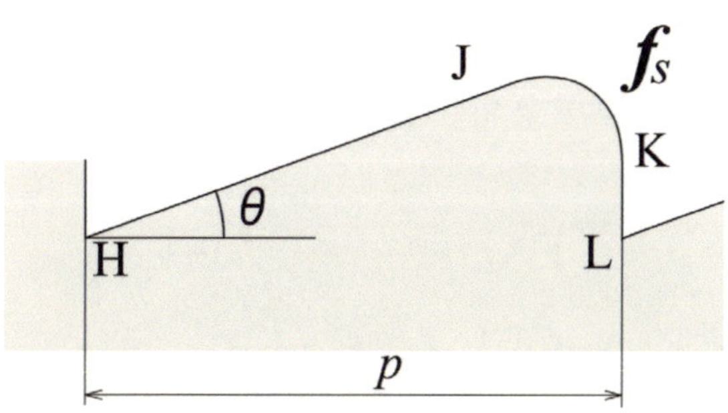

Fig. 13. Surface profile model of sawtooth

a_4	a_3	a_2	a_1	a_0
-0.772e-4	-0.370e-2	-0.611e-1	0.0	0.0

Table 1. Coefficients of approximation function

6. Analysis of contact between approximated models of both surfaces

6.1 Distance between two surfaces

Now we consider contact between two approximation functions represented by equations (2) and (5) as shown in Figure 14. Let us assume that these two functions share a tangent at the contact point $C(x_c, y_c)$, and also assume that adhesion acts perpendicular to the tangent.

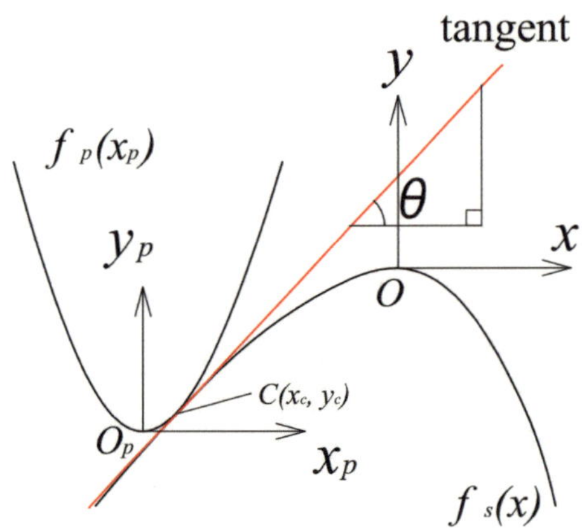

Fig. 14. Contact between two approximation models of micropart and sawtoothed surface

When the part origin O_p is located at ${}^0O_p(x_0, y_0)$ on the feeder coordinate, equation (2) can be rewritten as:

$$y = b(x - x_0)^2 + y_0. \tag{6}$$

Differentiating with respect to x and also substituating the contact point $C(x_c, y_c)$, we have the tangent as follows:

$$y = 2b(x_c - x_0)(x - x_0) + y_c. \tag{7}$$

When the incline of the tangent is defined as $y'(x_c) \equiv \tan\theta$, the following equations are obtained:

$$y'(x_c) = 2b(x_c - x_0) = f_s'(x_c), \tag{8}$$

$$f_s'(x_c) \equiv \frac{df_s(x)}{dx} = 4a_4 x_c^{\ 3} + 3a_3 x_c^{\ 2} + 2a_2 x_c^{\ 1} + a_1. \tag{9}$$

From these equations, the part origin ${}^0O_p(x_0, y_0)$ is calculated as:

$$x_0 = x_c - \frac{f_s'(x_c)}{2b}, \tag{10}$$

$$y_0 = y_c - \frac{\{f_s'(x_c)\}^2}{4b}. \tag{11}$$

Let us consider a normal equation against the tangent passing through a coordinate $Q(x_q, y_q)$. When the normal equation intersects two surfaces at the coorinates $Q_1(x_1, y_1)$ and $Q_2(x_2, y_2)$, respectively (Figure 15), distance of two surfaces can be represented as:

$$dl = Q_1Q_2 = \sqrt{(x_2 - x_1)^2 + (y_2 - y_1)^2}. \tag{12}$$

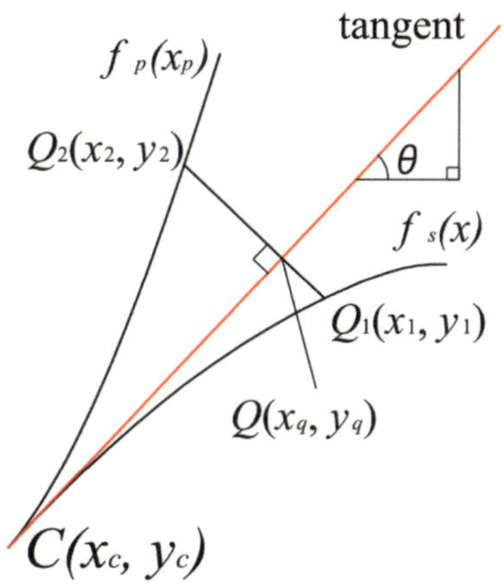

Fig. 15. Distance of two surface models

Now we formulate the coordinate $Q_2(x_2, y_2)$ assuming that the coordinate $Q_1(x_1, y_1)$ is already known. The normal equation is represented as:

$$\begin{cases} y = -\dfrac{1}{y_p'(x_c)}(x - x_1) + y_1 & (y_p'(x_c) \neq 0), \\ x = x_1 & (y_p'(x_c) = 0). \end{cases} \tag{13}$$

Then, substituting into equation (5), we have:

$$x_2 = \begin{cases} x_0 - \mathrm{x_a} & (y_p'(x_c) \neq 0), \\ x_1 & (y_p'(x_c) = 0), \end{cases} \tag{14}$$

$$y_2 = \begin{cases} y_0 + b\mathrm{x_a}^2 & (y_p'(x_c) \neq 0), \\ y_0 + b(x_1 - \mathrm{x_0}) & (y_p'(x_c) = 0), \end{cases} \tag{15}$$

where,

$$x_a \equiv \begin{cases} \dfrac{1}{2b}\left\{\dfrac{1}{y_p'(x_c)} - \sqrt{\dfrac{1}{y_p'(x_c)^2} - 4b\left(\dfrac{x_0 - x_1}{y_p'(x_c)} - (y_0 - y_1)\right)}\right\} & (y_p'(x_c) > 0), \\ \dfrac{1}{2b}\left\{\dfrac{1}{y_p'(x_c)} + \sqrt{\dfrac{1}{y_p'(x_c)^2} - 4b\left(\dfrac{x_0 - x_1}{y_p'(x_c)} - (y_0 - y_1)\right)}\right\} & (y_p'(x_c) < 0). \end{cases} \quad (16)$$

Here, when the square root in equation (16) is imaginary, equations (5) and (13) do not intersect each other, which means that $dl = \infty$.

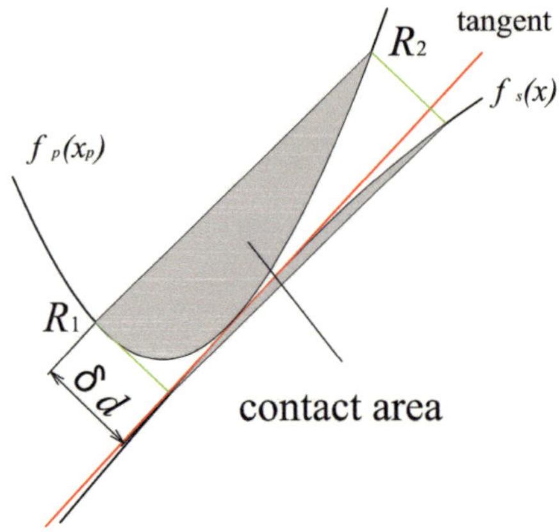

Fig. 15. Definition of contact area

6.2 Area of adhesion

Let as assume that adhesion acts when the distance dl is less than or equal to an adhesion limit δd. In Figure 16, area of adhesion can be defined as colored part between two lines satisfying $dl = \delta d$. Now we defined coordinates R_1 and R_2 as $R_1(x_{r1}, y_{r1})$ and $R_2(x_{r2}, y_{r2})$, (however, $x_{r1} < x_{r2}$), respectively. The equation that passes through R_1 and R_2 is described in the part coordinate system as:

$$y_p = c_r(x_p - x_{r1}) + x_{r1}^2, \quad (17)$$

where,

$$c_r = \frac{y_{r2} - y_{r1}}{x_{r2} - x_{r1}}.$$

When equation (17) is applied to the coordinate system $O_p - x_p y_p z_p$ as a plane parallel to the z_p axis, equation (17) cuts the hyperboloid represented in equation (4). In this study, the area of adhesion A is determined by the cut plane as shown in Figure 16. Substituting equation (17) into (4), equation of intersection is obtained:

$$(x_p - \frac{c_r}{2})^2 + z_p^2 = (x_{r1} - \frac{c_r}{2})^2. \quad (18)$$

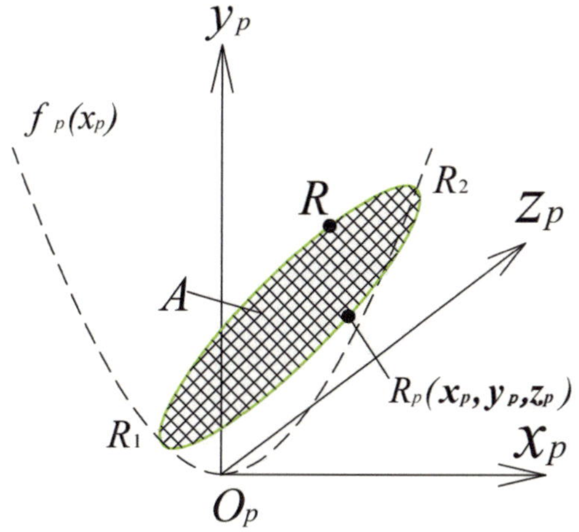

Fig. 16. Area of adhesion

Consequently, we have:

$$A = \pi(x_{r1} - \frac{c_r}{2})^2. \tag{19}$$

Figure 17 show calculation results of area of adhesion, assuming that the adhesion limit δl is determined by the Kelvin equation as follows:

$$\delta l = c_k r_k \equiv -c_k \frac{2\gamma V_m}{\ln \frac{P}{P_0} RT}, \tag{19}$$

where, T is the thermodynamic temperature, R the gas constant, γ the surface tension, P_0 the saturated vapor pressure, P vapor pressure, V_m molecular volume, r_k the Kelvin radius, and c_k proportionally coefficient.

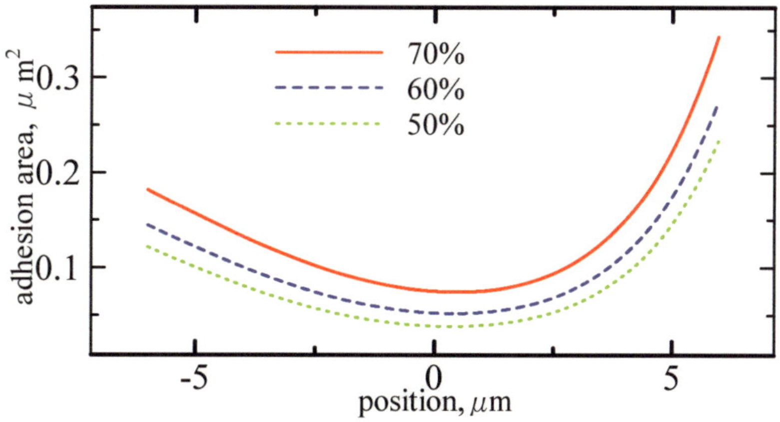

Fig. 17. Area of adhesion

Let F_a, D_A, n, and A_i be the adhesion force, the coefficient of adhesion, number of micropart convexity contacting with the sawtoothed surface, the area of adhesion of i-th micropart convexity ($i = 1,\cdots,n$), respectively . Assuming that adhesion force is proportional to the area of adhesion, the adhesion force is finally represented as follows:

$$F_a = D_A \sum_{i=1}^{n} A_i, \tag{19}$$

7. Identification of adhesion by angle of friction of microparts

Adhesion between microparts and a feeder surface is affected by surroundings such as temperature and ambient humidity. The Kelvin radius is getting larger as the ambient humidity increases, and then the adhesion force is also getting larger. In this section, we identified the adhesion force based on measurements of angle of friction of microparts under several conditions of ambient humidity.

7.1 Measurements of angle of friction of microparts

Angle of friction of microparts were measured under a temperature of 24^oC and an ambient humidity of 50, 60, or 70 %. We prepared sawtoothed silicon wafers with an elevation angle of $\theta = 20^o$ and various sawtooth pitches of $p = 0.01, 0.02, \cdots, 0.1$ mm . Experiments were conducted three times using 35 capacitors. Before experiments, all the experimental equipments were left in the sealed room with keeping constant temperature and ambient humidity for a day.

The averaged experimental data of each experimental condition were plotted in Figures 18 to 20. In these figures, 'positive' direction means that the sawtoothed surface was put as Figure 13, and then was turned around with the clockwise direction, whereas 'negative' direction means when it was turned around with the counter clockwise. Also, the averaged angle of friction at each ambient humidity is shown in Figure 21.

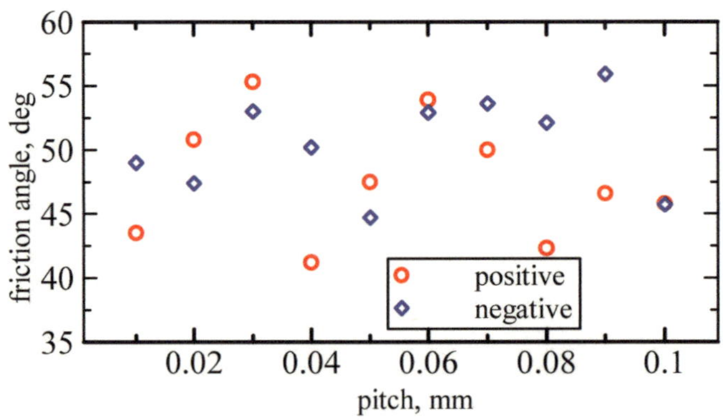

Fig. 18. Angle of friction of microparts with an ambient humidity of 50 %

Now we examine the directionality of friction. From Figures 18 to 20, experimental results at 'positive' direction were totally smaller than that of 'negative' direction, even opposite directions were appeared at on the surfaces of p=0.02, 0.03, 0.05, and 0.06 mm under an ambient humidity of 50 %, and on the surface of p=0.07, 0.08, and 0.09 mm under an ambient humidity of 60 %. The maximum directionality was 17.9 % realized on the surface of p=0.04 mm under an ambient humidity of 50 %, 26.6 % on the surface of p=0.05 mm under an ambient humidity of 60 %, and 15 % on the surface of p=0.06 mm under an ambient humidity of 70 %. From Figure 21, the angle of friction is getting larger according to ambient humidity, which indicates that the effect of adhesion increases as the increase of ambient humidity.

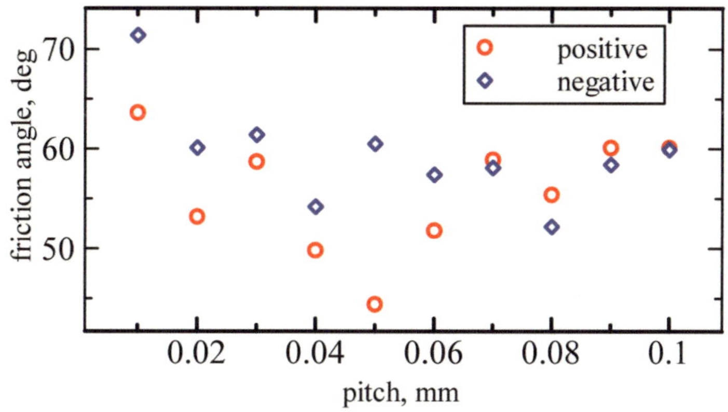

Fig. 19. Angle of friction of microparts with an ambient humidity of 60 %

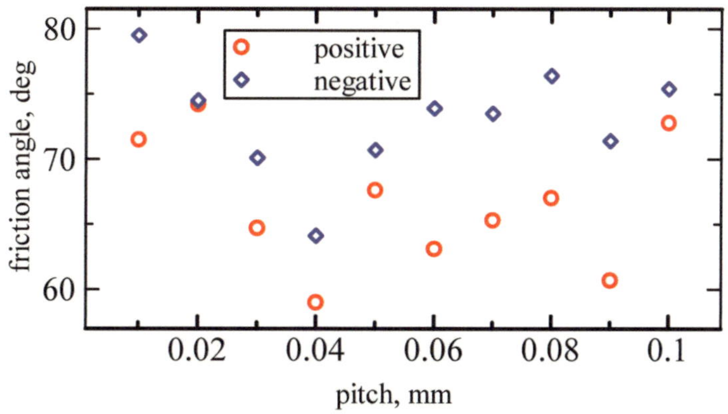

Fig. 20. Angle of friction of microparts with an ambient humidity of 70 %

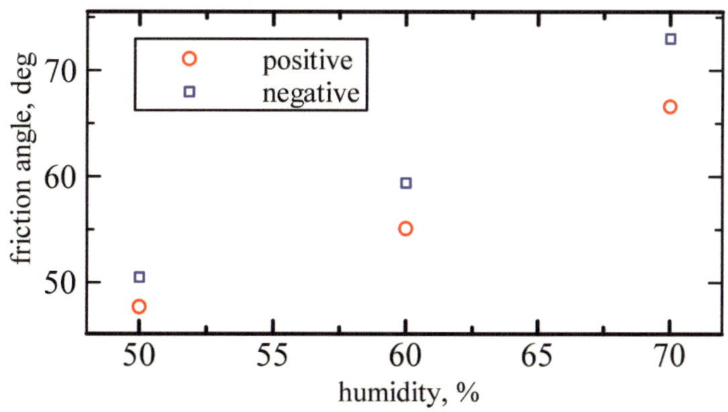

Fig. 21. Relationship between ambient humidity and angle of friction

7.2 Examination of friction coefficient

We consider the case that i-th convexity contacts a sawtooth at a position $x<0$, that is, $\theta_i>0$ (Figure 22). When the surface is inclined to the positive direction, adhesion acts as friction resistance against sliding motion, and also when inclined to the negative direction, adhesion acts as resistance against pull-off force. Let f_{si} be friction resistance against sliding motion, and f_{pi} be resistance against pull-off force, these resistances can be represented as:

$$f_{si}=\mu D_A A_i \cos\theta_i, \tag{20}$$

$$f_{pi}=D_A A_i \sin\theta_i. \tag{21}$$

Similarly, when contact at a position $x>0$ ($\theta_i<0$), these two resistance is rewritten as follows:

$$f_{si}=-\mu D_A A_i \cos\theta_i, \tag{22}$$

$$f_{pi}=D_A A_i \sin\theta_i. \tag{23}$$

On the other hand, when contact occurs at $x=0$ ($\theta_i=0$), adhesion acts as friction resistant against sliding motion according to the direction of incline. If ϕ is the incline of the sawtoothed surface, we have:

$$f_{si}=\begin{cases}-\mu D_A A_i & (\phi<0)\\ \mu D_A A_i & (\phi>0)\end{cases} \tag{24}$$

Let us assume that (m+n) convexities contact sawteeth, then each convexity numbered 1, 2, $\cdots$, m is shared a tangent with $\theta_{pi}>0,(i=1,2,\cdots,m)$, and also each convexity numbered (m+1), (m+2), $\cdots$, (m+n) is shared a tangent with $\theta_{nj}<0,(j=m+1,m+2,\cdots,m+n)$. Let F_p and F_n be the resistances at the positive and negative direction. Also, let A_{pi} and A_{nj} be adhesion area of the i-th convexity and j-th convexity, respectively, we obtained:

$$F_p=D_A\left(\sum_{i=1}^{m}A_{pi}\sin\theta_{pi}+\mu\sum_{j=1}^{n}A_{nj}\cos\theta_{nj}\right), \tag{25}$$

$$F_n=D_A\left(\mu\sum_{i=1}^{m}A_{pi}\cos\theta_{pi}-\sum_{j=1}^{n}A_{nj}\sin\theta_{nj}\right). \tag{26}$$

When the incline of the feeder surface is ϕ, inertia of micropart along the feeder surface is represented as:

$$F(\phi)=mg\sin\phi-\mu mg\cos\phi, \tag{27}$$

where, m is mass of micropart and g is gravity. Let as assume that micropart starts to move when the resistance caused by adhesion balances the inertia of micropart, $F(\phi)$. If ϕ_p and ϕ_n are angles of friction of positive and negative direction, respectively, we have:

$$F_p=mg\sin\phi_p-\mu mg\cos\phi_p, \tag{28}$$

$$F_n = mg\sin\phi_n - \mu mg\cos\phi_n. \tag{29}$$

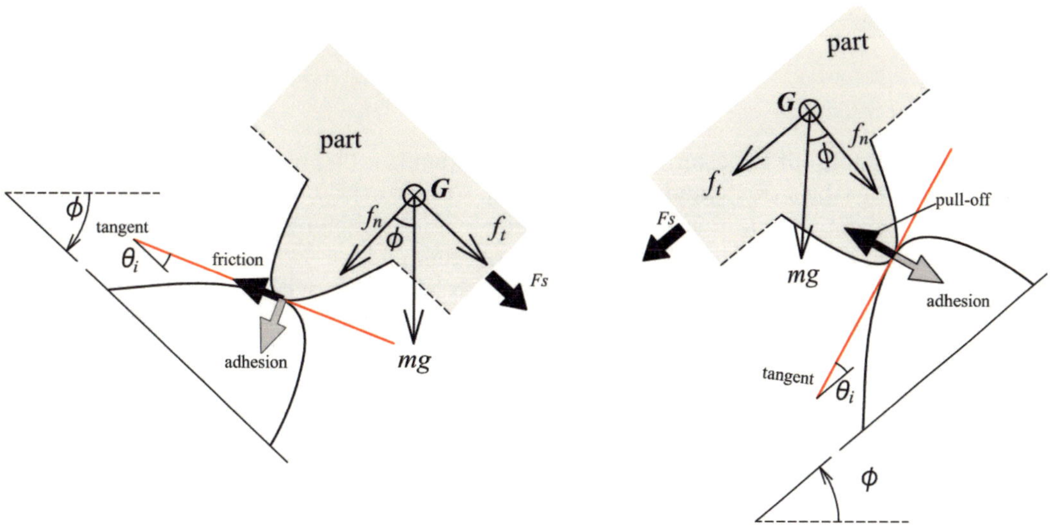

Fig. 22. Resistance caused by adhesion

7.3 Identification of friction and adhesion

First, we identified the coefficient of friction from experimental results in Figure 21. Assuming that adhesion is proportional to area adhesion, we decided the ratio of adhesion according to ambient humidity from Figure 17 as follows:

$$\frac{A_{(dir)}(60\%)}{A_{(dir)}(50\%)} = \frac{F_{(dir)}(60\%)}{F_{(dir)}(50\%)} = 1.18, \frac{A_{(dir)}(70\%)}{A_{(dir)}(50\%)} = \frac{F_{(dir)}(70\%)}{F_{(dir)}(50\%)} = 1.47, \tag{30}$$

where, either symbol 'p' or 'n' is substituted into the subscript '(dir)' according to direction. Substituting m=0.3 mg and g = 9.8 m/s^2 into equations (28) and (29), we identified the coefficient of friction so as to fit equation (30). From Figure 23, the identification results when $\mu = 0.28$ corresponds with simulations, error between both results is 0.96 %.

Next, we considered the identification of adhesion. In equations (25) and (26), we assumed that:

$$m = n, \tag{31}$$

$$\sum_{i=1}^{n} A_{(dir)i}\sin\theta_{(dir)i} \equiv A_{(dir)0}\sin\theta_{(dir)0}, \tag{32}$$

$$\sum_{i=1}^{n} A_{(dir)i}\cos\theta_{(dir)i} \equiv A_{(dir)0}\cos\theta_{(dir)0}. \tag{33}$$

Substituting equations (31), (32) and (33) into equations (25) and (26), we have:

$$F_p = D_A(A_{p0}\sin\theta_{p0} + \mu A_{n0}\cos\theta_{n0}), \tag{34}$$

$$F_n = D_A(\mu A_{p0}\cos\theta_{p0} - A_{n0}\sin\theta_{n0}). \tag{35}$$

Then, the ratio of adhesion of positive and negative direction was formulated as:

$$\frac{F_p}{F_n} = \frac{A_{p0}\sin\theta_{p0} + \mu A_{n0}\cos\theta_{n0}}{-A_{n0}\sin\theta_{n0} + \mu A_{p0}\cos\theta_{p0}}. \tag{36}$$

Substituting the ratio of adhesion calculated from equations (28) and (29) into equation (36), we identified variables $A_{(dir)0}$ and $\theta_{(dir)0}$ (Table 2). Consequently, the coefficient of adhesion was almost constant while there was 4 % error at each ambient humidity condition. We finally decided $D_A = 3.72\times10^2\ \mu N / \mu m^2$ averaging them.
To assess the identified results, we compared experiments with calculation using the identified results. From Figure 24, identification results were in well agreement with experiments.

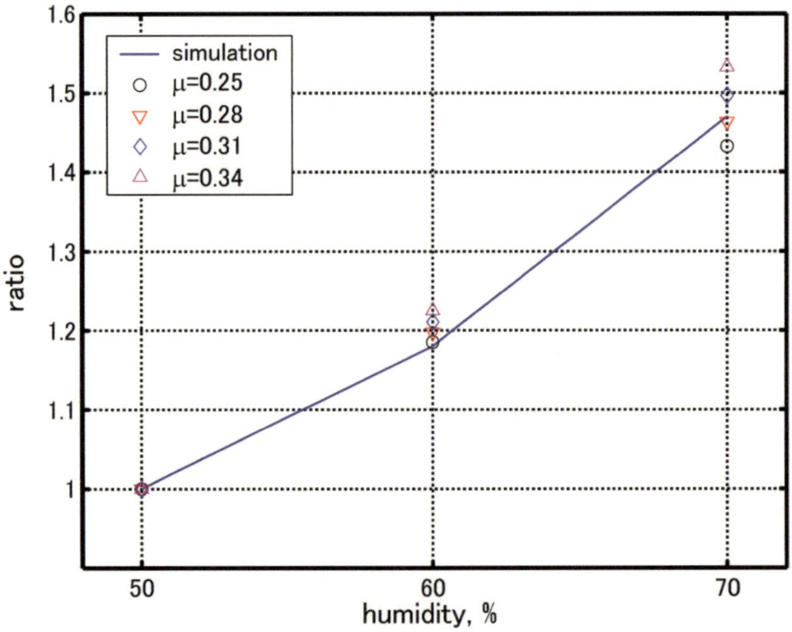

Fig. 23. Identification of coefficient of friction

7.4 Micropart dynamics including adhesion

When the feeder surface moves with sinusoidal vibration at an amplitude A_{vib} and an angular frequency ω (Figure 25), the inertia F_s transffered to a micropart is defined according to relative motion of the micropart and the feeder surface and its contact position as follows:

$$F_{vib} = -mAvib\omega^2\sin\omega t,$$
$$F_s = \begin{cases} F_{vib}\sin^2\theta & (\theta \neq 0) \\ 0 & (\theta = 0) \end{cases} \tag{37}$$

ambient humidity	50 %	60 %	70 %
x_c, μm	±0.913		
θ_{p0}, rad	0.102		
θ_{n0}, rad	−0.121		
A_{p0}, μm^2	$1.21e-2$	$1.42e-2$	$1.77e-2$
A_{n0}, μm^2	$1.12e-2$	$1.32e-2$	$1.65e-2$
D_A, $\mu N / \mu m^2$	$3.63e+2$	$3.80e+2$	$3.72e+2$

Table 2. Identification of adhesion

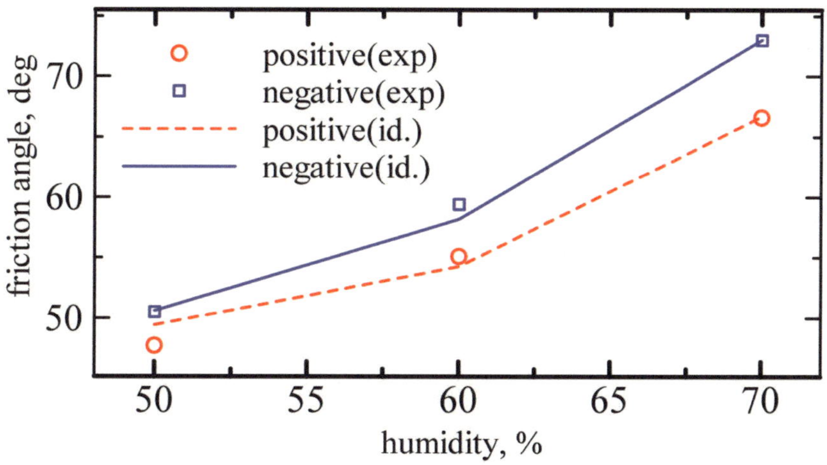

Fig. 24. Comparison of identfication and experiments

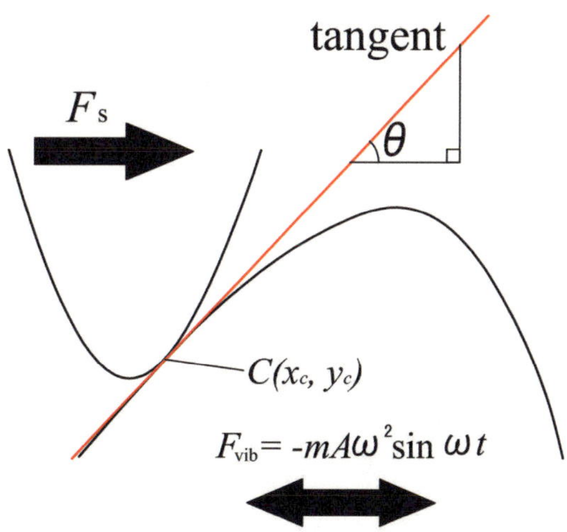

Fig. 25. Transferred force from feeder surface to micropart

Also, If x_p is micropart position, micropart dynamics is given by:

$$F_s = m\ddot{x}_p + c\dot{x}_p, \tag{38}$$

where, c is the coefficient of viscous attenuation, $\ddot{x}_p$ second order time differential, and $\dot{x}_p$ time differential.
Next we considered the effect of adhesion. Adhesion changes according to the relative motion of micropart on the feeder surface. If x is displacement of the feeder surface, velocity of the feeder surface is represented as:

$$\dot{x} = \frac{dx}{dt} = \omega A_{vib} \cos \omega t, \tag{39}$$

Then the micropart dynamics along the x axis can be expressed as:

$$m\ddot{x}_p + c\dot{x}_p = F_s - F_{(dir)}, \tag{40}$$

where,

$$F_{(dir)} = \begin{cases} F_p & (\dot{x}_p - \dot{x} > 0) \\ F_n & (\dot{x}_p - \dot{x} < 0) \end{cases}$$

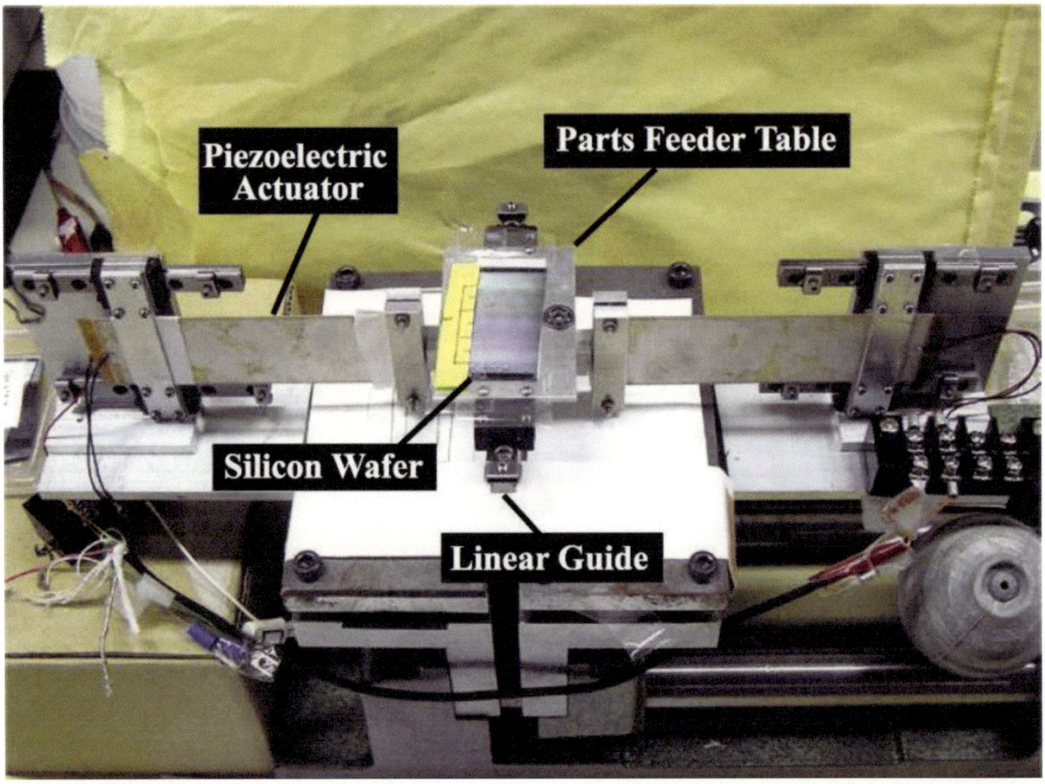

Fig. 26. Microparts feeder using bimorph piezoelectric actuators

8. Feeding experiments of micropart

8.1 Experimental equipment

In micropart feeder (Figure 26), a sawtoothed silicon wafer is placed at the top of the feeder table, which is driven back and forth in a track by a pair of piezoelectric bimorph elements, powered by a function generator and an amplifier that delivers peak-to-peak output voltage of up to 300 V.

8.2 Feeding experiments

Using this microparts feeder and sawtoothed silicon wafers mentioned in section 7.2, we conducted feeding experiments of microparts at a frequency of f=98 to 102 Hz with an interval of 0.2 Hz, and at an amplitude of A=0.5 mm under an ambient humidity of 60 % and a temperature of 24°C.

Each experimental result is the average of three trials using five microparts. Then the maximum feeding velocities of each feeder surface was recorded in Table 3.

When the pitch was 0.04 mm or less, the velocity was around 0.6 mm/s at a driving frequency f=98 to 100 Hz. The fastest feeding was 1.7 mm/s which was realized at a frequency f=101.4 Hz on p=0.05 mm surface. When the pitch was 0.06 mm or larger, the maximum velocities were around 1.0 mm/s at a frequency around f=101.4 Hz.

pitch, mm	velocity, mm/s	frequency, Hz
0.01	0.695	99.2
0.02	0.839	98.8
0.03	0.749	100.0
0.04	0.582	99.2
0.05	1.705	101.4
0.06	0.880	101.6
0.07	1.253	101.4
0.08	1.262	101.8
0.09	0.883	101.2
0.10	1.049	101.6

Table 3. Maximum feeding velocity on each feeder surface

8.3 Comparison of feeding simulation

Using equations (37) and (40), we simulated microparts feeding with the same conditions as experiments. In order to assess the effectiveness of adhesion, we conducted simulations when adhesion would be ignored. Experimental results and both simulation results were plotted simultaneously (Figure 27).

From this figure, both simulations were far from experimental results. These differences were caused by rotational motion around the axis along the sawtooth groove (Mitani, 2007).

9. Conclusion

We formulated feeding dynamics of microparts considering the effect of adhesion between sawtoothed silicon wafers and capacitors. Using a microscopy system, we obtained precise surface models of a micropart and sawtoothed silicon wafers. Contact between two surface models was analysed assuming that they shared a tangent at the contact point. Adhesion was then examined according to adhesion limit that both surfaces are near enough to adhere each other. Experiments of angle of friction of microparts were conducted in order to identify the coefficients of friction and adhesion. The feeding dynamics including the effect of adhesion were finally formulated.

Comparing simulation using the dynamics derived and experimental results, we found large differences between them because of rotation around the axis along to sawtooth groove.

In future studies, we will try to:

- Identify micropart dynamics including rotation, and
- Develop feeder surfaces with more precise profile.

This research was supported in part by a Grant-in-Aid for Young Scientists (B) (20760150) from the Ministry of Education, Culture, Sports, Science and Technology, Japan, and by a grant from the Electro-Mechanic Technology Advancing Foundation (**EMTAF**), Japan.

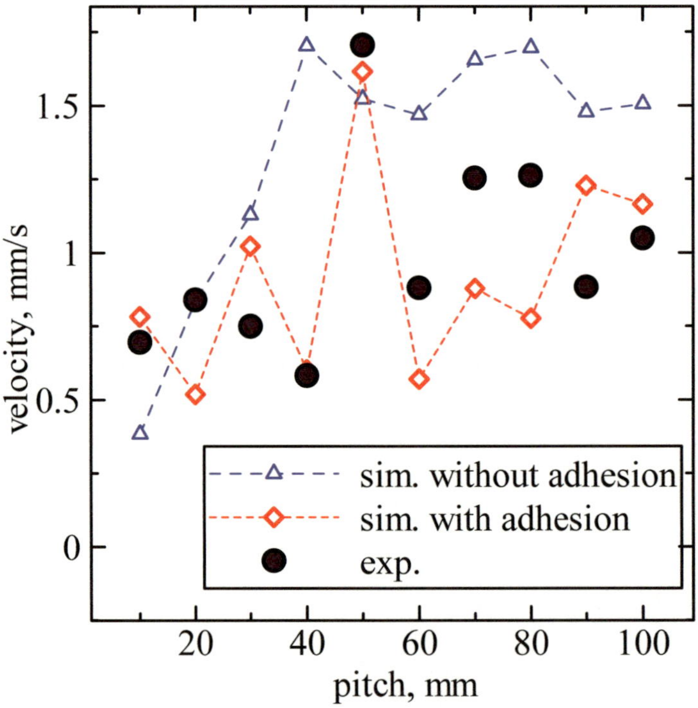

Fig. 27. Comparison of feeding experiments and simulations

10. References

Mitani, A., Sugano, N. & Hirai, S.(2006). Micro-parts Feeding by a Saw-tooth Surface, *IEEE/ASME Transactions on Mechatronics*, Vol. 11, No. 6, 671-681.

Ando, Y. & Ino, J. (1997). The effect of asperity array geometry on friction and pull-off force, *Transactions of the ASME Journal of Tribology*, Vol. 119, 781-787.

Maul, G. P. & Thomas, M. B. (1997). A systems model and simulation of the vibratory bowl feeder, *Journal of Manufacturing System*, Vol. 16, No. 5, 309-314.

Wolfsteiner, P. & Pfeiffer, F. (1999). The parts transportation in a vibratory feeder, *Procs. IUTAM Symposium on Unilateral Multibody Contacts*, 309-318.

Reznik, D. & Canny, J. (2001). C'mon part, do the local motion!, *Procs. 2001 International Conference on Robotics and Automation*, Vol. 3, 2235-2242.

Berkowitz, D.R. & Canny, J. (1997), A comparison of real and simulated designs for vibratory parts feeding, *Procs. 1997 IEEE International Conference on Robotics and Automation*, Vol. 3, 2377-2382.

Christiansen, A. & Edwards, A. & Coello, C. (1996). Automated design of parts feeders using a genetic algorithm, *Procs. 1996 IEEE International Conference on Robotics and Automation*, Vol. 1, 846-851

Doi, T, (2001), Feedback control for electromagnetic vibration feeder (Applications of two-degrees-of-freedom proportional plus integral plus derivative controller with nonlinear element), *JSME International Journal*, Series C, Vol. 44, No. 1, 44-52.

Konishi, S. (1997). Analysis of non-linear resonance phenomenon for vibratory feeder,
Procs. APVC '97, 854-859.

Fukuta, Y. (2004). Conveyor for pneumatic two-dimensional manipulation realized by arrayed MEMS and its control, *Journal of Robotics and Mechatronics*, Vol. 16, No. 2, 163-170.

Arai, M (2002). An air-flow actuator array realized by bulk micromachining technique, *Procs. IEEJ the 19th Sensor Symposium*, 447-450.

Ebefors, T. (2000), A robust micro conveyer realized by arrayed polyimide joint actuators, *Journal of Micromechanics and Microengineering*, Vol. 10, 337-349.

Böhringer, K.-F. (2003). Surface modification and modulation in microstructures: controlling protein adsorption, monolayer desorption and micro-self-assembly, *Journal of Micromechanics and microengineering*, Vol. 13, S1-S10.

Oyobe, H. & Hori, Y. (2001). Object conveyance system "Magic Carpet" consisting of 64 linear actuators-object position feedback control with object position estimation, *Procs. 2001 IEEE/ASME International Conference on Advanced Intelligent Mechatronics*, Vol. 2, 1307-1312.

Fuhr, G. (1999), Linear motion of dielectric particles and living cells in microfabricated structures induced by traveling electric fields, *Procs. 1999 IEEE Micro Electro Mechanical Systems*, 259-264.

Komori, M. & Tachihara, T. (2005). A magnetically driven linear microactuator with new driving method, *IEEE/ASME Transactions on Mechatronics*, Vol. 10, No. 3, 335-338.

Ting, Y. (2005), A new type of parts feeder driven by bimorph piezo actuator, *Ultrasonics*, Vol. 43, 566-573.

Codourey, A. (1995). A robot system for automated handling in micro-world, *Procs. 1995 IEEE/RSJ International Conference on Intelligent Robots and Systems*, Vol. 3, 185-190.

Mitani, A. & Hirai, S. (2007) Feeding of Submillimeter-sized Microparts along a Saw-tooth Surface Using Only Horizontal Vibration: Analysis of Convexities on the Surface of Microparts, Procs. IEEE 2007 3rd Conference on Automation Science and Engineering (CASE2007), Scottsdale, AZ, USA, Sep. 22-25, 2007.

19

Vibration Analysis of a Moving Probe with Long Cable for Defect Detection of Helical Tubes

Takumi Inoue and Atsuo Sueoka
Department of Mechanical Engineering, Kyushu University
Japan

1. Introduction

A defect detection of a heating tube installed in a power station is a very important process for avoidance of a serious disaster. The defect detection for the fast breeder reactor "Monju" in Japan is implemented by feeding an eddy current testing (ECT) probe (Isobe et al., 1995; Robinson, 1998) with a magnetic sensor, into the tube. The ECT probe (hereafter, simply called probe) is controlled so as to move in the heating tube at a constant velocity. A peculiar feature of the heating tubes in "Monju" is that each tube is mostly helical. An undesirable vibration of the probe always happened in the helical heating tube under a certain condition (Inoue et al., 2007). The vibration was considerably large and generated an obstructive noise in the signal of the magnetic sensor. It made the detection of defects difficult. Some papers reported similar problems (Bihan, 2002; Giguere et al., 2001; Tian and Sophian, 2005), but a large vibration of the probe was not involved. A key to the problem is that the noise in the signal was accompanied with the hard vibration. Several characteristics of the vibration became clear through some experiments by using a mock-up, and a countermeasure was taken by making use of the characteristics of the vibration (Inoue et al., 2007). However, an essential factor on the cause of the vibration was still unclear. Since the noise in the signal is highly correlated with the vibration, a thorough investigation of the vibration is needed. It is desirable to find out the cause of the vibration in order to remove or reduce the vibration and ensure the reliability of the inspection.

In this study, the cause of the vibration is assumed to be Coulomb friction between floats, which are attached to the probe, and the inner wall of the heating tube on the basis of the experimental results. An analytical model is obtained by taking Coulomb friction into account and numerical simulation is implemented by applying a step-by-step time integration scheme. However, the analytical model has a very large number of degree of freedom. Furthermore, there are many points on which Coulomb friction acts when the probe is fed into the tube under air pressure since many floats, which are in contact with the inner wall of the heating tube, are attached to the probe. It implies that a lot of strong non-linearities exist in the analytical model. There is no precedent for this kind of problem, and heavy computational costs are ordinarily required to carry out the numerical simulation.

Sueoka et al. (1985) presented the Transfer Influence Coefficient Method (Inoue et al., 1997; Kondou et al., 1989, hereafter: TICM), which is a computational method for a dynamic response of a structure and has advantages in computational accuracy and speed. The TICM is especially good at a longitudinally extended structure, such as a pipeline system and rotational machinery of a large plant. The advantages of the TICM are outstanding in an application to such structures. The probe can be regarded as a long cable, so that it exactly coincides with the structure suitable for the TICM. The TICM is applicable to various fields of the dynamic response, that is, free vibration analysis, forced vibration analysis, and time historical response analysis. The numerical simulation of the probe is efficiently implemented by applying the time historical response analysis of the TICM. The results of the numerical simulation qualitatively agree well with the experimental results. It confirms the validity of the assumption that the vibration is caused by Coulomb friction. In other words, the numerical simulation is regarded as an available tool to estimate a vibration of some modified probes. Based on this study, some improvements of probe sufficiently suppress the vibration, and a reliable inspection of helical tubes is realized.

2. The mock-up experimental equipment and analytical model of the probe

A mock-up experimental equipment is shown in Fig. 1. For the most part, the heating tube is helical. Six heating tubes with different helical diameters are mounted in the mock-up. The probe consists of a remote field (RF) sensor, cable and floats as shown in Fig. 2. The floats are attached to the cable at equal spaces. The probe is fed into the heating tube from the upper side of the steam generator. The RF sensor inspects the attenuation of the wall thickness of the heating tube by detecting the change of eddy current. The cable of the forward section from the sensor is called the guide cable and the aft section is called the carrier cable. A drag force which acts on the floats by means of dry compressed air flow is the driving force of the probe. The directions of the air flow and the movement of the probe are the same, that is, the direction of the air flow in the insertion process is opposite to the

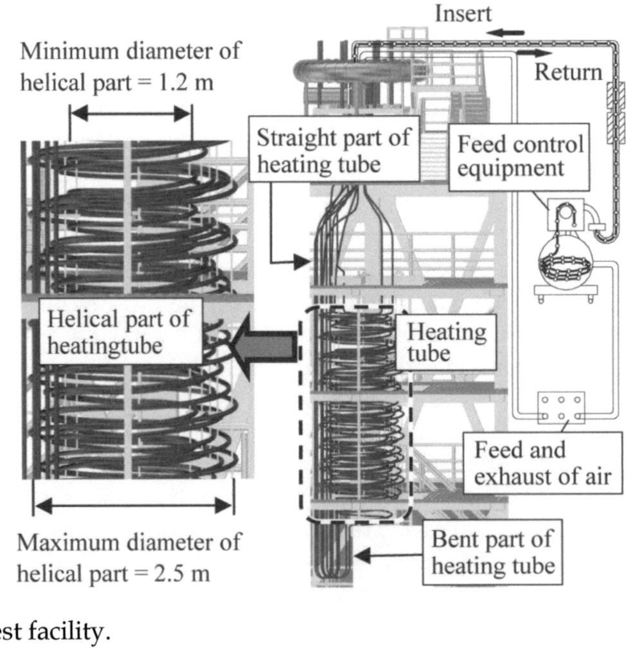

Fig. 1. Mock-up test facility.

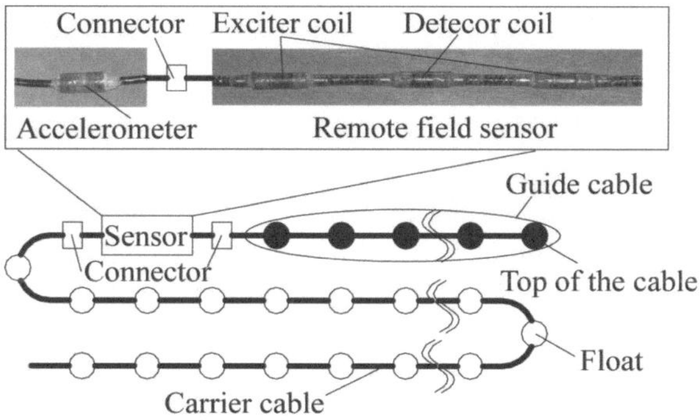

Fig. 2. ECT probe and accelerometer.

air flow of the return process. The probe passes through the heating tube very quickly unless the feed control equipment, which is shown in Fig. 1, regulates the feeding speed. An axial force of which direction is opposite to the moving direction acts on the probe from the feed control equipment. Thus, a tensile force acts on the probe in the insertion process on the average, while a compressive force acts on the probe in the return process. The detection of defects can be operated both in the insertion and the return processes, and inspections in both processes are desirable in order to ensure the reliability of the inspection.

2.1 Summary of the experimental results

Experimental results by using the mock-up (Inoue et al., 2007) are summarized as follows.

a. During the inspection, the RF sensor transmits two signals X and Y, which are output voltage from the detector coil. Their directions are perpendicular to each other, and also perpendicular to the axial direction of the helical tube as shown in Fig. 3(a). Usually, the directions of X and Y do not correspond to the normal and the binormal ones of the helical tube. Fig. 3(b) shows RF signal at the carrier velocity of 200 mm/s when the sensor part passes through the sensitivity test piece. Signals X and Y generate fluctuations in opposite directions at the same time, but the amplitudes are different from each other. In Fig. 3(c), the Lissajous' figures for signals X and Y are illustrated.

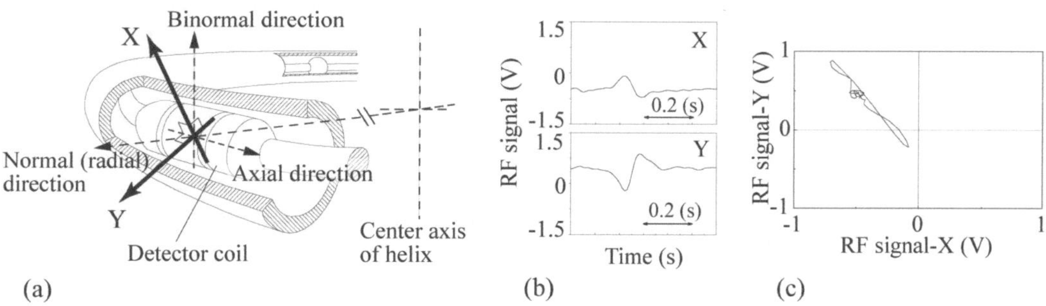

Fig. 3. (a) Two RF signals X and Y, (b) RF signals at the test piece and (c) its Lissajous' figure.

b. The total length of the heating tube is about 90 m. The length of the helical part is about 60m (see Fig. 1). RF signals of X, Y and accelerations nearby the sensor in the insertion process are shown in Fig. 4(a and b), respectively. The sensor passed the helical part of the heating tube in the shaded area of Fig. 4(a and b) and an approximate length of the probe inserted into the helical part is also indicated. Large impulsive signals at positions A and B shown in Fig. 4(a) were caused by metallic flanges to connect the both ends of the acrylic fluoroscopy tube. The acrylic fluoroscopy tube can be set up at either position A or B in order to observe the movement of probe by high-speed camera. Although the impulsive signals are large noises on the RF signals, we ignore them because the actual heating tubes are not equipped with the acrylic fluoroscopy tube and metallic flanges. On one hand, the small impulsive signals in the RF signals like short beards in the region of the helical tube occurring at equal intervals. These signals are generated as the sensor part passes through the metallic outer support of the heating tube. The small impulsive signal is called "support signal". Although the support signal is a kind of noise on the RF signals, the discrimination between the attenuation and the support signal is not discussed in this study, because the actual metallic outer supports are different from the ones of the mock-up. We focus on the relationship between the vibration and the RF signal noise.

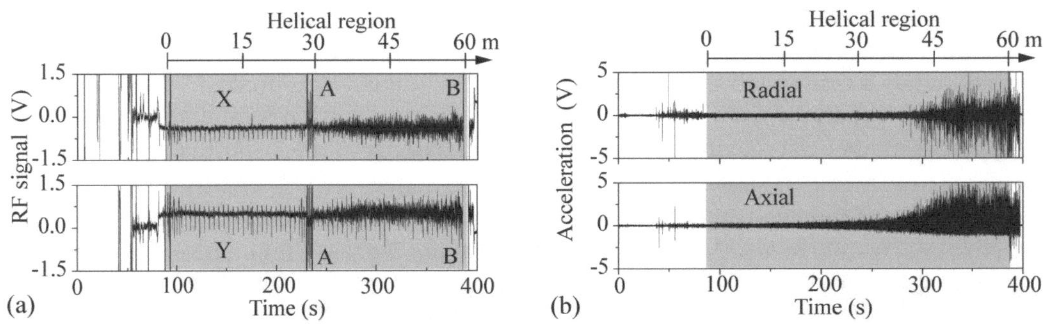

Fig. 4. (a and b) RF signal and acceleration in insertion process.

c. The accelerations shown in Fig. 4(b) were measured by an accelerometer, which was specially arranged for the experiment, located nearby the sensor as shown in Fig. 2. The directions of the acceleration were lateral and longitudinal of the probe and correspond to the radial and axial directions of the helical heating tube. From Fig. 4(b), the vibration of the probe rapidly increased after the sensor passed through the middle position of the helical part. At the same time, the noises were raised in the RF signals and kept a large value until the insertion process finished. It means that there was adequate correlation between the probe vibration and RF signal noise. In addition, we confirmed that a noticeable peak in the frequency analysis (about 20 Hz) appeared in both the axial and the radial vibrations of the probe. Both vibrations were weakly coupled and the probe showed an inchworm-like motion.
d. In the case of non-feeding, no vibration of the probe occurred even if the dry compressed air streamed into the heating tube. No RF signal noise was also appeared. It was expected that the vibration of the probe was mainly caused by a frictional force between the floats and the inner wall of the heating tube, and the fluid force was not an essential factor of the vibration.
e. The vibration of the probe in the return process was smaller than the one in the insertion process. There was no noticeable peak in the frequency analysis of the vibration in the return process.
f. The vibration of the probe became small in the case of low feeding speed, large helical diameter and low supply rate of the air flow.

g. It was found that the RF signal noise highly correlated with radial vibration of the probe. A long guide cable made the RF signal noise small because it was effective in suppressing the radial vibration. In addition, a large size of float attached to the guide cable was also effective in suppressing the vibration.

In this study, only the vibration of the probe is focused on because there was a certain correlation between the probe vibration and RF signal noise. The inspection of the attenuation of the wall thickness is operated in both the insertion and the return processes in order to perform a firm inspection. In this study, the vibration of the insertion process is focused on since it is larger than the one of the return process as mentioned above e.

2.2 Analytical model of probe

The analytical model is obtained under the following simplifications so that the numerical analysis can be implemented as easily as possible.

a. The heating tubes consist of straight, helical and bending parts as shown in Fig. 1. The vibration of the probe always occurred in the helical part, and it did not occur in the other parts of the heating tubes. Therefore, only the helical part of the heating tube is considered.

b. The length of the actual probe becomes longer as the insertion process goes on. However, it is difficult to treat a probe with time varying length. On one hand, if a vibrating probe, which is sufficiently inserted in the helical tube, stops feeding and restarts, the vibration of the probe is always reproduced. It follows that a probe with a constant length can be regarded as a momentary situation in which the actual time varying length of probe just reached the length. Hence, many probes with constant length (each length is different from one another) can be substitutes for the actual probe with time varying length. In this paper, the length of the probe is assumed to be constant and many probes with constant length are treated in order to cope with the actual probe with time varying length.

c. Contact points between the floats and the inner wall of the heating tube are always generated at the inside of the helical tube as shown in Fig. 5, because tensile force acts on the probe in the insertion process.

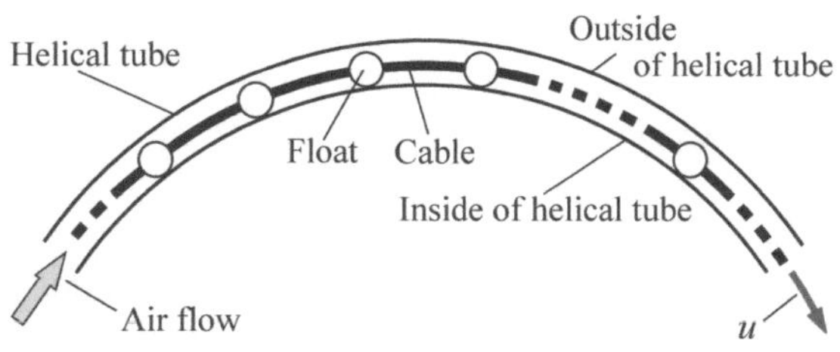

Fig. 5. Analytical model of probe in helical tube.

d. The vertical motion of the probe is disregarded. The motion of the probe is restricted within the horizontal plane. Thus, the probe moves in a circular tube placed in the horizontal plane as shown in Fig. 6.

e. The movement of the probe is modeled as illustrated in Fig. 5. The probe moves in the heating tube at a constant speed u from the left-hand side to the right-hand side of Fig. 5. The dry compressed air also flows inside the tube in the same direction of the movement of probe. Secondary flow around the floats and cable is neglected.

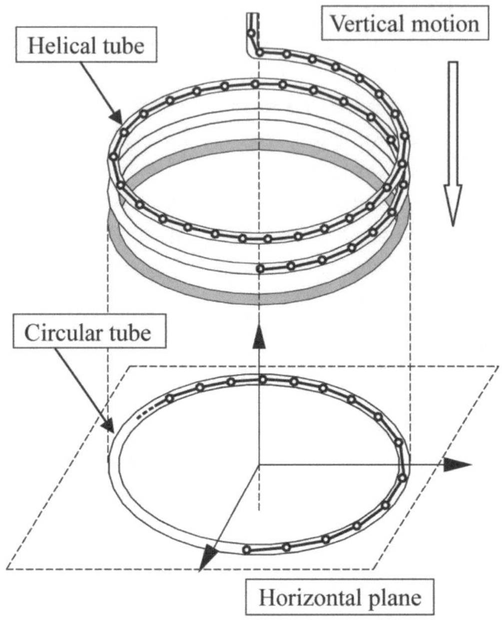

Fig. 6. Actual and analytical heating tube.

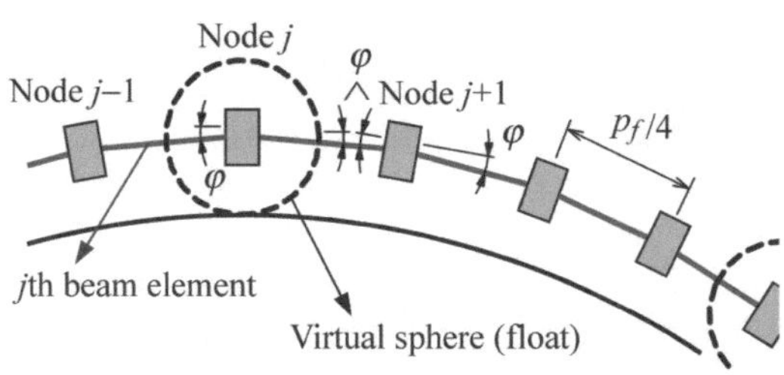

Fig. 7. Lumped mass modeling.

Based on the simplifications, the probe is modeled as a lumped mass system as shown in Fig. 7. The cable is equally divided, and rigid bodies which possess mass and moment of inertia, are put to each divided point. Each section spaced by floats is divided into four by taking a balance between the float pitch p_f and diameter of the cable d_c into consideration. The analytical model is formed by a connection of the rigid bodies and massless beams in series as shown in Fig. 7. The probe can be regarded as almost uniform because it was made by a continuous cable and lightweight spherical floats which are attached to the cable. Thus, the mass and moment of inertia of each rigid body are assumed to be identical and given as follows:

$$m = \rho_c \frac{p_f}{4}, \quad J = m\left[\frac{1}{12}\left(\frac{p_f}{4}\right)^2 + \frac{d_c^2}{16}\right] \tag{1}$$

where ρ_c is mass per unit length of probe, including the mass of the cable and floats. The moment of inertia J was obtained as a rigid column with diameter d_c and height $p_f/4$. Virtual spheres are assumed to be around the rigid bodies which occupy the place where the floats originally existed. The diameter of the virtual spheres is equal to one of the floats and is common to all spheres. The spheres fill the role of the floats, which are subjected to the drag force of air flow and are in contact with the inner wall of the heating tube. Contact forces and frictional forces from the inner wall of the heating tube also act on the virtual sphere. The forces are transmitted to the rigid bodies through the virtual sphere. The mass and the moment of inertia of the RF sensor are also assumed to be m and J without a special treatment.

Each rigid body is called "Node" and the left- and the righ-thand ends of the system are defined as node 0 and node n, respectively. The beam element between the node j and $j-1$ is called jth beam element. Each of the beam elements is assumed to be straight and slantingly connects with rigid bodies at both ends as shown in Fig. 7. The slant connection is due to the curvature of the helical heating tube and the slanting angle φ is given as:

$$\varphi = \sin^{-1}[p_f / (4d_h)] \tag{2}$$

where d_h is a diameter of the helix.

2.3 Equation of motion

In this paper, variables with head symbol and subscripts have following principles:

a. Variables with subscript j represent the physical quantities related to node j or the jth beam element.
b. Variables with and without head symbol "–" represent the physical quantities on the left- and the right-hand side of node, respectively.

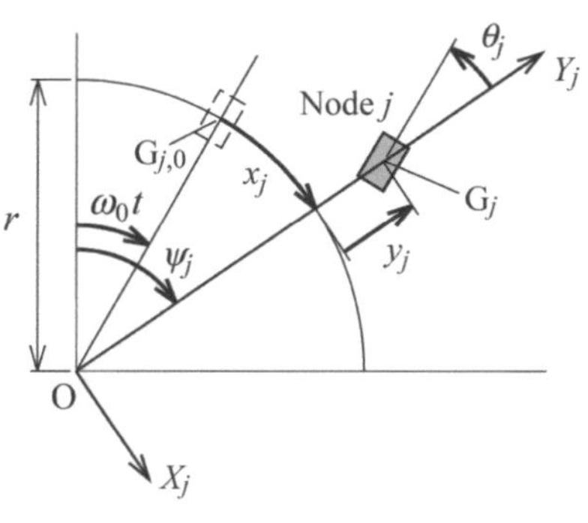

Fig. 8. Polar coordinate.

Since the probe goes into the helical (circular, under the assumption d of Section 2.2) tube at a constant speed, the motion of the rigid body at node j is represented in a polar coordinate O–X_jY_j as shown in Fig. 8. The point O in Fig. 8 corresponds to the center of the helix (or circle) and the X_j-axis points toward a center of gravity of the rigid body G_j. Supposing that a center of gravity of the rigid body without stretch and lateral motion of the probe is denoted $G_{j,0}$, the point of $G_{j,0}$ turns around the center O at a constant angular velocity ω_0 which is given as:

$$\omega_0 = u / r, \quad r = d_h / 2 \tag{3}$$

where r is the radius of the helix. The relative movement of the rigid body at node j with respect to the unstretched probe is represented as an axial displacement $x_j(t)$ (arc coordinate along the helix) from $G_{j,0}$, a radial displacement $y_j(t)$ from $G_{j,0}$ and a rotation $\theta_j(t)$ around G_j as shown in Fig. 8. Physical quantities such as $x_j(t)$, $y_j(t)$, $\theta_j(t)$, etc., are functions of time t in principle but the argument (t) is omitted in Fig. 8. The root of the probe, node 0, is assumed to move at the regulation speed u without lateral motion and rotation around $G_0 = G_{0,0}$. So, physical quantities on node 0 are fixed as: $x_0(t) = y_0(t) = \theta_0(t) = 0$.

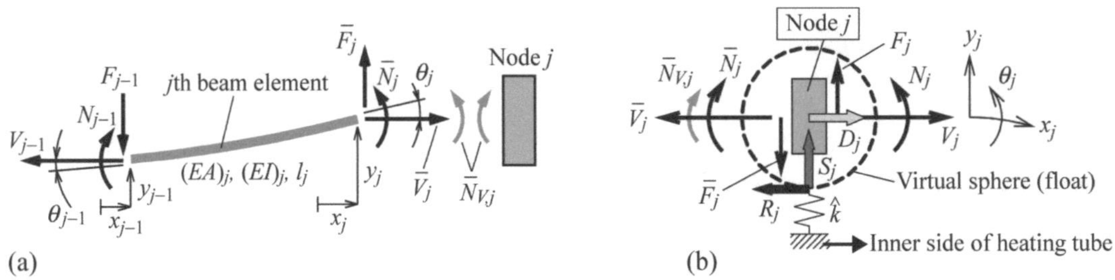

Fig. 9. State variables and external forces at (a) both ends of the jth beam element and (b) node j.

State variables at both ends of the jth beam element and the positive direction of the variables are shown in Fig. 9(a), where $\bar{V}_j(t)$, $V_{j-1}(t)$ are axial forces, $\bar{F}_j(t)$, $F_{j-1}(t)$ are shearing forces and $\bar{N}_j(t)$, $N_{j-1}(t)$ are moment of forces. Displacement vectors and force vectors are defined as follows:

$$\begin{aligned} \boldsymbol{d}_{j-1}(t) &= \{x(t), y(t), \theta(t)\}_{j-1}^{\mathrm{T}}, \quad \boldsymbol{d}_j(t) = \{x(t), y(t), \theta(t)\}_j^{\mathrm{T}} \\ \boldsymbol{f}_{j-1}(t) &= \{V(t), F(t), N(t)\}_{j-1}^{\mathrm{T}}, \quad \bar{\boldsymbol{f}}_j(t) = \{\bar{V}(t), \bar{F}(t), \bar{N}(t)\}_j^{\mathrm{T}} \end{aligned} \tag{4}$$

where superscript "T" denotes transpose. Argument (t) is omitted in Fig. 9 as same as in Fig. 8. The displacements $x_j(t)$, $y_j(t)$ and $x_{j-1}(t)$, $y_{j-1}(t)$ should be represented in a polar coordinate as shown in Fig. 8 because the jth beam element connects with the rigid bodies of node j and $j-1$. However, inertia forces of polar coordinate do not act on the jth beam element because the beam elements are assumed to be massless. In addition, the curvature of $x_j(t)$ and $x_{j-1}(t)$ are negligible since $x_j(t)$ and $x_{j-1}(t)$ are sufficiently smaller than the diameter of the helix d_h. Hence, the displacements are approximately treated in a local orthogonal coordinate, and the relationships of the state variables are represented in following simple forms.

$$\boldsymbol{d}_j(t) = \mathbf{L}_j^{\mathrm{T}} \boldsymbol{d}_{j-1}(t) + \mathbf{F}_j \bar{\boldsymbol{f}}_j(t) \tag{5}$$

$$\boldsymbol{f}_{j-1}(t) = \mathbf{L}_j \bar{\boldsymbol{f}}_j(t) \tag{6}$$

where

$$\mathbf{L}_j = \begin{bmatrix} 1 & 0 & 0 \\ 0 & 1 & 0 \\ 0 & l_j & 1 \end{bmatrix}, \quad \mathbf{F}_j = \begin{bmatrix} a_x & 0 & 0 \\ 0 & a_y & \gamma \\ 0 & \gamma & \beta \end{bmatrix}_j, \quad (a_x, a_y, \beta, \gamma)_j = \left(\frac{l}{EA}, \frac{l^3}{3EI}, \frac{l}{EI}, \frac{l^2}{2EI} \right)_j$$

where l_j, $(EA)_j$ and $(EI)_j$ are length, tensile rigidity and flexural rigidity of the jth beam element. Eqs. (5) and (6) are arranged as follows:

$$\begin{bmatrix} \boldsymbol{f}_{j-1}(t) \\ \overline{\boldsymbol{f}}_j(t) \end{bmatrix} = \begin{bmatrix} -\mathbf{L}_j\boldsymbol{F}_j^{-1}\mathbf{L}_j^{\mathrm{T}} & \mathbf{L}_j\boldsymbol{F}_j^{-1} \\ -\boldsymbol{F}_j^{-1}\mathbf{L}_j^{\mathrm{T}} & \boldsymbol{F}_j^{-1} \end{bmatrix} \begin{bmatrix} \boldsymbol{d}_{j-1}(t) \\ \boldsymbol{d}_j(t) \end{bmatrix} \tag{7a}$$

The coefficient matrix of the right-hand side of Eq. (7a) represents a stiffness matrix of jth beam element, and it also represents the rigidity of the cable. In addition, since the cable shows a considerable damping property, we introduce a damping term associated with the velocity vectors $\dot{\boldsymbol{d}}_j(t)$ and $\dot{\boldsymbol{d}}_{j-1}(t)$. In this study, a proportional viscous damping (proportional coefficient δ) is adopted. Thus, the relationship of the state variables at both ends of the jth beam element is represented as follows:

$$\begin{bmatrix} \boldsymbol{f}_{j-1}(t) \\ \overline{\boldsymbol{f}}_j(t) \end{bmatrix} = \begin{bmatrix} -\mathbf{L}_j\boldsymbol{F}_j^{-1}\mathbf{L}_j^{\mathrm{T}} & \mathbf{L}_j\boldsymbol{F}_j^{-1} \\ -\boldsymbol{F}_j^{-1}\mathbf{L}_j^{\mathrm{T}} & \boldsymbol{F}_j^{-1} \end{bmatrix} \begin{bmatrix} \boldsymbol{d}_{j-1}(t) \\ \boldsymbol{d}_j(t) \end{bmatrix} + \delta \begin{bmatrix} -\mathbf{L}_j\boldsymbol{F}_j^{-1}\mathbf{L}_j^{\mathrm{T}} & \mathbf{L}_j\boldsymbol{F}_j^{-1} \\ -\boldsymbol{F}_j^{-1}\mathbf{L}_j^{\mathrm{T}} & \boldsymbol{F}_j^{-1} \end{bmatrix} \begin{bmatrix} \dot{\boldsymbol{d}}_{j-1}(t) \\ \dot{\boldsymbol{d}}_j(t) \end{bmatrix} \tag{7b}$$

The second term of the right-hand side of Eq. (7b) corresponds to a proportional viscous damping.
A following couple (moment of force) due to the axial force is generated.

$$N_{V,j}(t) = \overline{V}_j(t)[y_j(t) - y_{j-1}(t)] = V_{j-1}(t)[y_j(t) - y_{j-1}(t)] \tag{8}$$

However, this couple is not considered in Eq. (7b). Since the axial force generated in the cable is very large, the effect of this moment of force is not negligible. It is substituted by a moment of force $\overline{N}_{V,j}(t)$ which equilibrates $N_{V,j}(t)$ and acts on the jth beam element from node j as shown in Fig. 9(a). A reaction of $\overline{N}_{V,j}(t)$ also acts on the node j. Assuming that the elastic deformation of the beam element is small, the following approximation is available:

$$\overline{N}_{V,j}(t) = \overline{V}_j(t)[y_j(t) - y_{j-1}(t)] \cong \overline{V}_j(t) l_j \theta_j(t) \tag{9}$$

State variables at node j are shown in Fig. 9(b), where the node is regarded as a float so that a virtual sphere and forces acting on the sphere are also depicted in Fig. 9(b). Since node j possesses mass and moment of inertia, inertia forces of polar coordinate, such as a centrifugal force and Coriolis force, act on node j. Hence, displacements $x_j(t)$, $y_j(t)$ and rotational angle $\theta_j(t)$ are treated in the polar coordinate as shown in Fig. 8.
Drag force $D_j(\dot{x}_j)$ acting on the float by air flow, which is the driving force of the probe, is represented as follows:

$$D_j(\dot{x}_j) = (1/2) c_D \rho_f (A_f - A_c) v_{rel}(\dot{x}_j) |v_{rel}(\dot{x}_j)| \tag{10}$$

where $v_{rel}(\dot{x}_j) = Q/(a - A_c) - u - \dot{x}_j(t)$, $c_D = 0.44$ is the coefficient of drag force, ρ_f is the air flow density, $A_f = \pi d_f^2/4$ is the cross-section of float, d_f is the diameter of float, $A_c = \pi d_c^2/4$ is the cross-section of cable, Q is the supply rate of air flow, $a = \pi d_{in}^2/4$ is the cross-section of heating tube and d_{in} is the inner diameter of heating tube.
The loss of the pressure head is not considered here.
A contact spring, of which spring constant is $\hat{k}$, is assumed to exist between the floats and the inner wall of the heating tube as shown in Fig. 9(b). The float contacts with the inner wall through the contact spring when the radial displacement $y_j(t) < 0$, and a reaction force $S_j(y_j)$ acts on the float from the inner wall of the heating tube. Positive direction of $S_j(y_j)$ is defined as shown in Fig. 9(b) and hence $S_j(y_j)$ is represented as:

$$S_j(y_j) = -k_j(y_j)y_j(t), \quad k_j(y_j) = \begin{cases} 0 & : y_j(t) \geq 0 \\ \hat{k} & : y_j(t) < 0 \end{cases} \tag{11}$$

where the spring constant $\hat{k}$ is assumed to be common to all floats. The Coulomb friction $R_j(y_j, \dot{x}_j, \theta_j)$, which is proportional to $S_j(y_j)$, also acts on the float. Since the direction of Coulomb friction is determined by the relative velocity at the contact point between the float and the inner wall, Coulomb friction is represented as follows:

$$\begin{aligned} R_j(y_j, \dot{x}_j, \dot{\theta}_j) &= \mu_j(\dot{x}_j, \dot{\theta}_j)S_j(y_j) = -\mu_j(\dot{x}_j, \dot{\theta}_j)k_j(y_j)y_j(t) \\ \mu_j(\dot{x}_j, \dot{\theta}_j) &= \begin{cases} \mu_d & : u + \dot{x}_j(t) + \dot{\theta}_j(t)d_f/2 \geq 0 \\ -\mu_d & : u + \dot{x}_j(t) + \dot{\theta}_j(t)d_f/2 < 0 \end{cases} \end{aligned} \tag{12}$$

where the coefficient of kinetic friction μ_d is also assumed to be common to all floats. In addition, node j is subjected to axial forces $\bar{V}_j(t), V_j(t)$, shearing forces $\bar{F}_j(t), F_j(t)$ and moment of forces $\bar{N}_j(t), N_j(t)$ from the beam elements connecting to the node j as shown in Fig. 9(b). Furthermore, the moment of force $\bar{N}_{V,j}(t)$ due to the axial force, which is treated as a time varying rotational spring [see Eq. (9)], also acts on the node j. Since the movement of the rigid body is represented in a polar coordinate (Fig. 8), accelerations of the rigid body at node j are represented as:

$$X_j\text{-direction}: r_j'\ddot{\psi}_j + 2\dot{r}_j'\dot{\psi}_j, \quad Y_j\text{-direction}: \ddot{r}_j' - r_j'\dot{\psi}_j^2, \quad \text{rotation}: \ddot{\theta}_j - \ddot{\psi}_j \tag{13}$$

where $r_j' = r + y_j(t)$, $\psi_j = \omega_0 t + x_j(t)/r$, $\dot{\psi}_j = \omega_j(\dot{x}_j) = \omega_0 + \dot{x}_j(t)/r$, $\ddot{\psi}_j = \dot{\omega}_j(\dot{x}_j) = \ddot{x}_j(t)/r$. Thus, the equation of motion of the rigid body at node j is expressed as follows with respect to X_j-, Y_j-directions and rotation.

$$\begin{aligned} & mr\left(1 + \frac{y_j(t)}{r}\right)\frac{\ddot{x}_j(t)}{r} + 2m\omega_j(\dot{x}_j)\dot{y}_j(t) = -R_j(y_j, \dot{x}_j, \dot{\theta}_j) + V_j(t) - \bar{V}_j(t) \\ & m\ddot{y}_j(t) - mr\left(1 + \frac{y_j(t)}{r}\right)\omega_j^2(\dot{x}_j) = S_j(y_j) + F_j(t) - \bar{F}_j(t) \\ & J_j\left(\ddot{\theta}_j - \frac{\ddot{x}_j(t)}{r}\right) = -R_j(y_j, \dot{x}_j, \dot{\theta}_j)\frac{d_f}{2} - \bar{N}_{V,j}(t) \end{aligned} \tag{14}$$

Eq. (14) is arranged as follows with an approximation: $1 + y_j(t)/r \cong 1$.

$$\boldsymbol{M}\ddot{\boldsymbol{d}}_j(t) + \boldsymbol{C}_j(\dot{x}_j)\dot{\boldsymbol{d}}_j(t) + \boldsymbol{K}_j(y_j, \dot{x}_j, \dot{\theta}_j, \bar{V}_j)\boldsymbol{d}_j(t) = \tilde{\boldsymbol{q}}_j(\dot{x}_j) + \boldsymbol{f}_j(t) - \bar{\boldsymbol{f}}_j(t) \tag{15}$$

where

$$\boldsymbol{M} = \begin{bmatrix} m & 0 & 0 \\ 0 & m & 0 \\ -J/r & 0 & J \end{bmatrix}, \quad \boldsymbol{C}_j(\dot{x}_j) = \begin{bmatrix} 0 & 2m\omega_j(\dot{x}_j) & 0 \\ 0 & 0 & 0 \\ 0 & 0 & 0 \end{bmatrix}$$

$$\boldsymbol{K}_j(y_j, \dot{x}_j, \dot{\theta}_j, \bar{V}_j) = \begin{bmatrix} 0 & -\mu_j(\dot{x}_j, \dot{\theta}_j)k_j(y_j) & 0 \\ 0 & k_j(y_j) & 0 \\ 0 & -\mu_j(\dot{x}_j, \dot{\theta}_j)k_j(y_j)d_f/2 & \bar{V}_j(t)l_j \end{bmatrix}, \quad \tilde{\boldsymbol{q}}_j(\dot{x}_j) = \{D_j(\dot{x}_j), mr\omega_j^2(\dot{x}_j), 0\}^{\mathrm{T}}$$

The matrix $C_j(\dot{x}_j)$ does not represent a damping matrix here, but a kind of Coriolis force as a term of $C_j(\dot{x}_j)\dot{d}_j(t)$. As for a damping, a proportional viscous damping, the second term of the right-hand side of Eq. (7b), is considered. An element $mr\omega_j^2(\dot{x}_j)$ of the vector $\tilde{q}_j(\dot{x}_j)$ represents a centrifugal force in a polar coordinate.
If the rigid body of node j does not correspond to a float, it does not contact with the inner wall of the heating tube and is free from the drag force $D_j(\dot{x}_j)$. The equation of motion of such a rigid body is similarly represented as Eq. (15) by setting $D_j(\dot{x}_j)$, $\hat{k}$ and μ_d to be zero.

3. Step-by-step time integration scheme by utilizing the Transfer Influence Coefficient Method

3.1 Step-by-step time integration scheme

Time historical response of the probe can be computed by applying a step-by-step time integration scheme to the equation of motion obtained in Section 2. Displacement $d_j(t)$, velocity $\dot{d}_j(t)$ and acceleration $\ddot{d}_j(t)$ vectors are successively computed at an interval of time step size Δt. For a large scale structure, the Newmark-β method and the Wilson-θ method (Belytschko and Hughes, 1983) are usually employed as the step-by-step time integration scheme. Velocity and acceleration vectors at time $t_i = i\Delta t$ are usually expressed as a linear function of displacement vector in an implicit method such as the Newmark-β and the Wilson-θ methods. The formulation is given as follows:

$$\ddot{d}_j(t_i)=B_a d_j(t_i)+h_{a,j}(t_{i-1}),\quad \dot{d}_j(t_i)=B_v d_j(t_i)+h_{v,j}(t_{i-1}) \tag{16}$$

where the coefficients B_a, B_v are the constants decided in each step-by-step integration scheme and $h_{a,j}(t_i-1)$, $h_{v,j}(t_i-1)$ are functions of displacement, velocity and acceleration vectors before the time t_{i-1} (see Table 1). The coefficients for the Newmark-β and the Wilson-θ methods are listed in Table 1.
Substituting Eq. (16) into Eq. (15), the equation of motion is transformed into following formula:

$$P_j(t_{i-1})d_j(t_i)=q_j(t_{i-1})+f_j(t_i)-\bar{f}_j(t_i) \tag{17}$$

where

$$P_j(t_{i-1})=B_a M+B_v C_j(t_{i-1})+K_j(t_{i-1}),\quad q_j(t_{i-1})=\tilde{q}_j(t_{i-1})-M h_{a,j}(t_{i-1})-C_j(t_{i-1})h_{v,j}(t_{i-1})$$

The state variables $y_j(t_i)$, $\dot{x}_j(t_i)$, $\dot{\theta}_j(t_i)$ and $\bar{V}_j(t_i)$ in the $C_j(\dot{x}_j)$, $K_j(y_j,\dot{x}_j,\dot{\theta}_j,\bar{V}_j)$ and $\tilde{q}_j(\dot{x}_j)$ of

	B_a	$\mathbf{h}_{a,j}(t_{i-1})$	B_v	$\mathbf{h}_{v,j}(t_{i-1})$
Newmark-β	$\frac{1}{\beta\Delta t^2}$	$\frac{-1}{\beta\Delta t^2}\left[\mathbf{d}_j(t_{i-1})+\Delta t\,\dot{\mathbf{d}}_j(t_{i-1})+\left(\frac{1}{2}-\beta\right)\Delta t^2\ddot{\mathbf{d}}_j(t_{i-1})\right]$	$\frac{\gamma}{\beta\Delta t}$	$\dot{\mathbf{d}}_j(t_{i-1})+\Delta t[(1-\gamma)\ddot{\mathbf{d}}_j(t_{i-1})+\gamma\,\mathbf{h}_{a,j}(t_{i-1})]$
Wilson-θ	$\frac{6}{(\theta\Delta t)^2}$	$\frac{-6}{(\theta\Delta t)^2}\left[\mathbf{d}_j(t_{i-1})+\theta\Delta t\,\dot{\mathbf{d}}_j(t_{i-1})+\frac{(\theta\Delta t)^2}{3}\ddot{\mathbf{d}}_j(t_{i-1})\right]$	$\frac{3}{\theta\Delta t}$	$\frac{-3}{\theta\Delta t}\left[\mathbf{d}_j(t_{i-1})+\frac{2\theta\Delta t}{3}\dot{\mathbf{d}}_j(t_{i-1})+\frac{(\theta\Delta t)^2}{6}\ddot{\mathbf{d}}_j(t_{i-1})\right]$

Table 1. Coefficients for step-by-step solution scheme.

Eq. (15) are replaced by the ones of one time step before $y_j(t_{i-1})$, $\dot{x}_j(t_{i-1})$, $\dot{\theta}_j(t_{i-1})$ and $\bar{V}_j(t_{i-1})$, respectively. Thus, the coefficients $\boldsymbol{C}_j(\dot{x}_j)$, $\boldsymbol{K}_j(y_j,\dot{x}_j,\dot{\theta}_j,\bar{V}_j)$ and vector $\tilde{\boldsymbol{q}}_j(\dot{x}_j)$ are known at the time t_i. These are expressed as $\boldsymbol{C}_j(t_{i-1})$, $\boldsymbol{K}_j(t_{i-1})$ and $\tilde{\boldsymbol{q}}_j(t_{i-1})$ in Eq. (17). It implies that time delay components are generated in Eq. (17).
Substituting Eq. (16) into Eq. (7b), we have transformed relationships of state variables between both ends of a beam element.

$$\boldsymbol{d}_j(t_i)=\boldsymbol{L}_j^{\mathrm{T}}\boldsymbol{d}_{j-1}(t_i)+\frac{1}{1+\delta B_v}\boldsymbol{F}_j\bar{\boldsymbol{f}}_j(t_i)+\frac{\delta}{1+\delta B_v}[\boldsymbol{L}_j^{\mathrm{T}}\boldsymbol{h}_{v,j-1}(t_{i-1})-\boldsymbol{h}_{v,j}(t_{i-1})] \tag{18}$$

Solving Eqs. (17), (18) and (6) simultaneously with respect to all nodes (j : 0 – n) yields displacement vectors $\boldsymbol{d}_j(t_i)$ of all nodes. Velocity and acceleration vectors $\dot{\boldsymbol{d}}_j(t_i)$ and $\ddot{\boldsymbol{d}}_j(t_i)$ are given by Eq. (16) according to the obtained displacement vector $\boldsymbol{d}_j(t_i)$. The obtained $\boldsymbol{d}_j(t_i)$, $\dot{\boldsymbol{d}}_j(t_i)$ and $\ddot{\boldsymbol{d}}_j(t_i)$ yield $\boldsymbol{h}_{a,j}(t_i)$, $\boldsymbol{h}_{v,j}(t_i)$, (see Table 1), then displacement, velocity and acceleration vectors at the next time step $\boldsymbol{d}_j(t_{i+1})$, $\dot{\boldsymbol{d}}_j(t_{i+1})$ and $\ddot{\boldsymbol{d}}_j(t_{i+1})$ are similarly obtained. Step-by-step time integration scheme proceeds in the same way.
However, numerical instability possibly happens in the step-by-step time integration because the reaction force $S_j(y_j)$ and the frictional force $R_j(y_j,\dot{x}_j,\dot{\theta}_j)$ involve the time delay components in addition to the strong non-linearity in Eq. (17). Although the Newmark-β method (β = 1/4, γ = 1/2) and theWilson-θ method (θ = 1.4) are absolutely stable to linear systems, numerical instability often happens in the both methods in the treatment of non-linear systems (Crisfield and Shi, 1996; Xie, 1996). A way to avoid the numerical instability is to use a small time step size but it takes a great deal of computational costs. Furthermore, Eqs. (17), (18) and (6) are solved as simultaneous equations of which the degree of freedom corresponds to that of the whole system. The total computational costs will be extremely large.
The TICM, which was developed by one of the authors, is a useful method to solve Eqs. (17), (18) and (6). The TICM does not solve Eqs. (17), (18) and (6) simultaneously, but solves them through a recursive algorithm. It makes the computation very fast. The concept of the TICM is based on a transmission of physical quantities from one end of structure to the other end of structure. This concept is similar to the Transfer Matrix Method (Pestel and Leckie, 1963, hereafter: TMM). However, the TICM has an advantage in computational speed since the dimensions of matrices and vectors used in the algorithm are smaller than the ones of the TMM. In addition, the TICM also has an advantage in computational accuracy. The TICM is free from a numerical instability, which sometimes happens in the TMM under a certain condition, for example, treatment of rigid supports, computation of high frequency range, etc. The TICM is good at a longitudinally extended structure, which includes the probe discussed here, according to its concept. Therefore, we employ the TICM as a device to compute the displacement vectors $\boldsymbol{d}_j(t_i)$. Even a standard computer can perform efficient computation by utilizing the TICM.

3.2 The Transfer Influence Coefficient Method

The concept of the TICM is illustrated in Fig. 10. The structure enclosed with a broken line in Fig. 10(a) schematically shows a connection of rigid bodies, from node 0 to node j−1, by means of beam elements. A curvature of the probe, spheres (floats) and contact springs are omitted in Fig. 10. Connections of jth beam element and node j follow as shown in Fig. 10(a). The repetition of the connection brings to the completion of the whole structure (probe).

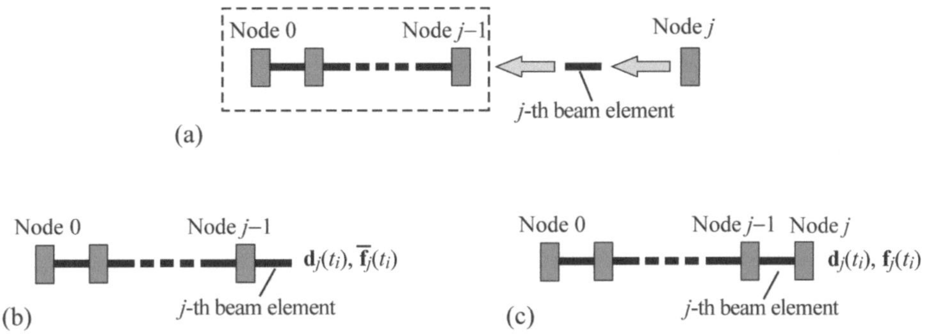

Fig. 10. Concept of the TICM: (a) connections of jth beam element and node j. (b) After the connection of jth beam element, and (c) after the following connection of node j.

This is the concept of the TICM. A structure after the connection of jth beam element, and following connection of node j are illustrated in Fig. 10(b and c), respectively. In the formulation of the TICM for a step-by-step time integration, a relationship between the displacement vector $\boldsymbol{d}_j(t_i)$ and the force vector $\overline{\boldsymbol{f}}_j(t_i)$ illustrated in Fig. 10(b), before the connection of node j, is defined as follows:

$$\boldsymbol{d}_j(t_i)=\overline{\boldsymbol{T}}_j(t_i)\,\overline{\boldsymbol{f}}_j(t_i)+\overline{\boldsymbol{s}}_j(t_i) \tag{19}$$

We call the 3×3 square matrix $\overline{\boldsymbol{T}}_j(t_i)$ and three-dimensional vector $\overline{\boldsymbol{s}}_j(t_i)$ a dynamic influence coefficient matrix and an additional vector of the left-hand side of node j, respectively. The additional vector $\overline{\boldsymbol{s}}_j(t_i)$ represents an influence of external forces, which act on the preceding nodes 0 to $j-1$, to displacement vector at node j.
Similarly, a relationship between $\boldsymbol{d}_j(t_i)$ and $\boldsymbol{f}_j(t_i)$ illustrated in Fig. 10(c), after the connection of node j, is defined as:

$$\boldsymbol{d}_j(t_i)=\boldsymbol{T}_j(t_i)\,\boldsymbol{f}_j(t_i)+\boldsymbol{s}_j(t_i) \tag{20}$$

where the matrix $\boldsymbol{T}_j(t_i)$ and the vector $\boldsymbol{s}_j(t_i)$ are called a dynamic influence coefficient matrix and an additional vector of the right-hand side of node j, respectively. The additional vector $\boldsymbol{s}_j(t_i)$ represents an influence of external forces, which act on the preceding nodes 0 to $j-1$ and newly connected node j.
In the algorithm of the TICM, the matrices $\overline{\boldsymbol{T}}_j(t_i)$, $\boldsymbol{T}_j(t_i)$ and vectors $\overline{\boldsymbol{s}}_j(t_i)$, $\boldsymbol{s}_j(t_i)$ are successively computed from node 0 (root of the probe) to node n (top of the probe) at first. Subsequently, the displacement vectors are computed in the reverse order from node n to node 0. Substituting Eq. (20) with subscript $j-1$ and Eq. (6) into Eq. (18) yields

$$\begin{aligned}\boldsymbol{d}_j(t_i)=&\left[\boldsymbol{L}_j^{\mathrm{T}}\boldsymbol{T}_{j-1}(t_i)\boldsymbol{L}_j\overline{\boldsymbol{f}}_j(t_i)+\frac{1}{1+\delta B_v}\boldsymbol{F}_j\right]\overline{\boldsymbol{f}}_j(t_i)\\&+\boldsymbol{L}_{j-1}^{\mathrm{T}}\boldsymbol{s}_j(t_i)+\frac{\delta}{1+\delta B_v}[\boldsymbol{L}_j^{\mathrm{T}}\boldsymbol{h}_{v,j-1}(t_{i-1})-\boldsymbol{h}_{v,j}(t_{i-1})]\end{aligned} \tag{21}$$

Comparing Eq. (21) with Eq. (19), we have

$$\overline{\boldsymbol{T}}_j(t_i)={}^{t}\boldsymbol{L}_j\boldsymbol{T}_{j-1}(t_i)\boldsymbol{L}_j+\frac{1}{1+\delta B_v}\boldsymbol{F}_j \tag{22a}$$

$$\bar{s}_j(t_i)={}^{t}\boldsymbol{L}_j\,\boldsymbol{s}_{j-1}(t_i)+\frac{\delta}{1+\delta B_v}[{}^{t}\boldsymbol{L}_j\boldsymbol{h}_{v,j-1}(t_{i-1})-\boldsymbol{h}_{v,j}(t_{i-1})] \tag{22b}$$

Multiplying both sides of Eq. (17) by $\bar{\boldsymbol{T}}_j(t_i)$ and utilizing the relationship $\bar{\boldsymbol{T}}_j(t_i)\bar{\boldsymbol{f}}_j(t_i)=\boldsymbol{d}_j(t_i)-\bar{s}_j(t_i)$ [Eq. (19)] yields

$$[\boldsymbol{I}_3+\bar{\boldsymbol{T}}_j(t_i)\boldsymbol{P}_j(t_{i-1})]\boldsymbol{d}_j(t_i)=\bar{\boldsymbol{T}}_j(t_i)\boldsymbol{f}_j(t_i)+\bar{s}_j(t_i)+\bar{\boldsymbol{T}}_j(t_i)\boldsymbol{q}_j(t_{i-1}) \tag{23}$$

Comparing Eq. (23) with Eq. (20), we obtain

$$[\boldsymbol{I}_3+\bar{\boldsymbol{T}}_j(t_i)\boldsymbol{P}_j(t_{i-1})]\boldsymbol{T}_j(t_i)=\bar{\boldsymbol{T}}_j(t_i) \tag{24a}$$

$$[\boldsymbol{I}_3+\bar{\boldsymbol{T}}_j(t_i)\boldsymbol{P}_j(t_{i-1})]\boldsymbol{s}_j(t_i)=\bar{s}_j(t_i)+\bar{\boldsymbol{T}}_j(t_i)\boldsymbol{q}_j(t_{i-1}) \tag{24b}$$

where $\boldsymbol{I}_3$ is a 3×3 unit matrix. We call Eqs. (22a), (22b) and (24a), (24b) "field transmission rule" and "point transmission rule", respectively. Supposing that the dynamic influence coefficient matrix and additional vector of the right-hand side of node $j-1$, $\boldsymbol{T}_{j-1}(t_i)$ and $s_{j-1}(t_i)$, are known, the ones of node j, that is $\boldsymbol{T}_j(t_i)$ and $s_j(t_i)$, are obtained through the field and point transmission rules Eqs. (22a), (22b) and (24a), (24b). In other words, if the dynamic influence coefficient matrix and additional vector of node 0 are known, the ones of other nodes are successively computed from node 1 to node n because the field and point transmission rules represent a recurrent formula to yield $\boldsymbol{T}_j(t_i)$ and $s_j(t_i)$. Since the root of the probe, node 0, is assumed to have no relative movement with respect to the unstretched probe, the displacement and force vectors at node 0 are regarded as $\boldsymbol{d}_0(t_i)=0$ and $\boldsymbol{f}_0(t_i)\neq 0$. Substituting the $\boldsymbol{d}_0(t_i)$ and the $\boldsymbol{f}_0(t_i)$ into Eq. (20) with subscript $j=0$, we obtain the dynamic influence coefficient matrix and additional vector of node 0.

$$T_0(t_i)=\boldsymbol{0}_3,\quad s_0(t_i)=\boldsymbol{0} \tag{25}$$

where $\boldsymbol{0}_3$ is a 3×3 zero matrix.

Node j slantingly connects with the jth and (j+1)th beam elements as shown in Fig. 7. Therefore, coordinate transform is necessary through the point transmission rule. The transform of coordinate from jth beam element to node j is operated as:

$$\boldsymbol{\Phi}\bar{\boldsymbol{T}}_j(t_i)\boldsymbol{\Phi}^{\mathrm{T}}\Rightarrow\bar{\boldsymbol{T}}_j(t_i),\quad \boldsymbol{\Phi}\bar{s}_j(t_i)\Rightarrow\bar{s}_j(t_i),\quad \boldsymbol{\Phi}=\begin{bmatrix}\cos\varphi & -\sin\varphi & 0\\ \sin\varphi & \cos\varphi & 0\\ 0 & 0 & 1\end{bmatrix} \tag{26a}$$

The transform of coordinate from node j to (j+1)th beam element is operated as:

$$\boldsymbol{\Phi}\boldsymbol{T}_j(t_i)\boldsymbol{\Phi}^{\mathrm{T}}\Rightarrow\boldsymbol{T}_j(t_i),\quad \boldsymbol{\Phi}\boldsymbol{s}_j(t_i)\Rightarrow\boldsymbol{s}_j(t_i) \tag{26b}$$

The dynamic influence coefficient matrix $\boldsymbol{T}_j(t_i)$ and additional vector $s_j(t_i)$ are successively computed from node 0 to node n through Eqs. (22a), (22b), (24a)-(26b).

The right-hand side of the system (top of the probe) is free, it follows that the force vector at the right-hand side of node n is zero, that is $\boldsymbol{f}_n(t_i)=\boldsymbol{0}$. Substituting $\boldsymbol{f}_n(t_i)=\boldsymbol{0}$ into Eq. (20), we obtain the displacement vector of node n as:

$$\boldsymbol{d}_n(t_i)=\boldsymbol{s}_n(t_i) \tag{27}$$

Displacement vectors of other nodes are recursively obtained from node $n-1$ to node 0 by applying the following equations, which are derived from Eqs. (17), (6) and (20).

$$\begin{aligned}&\bar{\boldsymbol{f}}_j(t_i)=\boldsymbol{q}_j(t_{i-1})+\boldsymbol{f}_j(t_i)-\boldsymbol{P}_j(t_{i-1})\boldsymbol{d}_j(t_i),\quad \boldsymbol{f}_{j-1}(t_i)=\boldsymbol{L}_j\,\bar{\boldsymbol{f}}_j(t_i)\\&\boldsymbol{d}_{j-1}(t_i)=\boldsymbol{T}_{j-1}(t_i)\,\boldsymbol{f}_{j-1}(t_i)+\boldsymbol{s}_{j-1}(t_i)\end{aligned} \tag{28}$$

where $j : n \rightarrow 1$. The following coordinate transform is also necessary for $\bar{\boldsymbol{f}}_j(t_i)$ and $\boldsymbol{f}_{j-1}(t_i)$ in the process of Eq. (28) because of the slanting connection of jth beam element with node $j-1$ and node j.

$$\boldsymbol{\Phi}^{\mathrm{T}}\bar{\boldsymbol{f}}_j(t_i)\Rightarrow\bar{\boldsymbol{f}}_j(t_i),\quad \boldsymbol{\Phi}^{\mathrm{T}}\boldsymbol{f}_{j-1}(t_i)\Rightarrow\boldsymbol{f}_{j-1}(t_i) \tag{29}$$

Velocity and acceleration vectors $\dot{\boldsymbol{d}}_j(t_i)$ and $\ddot{\boldsymbol{d}}_j(t_i)$ are given by Eq. (16) after the computation of displacement vectors $\boldsymbol{d}_j(t_i)$.

4. Numerical computations

4.1 Reproduction of the experimental results

Numerical simulations were implemented by using the analytical model obtained in Section 2. A standard computer (CPU 2.4 GHz, 512MB RAM) was used in the computation. The compiler was Fortran 95 and double precision variables were used. The Newmark-β method ($\beta=1/4$, $\gamma=1/2$) was employed as a step-by-step time integration scheme. We confirmed

d_h	: Diameter of helical tube	1.2 m
d_{in}	: Inner diameter of heating tube	24.2 mm
u	: Feed speed of probe	200 mm/s
ρ_f	: Air-flow density	1.023 kg/m^3
Q	: Supply rate of air-flow	63.0 m^3/h
d_c	: Diameter of cable	8.35 mm
l	: Length of carrier cable	10, 20, 30, 40, 50 m
l_G	: Length of guide cable	2.5 m
ρ_c	: Density of probe	0.111 kg/m
(EA)	:Tensile rigidity of cable	1.1×10^{5}N
(EI)	: Flexural rigidity of cable	4.8×10^{-2} Nm^2
d_f	: Diameter of float	19.0 mm
p_f	: Float pitch	100 mm
$\hat{k}$	: Contact rigidity between float and tube	10^5 N/m
μ	: Coefficient of kinetic friction	0.15
δ	: Coefficient of proportional viscous damping	5.0×10^{-5} s

Table 2. Parameters of numerical simulation

that the results by the Wilson-θ method (θ = 1.4) were almost the same as the ones by the Newmark-β method.

Parameters of the numerical simulation are listed in Table 2. Since probes of constant length are treated, five probes with different length are provided for numerical simulation. The five probes are different in length of carrier cable, $l = 10$, 20, 30, 40 and 50m as listed in Table 2. The total length of the cable L is the length of carrier cable l plus that of guide cable $l_G = 2.5$ m. As mentioned in Section 2.2 (b), numerical simulation of the probe is approximately regarded as a momentary situation in which the inserted length of the probe into the helical part of the heating tube reaches L. An initial condition was assumed to be static. The drag force of Eq. (10) simultaneously began to act on the all floats at the beginning of the simulation. At the same time, the probe began to move at a feeding speed u. Time step size Δt = 0.0001 s was chosen for the step-by-step integration and time historical responses during $t = 0 - 8$ s were computed. The numerical simulations were impossible because of a numerical divergence when the time step size was larger than 0.0001 s in both the Newmark-β and the Wilson-θ methods.

Displacements of the node corresponding to the sensor are shown in Fig. 11. Axial displacement $x_j(t)$ and radial displacement $y_j(t)$ are shown in Fig. 11(a and b), respectively. The vibration of the probe increases as the length of probe become longer. Particularly, the radial displacement rapidly increases between $l = 30$ and 40 m. Since the vibration of probe in experiment rapidly increased after the sensor passed through the middle point of the helical part (see Fig. 4), the results of the numerical simulation agree with the experimental results.

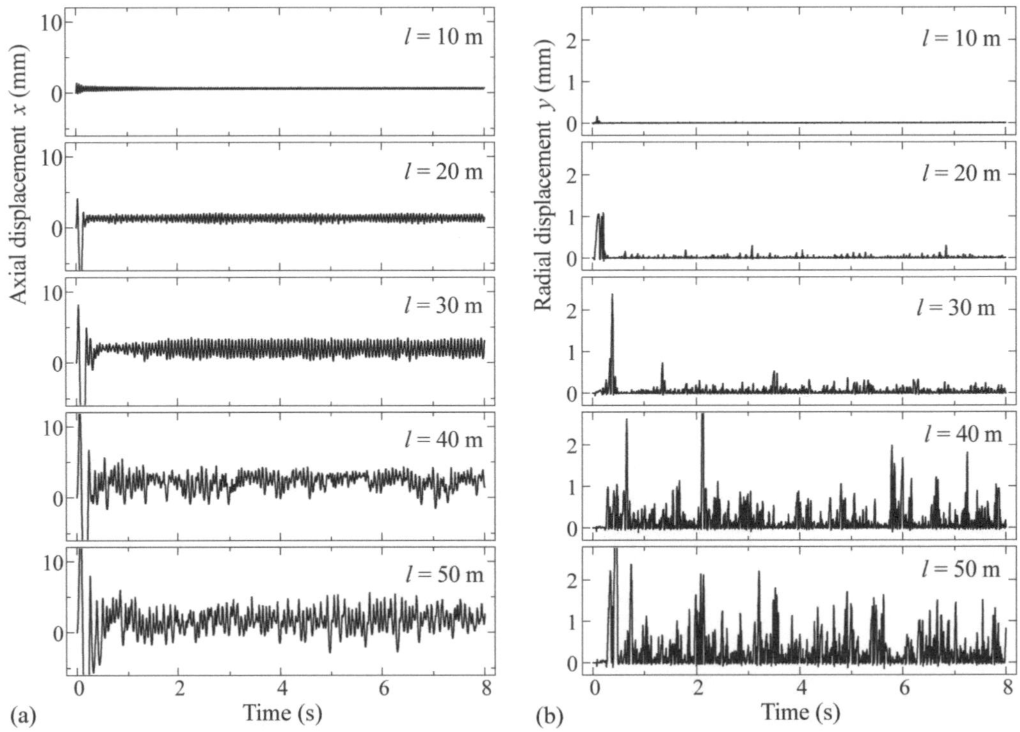

Fig. 11. Vibration of probe in insertion process: (a) axial and (b) radial displacements.

Finally, the inserted length of the probe into the helical heating tube reaches 55–60 m. Magnifications of the axial and the radial vibrations of $l = 55$ m (total length $L = l + l_G$ (2.5m) = 57.5 m) are shown in Fig. 12(a and b). Other parameters were the same as the ones listed in Table 2. The vibrations during $t = 1.0$–2.5 s are plotted. It is confirmed that the axial and the radial vibrations are weakly coupled. The locus of the vibration is plotted in Fig. 13(a). The horizontal axis indicates a fixed coordinate along the inner wall of the heating tube and the vertical axis shows the radial displacement. The probe is leaping around and shows an inchworm-like motion. The motion of the sensor in the experiment, where the inserted length of the probe into the helical part was about 57 m, is shown in Fig. 13(b). It was given by a tracing of the images of sensor, which was taken by a high-speed camera. Although both the axial and the radial motions in the experiment are larger than that of the simulation, the result of the simulation qualitatively agrees with the one of the experiment. The Fourier analysis of the axial and the radial vibrations of L = 57.5 m are shown in Fig. 14(a and b), respectively. The vibrations during t = 0.5–4.5 s, which are free from the transient response, are provided to the Fourier analysis. It is confirmed that the axial and the radial vibrations are coupled since an identical peak of 14 Hz appears in both vibrations. The frequency of the coupled vibration in the experiment was about 20 Hz, as mentioned in Section 2.1 c. There is a discrepancy between the experiment and the numerical simulation in this point. However, the results of numerical simulations are qualitatively similar to the ones of the experiment.

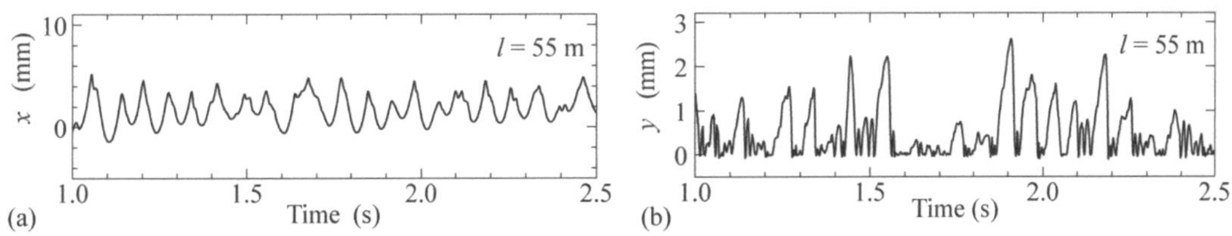

Fig. 12. Vibration of probe; $l = 55$ m, $t = 1.0$–2.5 s: (a) axial and (b) radial displacements.

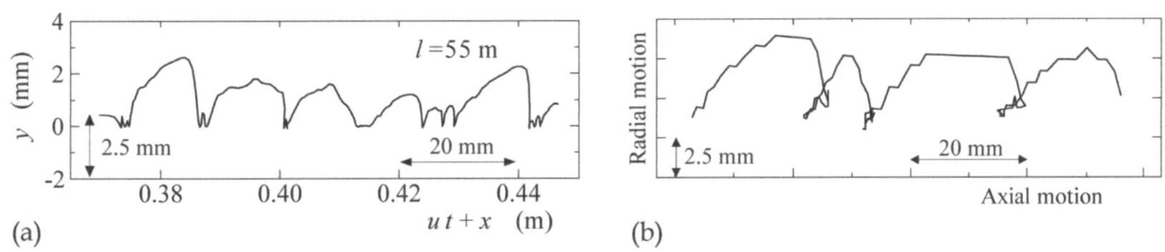

Fig. 13. Locus of probe; (a) numerical simulation of l=55 m, t=1.8–2.2 s and (b) in experiment, inserted length around 57 m.

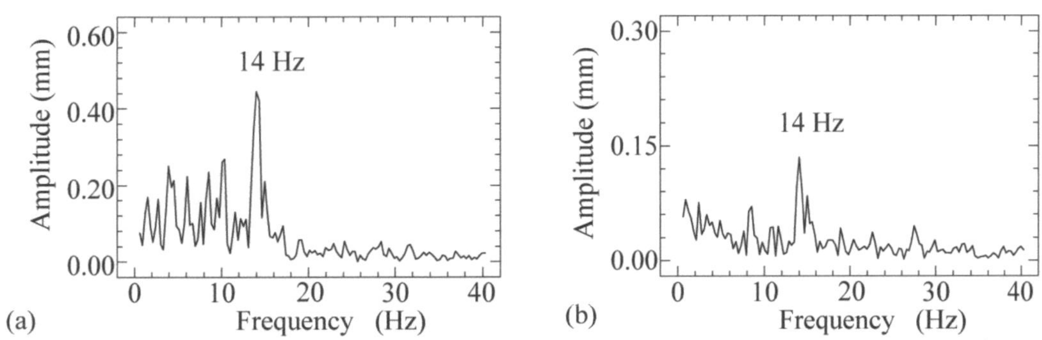

Fig. 14. Frequency analysis of vibration; $l = 55$m : (a) axial and (b) radial displacements.

A numerical simulation of the probe without feeding (feeding speed u = 0 mm/s) was implemented. The length of carrier cable was $l = 50$ m, which showed a severe vibration with feeding speed u = 200 mm/s as shown in Fig. 11. Other parameters were the same as the ones listed in Table 2. This simulation corresponds to the experiment that the dry compressed air streamed in the heating tube but the probe was not fed as mentioned in Section 2.1 d. Displacements of the node corresponding to the sensor are shown in Fig. 15. Both the axial and the radial displacements converged at constant values after an initial transient response. This result is similar to the experiment. It follows that the experimental result without feeding is also supported by the numerical simulation.

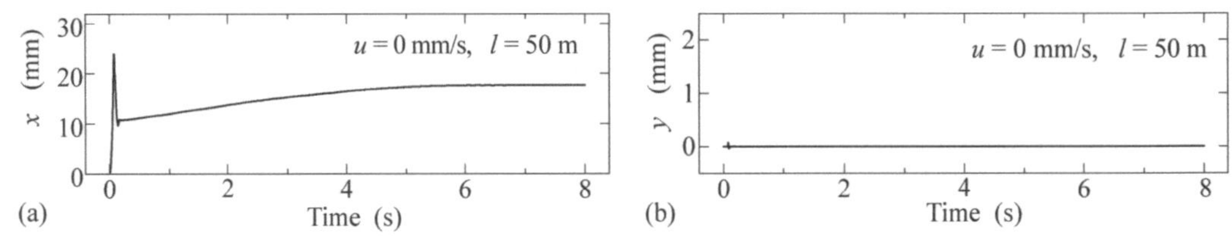

Fig. 15. Response at $u = 0$ mm/s; $l = 50$m : (a) axial and (b) radial displacements.

More numerical simulations were implemented in order to enhance the validity of the analytical model. Numerical simulations with variation of feeding speed, diameter of the helix and air supply rate were implemented. Only one parameter (feeding speed, diameter of the helix or air supply rate) was changed, and the other parameters were the same as Table 2. The length of carrier cable was $l = 50$ m as well as the simulation of the non-feeding probe, Fig. 15. The simulations of feeding speed $u = 100$ and 400 mm/s, diameter of the helix d_h = 2.5 m and air supply rate Q = 40m^3/h are shown in Figs. 16–18, respectively. In Fig. 16,the vibration of the probe became small at low feeding speed u = 100 mm/s, but large at high feeding speed u = 400 mm/s, compared with the result of l = 50 m in Fig. 11 (u = 200 mm/s). The vibration also became small in the case of large helical diameter (Fig. 17) and low supply rate of the air flow (Fig. 18). These results are similar to the experiments mentioned in Section 2.1 f. Note that in the case of $Q = 40$m^3/h, an ability to insert the actual probe is not guaranteed for lack of a drag force (Inoue et al., 2007).

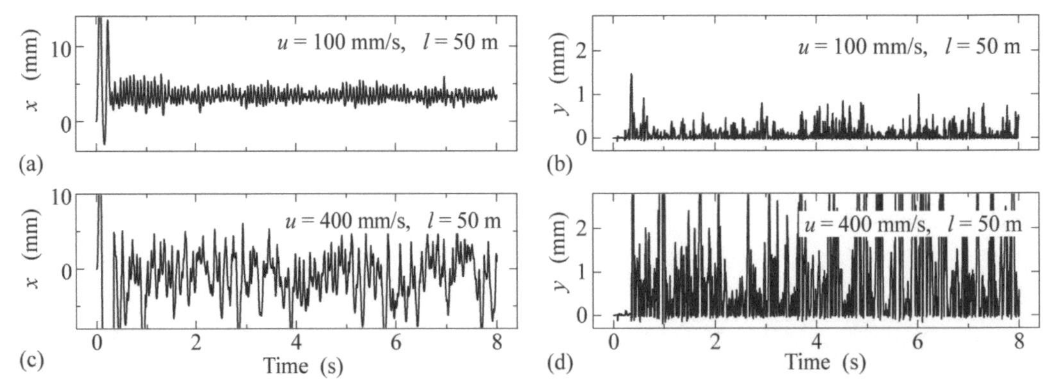

Fig. 16. Vibration of probe; $l = 50$m: (a) axial and (b) radial displacements at feeding speed $u = 100$ mm/s, (c) axial and (d) radial displacements at $u = 400$ mm/s.

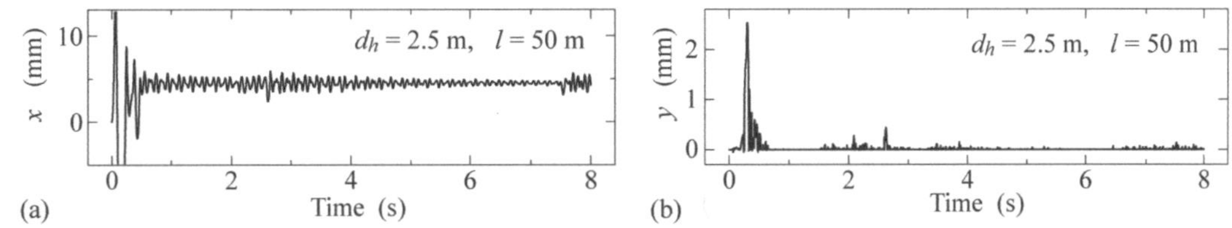

Fig. 17. Vibration of probe; diameter of helix d_h = 2.5 m, l = 50m: (a) axial and (b) radial displacements.

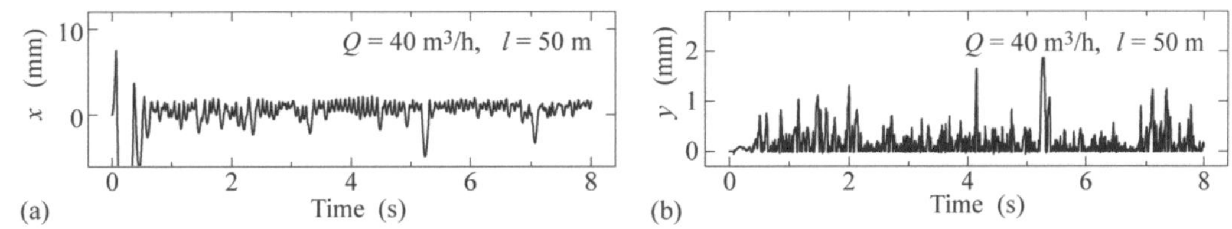

Fig. 18. Vibration of probe; air supply rate $Q = 40\,m^3/h$, $l = 50\,m$: (a) axial and (b) radial displacements.

The numerical simulation was qualitatively able to reproduce the experimental results. Thus, the validity of the analytical model obtained in this study was confirmed through the numerical simulations. It was demonstrated that the vibration of probe was caused by Coulomb friction between the floats and the inner wall of the heating tube.

4.2 Entire behavior of probe

A numerical simulation of the insertion process to the length of carrier cable l = 55 m is implemented, and the entire probe behavior is shown in Fig. 19. The other parameters are the same as the ones in Table 2. The total length of the cable is $L = l$ (55 m) + l_G (2.5 m) = 57.5 m. Momentary shapes of the entire probe during 1.56–1.65 s are displayed at an interval of 0.01 s. Axial and radial displacements are shown in Fig. 19(a and b), respectively. Each of the horizontal axes in Fig. 19(a and b) indicates a distance from the entrance of the helical heating tube. It is a fixed coordinate along the helical heating tube. The root of the probe, which is supposed to be located at the entrance of the helical heating tube, corresponds to L = 0 m, and the top of the cable is situated at L = 57.5 m. The vertical axes in Fig. 19(a) indicate the axial displacements, and the ones in Fig. 19(b) indicate the radial displacements. Although the direction of the axial displacement in the ordinate of Fig. 19(a) is the same as the coordinate along the heating tube L, it is displayed at right angles with the coordinate L. The sensor position is indicated as broken lines both in Fig. 19(a and b). The following characteristics are found in Fig. 19.

a. A shaded area in Fig. 19(a) indicates a segment in which a gradient of the axial displacement along the heating tube (dx/dL) obviously shows a negative value. The identical areas are also shaded in Fig. 19(b). We are able to observe a radial displacement in the shaded area. Furthermore, it becomes larger as the negative gradient of the axial displacement ($dx/dL < 0$) becomes steeper.
b. Local maxima of the axial displacement, points "A" and "B" in Fig. 19(a), move toward the top of the probe as the time step goes forward. This is a wave-like motion rather than a vibration. A reflection of the wave is not clearly observed in Fig. 19(a and b). It

seems that the noticeable peak at 14 Hz in Fig. 14 signifies the frequency of repetitiveness of the wave-like motion.

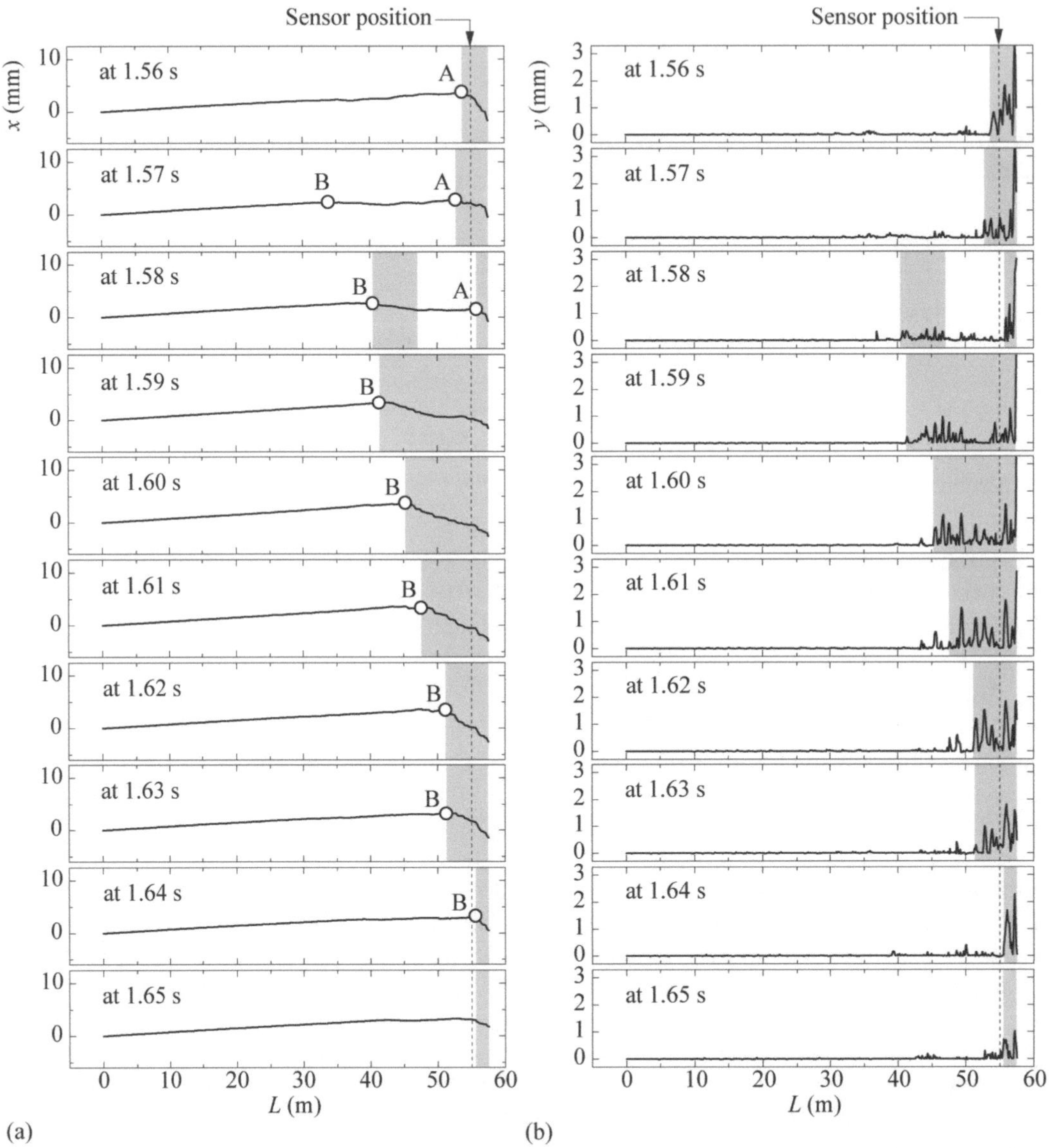

Fig. 19. Entire behavior of probe in the insertion process: (a) axial and (b) radial displacements.

c. Large amplitudes in the radial displacement are limited in the area near the top of the cable.

The countermeasures against vibration, which include a long guide cable and a large float of guide cable, were devised in order to reduce the RF sensor noise. It was confirmed that the countermeasures are effective in suppressing the vibration in the experiments. Although the countermeasures were empirically obtained, the entire behavior of the probe shown in Fig. 19 implies the mechanism of the countermeasures as follows:

a. The amplitude in the radial displacement is small at a position away from the top of the cable as shown in Fig. 19(b). The long guide cable keeps the sensor part away from the top of the cable, and the radial (displacement) vibration at the sensor position becomes small. Since the RF sensor noise is highly correlated to the radial vibration, it is reduced by means of the long guide cable. This effect has been also confirmed in the experiments (Inoue et al., 2007a).
b. In the shaded area in Fig. 19, where the gradient $dx/dL<0$, the driving force (drag force) acting on the probe is smaller than that of the non-shaded area. Originally, a tensile force acts on the probe in the insertion process. However, a "compressive force" is generated in the shaded area because of the lack of driving force, and the shaded area is pushed from the backward non-shaded area. Consequently, a kind of buckling happens and the probe in the shaded area, which is supposed to move in contact with the inside of the helical tube, rises off the inner wall of the heating tube. This phenomenon travels toward the top of the cable and makes the wave-like motion. At a fixed point, for example the sensor position, it appears as a vibration. This is the mechanism of the probe vibration. Similar rising (lift-off) phenomena were reported in previous studies (Bihan, 2002; Giguere et al., 2001; Tian and Sophian, 2005), but significant vibration was not reported in these studies. Relatively severe vibration induced by this rising phenomenon is a peculiar characteristic of this study. Since the shaded area is generated in the forward section of the probe, the large float of guide cable makes the driving force acting on the forward section large, and it reduces the "compressive force" acting on the shaded area. As a result, the large float of guide cable works to suppress the vibration at the sensor part.

4.3 Improvement of the countermeasure

The empirical countermeasures to suppress the vibration at the sensor part are supported by the numerical simulations. On the basis of the mechanism which suppresses the vibration, the following improvements are suggested:

a. Use of a longer guide cable. This acts on the principle that the vibration becomes smaller as the length between the sensor position and the top of cable becomes longer.
b. Further increase of the driving force of the guide cable. This makes the "compressive force" acting on the forward section of the probe relatively weak, and prevents the probe from rising off the inner wall of the heating tube.
c. Decrease the driving force of the carrier cable. This is similar to suggestion b. It directly reduces the "compressive force" toward the forward section of the probe by reducing the driving force of the backward section.

In reference to suggestion a, it makes the probe length inserted into the heating tube longer. Since the steam generator of the "Monju" has 140-layered heating tubes, use of an excessively long guide cable would negatively affect maintenance efficiency. Thus, a guide cable longer than 10m is undesirable in actual use. Suggestions b and c involve control of the drag force acting on the floats. There are two means to vary the drag force: One is to alter the float size, where the float is spherical. The other is to replace the float shape. However, it is difficult to practicably use a non-spherical float as it would compromise the smooth passage of the probe. Hence, control of the drag force by alteration of the float size is considered here.

The inner diameter of the heating tube is 24.2 mm, and some points are smaller than 24.2mm because of projections caused by welding. Consequently, a float diameter of 20 mm, which has been utilized in the countermeasure, seems to be the upper limit since a larger float would probably clog the heating tube. Thus, only suggestion c is adopted. The probe is fed into the upper side of the steam generator (see Fig. 1), goes down the heating tube, passes the helical part, goes up the straight part and reaches the upper side again. A strong driving force is needed when the probe passes the helical part and goes up the straight part of heating tubes. Thus, there is also a minimum float diameter in order to guarantee the driving force needed to propel the probe to achieve the inspection of the heating tubes. We choose the diameter for the float attached to carrier cable $d_f = 16$ mm.
The numerical simulation with these improvements, where the length of guide cable $l_G = 10$ m, the diameter of the float attached to guide cable $d_f = 20$ mm and the one to carrier cable $d_f = 16$ mm, is implemented. The length of carrier cable $l = 50$ m, (total length L is 60 m) and the other parameters are the same as the ones in Table 2. The vibration at the sensor part is shown in Fig. 20. Suppression of the vibration at the sensor part is almost accomplished in the radial direction. Comparing this result with the one of $l = 50$ m in Fig. 11, the validity of this improvement is indisputable. We can assess that the performance of the improved probe is satisfactory to suppressing the vibration.

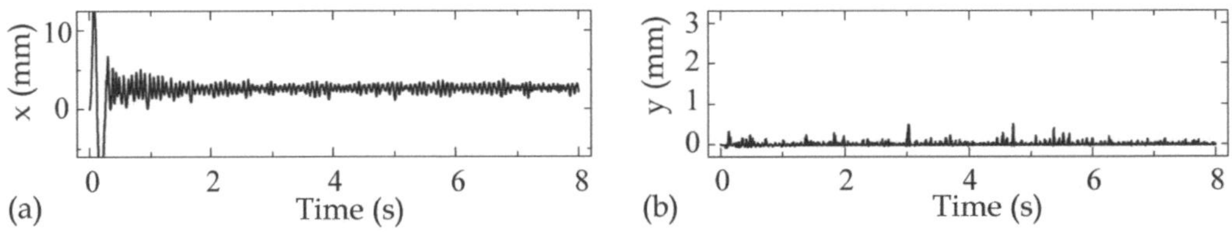

Fig. 20. Vibration of probe in the insertion process with the proposed improvement, diameter of the float attached to the guide cable 20 mm, carrier cable 16mm and length of the guide cable $l_G = 10$ m : (a) axial and (b) radial displacements.

In 2010, the fast breeder reactor "Monju" in Japan resumed work after a long time tie-up of operation. The tie-up was cause by a leakage accident of sodium in a heat exchanging system. The resumption of "Monju" was the target of public attention. An improved probe based on this study practically come into service for the defect detection of heating helical tubes installed in "Monju". A reliable inspection is performed and it has kept a safe operation of "Monju".

5. Conclusions

A defect detection of a helical heating tube installed in a fast breeder reactor "Monju" in Japan is operated by a feeding of an eddy current testing probe. A problem that the eddy current testing probe vibrates in the helical heating tubes happened and it makes the detection of defect difficult. In this study, the cause of the vibration of the eddy current testing probe was investigated. The results are summarized as follows:

a. The cause of the vibration was assumed to be Coulomb friction and an analytical model of the vibration incorporating Coulomb friction was obtained.
b. An effectual algorithm for the numerical simulation of the eddy current testing probe was formulated by applying the Transfer Influence Coefficient Method to the equation of motion derived from the analytical model.

c. The results of numerical simulations qualitatively reproduced the several characteristics of the vibration of the eddy current testing probe, which were obtained by experiments. The validity of the assumption that the vibration is cause by Coulomb friction was confirmed by an agreement between the results of experiments and numerical simulations.
d. The probe's motion in its entirety under the vibration conditions was obtained by the numerical simulation. The mechanism of the vibration and the countermeasures were revealed through a discussion on the probe's entire motion.
e. An improvement of the countermeasure was proposed based on the probe's entire motion. The validity of the proposed improvement was demonstrated through a numerical simulation. The improvement was effective both in the insertion and the return processes.

6. Acknowledgements

This investigation was performed through collaboration between Kyushu University and Japan Atomic Energy Agency (JAEA) as public research of Japan Nuclear Cycle. Here the authors would like to acknowledge the authorities concerned.

7. References

Belytschko, T. & Hughes, T.J.R. (1983). *Computational methods for transient analysis*, Belytschko, T. & Bathe, K.J. (Eds.), Computational Methods in Mechanics, Vol. 1, (417–471), Elsevier Science Publishers B.V., ISBN 0444864792, Amsterdam.

Bihan, Y.L. (2002). Lift-off and tilt effects on eddy current sensor measurements: a 3-D finite element study. *Eur. Phys. J. Appl. Phys.*, Vol. 17, (25–28), ISSN 0021-8979.

Crisfield, M.A. & Shi, J. (1996). An energy conserving co-rotation procedure for non-linear dynamics with finite elements. *Nonlinear Dyn.*, Vol. 9, (37–52), ISSN 1090-0578.

Giguere, S.; Lepine, B. & Dubois, J.M.S. (2001). Pulsed eddy current technology: characterizing material loss with gap and lift-off variations. *Res. Nondestructive Eval.*, Vol. 13, (119–129), ISSN 1075-4862.

Inoue, T.; Sueoka, A.; Nakano, Y.; Kanemoto, H.; Imai, Y. & Yamaguchi, T. (2007). Vibrations of probe used for the defect detection of helical heating tubes in a fast breeder reactor. Part 1. Experimental results by using mock-up. *Nucl. Eng. Des.*, Vol. 237, (858–867), ISSN 0029-5493.

Inoue, T.; Sueoka, A. & Shimokawa, Y. (1997). Time historical response analysis by applying the Transfer Influence Coefficient Method, *Proceedings of the Asia-Pacific Vibration Conference '97*, Vol. 1, pp. 471–476, Kyongju (Korea), November 1997.

Isobe, M.; Iwata, R. & Nishikawa, M. (1995). *High sensitive remote field eddy current testing by using dual exciting coils*, Collins, R.; Dover, W.D.; Bowler, J.R. & Miya, K. (Eds.), Nondestructive Testing Mater. (Studies in Applied Electromagnetics and Mechanics), Vol. 8, (145–152). IOS Press, ISBN 9051992394, Amsterdam.

Kondou, T.; Sueoka, A.; Moon, D.H.; Tamura, H. & Kawamura, T. (1989). Free vibration analysis of a distributed flexural vibrational system by the Transfer Influence Coefficient Method. *Theor. Appl. Mech.*, Vol.37, (289–304), ISSN 0285-6024.

Pestel, E.C. & Leckie, F.A. (1963). *Matrix Methods in Elastomechanics*, McGraw-Hill Publishers, ISBN 0070495203, New York.

Robinson, D. (1998). Identification and sizing of defects in metallic pipes by remote field eddy current inspection. *Tunnel. Underground Space Technol*, Vol. 13, (17–27), ISSN 0886-7798.

Sueoka, A.; Tamura, H.; Ayabe, T. & Kondou, T. (1985). A method of high speed structural analysis using a personal computer. *Bull. JSME*, Vol. 28, (924–930), ISSN 1344-7653.

Tian, G.Y. & Sophian, A. (2005). Reduction of lift-off effects for pulsed eddy current NDT. *NDT&E Int.*, Vol. 38, (319–324), ISSN 0963-8695.

Xie, Y.M. (1996). An assessment of time integration schemes for non-linear dynamic equations. *J. Sound Vibrat.*, Vol. 192, (321–331), ISSN 0022-460X.

20

Vibration and Sensitivity Analysis of Spatial Multibody Systems Based on Constraint Topology Transformation

Wei Jiang, Xuedong Chen and Xin Luo
Huazhong University of Science and Technology
P.R.China

1. Introduction

Many kinds of mechanical systems are often modeled as spatial multibody systems, such as robots, machine tools, automobiles and aircrafts. A multibody system typically consists of a set of rigid bodies interconnected by kinematic constraints and force elements in spatial configuration (Flores et al., 2008). Each flexible body can be further modeled as a set of rigid bodies interconnected by kinematic constraints and force elements (Wittbrodt et al., 2006). Dynamic modeling and vibration analysis based on multibody dynamics are essential to design, optimization and control of these systems (Wittenburg, 2008 ; Schiehlen et al., 2006). Vibration calculation of multibody systems is usually started by solving large-scale nonlinear equations of motion combined with constraint equations (Laulusa & Bauchau, 2008), and then linearization is carried out to obtain a set of linearized differential-algebraic equations (DAEs) or second-order ordinary differential equations (ODEs) (Cruz et al., 2007; Minaker & Frise, 2005; Negrut & Ortiz, 2006; Pott et al., 2007; Roy & Kumar, 2005). This kind of method is necessary for solving the dynamics of nonlinear systems with large deformation.
However, there are two major disadvantages for vibration calculation of multibody systems by using the conventional methods. On one hand, the computational efficiency is very low due to a large amount of efforts usually required for computation of trigonometric functions, derivation and linearization. Many approaches have been proposed to simplify the formulation, such as proper selection of reference frames (Wasfy & Noor, 2003), generalized coordinates (Attia, 2006; Liu et al., 2007; McPhee & Redmond, 2006; Valasek et al., 2007), mechanics principles (Amirouche, 2006; Eberhard & Schiehlen, 2006), and other methods (Richard et al., 2007; Rui et al., 2008). On the other hand, despite sensitivity analysis of multibody systems based on the conventional methods are well documented (Anderson & Hsu, 2002; Choi et al., 2004; Ding et al., 2007; Sliva et al. 2010; Sohl & Bobrow, 2001; Van Keulen et al. 2005; Xu et al., 2009), the formulation is quite complicated because the resulting equations are implicit functions of the design parameters.
Actually, what people concern, for many kinds of mechanical systems under working conditions, are eigenvalue problems and the relationship between the modal parameters and the design parameters. And the designer needs to know the results as quickly as possible so as to perform optimal design. From this point of view, fast algorithm for

vibration calculation and sensitivity analysis with easiness of application is critical to the design of a complex mechanical system. A novel formulation based on matrix transformation for open-loop multibody systems has been proposed recently (Jiang et al., 2008a). The algorithm has been further improved to directly generate the open-loop constraint matrix instead of matrix multiplication (Jiang et al., 2008b). The computational efficiency has been significantly improved, and the resulting equations are explicit functions of the design parameters that can be easily applied for sensitivity analysis. Particularly, the proposed method can be used to directly obtain sensitivity of system matrices about design parameters which are required to perform mode shape sensitivity analysis (Lee et al., 1999a; 1999b).

Vibration calculation of general multibody system containing closed-loop constraints is investigated in this article. Vibration displacements of bodies are selected as generalized coordinates. The translational and rotational displacements are integrated in spatial notation. Linear transformation of vibration displacements between different points on the same rigid body is derived. Absolute joint displacement is introduced to give mathematical definition for ideal joint in a new form. Constraint equations written in this way can be solved easily via the proposed linear transformation. A new formulation based on constraint-topology transformation is proposed to generate oscillatory differential equations for a general multibody system, by matrix generation and quadric transformation in three steps:

1. Linearized ODEs in terms of absolute displacements are firstly derived by using Lagrangian method for free multibody system without considering any constraint.
2. An open-loop constraint matrix $\boldsymbol{B}'$ is derived to formulate linearized ODEs via quadric transformation $\boldsymbol{E}' = \boldsymbol{B}'^{\mathrm{T}}\boldsymbol{E}\boldsymbol{B}'$ $(\boldsymbol{E} = \boldsymbol{M},\boldsymbol{K},\boldsymbol{C})$ for open-loop multibody system, which is obtained from closed-loop multibody system by using cut-joint method.
3. A constraint matrix $\boldsymbol{B}''$ corresponding to all cut-joints is finally derived to formulate a minimal set of ODEs via quadric transformation $\boldsymbol{E}'' = \boldsymbol{B}''^{\mathrm{T}}\boldsymbol{E}'\boldsymbol{B}''$ $(\boldsymbol{E} = \boldsymbol{M},\boldsymbol{K},\boldsymbol{C})$ for closed-loop multibody system.

Complicated solving for constraints and linearization are unnecessary for the proposed method, therefore the procedure of vibration calculation can be greatly simplified. In addition, since the resulting equations are explicit functions of the design parameters, the suggested method is particularly suitable for sensitivity analysis and optimization for large-scale multibody system, which is very difficult to be achieved by using conventional approaches.

Large-scale spatial multibody systems with chain, tree and closed-loop topologies are taken as case studies to verify the proposed method. Comparisons with traditional approaches show that the results of vibration calculation by using the proposed method are accurate with improved computational efficiency. The proposed method has also been implemented in dynamic analysis of a quadruped robot and a Stewart isolation platform.

2. Fundamentals of multibody dynamics

2.1 Description of multibody system

As shown in Fig. 1, considering a multibody system which consists of n rigid bodies and the ground B_0, each two bodies are probably interconnected by at most one joint and arbitrary number of spatial spring-dampers. A spatial spring-damper means an integration of three spring-dampers and three torsional spring-dampers. Each joint contains at least one and at most six holonomic constraints. B_i denotes the ith rigid body, and J_{ij} is the joint between B_i and B_j, where $i,j=1,2,\cdots,n$ and $i \neq j$. s_{ij} denotes the total number of spring-dampers between B_i and B_j, among which K_{ijs} is the sth one, where $s=0,1,2,\cdots,s_{ij}$. $s_{ij}=0$ means there is no spring-damper between B_i and B_j.

Four kinds of reference frames are used in the formulation. The global reference frame, namely the inertial frame, i.e., $o\text{-}xyz$, is fixed on the ground. The body reference frame, e.g., $c_i\text{-}xyz$ for B_i, is fixed in the space with its origin coinciding with the center of mass (CM) of the body. For simplicity without loss of generality, all body reference frames are set to be parallel to $o\text{-}xyz$ in this paper. The spring reference frame, e.g., $u_{ijs}\text{-}x'y'z'$ for K_{ijs}, is located at one of the spring acting points. The joint reference frame, e.g., $v_{ij}\text{-}x''y''z''$ for J_{ij}, is located at one of the joint acting points.

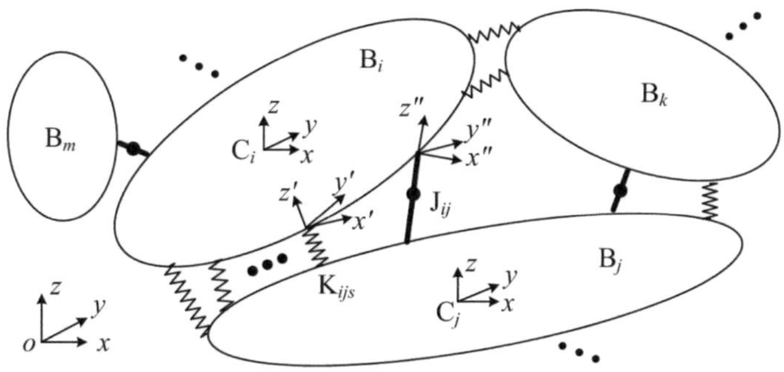

Fig. 1. Elements and reference frames in multibody system

Define m_i the mass of B_i, $\boldsymbol{J}_i$ the inertia tensor of B_i with respect to $c_i\text{-}xyz$, and $\boldsymbol{I}$ the 3×3 identity matrix. Then the mass matrix of body B_i with respect to $c_i\text{-}xyz$ is given by

$$\boldsymbol{M}_i = \text{diag}(m_i\boldsymbol{I} \;\; \boldsymbol{J}_i) \tag{1}$$

The mass matrix of the free multibody system can be organized as

$$\boldsymbol{M} = \text{diag}(\boldsymbol{M}_1 \;\; \boldsymbol{M}_2 \;\; \cdots \;\; \boldsymbol{M}_n) \tag{2}$$

The translation of CM of B_i is specified via vector $\boldsymbol{r}_i = [x_i \;\; y_i \;\; z_i]^{\mathrm{T}}$. The rotation of B_i is specified via Bryan angles $\boldsymbol{\theta}_i = [\alpha_i \;\; \beta_i \;\; \gamma_i]^{\mathrm{T}}$. The absolute angular velocities can be written as (Wittenburg, 2008)

$$\boldsymbol{\omega}_i = \begin{bmatrix} \omega_{ix} \\ \omega_{iy} \\ \omega_{iz} \end{bmatrix} = \begin{bmatrix} C_{\beta i}C_{\gamma i} & S_{\gamma i} & 0 \\ -C_{\beta i}S_{\gamma i} & C_{\gamma i} & 0 \\ S_{\beta i} & 0 & 1 \end{bmatrix} \begin{bmatrix} \dot{\alpha}_i \\ \dot{\beta}_i \\ \dot{\gamma}_i \end{bmatrix} \tag{3}$$

where $S_\mu = \sin\mu$, $C_\mu = \cos\mu$ ($\mu = \alpha_i, \beta_i, \gamma_i$).

Due to small angular displacements of bodies, i.e., $\alpha_i, \beta_i, \gamma_i \approx 0$, the absolute angular velocities and displacements can be linearized as (Wittenburg, 2008)

$$\boldsymbol{\omega}_i \approx [\dot{\alpha}_i \;\; \dot{\beta}_i \;\; \dot{\gamma}_i]^{\mathrm{T}} = \dot{\boldsymbol{\theta}}_i \tag{4}$$

$$\Theta = \int \omega_i \mathrm{d}t \approx \int \dot{\theta}_i \mathrm{d}t = \theta_i \tag{5}$$

The spatial displacements of B_i can be unified as

$$q_i = [r_i^{\mathrm{T}} \ \theta_i^{\mathrm{T}}]^{\mathrm{T}} = [x_i \ y_i \ z_i \ \alpha_i \ \beta_i \ \gamma_i]^{\mathrm{T}} \tag{6}$$

The displacements and velocities for free multibody system can be organized as $q = [q_1^{\mathrm{T}} \ q_2^{\mathrm{T}} \ \cdots \ q_n^{\mathrm{T}}]^{\mathrm{T}}$ and $\dot{q} = [\dot{q}_1^{\mathrm{T}} \ \dot{q}_2^{\mathrm{T}} \ \cdots \ \dot{q}_n^{\mathrm{T}}]^{\mathrm{T}}$.
The stiffness and damping coefficients of K_{ijs} are defined in spring reference frame $u_{ijs}\text{-}x'y'z'$ as $K_{ijs}^{u} = \mathrm{diag}(k_x \ k_y \ k_z \ k_\alpha \ k_\beta \ k_\gamma)$, $C_{ijs}^{u} = \mathrm{diag}(c_x \ c_y \ c_z \ c_\alpha \ c_\beta \ c_\gamma)$. P_{ijs} and P_{jis} are the acting points of K_{ijs} on B_i and B_j. $\bar{r}_{ijs} = [\bar{x}_{ijs} \ \bar{y}_{ijs} \ \bar{z}_{ijs}]^{\mathrm{T}}$ denotes the original position of P_{ijs} relative to $c_i\text{-}xyz$. $\bar{r}_{jis} = [\bar{x}_{jis} \ \bar{y}_{jis} \ \bar{z}_{jis}]^{\mathrm{T}}$ denotes the original position of P_{jis} relative to $c_j\text{-}xyz$. $\bar{\theta}_{ijs} = [\bar{\alpha}_{ijs} \ \bar{\beta}_{ijs} \ \bar{\gamma}_{ijs}]^{\mathrm{T}}$ denotes the original orientation of K_{ijs} relative to $c_i\text{-}xyz$.
Most of the joints that used for practical applications can be modeled in terms of the so-called lower pairs, including revolute, prismatic, cylindrical, universal, spherical, and planar joints. Each joint reduces corresponding number of degrees of freedom (DOFs) of the distal body (Pott et al., 2007; Müller, 2004) between two connected bodies. Assume there is an ideal joint J_{ij} between body B_i and B_j. The acting points of J_{ij} on B_i and B_j are marked as Q_{ij} and Q_{ji}, respectively. $\bar{r}_{ijq} = [\bar{x}_{ijq} \ \bar{y}_{ijq} \ \bar{z}_{ijq}]^{\mathrm{T}}$ denotes the original position of Q_{ij} relative to $c_i\text{-}xyz$. $\bar{r}_{jiq} = [\bar{x}_{jiq} \ \bar{y}_{jiq} \ \bar{z}_{jiq}]^{\mathrm{T}}$ denotes the original position of Q_{ji} relative to $c_j\text{-}xyz$. $\bar{\theta}_{ij} = [\bar{\alpha}_{ij} \ \bar{\beta}_{ij} \ \bar{\gamma}_{ij}]^{\mathrm{T}}$ denotes the original orientation of J_{ij} relative to $c_i\text{-}xyz$. q_{ij}^{v} and q_{ji}^{v} are absolute joint displacements of Q_{ij} and Q_{ji} with respect to $v_{ij}\text{-}x''y''z''$. A 6×6 diagonal matrix H is introduced for each kind of joint to formulate the constraint equations in terms of absolute joint displacements. For example, the constraint equations for joint J_{ij} can be written as

$$H_{ij} q_{ij}^{v} = H_{ij} q_{ji}^{v} \tag{7}$$

The meaning of matrix H can be explained as follows: the value of each diagonal element in H is either one or zero, representing whether the DOF along the corresponding axis is constrained or not. In order to reduce the number of constraint equations, another matrix D is introduced for each kind of joint to extract the independent variables, e.g., for joint J_{ij} it turns to be $q_j' = D_{ij} q_{qji}^{v}$. Matrix D is obtained from matrix $I - H$ by removing those rows whose elements are all zero. Matrices for some common joints are shown in Table 1.
Transmission mechanisms are another kind of constraints widely used in mechanical systems, such as gear pair, rackandpinion, worm gear pair, screw pair, etc. They are usually related to a pair of joints, therefore the constraint equations can be written in terms of absolute joint displacements. Suppose there is a transmission mechanism T_{kr} between body B_k and B_r, T_{kr} is related to joint J_{jk} and J_{mr}. The joint acting point of J_{jk} on B_k is marked as Q_{jk}, and that of J_{mr} on B_r is marked as Q_{mr}. The constraint equations for T_{kr} can be expressed as

$$G_k q_{jk}^{v} + G_r q_{mr}^{v} = 0 \tag{8}$$

where q_{jk}^{v} is the absolute joint displacement of Q_{jk} with respect to $v_{jk}\text{-}x''y''z''$, and q_{mr}^{v} is that of Q_{mr} with respect to $v_{mr}\text{-}x''y''z''$. Matrices G_k and G_r are used to extract variables relative to transmission mechanism. Matrices for some common transmission mechanisms are shown in Table 2, in which i is the transmission ratio.

Joint type	Free axes	Matrix $\boldsymbol{H}$	Matrix $\boldsymbol{D}$
Fixed	none	$\boldsymbol{I}_6$	null matrix
revolute	γ	$\mathrm{diag}(1\ 1\ 1\ 1\ 1\ 0)$	$[0\ 0\ 0\ 0\ 0\ 1]$
prismatic	z	$\mathrm{diag}(1\ 1\ 0\ 1\ 1\ 1)$	$[0\ 0\ 1\ 0\ 0\ 0]$
cylindrical	z,γ	$\mathrm{diag}(1\ 1\ 0\ 1\ 1\ 0)$	$\begin{bmatrix}0&0&1&0&0&0\\0&0&0&0&0&1\end{bmatrix}$
universal	α,β	$\mathrm{diag}(1\ 1\ 1\ 0\ 0\ 1)$	$\begin{bmatrix}0&0&0&1&0&0\\0&0&0&0&1&0\end{bmatrix}$
spherical	α,β,γ	$\mathrm{diag}(1\ 1\ 1\ 0\ 0\ 0)$	$\begin{bmatrix}0&0&0&1&0&0\\0&0&0&0&1&0\\0&0&0&0&0&1\end{bmatrix}$
planar	x,y,γ	$\mathrm{diag}(0\ 0\ 1\ 1\ 1\ 0)$	$\begin{bmatrix}1&0&0&0&0&0\\0&1&0&0&0&0\\0&0&0&0&0&1\end{bmatrix}$
…	…	…	…

Table 1. Mathematical definition of some common joints

Transmission	Constraint equation	Matrix $\boldsymbol{G}_1$	Matrix $\boldsymbol{G}_2$
Gear pair	$\hat{\gamma}_1 + i\,\hat{\gamma}_2 = 0$	$[0\ 0\ 0\ 0\ 0\ 1]$	$[0\ 0\ 0\ 0\ 0\ i]$
Worm gear pair	$\hat{\gamma}_1 + i\,\hat{\gamma}_2 = 0$	$[0\ 0\ 0\ 0\ 0\ 1]$	$[0\ 0\ 0\ 0\ 0\ i]$
Rackandpinion	$\hat{\gamma}_1 + i\,\hat{z}_2 = 0$	$[0\ 0\ 0\ 0\ 0\ 1]$	$[0\ 0\ i\ 0\ 0\ 0]$
Screw pair	$\hat{\gamma}_1 + i\,\hat{z}_1 - i\,\hat{z}_2 = 0$	$[0\ 0\ i\ 0\ 0\ 1]$	$[0\ 0\ -i\ 0\ 0\ 0]$
…	…	…	…

Table 2. Mathematical definition of some transmission mechanisms

2.2 Linear transformation of vibration displacements

Transformation of displacements of two points on a same rigid body is fundamental to the dynamics of a multibody system. The transformation can be divided into two steps. Firstly, the displacements of spring acting point are formulated by using the displacements of CM on the same body, with respect to the same reference frame. And then the resulting displacements are transformed from body reference frame to spring reference frame. A linear transformation is proposed for vibration displacements based on homogeneous transformation.

Assume that there are two reference frames, $c\text{-}xyz$ and $u\text{-}x'y'z'$. The direction cosine matrix from $c\text{-}xyz$ to $u\text{-}x'y'z'$ is determined by $\boldsymbol{\theta} = [\alpha\ \ \beta\ \ \gamma]^{\mathrm{T}}$ as follows

$$\boldsymbol{A}^{cu} = \begin{bmatrix} \mathrm{C}_\beta \mathrm{C}_\gamma & \mathrm{C}_\alpha \mathrm{S}_\gamma + \mathrm{S}_\alpha \mathrm{S}_\beta \mathrm{C}_\gamma & \mathrm{S}_\alpha \mathrm{S}_\gamma - \mathrm{C}_\alpha \mathrm{S}_\beta \mathrm{C}_\gamma \\ -\mathrm{C}_\beta \mathrm{S}_\gamma & \mathrm{C}_\alpha \mathrm{C}_\gamma - \mathrm{S}_\alpha \mathrm{S}_\beta \mathrm{S}_\gamma & \mathrm{S}_\alpha \mathrm{C}_\gamma + \mathrm{C}_\alpha \mathrm{S}_\beta \mathrm{S}_\gamma \\ \mathrm{S}_\beta & -\mathrm{S}_\alpha \mathrm{C}_\beta & \mathrm{C}_\alpha \mathrm{C}_\beta \end{bmatrix} \tag{9}$$

where $S_\mu = \sin\mu$, $C_\mu = \cos\mu$ $(\mu = \alpha, \beta, \gamma)$.
The translational and rotational displacements of a same rigid body can be integrated as a spatial vector, as shown in Fig. 2. And its transformation between different reference frames can be expressed as

$$\boldsymbol{q}_{\mathrm{C}}^{u} = \begin{bmatrix} \boldsymbol{r}_{\mathrm{C}}^{u} \\ \boldsymbol{\theta}_{\mathrm{C}}^{u} \end{bmatrix} = \begin{bmatrix} \boldsymbol{A}^{cu} & \boldsymbol{0} \\ \boldsymbol{0} & \boldsymbol{A}^{cu} \end{bmatrix} \begin{bmatrix} \boldsymbol{r}_{\mathrm{C}}^{c} \\ \boldsymbol{\theta}_{\mathrm{C}}^{c} \end{bmatrix} = \boldsymbol{R}^{cu} \boldsymbol{q}_{\mathrm{C}}^{c} \tag{10}$$

Suppose C and P are two different points on a same rigid body. As shown in Fig. 3, $\overline{\boldsymbol{r}}_{CP} = [\overline{x}_{CP}\ \ \overline{y}_{CP}\ \ \overline{z}_{CP}]^{\mathrm{T}}$ denotes the position of P relative to C. $\boldsymbol{q}_{\mathrm{C}} = [\boldsymbol{r}_{\mathrm{C}}^{\mathrm{T}}\ \ \boldsymbol{\theta}_{\mathrm{C}}^{\mathrm{T}}]^{\mathrm{T}}$ denotes the vector of displacements of point C. Notice that point mentioned in this paper is actually mark that has angular displacements. The translational displacements of point P can be expressed as

$$\begin{aligned} \boldsymbol{r}_{P} &= \overline{\boldsymbol{r}}_{OP'} - \overline{\boldsymbol{r}}_{OP} \\ &= \overline{\boldsymbol{r}}_{OC} + \boldsymbol{r}_{\mathrm{C}} + \overline{\boldsymbol{r}}_{C'P'} - (\overline{\boldsymbol{r}}_{OC} + \overline{\boldsymbol{r}}_{CP}) \\ &= \boldsymbol{r}_{\mathrm{C}} + \boldsymbol{A}^{-1}\, \overline{\boldsymbol{r}}_{CP} - \overline{\boldsymbol{r}}_{CP} \\ &= \boldsymbol{r}_{\mathrm{C}} + (\boldsymbol{A}^{\mathrm{T}} - \boldsymbol{I})\overline{\boldsymbol{r}}_{CP} \end{aligned} \tag{11}$$

The rotational displacements of different points on a same rigid body are equal to each other, i.e., $\boldsymbol{\theta}_P = \boldsymbol{\theta}_{\mathrm{C}}$. It means that the translational and rotational displacements of point P can be integrated as

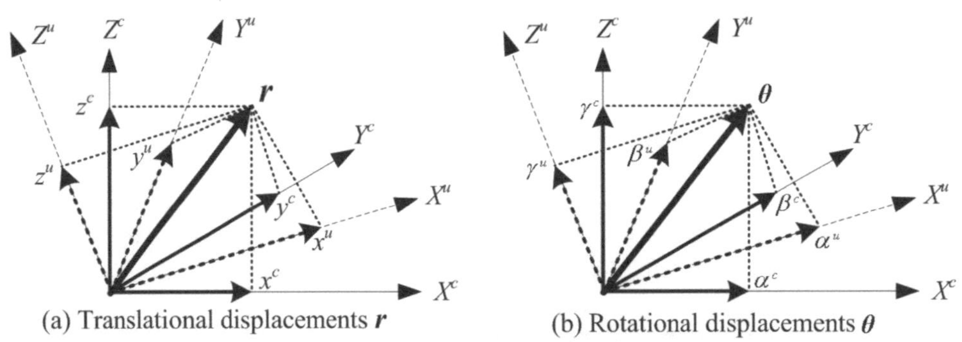

Fig. 2. Finite displacements of the same rigid body in two frames

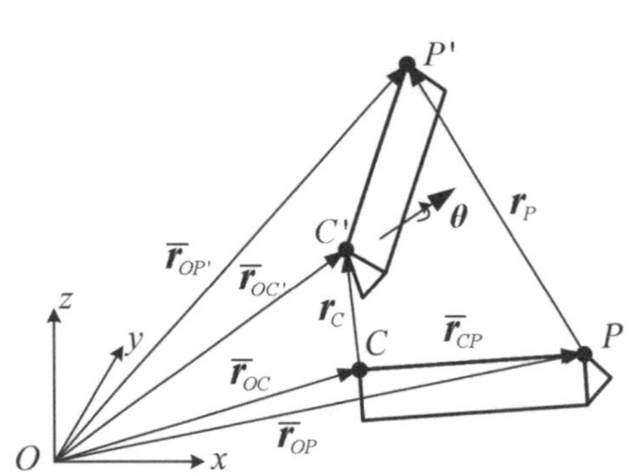

Fig. 3. Finite displacements of two points on a same rigid body

$$q_P = \begin{bmatrix} r_P \\ \theta_P \end{bmatrix} = \begin{bmatrix} r_C + (A^{\mathrm{T}} - I)\overline{r}_{CP} \\ \theta_C \end{bmatrix} \tag{12}$$

Due to small angular displacements for vibration analysis, i.e., $\alpha,\beta,\gamma \approx 0$, the direction cosine matrix in Eq. (9) can be linearized as (Wittenburg, 2008)

$$A \approx \begin{bmatrix} 1 & \gamma & -\beta \\ -\gamma & 1 & \alpha \\ \beta & -\alpha & 1 \end{bmatrix} \tag{13}$$

Substitute Eq. (13) into Eq.(11), it yields

$$(A^{\mathrm{T}} - I)\overline{r}_{CP} \approx \begin{bmatrix} 0 & -\gamma & \beta \\ \gamma & 0 & -\alpha \\ -\beta & \alpha & 0 \end{bmatrix} \begin{bmatrix} \overline{x}_{CP} \\ \overline{y}_{CP} \\ \overline{z}_{CP} \end{bmatrix} = \begin{bmatrix} 0 & -\overline{z}_{CP} & \overline{y}_{CP} \\ \overline{z}_{CP} & 0 & -\overline{x}_{CP} \\ -\overline{y}_{CP} & \overline{x}_{CP} & 0 \end{bmatrix} \begin{bmatrix} \alpha \\ \beta \\ \gamma \end{bmatrix} = U_{CP}\theta_C \tag{14}$$

Therefore Eq. (12) can be linearized to formulate the relationship between fine displacements of two points on a same rigid body as follows

$$q_P \approx \begin{bmatrix} I & U_{CP} \\ 0 & I \end{bmatrix} q_C = T_{CP} q_C \tag{15}$$

According to description in Section 2, the displacements of spring acting point P_{ijs} in u_{ijs}-$x'y'z'$ can be figured out using fine displacements of CM of the body in c-xyz as follows

$$q_{ijs}^{u} = R_{ijs}^{cu} T_{ijs} q_i \tag{16}$$

where R_{ijs}^{cu} can be formulated using $\overline{\theta}_{ijs}$ according to Eqs. (9) and (10), and T_{ijs} can be formulated using $\overline{r}_{ijs}$ according to Eqs. (14) and (15).
Similarly, displacements of joint acting point Q_{ij} in v_{ij}-$x''y''z''$ can be expressed as

$$q_{ij}^{v} = R_{ij}^{cv} T_{ij} q_i \tag{17}$$

where R_{ij}^{cv} can be formulated using $\overline{\theta}_{ij}$ according to Eqs. (9) and (10), and T_{ij} can be formulated using $\overline{r}_{ij}$ according to Eqs. (14) and (15).

3. Topology-based vibration formulation of multibody systems

Generally, there might be none or more then one joint in a multibody system. As shown in Fig. 4, the topologies of constraints in multibody systems can be classified into five groups: (a) free, (b) scattered, (c) chain, (d) tree, and (e) closed-loop. Free multibody system means that there is no constraint in the system. Groups (b), (c) and (d) can all be regarded as general open-loop multibody system. Since the spring-dampers do not change the topology of constraints in a multibody system, spring-dampers between two nonadjacent bodies are not displayed in the figure.
Considering a general closed-loop multibody system as shown in Fig. 4(e), body B_i, B_j, B_k and B_r are connected with joints J_{ij}, J_{jk} and J_{rk}, whereas B_j, B_m and B_r are connected with joints J_{jm} and J_{mr}. Without loss of generality, assume that $1 \le i < j < k < m < r \le n$. Firstly,

linearized ODEs in terms of absolute displacements are derived by using Lagrangian method for free multibody system without considering any constraint, as shown in Fig. 4(a). Secondly, an open-loop constraint matrix is derived to formulate linearized ODEs via quadric transformation for open-loop multibody system, which is obtained by ignoring all cut-joints (Müller, 2004 ; Pott et al., 2007), e.g., if J_{kr} is chosen as cut-joint and one can obtain open-loop multibody system as shown in Fig. 4(d). Finally, a cut-joint constraint matrix corresponding to all cut-joints is solved to formulate a minimal set of ODEs via quadric transformation for closed-loop multibody system.

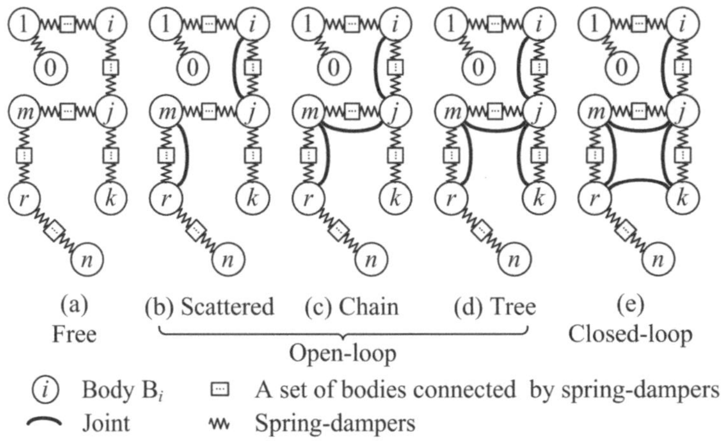

Fig. 4. Topologies of constraints in multibody system

3.1 Vibration formulation of free multibody system

The total kinetic energy of the system as shown in Fig. 4(a) is the summation of translational energy and rotational energy of all bodies, i.e.,

$$T = \sum_{i=1}^{n}\left(\frac{1}{2}\dot{\boldsymbol{r}}_i^{\mathrm{T}} m_i \dot{\boldsymbol{r}}_i + \frac{1}{2}\boldsymbol{\omega}_i^{\mathrm{T}} \boldsymbol{J}_i \boldsymbol{\omega}_i\right) \approx \sum_{i=1}^{n}\frac{1}{2}\dot{\boldsymbol{q}}_i^{\mathrm{T}} \boldsymbol{M}_i \dot{\boldsymbol{q}}_i \tag{18}$$

The fine deformation of spring K_{ijs} can be formulated as difference of displacements between P_{ijs} and P_{jis} in u_{ijs}-$x'y'z'$

$$\Delta \boldsymbol{q}_{ijs}^{u} = \boldsymbol{q}_{jis}^{u} - \boldsymbol{q}_{ijs}^{u} = \boldsymbol{R}_{ijs}^{cu} \boldsymbol{T}_{jis} \boldsymbol{q}_j - \boldsymbol{R}_{ijs}^{cu} \boldsymbol{T}_{ijs} \boldsymbol{q}_i \tag{19}$$

Set the potential energy of the system at equilibrium positions to be zero. Then the potential energy of spring K_{ijs} can be formulated as

$$V_{ijs} = \frac{1}{2}\left(\Delta \boldsymbol{q}_{ijs}^{u}\right)^{\mathrm{T}} \boldsymbol{K}_{ijs}^{u} \Delta \boldsymbol{q}_{ijs}^{u} \tag{20}$$

The potential energy of the entire system is the sum of gravitational potential V_g and elastic potential V_k, i.e.,

$$V = V_g + V_k = \sum_{i=0}^{n} \boldsymbol{q}_i^{\mathrm{T}} \boldsymbol{M}_i \boldsymbol{g} + \sum_{i=0}^{n-1} \sum_{j=i+1}^{n} \sum_{s=0}^{s_{ij}} V_{ijs} \tag{21}$$

where $g = [0\ 0\ g\ 0\ 0\ 0]^{\mathrm{T}}$ is the vector of gravitational acceleration. Since there might be no spring-damper between two bodies, a "virtual spring-damper" which has no effect on the system is introduced between each two bodies for consistency in formula. For example, K_{ij0} is the "virtual spring-damper" between body B_i and B_j, and $\boldsymbol{K}_{ij0}^{u} = \boldsymbol{0}$, $\boldsymbol{C}_{ij0}^{u} = \boldsymbol{0}$.
The Lagrangian equations of the system take the form

$$\frac{d}{dt}\left(\frac{\partial T}{\partial \dot{\boldsymbol{q}}_i^{\mathrm{T}}}\right) - \frac{\partial V}{\partial \boldsymbol{q}_i^{\mathrm{T}}} = \boldsymbol{f}_{di} + \boldsymbol{f}_{ei} \tag{22}$$

where $i = 1, 2, \cdots, n$, $\boldsymbol{f}_{di}$ and $\boldsymbol{f}_{ei}$ denote the damping forces and other non-potential forces acting on body B_i.
Due to property $\boldsymbol{M}_i^{\mathrm{T}} = \boldsymbol{M}_i$, it yields

$$\frac{\mathrm{d}}{\mathrm{d}t}\left(\frac{\partial T}{\partial \dot{\boldsymbol{q}}_i^{\mathrm{T}}}\right) = \frac{1}{2}(\boldsymbol{M}_i + \boldsymbol{M}_i^{\mathrm{T}})\ddot{\boldsymbol{q}}_i = \boldsymbol{M}_i \ddot{\boldsymbol{q}}_i \tag{23}$$

Substitute Eqs. (19) and (20) into Eq. (21), and derivate V with respect to $\boldsymbol{q}_i^{\mathrm{T}}$, it yields

$$\begin{aligned}
\frac{\partial V}{\partial \boldsymbol{q}_i^{\mathrm{T}}} &= \sum_{k=0,k\neq i}^{n} \frac{\partial \boldsymbol{q}_k^{\mathrm{T}}}{\partial \boldsymbol{q}_i^{\mathrm{T}}} \boldsymbol{M}_k \boldsymbol{g} + \frac{\partial \boldsymbol{q}_i^{\mathrm{T}}}{\partial \boldsymbol{q}_i^{\mathrm{T}}} \boldsymbol{M}_i \boldsymbol{g} + \sum_{k=0}^{i-1}\sum_{s=0}^{s_{ki}} \frac{\partial V_{kis}}{\partial \boldsymbol{q}_i^{\mathrm{T}}} + \sum_{j=i+1}^{n}\sum_{s=0}^{s_{ij}} \frac{\partial V_{ijs}}{\partial \boldsymbol{q}_i^{\mathrm{T}}} + \sum_{k=i+1}^{n-1}\sum_{j=k+1}^{n}\sum_{s=0}^{s_{kj}} \frac{\partial V_{kjs}}{\partial \boldsymbol{q}_i^{\mathrm{T}}} \\
&= \boldsymbol{0} + \boldsymbol{M}_i \boldsymbol{g} + \sum_{j=0}^{i-1}\sum_{s=0}^{s_{ij}} \frac{\partial V_{ijs}}{\partial \boldsymbol{q}_i^{\mathrm{T}}} + \sum_{j=i+1}^{n}\sum_{s=0}^{s_{ij}} \frac{\partial V_{ijs}}{\partial \boldsymbol{q}_i^{\mathrm{T}}} + \boldsymbol{0} \\
&= \boldsymbol{M}_i \boldsymbol{g} + \sum_{j=0,j\neq i}^{n}\sum_{s=0}^{s_{ij}} \frac{\partial V_{ijs}}{\partial \boldsymbol{q}_i^{\mathrm{T}}} \\
&= \boldsymbol{M}_i \boldsymbol{g} + \sum_{j=0,j\neq i}^{n}\sum_{s=0}^{s_{ij}} \{(\boldsymbol{T}_{ijs})^{\mathrm{T}} (\boldsymbol{R}_{ijs}^{cu})^{\mathrm{T}} \boldsymbol{K}_{ijs}^{u} \boldsymbol{R}_{ijs}^{cu} (\boldsymbol{T}_{ijs}\boldsymbol{q}_i - \boldsymbol{T}_{jis}\boldsymbol{q}_j)\} \\
&= \boldsymbol{M}_i \boldsymbol{g} + \left[\sum_{j=0,j\neq i}^{n}\sum_{s=0}^{s_{ij}} (\boldsymbol{T}_{ijs})^{\mathrm{T}} (\boldsymbol{R}_{ijs}^{cu})^{\mathrm{T}} \boldsymbol{K}_{ijs}^{u} \boldsymbol{R}_{ijs}^{cu} \boldsymbol{T}_{ijs}\right]\boldsymbol{q}_i - \sum_{j=0,j\neq i}^{n}\left[\sum_{s=0}^{s_{ij}} (\boldsymbol{T}_{ijs})^{\mathrm{T}} (\boldsymbol{R}_{ijs}^{cu})^{\mathrm{T}} \boldsymbol{K}_{ijs}^{u} \boldsymbol{R}_{ijs}^{cu} \boldsymbol{T}_{jis}\boldsymbol{q}_j\right]
\end{aligned} \tag{24}$$

Denote

$$\boldsymbol{E}_{ii} = \sum_{j=0,j\neq i}^{n}\sum_{s=0}^{s_{ij}} (\boldsymbol{T}_{ijs})^{\mathrm{T}} (\boldsymbol{R}_{ijs}^{cu})^{\mathrm{T}} \boldsymbol{E}_{ijs}^{u} \boldsymbol{R}_{ijs}^{cu} \boldsymbol{T}_{ijs} \tag{25}$$

$$\boldsymbol{E}_{ij} = \sum_{s=0}^{s_{ij}} (\boldsymbol{T}_{ijs})^{\mathrm{T}} (\boldsymbol{R}_{ijs}^{cu})^{\mathrm{T}} \boldsymbol{E}_{ijs}^{u} \boldsymbol{R}_{ijs}^{cu} \boldsymbol{T}_{jis} \tag{26}$$

Let $\boldsymbol{E} = \boldsymbol{K}$, then Eq. (24) can be rewritten as

$$\frac{\partial V}{\partial \boldsymbol{q}_i^{\mathrm{T}}} = \boldsymbol{K}_{ii}\boldsymbol{q}_i - \sum_{j=0,j\neq i}^{n} \boldsymbol{K}_{ij}\boldsymbol{q}_j + \boldsymbol{M}_i \boldsymbol{g} \tag{27}$$

The dissipation power due to damping forces can be formulated as (Wittbrodt, 2006)

$$P = -\sum_{i=0}^{n-1}\sum_{j=i+1}^{n}\sum_{s=0}^{s_{ij}} \frac{1}{2}(\Delta\dot{\boldsymbol{q}}_{ijs}^{u})^{\mathrm{T}} \boldsymbol{C}_{ijs}^{u} \Delta\dot{\boldsymbol{q}}_{ijs}^{u} \tag{28}$$

Similarly, the damping forces acting on B_i with respect to c_i-xyz can be evaluated as

$$\boldsymbol{f}_{di} = \frac{\partial P}{\partial \dot{\boldsymbol{q}}_i^{\mathrm{T}}} = -\boldsymbol{C}_{ii}\dot{\boldsymbol{q}}_i + \sum_{j=0,j\neq i}^{n} \boldsymbol{C}_{ij}\dot{\boldsymbol{q}}_j \tag{29}$$

It can be proved that C_{ii} and C_{ij} are also determined by Eqs. (25) and (26) for $E = C$.
The linearized ODEs for a free multibody system turn to be

$$M\ddot{q} + C\dot{q} + Kq = f_e - f_g \tag{30}$$

where quantities $f_g = [(M_1 g)^{\mathrm{T}}\ (M_2 g)^{\mathrm{T}}\ \cdots\ (M_n g)^{\mathrm{T}}]^{\mathrm{T}}$ and $f_e = [f_{e1}{}^{\mathrm{T}}\ f_{e2}{}^{\mathrm{T}}\ \cdots\ f_{en}{}^{\mathrm{T}}]^{\mathrm{T}}$ denote gravity forces and other non-potential forces. The damping matrix C and stiffness matrix K in Eq. (30) take the same form

$$E = \begin{bmatrix} E_{11} & -E_{12} & \cdots & -E_{1n} \\ -E_{21} & E_{22} & \cdots & \vdots \\ \vdots & \vdots & \ddots & -E_{n-1,n} \\ -E_{n1} & \cdots & -E_{n,n-1} & E_{nn} \end{bmatrix} \quad (E = C, K) \tag{31}$$

The block matrices K_{ii} and C_{ii} contain parameters of all springs and dampers that connected with B_i. K_{ij} and C_{ij} contain parameters of all springs and dampers that connected between B_i and B_j. Matrices C and K contain explicitly damping coefficients and stiffness coefficients, and reveal clearly the topology of spring-dampers.
By using the system matrices M, C and K, Eqs (18), (21) and (28) can be reformed as

$$T = \frac{1}{2}\dot{q}^{\mathrm{T}} M \dot{q} \tag{32}$$

$$V = \frac{1}{2}q^{\mathrm{T}} K q + q^{\mathrm{T}} f_g \tag{33}$$

$$P = \frac{1}{2}\dot{q}^{\mathrm{T}} C \dot{q} \tag{34}$$

3.2 Vibration formulation of open-loop multibody system

Select J_{rk} in Fig. 4(e) as cut-joint and one can obtain open-loop multibody system as shown in Fig. 4(d). The constraint equations for joint J_{ij} can be written as

$$H_{ij} q_{ij}^{v} = H_{ij} R_{ij}^{cv} T_{ij} q_i = H_{ij} q_{ji}^{v} \tag{35}$$

where q_{ij}^{v} and q_{ji}^{v} denote the displacements of joint acting points Q_{ij} and Q_{ji} with respect to $v_{ij}\text{-}x''y''z''$, respectively. R_{ij}^{cv} is determined by $\bar{\theta}_{ij}$ according to Eqs. (9) and (10). T_{ij} is determined by $\bar{r}_{ij}$ according to Eqs. (14) and (15).
Due to properties $(I - H_{ij}) D_{ij}^{\mathrm{T}} D_{ij} = I - H_{ij}$ and $(R^{cv})^{-1} = R^{vc}$, Eq. (35) can be reformed as

$$\begin{aligned} q_j &= (T_{ji})^{-1} R_{ij}^{vc} H_{ij} R_{ij}^{cv} T_{ij} q_i + (T_{ji})^{-1} R_{ij}^{vc} (I - H_{ij}) q_{ji}^{v} \\ &= (T_{ji})^{-1} R_{ij}^{vc} H_{ij} R_{ij}^{cv} T_{ij} q_i + (T_{ji})^{-1} R_{ij}^{vc} (I - H_{ij}) D_{ij}^{\mathrm{T}} D_{ij} q_{ji}^{v} \end{aligned} \tag{36}$$

Define

$$P_{ij} = (T_{ji})^{-1} R_{ij}^{vc} H_{ij} R_{ij}^{cv} T_{ij} \tag{37}$$

$$Q_{ij} = (T_{ji})^{-1} R_{ij}^{vc} (I - H_{ij}) D_{ij}^{\mathrm{T}} \tag{38}$$

Considering that $\boldsymbol{q}'_j = \boldsymbol{D}_{ij}\boldsymbol{q}^v_{ji}$, Eq. (36) can be written as

$$\boldsymbol{q}_j = \boldsymbol{P}_{ij}\boldsymbol{q}_i + \boldsymbol{Q}_{ij}\boldsymbol{q}'_j \tag{39}$$

Similarly, the constraint equations for joint J_{jk} are

$$\boldsymbol{q}_k = \boldsymbol{P}_{jk}\boldsymbol{P}_{ij}\boldsymbol{q}_i + \boldsymbol{P}_{jk}\boldsymbol{Q}_{ij}\boldsymbol{q}'_j + \boldsymbol{Q}_{jk}\boldsymbol{q}'_k \tag{40}$$

The constraint equations for all the rest joints can be formulated similar to Eq. (40). The constraint equations for the entire open-loop system can thus be integrated as

$$\boldsymbol{q} = \boldsymbol{B}'\boldsymbol{q}' \tag{41}$$

The open-loop constraint matrix $\boldsymbol{B}'$ corresponding to system shown in Fig. 4(d) takes the form

$$\boldsymbol{B}' = \begin{bmatrix} \boldsymbol{I}_a & \boldsymbol{0} & \boldsymbol{0} & \boldsymbol{0} & \boldsymbol{0} & \boldsymbol{0} & \boldsymbol{0} & \boldsymbol{0} & \boldsymbol{0} & \boldsymbol{0} & \boldsymbol{0} \\ \boldsymbol{0} & \boldsymbol{I}_6 & \boldsymbol{0} & \boldsymbol{0} & \boldsymbol{0} & \boldsymbol{0} & \boldsymbol{0} & \boldsymbol{0} & \boldsymbol{0} & \boldsymbol{0} & \boldsymbol{0} \\ \boldsymbol{0} & \boldsymbol{0} & \boldsymbol{I}_b & \boldsymbol{0} & \boldsymbol{0} & \boldsymbol{0} & \boldsymbol{0} & \boldsymbol{0} & \boldsymbol{0} & \boldsymbol{0} & \boldsymbol{0} \\ \boldsymbol{0} & \boldsymbol{P}_{ij} & \boldsymbol{0} & \boldsymbol{Q}_{ij} & \boldsymbol{0} & \boldsymbol{0} & \boldsymbol{0} & \boldsymbol{0} & \boldsymbol{0} & \boldsymbol{0} & \boldsymbol{0} \\ \boldsymbol{0} & \boldsymbol{0} & \boldsymbol{0} & \boldsymbol{0} & \boldsymbol{I}_c & \boldsymbol{0} & \boldsymbol{0} & \boldsymbol{0} & \boldsymbol{0} & \boldsymbol{0} & \boldsymbol{0} \\ \boldsymbol{0} & \boldsymbol{P}_{jk}\boldsymbol{P}_{ij} & \boldsymbol{0} & \boldsymbol{P}_{jk}\boldsymbol{Q}_{ij} & \boldsymbol{0} & \boldsymbol{Q}_{jk} & \boldsymbol{0} & \boldsymbol{0} & \boldsymbol{0} & \boldsymbol{0} & \boldsymbol{0} \\ \boldsymbol{0} & \boldsymbol{0} & \boldsymbol{0} & \boldsymbol{0} & \boldsymbol{0} & \boldsymbol{0} & \boldsymbol{I}_d & \boldsymbol{0} & \boldsymbol{0} & \boldsymbol{0} & \boldsymbol{0} \\ \boldsymbol{0} & \boldsymbol{P}_{jm}\boldsymbol{P}_{ij} & \boldsymbol{0} & \boldsymbol{P}_{jm}\boldsymbol{Q}_{ij} & \boldsymbol{0} & \boldsymbol{0} & \boldsymbol{0} & \boldsymbol{Q}_{jm} & \boldsymbol{0} & \boldsymbol{0} & \boldsymbol{0} \\ \boldsymbol{0} & \boldsymbol{0} & \boldsymbol{0} & \boldsymbol{0} & \boldsymbol{0} & \boldsymbol{0} & \boldsymbol{0} & \boldsymbol{0} & \boldsymbol{I}_e & \boldsymbol{0} & \boldsymbol{0} \\ \boldsymbol{0} & \boldsymbol{P}_{mr}\boldsymbol{P}_{jm}\boldsymbol{P}_{ij} & \boldsymbol{0} & \boldsymbol{P}_{mr}\boldsymbol{P}_{jm}\boldsymbol{Q}_{ij} & \boldsymbol{0} & \boldsymbol{0} & \boldsymbol{0} & \boldsymbol{P}_{mr}\boldsymbol{Q}_{jm} & \boldsymbol{0} & \boldsymbol{Q}_{mr} & \boldsymbol{0} \\ \boldsymbol{0} & \boldsymbol{0} & \boldsymbol{0} & \boldsymbol{0} & \boldsymbol{0} & \boldsymbol{0} & \boldsymbol{0} & \boldsymbol{0} & \boldsymbol{0} & \boldsymbol{0} & \boldsymbol{I}_h \end{bmatrix} \tag{42}$$

where $a = 6i-6$, $b = 6(j-i-1)$, $c = 6(k-j-1)$, $d = 6(m-k-1)$, $e = 6(r-m-1)$, and $h = 6(n-r)$. The subscript of each identity matrix $\boldsymbol{I}$ denotes its dimension. Obviously, matrix $\boldsymbol{B}'$ contains information about all joints and reveals constraint topology of open-loop multibody system. In Eq. (41), $\boldsymbol{q}'$ are the general displacements of open-loop multibody system, which are the combination of absolute displacements of CM of unconstrained bodies and absolute joint displacements of constrained bodies, i.e.,

$$\boldsymbol{q}' = [(\boldsymbol{q}'_1)^{\mathrm{T}} \ (\boldsymbol{q}'_2)^{\mathrm{T}} \ \cdots \ (\boldsymbol{q}'_n)^{\mathrm{T}}]^{\mathrm{T}} \tag{43}$$

where $\boldsymbol{q}'_j = \boldsymbol{D}_{ij}\boldsymbol{q}^v_{ji}$, $\boldsymbol{q}'_k = \boldsymbol{D}_{jk}\boldsymbol{q}^v_{kj}$, $\boldsymbol{q}'_m = \boldsymbol{D}_{jm}\boldsymbol{q}^v_{mj}$, $\boldsymbol{q}'_r = \boldsymbol{D}_{mr}\boldsymbol{q}^v_{rm}$, $\boldsymbol{q}'_\varepsilon = \boldsymbol{q}_\varepsilon$ ($\varepsilon = 1,2,\cdots,n$ and $\varepsilon \neq j,k,m,r$). Substitute Eq. (41) and its time derivation, i.e., $\dot{\boldsymbol{q}} = \boldsymbol{B}'\dot{\boldsymbol{q}}'$, into Eqs. (32)-(34), it yields

$$\frac{\mathrm{d}}{\mathrm{d}t}\left(\frac{\partial T}{\partial \dot{\boldsymbol{q}}'^{\mathrm{T}}}\right) = \boldsymbol{B}'^{\mathrm{T}}\boldsymbol{M}\boldsymbol{B}'\ddot{\boldsymbol{q}}' = \boldsymbol{M}'\ddot{\boldsymbol{q}}' \tag{44}$$

$$\frac{\partial V}{\partial \boldsymbol{q}'^{\mathrm{T}}} = \boldsymbol{B}'^{\mathrm{T}}\boldsymbol{K}\boldsymbol{B}'\boldsymbol{q}' + \boldsymbol{B}'^{\mathrm{T}}\boldsymbol{f}_g = \boldsymbol{K}'\boldsymbol{q}' + \boldsymbol{B}'^{\mathrm{T}}\boldsymbol{f}_g \tag{45}$$

$$\boldsymbol{f}'_d = \frac{\partial P}{\partial \dot{\boldsymbol{q}}'^{\mathrm{T}}} = \boldsymbol{B}'^{\mathrm{T}}\boldsymbol{C}\boldsymbol{B}'\dot{\boldsymbol{q}}' = \boldsymbol{C}'\dot{\boldsymbol{q}}' \tag{46}$$

It then follows a minimal set of linearized ODEs for an open-loop multibody system

$$M'\ddot{q}' + C'\dot{q}' + K'q' = B'^{\mathrm{T}}(f_e - f_g) \tag{47}$$

where M', C' and K' are determined via the same quadric transformation

$$E' = B'^{\mathrm{T}}EB' \ \ (E = M, K, C) \tag{48}$$

Eq. (47) can be regarded as obtained by multiplying Eq. (30) with B'^{T} and replacing q by $B'q'$. It indicates that the solution of constraint equations for open-loop multibody system can be directly obtained via quadric transformation upon system matrices for free multibody system, by using the corresponding open-loop constraint matrix B'.

3.3 Vibration formulation of closed-loop multibody system

Considering closed-loop multibody system as shown in Fig. 4(e), similar to Eq. (35), the constraint equations for joint J_{kr} can be expressed as

$$H_{kr}q_{kr}^{v} = H_{kr}q_{rk}^{v} \tag{49}$$

where q_{kr}^{v} and q_{rk}^{v} denote the displacements of points Q_{kr} and Q_{rk} with respect to v_{kr}-$x''y''z''$, respectively.

Rewrite matrix B' with each six rows as a block, i.e., $B' = [B_1'^{\mathrm{T}} \ B_2'^{\mathrm{T}} \ \cdots \ B_n'^{\mathrm{T}}]^{\mathrm{T}}$. According to Eqs. (41) and (17) one can obtain $q_{kr}^{v} = R_{kr}^{cv}T_{kr}B_k'$ and $q_{rk}^{v} = R_{kr}^{cv}T_{rk}B'$. Then Eq. (49) can be rewritten as

$$H_{kr}R_{kr}^{cv}(T_{kr}B_k' - T_{rk}B_r')q' = 0 \tag{50}$$

If the number of cut-joints in a general spatial closed-loop multibody system is c, the constraint equations for all cut-joints can be integrated as

$$Bq' = 0 \tag{51}$$

where $B = [B_1^{\mathrm{T}} \ B_2^{\mathrm{T}} \ \cdots \ B_c^{\mathrm{T}}]^{\mathrm{T}}$, and B_i is the coefficient matrix of constraint equations for the ith cut-joint.

Transmission mechanism can be treated as cut-joint. Suppose the constraints between body B_k and B_r in Fig. 4(e) is not a joint J_{kr} as mentioned before but a transmission mechanism T_{kr}. The details of T_{kr} can be seen in section 1. Similar to Eq. (50), constraint equations specified as Eq. (8) can be rewritten as

$$(G_k R_{jk}^{ck} T_{kj} B_k' + G_r R_{mr}^{cr} T_{rm} B_r')q' = 0 \tag{52}$$

If the number of transmission mechanisms in a general multibody system is t, the constraint equations for all transmission mechanisms can be integrated as

$$Zq' = 0 \tag{53}$$

where $Z = [Z_1^{\mathrm{T}} \ Z_2^{\mathrm{T}} \ \cdots \ Z_t^{\mathrm{T}}]^{\mathrm{T}}$, and Z_j is the coefficient matrix of constraint equations for the jth transmission mechanism.

Equation (51) and (53) can be integrated as constraint equations for cut-joints as follows

$$\begin{bmatrix} \boldsymbol{B} \\ \boldsymbol{Z} \end{bmatrix} \boldsymbol{q}' = \boldsymbol{0} \tag{54}$$

Since there might be redundant constraints in closed-loop system, Eq. (54) can be solved to form independent constraint equations

$$\tilde{\boldsymbol{q}}' = \tilde{\boldsymbol{B}}' \boldsymbol{q}'' \tag{55}$$

where $\boldsymbol{q}''$ is a vector of all independent variables in $\boldsymbol{q}'$, and $\tilde{\boldsymbol{q}}'$ is that of dependent ones.

Considering that the elements in $\boldsymbol{q}''$ or $\tilde{\boldsymbol{q}}'$ are not necessarily consecutive variables in $\boldsymbol{q}'$, they are reordered by introducing a matrix S as

$$\boldsymbol{q}' = \boldsymbol{S}[\boldsymbol{q}''^{\mathrm{T}} \;\; \tilde{\boldsymbol{q}}'^{\mathrm{T}}]^{\mathrm{T}} \tag{56}$$

Substituting Eq. (55) into Eq. (56), and let $\boldsymbol{B}'' = \boldsymbol{S}[\boldsymbol{I} \;\; (\tilde{\boldsymbol{B}}')^{\mathrm{T}}]^{\mathrm{T}}$, it yields

$$\boldsymbol{q}' = \boldsymbol{B}'' \boldsymbol{q}'' \tag{57}$$

Here we call matrix $\boldsymbol{B}''$ the cut-joint constraint matrix. Considering Eq. (41), one can obtain

$$\boldsymbol{q} = \boldsymbol{B}' \boldsymbol{q}' = \boldsymbol{B}' \boldsymbol{B}'' \boldsymbol{q}'' \tag{58}$$

Similar to formulation of open-loop multibody system, substitute Eq. (58) and its time derivation, i.e., $\dot{\boldsymbol{q}} = \boldsymbol{B}' \boldsymbol{B}'' \dot{\boldsymbol{q}}''$, into Eqs. (32)-(34), a minimal set of linearized ODEs for closed-loop multibody system can be expressed as

$$\boldsymbol{M}'' \ddot{\boldsymbol{q}}'' + \boldsymbol{C}'' \dot{\boldsymbol{q}}'' + \boldsymbol{K}'' \boldsymbol{q}'' = \boldsymbol{B}''^{\mathrm{T}} \boldsymbol{B}'^{\mathrm{T}} (\boldsymbol{f}_e - \boldsymbol{f}_g) \tag{59}$$

where $\boldsymbol{M}''$, $\boldsymbol{C}''$ and $\boldsymbol{K}''$ are determined via the same quadric transformation

$$\boldsymbol{E}'' = \boldsymbol{B}''^{\mathrm{T}} \boldsymbol{E}' \boldsymbol{B}'' = \boldsymbol{B}''^{\mathrm{T}} \boldsymbol{B}'^{\mathrm{T}} \boldsymbol{E} \boldsymbol{B}' \boldsymbol{B}'' \quad (\boldsymbol{E} = \boldsymbol{M}, \boldsymbol{K}, \boldsymbol{C}) \tag{60}$$

Equation (59) can be regarded as obtained by multiplying Eq. (47) with the transposed cut-joint constraint matrix $\boldsymbol{B}''^{\mathrm{T}}$ and replacing $\boldsymbol{q}'$ by $\boldsymbol{B}''\boldsymbol{q}''$. It indicates that the solution of constraint equations for cut-joints can be directly obtained via quadric transformation upon system matrices for open-loop system, by using the corresponding cut-joint constraint matrix $\boldsymbol{B}''$.
Complicated solving for constraints and linearization are unnecessary in this method, and the resulting equations contain explicitly the design parameters. The suggested method can be used to greatly simplify the procedure of vibration calculation. Furthermore, the suggested method is particularly suitable for sensitivity analysis and optimization for large-scale multibody system.
The proposed algorithm has been implemented in MATLAB, and is named as AMVA (Automatic Modeling for Vibration Analysis). The eigenvalue problem is solved using standard LAPACK routines. The flowchart of the proposed algorithm is illustrated in Fig. 5.

3.4 Comparison with the traditional methods

The procedure of most of the conventional methods for vibration calculation can be concluded as follows. Firstly, the general-purpose nonlinear equations of motion, in most

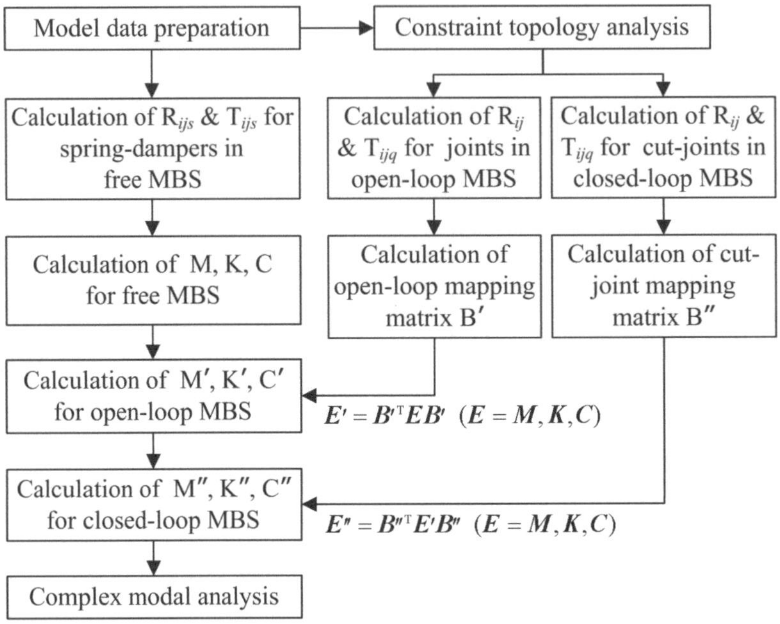

Fig. 5. Flowchart of the proposed formulation

cases DAEs, are formulated in terms of coordinates of all bodies. Secondly, the Jacobian of constraint equations is calculated to transform DAEs into ODEs by eliminating the Lagrange's Multipliers. Thirdly, a minimal set of nonlinear ODEs in terms of independent generalized coordinates are obtained. Finally, the resulting equations are linearized at small vicinity near the equilibrium position. A large amount of computational efforts are required for computation of trigonometric functions, derivation and linearization. Many kinds of software such as ADAMS employ this kind of method for obtaining a minimal set of linear ODEs for vibration analysis.

As shown in Fig. 5, there are three steps in the proposed method to generate a minimal set of second-order linear ODEs for vibration calculation. Firstly, system matrices for linear ODEs of free system are directly generated by using linear transformation. Secondly, an open-loop constraint matrix is formulated to obtain linear ODEs for open-loop system. Finally, a cut-joint constraint matrix is solved to formulate a minimal set of second-order linear ODEs for closed-loop system.

Considering the definitions for vibration calculation, the major difference between the proposed method and previous studies lies in the definition and formulation of constraint equations. Conventionally, the constraint equations are defined in terms of coordinates of bodies or joints. The constraint equations and the Jacobian of constraint matrix are usually nonlinear ones. It is difficult, particularly for large-scale multibody system, to obtain the transformation matrix from the generalized coordinates to the independent coordinates. In this paper, however, the constraint equations are defined in terms of fine displacements of two acting points of the joint. The resulting linear constraint equations can be easily resolved to obtain the transformation matrix, i.e., the open-loop constraint matrix and the cut-joint constraint matrix.

There are two major differences between the proposed method and most of the traditional methods. One is that the linearization is carried out before generating ODEs with small motion assumption which is satisfied for vibration. The other is that the formulation of a minimal set of second-order linear ODEs for constrained system is achieved by directly generating five matrices, i.e., mass matrix, stiffness matrix and damping matrix for free system, an open-loop constraint matrix $\boldsymbol{B}'$ for open-loop system, and a cut-joint constraint matrix $\boldsymbol{B}''$ for closed-loop system.

Notice that Kang et al. have also proposed a similar method in which the linearization is carried out before generating ODEs with small motion assumption (Kang, 2003). The results of system matrices for free system are actually the same as those derived by our method. The difference between Kang's method and ours lies in the formulation of a minimal set of ODEs for constrained system. They employ the partition of the Jacobian of constraint matrix, which is time-consuming to be obtained for multibody system with a large amount of constraints, to derive the relationship between generalized coordinates and the independent coordinates. We use the linear transformation matrix to directly formulate linearized constraint equations and then derive the relationship between generalized coordinates and the independent coordinates. Most of all, since the final system matrices can be directly obtained by only a few steps of matrices generation and multiplication, the computational efficiency can be significantly improved for large-scale multibody system with a large amount of constraints.

4. Topology-based sensitivity formulation of multibody systems

Besides the promise in improving the computational efficiency, the proposed method can be applied in sensitivity analysis because the resulting equations depend on the design parameters explicitly. As is known to all, the eigen-sensitivity is based on the derivatives of the system matrices, which are denoted as $\boldsymbol{M}''$, $\boldsymbol{C}''$ and $\boldsymbol{K}''$ in this paper, with respect to the design parameters (Lee et al., 1999a; 1999b). Conventionally, the system matrices are solved numerically and they depend on the design parameters implicitly. Therefore the derivatives of the system matrices with respect to a certain parameter p are usually obtained by using finite difference method. However, it can be seen that each kind of design parameters can be easily traced in different system matrices obtained by using the proposed method. For example, the stiffness coefficients of spatial spring K_{ijs} only exist in matrix $\boldsymbol{K}^u_{ijs}$ in Eqs. (25) and (26) ($\boldsymbol{E}^u_{ijs}$ refers to $\boldsymbol{K}^u_{ijs}$ for spring). The position parameters of K_{ijs} exist in $\boldsymbol{T}_{ijs}$ and $\boldsymbol{T}_{jis}$, and its orientation parameters exist in $\boldsymbol{R}^{cu}_{ijs}$. Similarly, the position and orientation parameters of joint exist in $\boldsymbol{B}'$ and $\boldsymbol{B}''$. Therefore the derivatives $d\boldsymbol{M}''/dp$, $d\boldsymbol{C}''/dp$ and $d\boldsymbol{K}''/dp$ can be further derived analytically.

4.1 Conventional sensitivity formulation

The eigenvalue sensitivity can be expressed as

$$\frac{\partial \lambda_r}{\partial p} = -\lambda_r^2 \boldsymbol{\psi}_r^{\mathrm{T}} \frac{\partial \boldsymbol{M}''}{\partial p} \boldsymbol{\psi}_r - \lambda_r \boldsymbol{\psi}_r^{\mathrm{T}} \frac{\partial \boldsymbol{C}''}{\partial p} \boldsymbol{\psi}_r - \boldsymbol{\psi}_r^{\mathrm{T}} \frac{\partial \boldsymbol{K}''}{\partial p} \boldsymbol{\psi}_r \tag{61}$$

where λ_r is the rth eigenvalue, $\boldsymbol{\psi}_r = [\varphi_{1r} \;\; \varphi_{1r} \;\; \cdots \;\; \varphi_{Nr}]^{\mathrm{T}}$ ($N = \mathrm{rank}(\boldsymbol{M}'')$) is the rth unitary eigenvector, and p represents the considered parameter. Denote m''_{ij}, c''_{ij} and k''_{ij} the elements at row i and column j in matrices $\boldsymbol{M}''$, $\boldsymbol{C}''$ and $\boldsymbol{K}''$, respectively, eigenvalue sensitivity can be formulated as

$$\frac{\partial \lambda_r}{\partial m''_{ij}} = \begin{cases} -2\lambda_r^2 \varphi_{ir}\varphi_{jr} & (i \neq j) \\ \lambda_r^2 \varphi_{ir}^2 & (i = j) \end{cases}$$

$$\frac{\partial \lambda_r}{\partial c''_{ij}} = \begin{cases} -2\lambda_r \varphi_{ir}\varphi_{jr} & (i \neq j) \\ \lambda_r \varphi_{ir}^2 & (i = j) \end{cases}$$

$$\frac{\partial \lambda_r}{\partial k''_{ij}} = \begin{cases} -2\varphi_{ir}\varphi_{jr} & (i \neq j) \\ \varphi_{ir}^2 & (i = j) \end{cases}$$

The formulation is very simple. However, matrices $\boldsymbol{M}''$, $\boldsymbol{C}''$ and $\boldsymbol{K}''$ generated by using conventional methods are implicit functions of design parameters, such as mass and inertia of bodies, stiffness coefficients and damping coefficients of spring-dampers, position and orientation of spring-dampers and joints, and etc. That is to say, m''_{ij}, c''_{ij} and k''_{ij} are intermediate quantities instead of original design parameters. Therefore, the existing sensitivity formula can not be directly used for optimization.

4.2 Proposed sensitivity formulation about physical design parameters

Since matrices $\boldsymbol{M}''$, $\boldsymbol{C}''$ and $\boldsymbol{K}''$ generated by using the proposed method are explicit functions of design parameters, sensitivity analysis about design parameters can be easily carried out. Considering that $\boldsymbol{E}'' = \boldsymbol{B}''^{\mathrm{T}}\boldsymbol{B}'^{\mathrm{T}}\boldsymbol{E}\boldsymbol{B}'\boldsymbol{B}''$ $(\boldsymbol{E} = \boldsymbol{M}, \boldsymbol{K}, \boldsymbol{C})$, eigenvalue sensitivity about design parameter p in Eq. (61) can be expressed as follows

$$\begin{aligned}\frac{\partial \lambda_r}{\partial p} &= -\lambda_r^2 \boldsymbol{\psi}_r^{\mathrm{T}} \frac{\partial \boldsymbol{M}''}{\partial p}\boldsymbol{\psi}_r - \lambda_r \boldsymbol{\psi}_r^{\mathrm{T}} \frac{\partial \boldsymbol{C}''}{\partial p}\boldsymbol{\psi}_r - \boldsymbol{\psi}_r^{\mathrm{T}} \frac{\partial \boldsymbol{K}''}{\partial p}\boldsymbol{\psi}_r \\ &= -\boldsymbol{\psi}_r^{\mathrm{T}}\boldsymbol{B}''^{\mathrm{T}}\boldsymbol{B}'^{\mathrm{T}}\left(\lambda_r^2 \frac{\partial \boldsymbol{M}}{\partial p} + \lambda_r \frac{\partial \boldsymbol{C}}{\partial p} + \frac{\partial \boldsymbol{K}}{\partial p}\right)\boldsymbol{B}'\boldsymbol{B}''\boldsymbol{\psi}_r - 2\boldsymbol{\psi}_r^{\mathrm{T}}\boldsymbol{B}''^{\mathrm{T}}\boldsymbol{B}'^{\mathrm{T}}(\lambda_r^2\boldsymbol{M} + \lambda_r\boldsymbol{C} + \boldsymbol{K})\frac{\partial(\boldsymbol{B}'\boldsymbol{B}'')}{\partial p}\boldsymbol{\psi}_r \end{aligned} \quad (62)$$

As pointed out in previous derivation, the mass matrix $\boldsymbol{M}$ of free system contains only mass and inertia parameters of each body. The damping matrix $\boldsymbol{C}$ of free system contains only damping coefficients and position and orientation of dampers. The stiffness matrix $\boldsymbol{K}$ of free system contains only stiffness coefficients and position and orientation of springs. Matrices $\boldsymbol{B}'$ and $\boldsymbol{B}''$ contain information such as position and orientation of all joints. Therefore eigenvalue sensitivity about specific design parameter can be obtained.

a. Eigenvalue sensitivity about mass or inertia parameter

If p is the mass or inertia parameter of body B_i, one can obtain that

$$\frac{\partial \boldsymbol{M}}{\partial p} = \mathrm{diag}(\boldsymbol{0} \ \cdots \ \boldsymbol{0} \ \frac{\partial \boldsymbol{M}_i}{\partial p} \ \boldsymbol{0} \ \cdots \ \boldsymbol{0}) = \boldsymbol{M}\big|_{p=1, p_{rest}=0} = \boldsymbol{M}_{sp} \quad (63)$$

where p_{rest} stands for all parameters except p in the system. It means that sensitivity of mass matrix $\boldsymbol{M}$ about mass or inertia parameter p can be directly obtained by reevaluating $\boldsymbol{M}$ under condition that all parameters being equal to zero except $p = 1$. There is no need for calculating derivatives. Accordingly, eigenvalue sensitivity can be formulated as

$$\frac{\partial \lambda_r}{\partial p} = -\lambda_r^2(\boldsymbol{B}'\boldsymbol{B}''\boldsymbol{\psi}_r)^{\mathrm{T}}\boldsymbol{M}_{sp}\boldsymbol{B}'\boldsymbol{B}''\boldsymbol{\psi}_r \quad (64)$$

Considering that $\boldsymbol{M}_{sp}$ is a sparse matrix because most elements in $\boldsymbol{M}$ are irrelative to parameter p, eigenvalue sensitivity can be significantly simplified by reducing dimension in matrix multiplication. Denote $\boldsymbol{\varphi}_r = \boldsymbol{B}'\boldsymbol{B}''\boldsymbol{\psi}_r$, and rewrite it by integrating each six rows as a block, i.e., $\boldsymbol{\varphi}_{i,r} = \boldsymbol{B}_i'\boldsymbol{B}''\boldsymbol{\psi}_r$, it yields

$$\boldsymbol{\varphi}_r = \left[\boldsymbol{\varphi}_{1,r}^{\mathrm{T}} \ \ \boldsymbol{\varphi}_{2,r}^{\mathrm{T}} \ \ \cdots \ \ \boldsymbol{\varphi}_{n,r}^{\mathrm{T}}\right]^{\mathrm{T}} \tag{65}$$

where n is the number of bodies in the system.
Eigenvalue sensitivity specified by Eq. (62) can be simplified as

$$\frac{\partial \lambda_r}{\partial p} = -\boldsymbol{\varphi}_r^{\mathrm{T}} \lambda_r^2 \frac{\partial \boldsymbol{M}}{\partial p} \boldsymbol{\varphi}_r = -\lambda_r^2 \boldsymbol{\varphi}_{i,r}^{\mathrm{T}} \frac{\partial \boldsymbol{M}_i}{\partial p} \boldsymbol{\varphi}_{i,r} \tag{66}$$

It can be seen that computational cost in Eq. (66) has been reduced by n^2 times in compare with that in Eq. (64).
Generally, there might be several components with identical structure used in a multibody system. That is to say, p is used as mass or inertia parameter for a set of bodies numbered as $\boldsymbol{e} = [e_1 \ \ e_2 \ \ \cdots \ \ e_k] \in R^n$. Eigenvalue sensitivity is difficult to be resolved by using traditional method because many elements in $\boldsymbol{M}''$ are determined by p and therefore they are correlative with each other. However, it can be directly formulated similar to Eq. (62)

$$\frac{\partial \lambda_r}{\partial p} = -\boldsymbol{\varphi}_r^{\mathrm{T}} \lambda_r^2 \frac{\partial \boldsymbol{M}}{\partial p} \boldsymbol{\varphi}_r = -\lambda_r^2 \sum_{s=1}^{k} \boldsymbol{\varphi}_{e_s,r}^{\mathrm{T}} \frac{\partial \boldsymbol{M}_{e_s}}{\partial p} \boldsymbol{\varphi}_{e_s,r} \tag{67}$$

b. Eigenvalue sensitivity about stiffness parameter
Eigenvalue sensitivity about stiffness and damping coefficient can be calculated in the same way. If p is the stiffness coefficient of spring-dampers interconnected between B_i and B_j, one can obtain that

$$\frac{\partial \lambda_r}{\partial p} = -\boldsymbol{\psi}_r^{\mathrm{T}} \boldsymbol{B}''^{\mathrm{T}} \boldsymbol{B}'^{\mathrm{T}} \frac{\partial \boldsymbol{K}}{\partial p} \boldsymbol{B}'\boldsymbol{B}''\boldsymbol{\psi}_r = -\boldsymbol{\varphi}_r^{\mathrm{T}} \frac{\partial \boldsymbol{K}}{\partial p} \boldsymbol{\varphi}_r \tag{68}$$

The variation of p affects only $\boldsymbol{K}_{ii}$, $\boldsymbol{K}_{jj}$, $\boldsymbol{K}_{ij}$ and $\boldsymbol{K}_{ji}$, it can be obtained that

$$\begin{aligned}\frac{\partial \boldsymbol{K}_{ii}}{\partial p} &= \sum_{j=0,j\neq i}^{n} \sum_{s=0}^{s_{ij}} (\boldsymbol{T}_{ijs})^{\mathrm{T}} (\boldsymbol{R}_{ijs}^{cu})^{\mathrm{T}} \frac{\partial \boldsymbol{K}_{ijs}^{u}}{\partial p} \boldsymbol{R}_{ijs}^{cu} \boldsymbol{T}_{ijs} \\ &= \sum_{j=0,j\neq i}^{n} \sum_{s=0}^{s_{ij}} (\boldsymbol{T}_{ijs})^{\mathrm{T}} (\boldsymbol{R}_{ijs}^{cu})^{\mathrm{T}} \left(\boldsymbol{K}_{ijs}^{u}\big|_{p=1,p_{rest}=0}\right) \boldsymbol{R}_{ijs}^{cu} \boldsymbol{T}_{ijs} \\ &= \boldsymbol{K}_{ii}\big|_{p=1,p_{rest}=0}\end{aligned} \tag{69}$$

$$\begin{aligned}\frac{\partial \boldsymbol{K}_{ij}}{\partial p} &= \sum_{s=0}^{s_{ij}} (\boldsymbol{T}_{ijs})^{\mathrm{T}} (\boldsymbol{R}_{ijs}^{cu})^{\mathrm{T}} \frac{\partial \boldsymbol{K}_{ijs}^{u}}{\partial p} \boldsymbol{R}_{ijs}^{cu} \boldsymbol{T}_{jis} \\ &= \sum_{s=0}^{s_{ij}} (\boldsymbol{T}_{ijs})^{\mathrm{T}} (\boldsymbol{R}_{ijs}^{cu})^{\mathrm{T}} \left(\boldsymbol{K}_{ijs}^{u}\big|_{p=1,p_{rest}=0}\right) \boldsymbol{R}_{ijs}^{cu} \boldsymbol{T}_{jis} \\ &= \boldsymbol{K}_{ij}\big|_{p=1,p_{rest}=0}\end{aligned} \tag{70}$$

$$\frac{\partial \boldsymbol{K}_{aa}}{\partial p} = \begin{cases} \boldsymbol{K}_{aa}\big|_{p=1,p_{rest}=0} & (a = i, j) \\ \boldsymbol{0} & (a \neq i, j) \end{cases} \tag{71}$$

$$\frac{\partial \mathbf{K}_{ab}}{\partial p} = \begin{cases} \mathbf{K}_{ab}\big|_{p=1,p_{rest}=0} & (a = i \,\&\, b = j, \text{ or } a = j \,\&\, b = i) \\ \mathbf{0} & (a \neq i \text{ or } b \neq j) \end{cases} \tag{72}$$

Combine Eq. (71) with Eq. (72) and it yields

$$\frac{\partial \mathbf{K}}{\partial p} = \mathbf{K}\big|_{p=1,p_{rest}=0} = \mathbf{K}_{sp} \tag{73}$$

Considering that $\mathbf{K}_{sp}$ is usually a sparse matrix, eigenvalue sensitivity about stiffness parameter used in springs between B_i and B_j can be formulated as

$$\frac{\partial \lambda_r}{\partial p} = -\boldsymbol{\varphi}_r^{\mathrm{T}} \frac{\partial \mathbf{K}}{\partial p} \boldsymbol{\varphi}_r = -\begin{bmatrix} \boldsymbol{\varphi}_{i,r}^{\mathrm{T}} & \boldsymbol{\varphi}_{j,r}^{\mathrm{T}} \end{bmatrix} \begin{bmatrix} \frac{\partial \mathbf{K}_{ii}}{\partial p} & -\frac{\partial \mathbf{K}_{ij}}{\partial p} \\ -\frac{\partial \mathbf{K}_{ji}}{\partial p} & \frac{\partial \mathbf{K}_{jj}}{\partial p} \end{bmatrix} \begin{bmatrix} \boldsymbol{\varphi}_{i,r} \\ \boldsymbol{\varphi}_{j,r} \end{bmatrix} \tag{74}$$

Generally, there might be several spring-dampers sharing the same stiffness or damping coefficient p in a multibody system. If p is the stiffness coefficient of spring-dampers interconnected between B_i and B_j, and B_j and B_k, it can be obtained that

$$\frac{\partial \lambda_r}{\partial p} = -\begin{bmatrix} \boldsymbol{\varphi}_{i,r}^{\mathrm{T}} & \boldsymbol{\varphi}_{j,r}^{\mathrm{T}} & \boldsymbol{\varphi}_{k,r}^{\mathrm{T}} \end{bmatrix} \begin{bmatrix} \frac{\partial \mathbf{K}_{ii}}{\partial p} & -\frac{\partial \mathbf{K}_{ij}}{\partial p} & \mathbf{0} \\ -\frac{\partial \mathbf{K}_{ji}}{\partial p} & \frac{\partial \mathbf{K}_{jj}}{\partial p} & -\frac{\partial \mathbf{K}_{jk}}{\partial p} \\ \mathbf{0} & -\frac{\partial \mathbf{K}_{kj}}{\partial p} & \frac{\partial \mathbf{K}_{kk}}{\partial p} \end{bmatrix} \begin{bmatrix} \boldsymbol{\varphi}_{i,r} \\ \boldsymbol{\varphi}_{j,r} \\ \boldsymbol{\varphi}_{k,r} \end{bmatrix} \tag{75}$$

If p is the stiffness coefficient of spring-dampers interconnected between B_i and B_j, and B_k and B_l, it can be obtained that

$$\frac{\partial \lambda_r}{\partial p} = -\begin{bmatrix} \boldsymbol{\varphi}_{i,r}^{\mathrm{T}} & \boldsymbol{\varphi}_{j,r}^{\mathrm{T}} \end{bmatrix} \begin{bmatrix} \frac{\partial \mathbf{K}_{ii}}{\partial p} & -\frac{\partial \mathbf{K}_{ij}}{\partial p} \\ -\frac{\partial \mathbf{K}_{ji}}{\partial p} & \frac{\partial \mathbf{K}_{jj}}{\partial p} \end{bmatrix} \begin{bmatrix} \boldsymbol{\varphi}_{i,r} \\ \boldsymbol{\varphi}_{j,r} \end{bmatrix} - \begin{bmatrix} \boldsymbol{\varphi}_{k,r}^{\mathrm{T}} & \boldsymbol{\varphi}_{l,r}^{\mathrm{T}} \end{bmatrix} \begin{bmatrix} \frac{\partial \mathbf{K}_{kk}}{\partial p} & -\frac{\partial \mathbf{K}_{kl}}{\partial p} \\ -\frac{\partial \mathbf{K}_{lk}}{\partial p} & \frac{\partial \mathbf{K}_{ll}}{\partial p} \end{bmatrix} \begin{bmatrix} \boldsymbol{\varphi}_{k,r} \\ \boldsymbol{\varphi}_{l,r} \end{bmatrix} \tag{76}$$

c. Eigenvalue sensitivity about damping parameter

Similarly, if p is the damping coefficient of spring-dampers interconnected between B_i and B_j, eigenvalue sensitivity about p can be formulated as

$$\frac{\partial \lambda_r}{\partial p} = -\lambda_r \boldsymbol{\varphi}_r^{\mathrm{T}} \frac{\partial \mathbf{C}}{\partial p} \boldsymbol{\varphi}_r = -\lambda_r \begin{bmatrix} \boldsymbol{\varphi}_{i,r}^{\mathrm{T}} & \boldsymbol{\varphi}_{j,r}^{\mathrm{T}} \end{bmatrix} \begin{bmatrix} \frac{\partial \mathbf{C}_{ii}}{\partial p} & -\frac{\partial \mathbf{C}_{ij}}{\partial p} \\ -\frac{\partial \mathbf{C}_{ji}}{\partial p} & \frac{\partial \mathbf{C}_{jj}}{\partial p} \end{bmatrix} \begin{bmatrix} \boldsymbol{\varphi}_{i,r} \\ \boldsymbol{\varphi}_{j,r} \end{bmatrix} \tag{77}$$

If p is the damping coefficient of spring-dampers interconnected between B_i and B_j, and B_j and B_k, it can be obtained that

$$\frac{\partial \lambda_r}{\partial p} = -\lambda_r [\boldsymbol{\varphi}_{i,r}^{\mathrm{T}} \quad \boldsymbol{\varphi}_{j,r}^{\mathrm{T}} \quad \boldsymbol{\varphi}_{k,r}^{\mathrm{T}}] \begin{bmatrix} \frac{\partial \boldsymbol{C}_{ii}}{\partial p} & -\frac{\partial \boldsymbol{C}_{ij}}{\partial p} & \boldsymbol{0} \\ -\frac{\partial \boldsymbol{C}_{ji}}{\partial p} & \frac{\partial \boldsymbol{C}_{jj}}{\partial p} & -\frac{\partial \boldsymbol{C}_{ik}}{\partial p} \\ \boldsymbol{0} & -\frac{\partial \boldsymbol{C}_{kj}}{\partial p} & \frac{\partial \boldsymbol{C}_{kk}}{\partial p} \end{bmatrix} \begin{bmatrix} \boldsymbol{\varphi}_{i,r} \\ \boldsymbol{\varphi}_{j,r} \\ \boldsymbol{\varphi}_{k,r} \end{bmatrix} \tag{78}$$

If p is the damping coefficient of spring-dampers interconnected between B_i and B_j , and B_k and B_l , it can be obtained that

$$\frac{\partial \lambda_r}{\partial p} = -\lambda_r [\boldsymbol{\varphi}_{i,r}^{\mathrm{T}} \quad \boldsymbol{\varphi}_{j,r}^{\mathrm{T}}] \begin{bmatrix} \frac{\partial \boldsymbol{C}_{ii}}{\partial p} & -\frac{\partial \boldsymbol{C}_{ij}}{\partial p} \\ -\frac{\partial \boldsymbol{C}_{ji}}{\partial p} & \frac{\partial \boldsymbol{C}_{jj}}{\partial p} \end{bmatrix} \begin{bmatrix} \boldsymbol{\varphi}_{i,r} \\ \boldsymbol{\varphi}_{j,r} \end{bmatrix} - \lambda_r [\boldsymbol{\varphi}_{k,r}^{\mathrm{T}} \quad \boldsymbol{\varphi}_{l,r}^{\mathrm{T}}] \begin{bmatrix} \frac{\partial \boldsymbol{C}_{kk}}{\partial p} & -\frac{\partial \boldsymbol{C}_{kl}}{\partial p} \\ -\frac{\partial \boldsymbol{C}_{lk}}{\partial p} & \frac{\partial \boldsymbol{C}_{ll}}{\partial p} \end{bmatrix} \begin{bmatrix} \boldsymbol{\varphi}_{k,r} \\ \boldsymbol{\varphi}_{l,r} \end{bmatrix} \tag{79}$$

4.3 Proposed sensitivity formulation about geometrical design parameters

The position and orientation of connection such as spring-damper and joint affect the dynamics of multibody system too. Eigenvalue sensitivity about these geometrical design parameters will be derived in this section.

If p is the position and orientation of spring-dampers, eigenvalue sensitivity can be formulated as

$$\frac{\partial \lambda_r}{\partial p} = -\boldsymbol{\varphi}_r^{\mathrm{T}} \left(\lambda_r \frac{\partial \boldsymbol{C}}{\partial p} + \frac{\partial \boldsymbol{K}}{\partial p} \right) \boldsymbol{\varphi}_r \tag{80}$$

If p is the position and orientation of spring-dampers interconnected between B_i and B_j , similar to Eq. (74), it can be obtained that

$$\frac{\partial \lambda_r}{\partial p} = -[\boldsymbol{\varphi}_{i,r}^{\mathrm{T}} \quad \boldsymbol{\varphi}_{j,r}^{\mathrm{T}}] \begin{bmatrix} \lambda_r \frac{\partial \boldsymbol{C}_{ii}}{\partial p} + \frac{\partial \boldsymbol{K}_{ii}}{\partial p} & -\lambda_r \frac{\partial \boldsymbol{C}_{ij}}{\partial p} - \frac{\partial \boldsymbol{K}_{ij}}{\partial p} \\ -\lambda_r \frac{\partial \boldsymbol{C}_{ji}}{\partial p} - \frac{\partial \boldsymbol{K}_{ji}}{\partial p} & \lambda_r \frac{\partial \boldsymbol{C}_{jj}}{\partial p} + \frac{\partial \boldsymbol{K}_{jj}}{\partial p} \end{bmatrix} \begin{bmatrix} \boldsymbol{\varphi}_{i,r} \\ \boldsymbol{\varphi}_{j,r} \end{bmatrix} \tag{81}$$

In addition, if p is the position of spring-dampers interconnected between B_i and B_j , it can be obtained that

$$\frac{\partial \boldsymbol{E}_{ii}}{\partial p} = \sum_{s=0}^{s_{ij}} \left[\frac{\partial (\boldsymbol{T}_{ijs})^{\mathrm{T}}}{\partial p} (\boldsymbol{R}_{ijs}^{cu})^{\mathrm{T}} \boldsymbol{E}_{ijs}^{u} \boldsymbol{R}_{ijs}^{cu} \boldsymbol{T}_{ijs} + (\boldsymbol{T}_{ijs})^{\mathrm{T}} (\boldsymbol{R}_{ijs}^{cu})^{\mathrm{T}} \boldsymbol{E}_{ijs}^{u} \boldsymbol{R}_{ijs}^{cu} \frac{\partial (\boldsymbol{T}_{ijs})}{\partial p} \right] \quad (\boldsymbol{E} = \boldsymbol{K}, \boldsymbol{C}) \tag{82}$$

$$\frac{\partial \boldsymbol{E}_{ij}}{\partial p} = \sum_{s=0}^{s_{ij}} \left[\frac{\partial (\boldsymbol{T}_{ijs})^{\mathrm{T}}}{\partial p} (\boldsymbol{R}_{ijs}^{cu})^{\mathrm{T}} \boldsymbol{E}_{ijs}^{u} \boldsymbol{R}_{ijs}^{cu} \boldsymbol{T}_{jis} + (\boldsymbol{T}_{ijs})^{\mathrm{T}} (\boldsymbol{R}_{ijs}^{cu})^{\mathrm{T}} \boldsymbol{E}_{ijs}^{u} \boldsymbol{R}_{ijs}^{cu} \frac{\partial (\boldsymbol{T}_{jis})}{\partial p} \right] \quad (\boldsymbol{E} = \boldsymbol{K}, \boldsymbol{C}) \tag{83}$$

If p is the orientation of spring-dampers interconnected between B_i and B_j , it can be obtained that

$$\frac{\partial \boldsymbol{E}_{ii}}{\partial p} = \sum_{s=0}^{s_{ij}} \left[(\boldsymbol{T}_{ijs})^{\mathrm{T}} \frac{\partial (\boldsymbol{R}_{ijs}^{cu})^{\mathrm{T}}}{\partial p} \boldsymbol{E}_{ijs}^{u} \boldsymbol{R}_{ijs}^{cu} \boldsymbol{T}_{ijs} + (\boldsymbol{T}_{ijs})^{\mathrm{T}} (\boldsymbol{R}_{ijs}^{cu})^{\mathrm{T}} \boldsymbol{E}_{ijs}^{u} \frac{\partial (\boldsymbol{R}_{ijs}^{cu})}{\partial p} \boldsymbol{T}_{ijs} \right] \quad (\boldsymbol{E} = \boldsymbol{K}, \boldsymbol{C}) \tag{84}$$

$$\frac{\partial \boldsymbol{E}_{ij}}{\partial p}=\sum_{s=0}^{s_{ij}}\left[(\boldsymbol{T}_{ijs})^{\mathrm{T}}\frac{\partial(\boldsymbol{R}_{ijs}^{cu})^{\mathrm{T}}}{\partial p}\boldsymbol{E}_{ijs}^{u}\boldsymbol{R}_{ijs}^{cu}\boldsymbol{T}_{jis}+(\boldsymbol{T}_{ijs})^{\mathrm{T}}(\boldsymbol{R}_{ijs}^{cu})^{\mathrm{T}}\boldsymbol{E}_{ijs}^{u}\frac{\partial(\boldsymbol{R}_{ijs}^{cu})}{\partial p}\boldsymbol{T}_{jis}\right]\ (\boldsymbol{E}=\boldsymbol{K},\boldsymbol{C}) \tag{85}$$

Generally, p may be used as position and orientation of spring-dampers among a set of bodies in a multibody system. For example, if p is the position and orientation of spring-dampers interconnected between B_i and B_j, and B_j and B_k, it can be obtained that

$$\frac{\partial \lambda_r}{\partial p}=-\left[\boldsymbol{\varphi}_{i,r}^{\mathrm{T}}\ \boldsymbol{\varphi}_{j,r}^{\mathrm{T}}\ \boldsymbol{\varphi}_{k,r}^{\mathrm{T}}\right]\begin{bmatrix}\lambda_r\frac{\partial \boldsymbol{C}_{ii}}{\partial p}+\frac{\partial \boldsymbol{K}_{ii}}{\partial p} & -\lambda_r\frac{\partial \boldsymbol{C}_{ij}}{\partial p}-\frac{\partial \boldsymbol{K}_{ij}}{\partial p} & \boldsymbol{0}\\ -\lambda_r\frac{\partial \boldsymbol{C}_{ji}}{\partial p}-\frac{\partial \boldsymbol{K}_{ji}}{\partial p} & \lambda_r\frac{\partial \boldsymbol{C}_{jj}}{\partial p}+\frac{\partial \boldsymbol{K}_{jj}}{\partial p} & -\lambda_r\frac{\partial \boldsymbol{C}_{jk}}{\partial p}-\frac{\partial \boldsymbol{K}_{jk}}{\partial p}\\ \boldsymbol{0} & -\lambda_r\frac{\partial \boldsymbol{C}_{kj}}{\partial p}-\frac{\partial \boldsymbol{K}_{kj}}{\partial p} & \lambda_r\frac{\partial \boldsymbol{C}_{kk}}{\partial p}+\frac{\partial \boldsymbol{K}_{kk}}{\partial p}\end{bmatrix}\begin{bmatrix}\boldsymbol{\varphi}_{i,r}\\ \boldsymbol{\varphi}_{j,r}\\ \boldsymbol{\varphi}_{k,r}\end{bmatrix} \tag{86}$$

If p is the position and orientation of spring-dampers interconnected between B_i and B_j, and B_k and B_l, it can be obtained that

$$\begin{aligned}\frac{\partial \lambda_r}{\partial p}=&-\left[\boldsymbol{\varphi}_{i,r}^{\mathrm{T}}\ \boldsymbol{\varphi}_{j,r}^{\mathrm{T}}\right]\begin{bmatrix}\lambda_r\frac{\partial \boldsymbol{C}_{ii}}{\partial p}+\frac{\partial \boldsymbol{K}_{ii}}{\partial p} & -\lambda_r\frac{\partial \boldsymbol{C}_{ij}}{\partial p}-\frac{\partial \boldsymbol{K}_{ij}}{\partial p}\\ -\lambda_r\frac{\partial \boldsymbol{C}_{ji}}{\partial p}-\frac{\partial \boldsymbol{K}_{ji}}{\partial p} & \lambda_r\frac{\partial \boldsymbol{C}_{jj}}{\partial p}+\frac{\partial \boldsymbol{K}_{jj}}{\partial p}\end{bmatrix}\begin{bmatrix}\boldsymbol{\varphi}_{i,r}\\ \boldsymbol{\varphi}_{j,r}\end{bmatrix}\\ &-\left[\boldsymbol{\varphi}_{k,r}^{\mathrm{T}}\ \boldsymbol{\varphi}_{l,r}^{\mathrm{T}}\right]\begin{bmatrix}\lambda_r\frac{\partial \boldsymbol{C}_{kk}}{\partial p}+\frac{\partial \boldsymbol{K}_{kk}}{\partial p} & -\lambda_r\frac{\partial \boldsymbol{C}_{kl}}{\partial p}-\frac{\partial \boldsymbol{K}_{kl}}{\partial p}\\ -\lambda_r\frac{\partial \boldsymbol{C}_{lk}}{\partial p}-\frac{\partial \boldsymbol{K}_{lk}}{\partial p} & \lambda_r\frac{\partial \boldsymbol{C}_{ll}}{\partial p}+\frac{\partial \boldsymbol{K}_{ll}}{\partial p}\end{bmatrix}\begin{bmatrix}\boldsymbol{\varphi}_{k,r}\\ \boldsymbol{\varphi}_{l,r}\end{bmatrix}\end{aligned} \tag{87}$$

The above-mentioned sensitivity formulations are based on the topology of the multibody systems. Particularly, eigen-sensitivity with respect to design parameters of mass and inertia, coefficients of stiffness and damping, position and orientation of connections are all derived analytically in detail. These results can be directly applied for sensitivity analysis of general mechanical systems and complex structures which are modelled as multibody systems.

5. Numerical examples and applications

5.1 Numerical verification

The computational efficiency for vibration calculation can be significantly improved by using the proposed method, in comparison with most of the traditional approaches. A multibody system with n rigid bodies and m DOFs is taken as an example to demonstrate it. Suppose there are p constraints for the open-loop system and q ($p \le 6n-m \le q$) constraints for the entire system. There are mainly four factors that can help to improve the computational efficiency.

1. Relative small scale of matrix computation. Traditionally, a matrix with size $(12n-m)\times(12n-m)$ must be generated and solved to obtain system matrices with size $m\times m$. In addition, in order to express the $6n-m$ dependent coordinates in terms of m independent coordinates, it is necessary to get the inverse of a matrix with size $6n-m$, according to the Kang's method (Kang et al., 2003). However, there are only matrices

$\boldsymbol{M}$, $\boldsymbol{C}$, $\boldsymbol{K}$ with size $6n \times 6n$ and an open-loop constraint matrix $\boldsymbol{B}'$ with size $6n \times (6n-p)$ need to be easily generated for the proposed method. And then a cut-joint constraint matrix $\boldsymbol{B}''$ with size $(6n-p) \times m$ needs to be resolved to perform simple matrix multiplication for obtaining the final system matrices. In addition, there are only $6n-p-m$ dependent coordinates in terms of m independent coordinates, the size of matrix to be inversed is $6n-p-m$. It can be easily concluded that less computational efforts are required for the proposed method.

2. Reduction of trigonometric functions computing. Conventionally, the variations of coordinates and postures between two acting points of a connection, such as spring-damper or joint, are computed based on homogeneous transformation. Instead, the linear transformation in the proposed method can significantly reduce computational efforts due to calculation of trigonometric functions. Obviously, the more connections there are, the more computational efforts can be reduced.
3. Avoidance of complex calculation of Jacobian of constraint equation which usually contains many trigonometric functions. It is time-consuming for the calculation of Jacobian of a matrix with size $(6n-m) \times (6n-m)$. Instead, the constraint matrices $\boldsymbol{B}'$ and $\boldsymbol{B}''$ can be easily obtained by using the presented definition of constraints for the proposed method.
4. Avoidance of linearization of nonlinear equations of motion. The ODEs generated by conventional methods are nonlinear ones that need to be linearized before perform vibration calculation (Cruz et al., 2007; Minaker & Frise, 2005; Negrut & Ortiz, 2006; Pott et al., 2007; Roy & Kumar, 2005). Instead, the ODEs obtained by using the proposed method are a minimal set of second-order linear ODEs which can be directly used for vibration calculation.

In this section, numerical experiments were carried out to verify the correctness and efficiency of the proposed method. It is unsuitable to compare straightforwardly the results of system matrices with theoretical solutions for they are usually very large in size. Normal mode analysis (NMA) and transfer function analysis (TFA) for the same model were performed in AMVA and commercial software ADAMS. The results of natural frequencies, the damping ratios, and the transfer function were compared to verify the correctness of the proposed method. Solution time was compared to testify the efficiency of the proposed method. The experiments were performed on a PC with CPU Pentium IV of 2.0 GHz and memory of 2.0 GB. Models with chain, tree, and closed-loop topology were taken as case studies, as shown in Fig. 6.

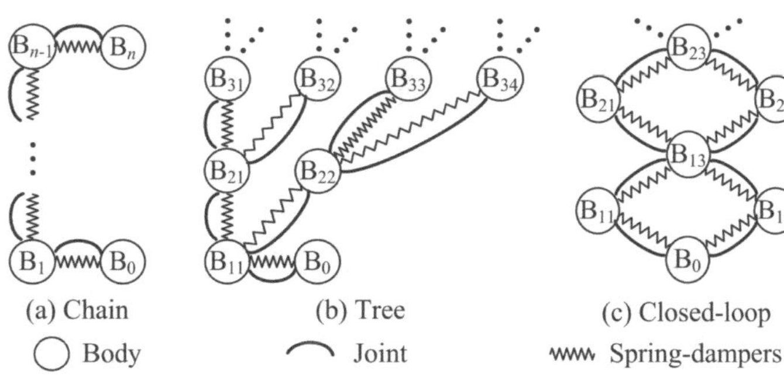

Fig. 6. Topologies of models used for numerical test

A. Chain topology MBS. As shown in Fig. 6(a), n moving bodies and the ground B_0 are connected by joints and spatial spring-dampers in a chain. The position and orientation of CM of body B_i are $[0\ 0\ 0.2i-0.1\ 0\ 0\ 0]$. The position and orientation of joint $J_{i-1,i}$ are $[0\ 0\ 0.2i-0.2\ 0\ 0\ 0]$.
B. Tree topology MBS. As shown in Fig. 6(b), the bodies are connected by joints and spatial spring-dampers in form of binary tree with N layers. There are $n_i = 2^{i-1}$ bodies in the ith layer, among which the jth one is denoted as B_{ij}. The position and orientation of CM of body B_{ij} are $[j\ i\ 0\ 0\ 0\ 0]$. The position and orientation of joint between body $B_{i+1,2j-1}$ and B_{ij} are $[(3j-1)/2\ \ i+0.5\ 0\ 0\ 0\ \mathrm{arccot}(j-1)]$, and that between body $B_{i+1,2j}$ and B_{ij} are $[3j/2\ \ i+0.5\ 0\ 0\ 0\ \mathrm{arccot}(j)]$.
C. Closed-loop topology MBS. As shown in Fig. 6(c), the bodies are connected by joints and spatial spring-dampers in form of ladder with N layers. There are three bodies in the ith layer, among which the jth one is denoted as B_{ij}. The position and orientation of CM of B_{ij} are $[0.2j-0.3\ \ 0.2i-0.1\ 0\ 0\ 0\ 0]$ (for $j=1,2$) or $[0\ 0.2i\ 0\ 0\ 0\ \pi/2]$ (for $j=3$). The position and orientation of joint between $B_{i,3}$ and $B_{i,u}$ ($u=1,2$) are $[0.2u-0.3\ \ 0.2i-0.1\ 0\ 0\ 0\ 0]$. The position and orientation of joint between $B_{i,3}$ and $B_{i+1,u}$ ($u=1,2$) are $[0.2u-0.3\ \ 0.2i+0.1\ 0\ 0\ 0\ 0]$.

The rule of name for each kind of models is specified as follows. The first letter, i.e., 'C', 'T', and 'L', means model with chain, tree, and closed-loop topology, respectively. It then follows the number of bodies (for models with chain topology) or layers (for models with tree or closed-loop topology). The letter before 'F' means the type of joint in the model, e.g., 'R' , 'P', 'C' and 'S' means revolute, prismatic, cylindrical and spherical joint. The figure at the end means the number of spring-dampers between two bodies connected by joint.

For simplicity without loss of generality, the mass and inertia tensor of all bodies, the stiffness and damping coefficients of all spring-dampers, as well as the position and orientation of joint and spring-dampers between each two bodies were set to be equal to each other, as specified in Table 2, where s is the number of spring-dampers between the two bodies considered.. The results of NMA and TFA (force input at CM of body $B_{6,1}$ in X-direction, displacement output at CM of body $B_{6,32}$ in Y-direction) for model TL7SF1 are shown in Fig.7 and Fig.8, respectively.

Parameter	Symbol	Value
Mass (kg)	m	1.0
Inertia ($\mathrm{kg \cdot m^2}$)	$[I_{xx}\ I_{yy}\ I_{zz}\ I_{xy}\ I_{xz}\ I_{yz}]$	$[1.0\ 1.0\ 1.0\ 0\ 0\ 0]$
Stiffness ($\mathrm{N \cdot m^{-1}}$)	$[k_x^k\ k_y^k\ k_z^k]$	$[1.0\ 1.0\ 1.0]\times 10^4/s$
Torsion stiffness ($\mathrm{N \cdot m \cdot deg^{-1}}$)	$[k_\alpha^k\ k_\beta^k\ k_\gamma^k]$	$[1.0\ 1.0\ 1.0]\times 10^4/s$
Damping ($\mathrm{N \cdot s \cdot m^{-1}}$)	$[c_x^k\ c_y^k\ c_z^k]$	$[1.0\ 1.0\ 1.0]\times 10^1/s$
Torsion damping ($\mathrm{N \cdot m \cdot s \cdot deg^{-1}}$)	$[c_\alpha^k\ c_\beta^k\ c_\gamma^k]$	$[1.0\ 1.0\ 1.0]\times 10^1/s$

Table 3. Parameters of bodies and spring-dampers in all case studies

Solutions in Fig.7 indicate that the results of eigenvalue calculated using AMVA are identical to those in ADAMS. The mean and maximal errors of natural frequencies between the two groups of results are 1.02×10^{-6} Hz and 5.00×10^{-5} Hz. The mean and maximal errors of damping ratios of the two groups of results are 1.73×10^{-10} and 5.00×10^{-8}. Comparisons in

Fig.8 indicate that solutions of transfer function calculated using AMVA coincide well with those in ADAMS.

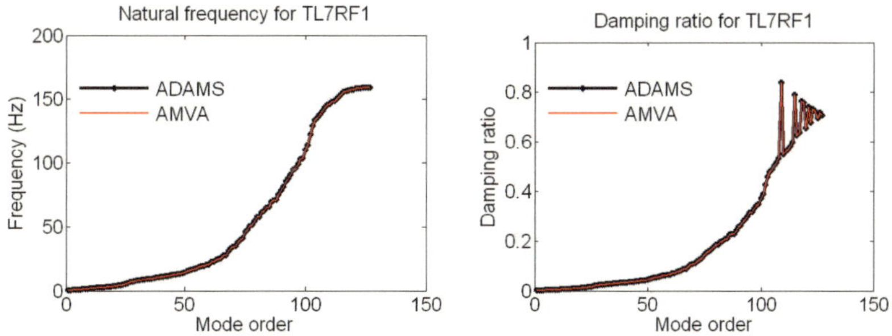

Fig. 7. Comparison of NMA results for model TL7RF1

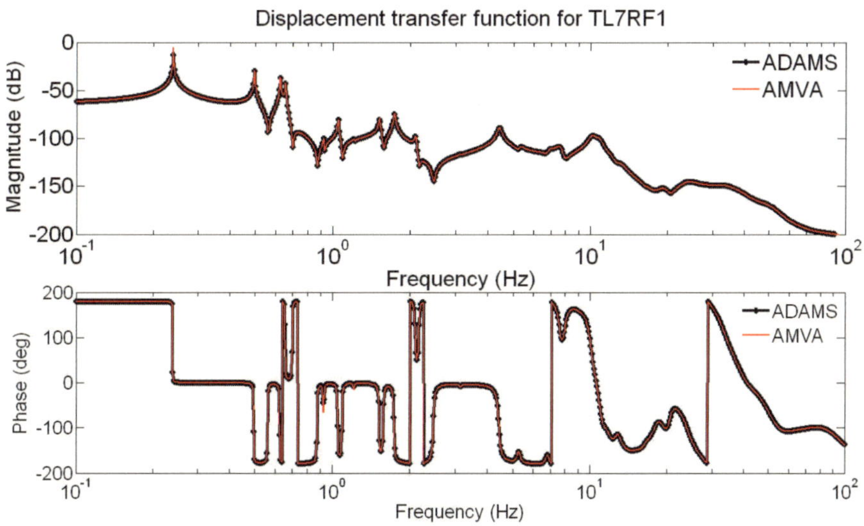

Fig. 8. Comparison of TFA solutions for model TL7RF1

5.2 Applications in engineering

A quadruped robot and a Stewart platform were taken as case studies to verify the effectiveness of the proposed method for both open-loop and closed-loop spatial mechanism systems, respectively. Simulations and experiments were further carried out on a wafer stage to justify the presented method.

a. Quadruped robot

The proposed method has been applied in linear vibration analysis of a quadruped robot, which is an open-loop spatial mechanism system. As shown in Fig. 9, the body is connected with four legs via revolute joints along z direction. Each leg consists of three parts which are connected by two turbine worm gears. The leg mechanism can be modeled as three rigid bodies connected by two revolute joints and torsion springs along x direction. Each flexible foot is modeled as a three dimensional linear spring-damper, then the quadruped robot becomes an open-loop spatial mechanism system with 13 bodies and 18 DOFs.

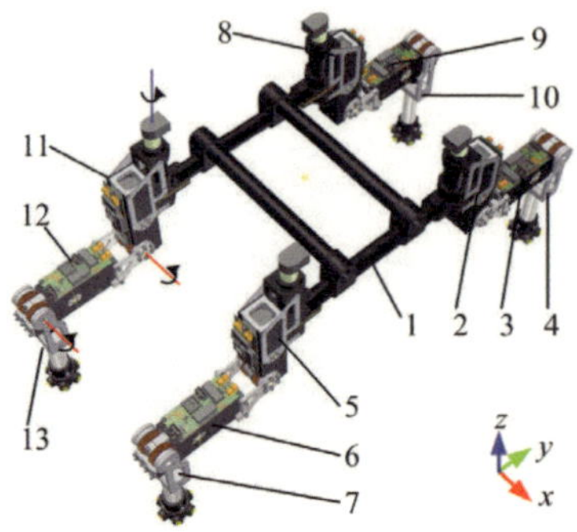

Fig. 9. Quadruped robot

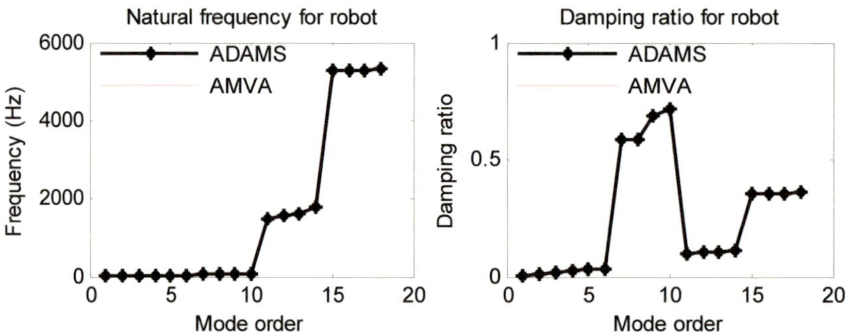

Fig. 10. Comparison of NMA results for quadruped robot

Normal mode analysis and transfer function analysis were both performed in ADAMS and AMVA for such a quadruped robot. As shown in Fig. 10, natural frequencies and damping ratio solved in two tools are equal to each other. Fig. 11 shows that results of transfer function computed in two packages are identical. It indicates that dynamic analysis of open-loop spatial mechanism system can also be solved using the proposed method.

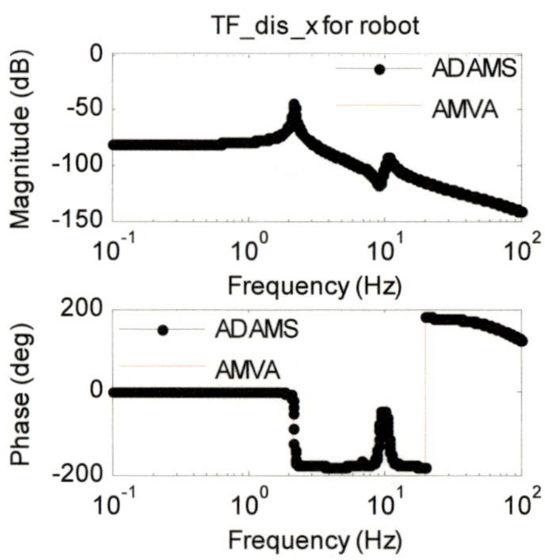

Fig. 11. Comparison of TFA results for quadruped robot

b. Stewart platform

The proposed method has also been applied in linear vibration analysis of a Stewart isolation platform, which is a closed-loop spatial mechanism system with six parallel linear actuators, as shown in Fig. 12. The isolated platform on the top layer is connected with linear actuators via flexible joints. The lower end of each actuator is also connected with the base via flexible joint. Based on previous finite element analysis, each flexible joint is modeled as spherical joint together with three-dimensional torsion spring-damper. And each linear actuator is modeled as two rigid bodies connected with a translational joint together with a linear spring-damper along the relative moving direction. Therefore the system can be modeled as a closed-loop spatial mechanism system with 14 rigid bodies and 12 DOFs.

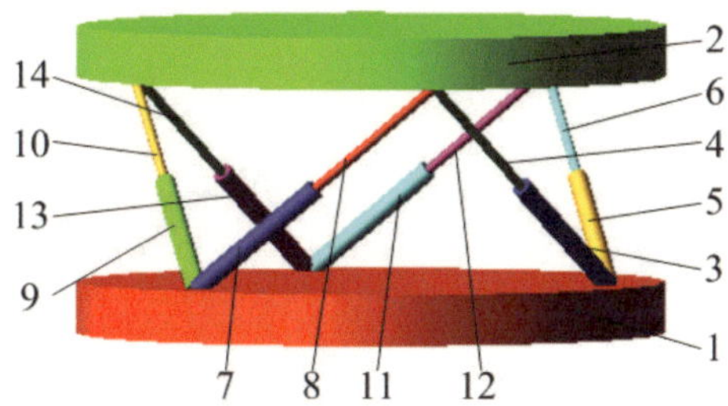

Fig. 12. Stewart platform

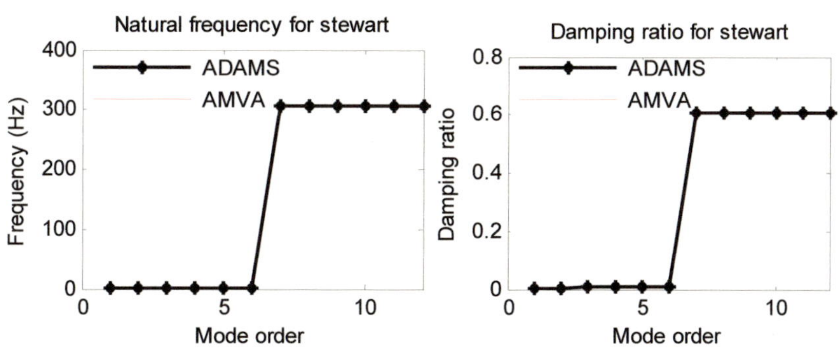

Fig. 13. Comparison of NMA results for Stewart platform

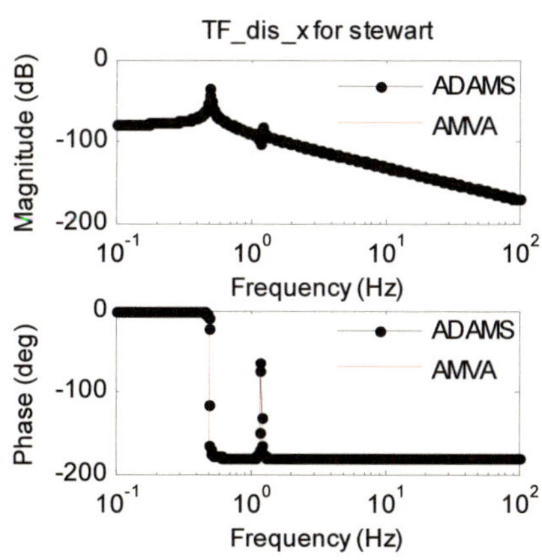

Fig. 14. Comparison of TFA results for Stewart platform

Normal mode analysis and transfer function analysis were both performed in ADAMS and AMVA to acquire vibration isolation performance of such a Stewart platform. As shown in Fig. 13, natural frequencies and damping ratio solved in two tools are equal to each other. Fig. 14 shows that results of transfer function of displacement computed in two packages are identical. Fig. 15 shows that results of time response of displacement computed in two packages are identical. It indicates that dynamic analysis of closed-loop spatial mechanism system can also be solved using the proposed method.

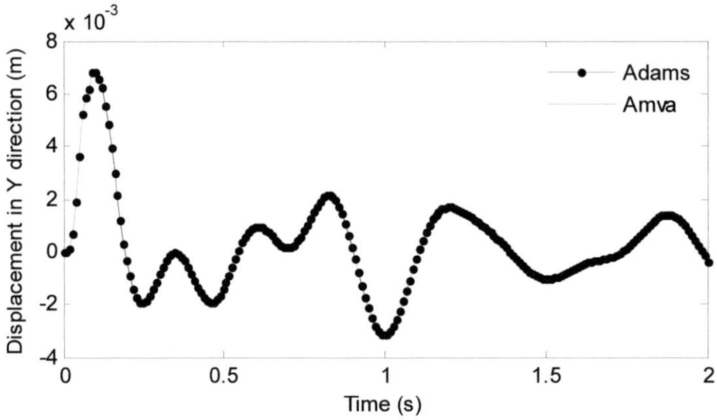

Fig. 15. Comparison of TRA solutions for the Stewart platform

7. Conclusion

A new formulation based on constraint-topology transformation is proposed to generate oscillatory differential equations for a general multibody system. Vibration displacements of bodies are selected as generalized coordinates. The translational and rotational displacements are integrated in spatial notation. Linear transformation of vibration displacements between different points on the same rigid body is derived. Absolute joint displacement is introduced to give mathematical definition for ideal joint in a new form. Constraint equations written in this way can be solved easily via the proposed linear transformation. The oscillatory differential equations for a general multibody system are derived by matrix generation and quadric transformation in three steps:

1. Linearized ODEs in terms of absolute displacements are firstly derived by using Lagrangian method for free multibody system without considering any constraint.
2. An open-loop constraint matrix is derived to formulate linearized ODEs via quadric transformation for open-loop multibody system, which is obtained from closed-loop multibody system by using cut-joint method.
3. A cut-joint constraint matrix corresponding to all cut-joints is finally derived to formulate a minimal set of ODEs via quadric transformation for closed-loop multibody system.

Sensitivity of the mass, stiffness and damping matrix about each kind of design parameters are derived based on the proposed algorithm for vibration calculation. The results show that they can be directly obtained by matrix generation and multiplication without derivatives. Eigen-sensitivity about design parameters are then carried out.

Several kinds of mechanical systems are taken as case studies to illustrate the presented method. The correctness of the proposed method has been verified via numerical experiments on multibody system with chain, tree, and closed-loop topology. Results show that the vibration calculation and sensitivity analysis have been greatly simplified because complicatedly solving for constraints, linearization and derivatives are unnecessary. Therefore the proposed method can be used to greatly improve the computational efficiency for vibration calculation and sensitivity analysis of large-scale multibody system. Sensitivity of the dynamic response with respect to the design parameters, and the computational efficiency of the proposed method will be investigated in the future.

8. References

Amirouche, F., (2006). *Fundamentals of multibody dynamics: theory and applications*, Birkhauser, 9780817642365, Boston

Anderson, KS & Hsu, Y., (2002). Analytical Fully-Recursive Sensitivity Analysis for Multibody Dynamic Chain Systems, *Multibody Syst. Dyn.*, Vol. 8, No. 1, (1-27), 1384-5640

Attia, HA, (2008). Modelling of three-dimensional mechanical systems using point coordinates with a recursive approach, *Appl. Math. Model.*, Vol. 32, No. 3, (315-326), 0307-904X

Choi, KM, Jo, HK, Kim, WH, et al., (2004). Sensitivity analysis of non-conservative eigensystems, *J. Sound Vib.*, Vol. 274, (997-1011), 0022-460X

Cruz, HD, Biscay, RJ, Carbonell, F., et al., (2007). A higher order local linearization method for solving ordinary differential equations, *Appl. Math. Comput.*, Vol. 185, No. 1, (197-212), 0096-3003

Ding, JY, Pan, ZK & Chen, LQ, (2007). Second order adjoint sensitivity analysis of multibody systems described by differential-algebraic equations, *Multibody Syst. Dyn.*, Vol. 18, (599–617), 1384-5640

Eberhard, P. & Schiehlen, W., (2006). Computational dynamics of multibody systems: history, formalisms, and applications, *J. Comput. Nonlin. Dyn.*, Vol. 1, (3-12), 1555-1415

Flores, P., Ambrósio, J., Claro, P., et al., (2008). *Kinematics and dynamics of multibody systems with imperfect joints: models and case studies*, Springer-Verlag, 9783540743590, Berlin

Jiang, W., Chen, XD & Yan, TH, (2008a). Symbolic formulation of multibody systems for vibration analysis based on matrix transformation, *Chinese J. Mech. Eng. (Chinese Ed.)*, Vol. 44, No. 6, (54-60), 0577-6686

Jiang, W., Chen, XD, Luo, X. & Huang, QJ, (2008b). Symbolic formulation of large-scale open-loop multibody systems for vibration analysis using absolute joint coordinates, *JSME J. Syst. Design Dyn.*, Vol. 2, No. 4, (1015-1026), 1881-3046

Kang, JS, Bae S., Lee JM & Tak TO, (2003). Force equilibrium approach for linearization of constrained mechanical system dynamics, *ASME J. Mech. Design*, Vol. 125, (143-149), 1050-0472

Laulusa, A. & Bauchau, OA, (2008). Review of classical approaches for constraint enforcement in multibody Systems, *J. Comput. Nonlin. Dyn.*, Vol. 3, No. 1, (011004), 1555-1415

Lee, IW, Kim, DO & Jung, GH, (1999a). Natural frequency and mode shape sensitivities of damped systems: part i, distinct natural frequencies, *J. Sound Vib.*, Vol. 223, No. 3, (399-412), 0022-460X

Lee, IW, Kim, DO & Jung, GH, (1999). Natural frequency and mode shape sensitivities of damped systems: part ii, multiple natural frequencies, *J. Sound Vib.*, Vol. 223, No. 3, (413-424), 0022-460X

Liu, JY, Hong, JZ & Cui, L., (2007). An exact nonlinear hybrid-coordinate formulation for flexible multibody systems, *Acta Mech. Sinica*, Vol. 23, No. 6, (699-706), 0567-7718

McPhee, JJ & Redmond, SM, (2006). Modelling multibody systems with indirect coordinates, *Comput. Method. Appl. Mech. Eng.*, Vol. 195, No. 50-51, (6942-6957), 0045-7825

Minaker, B. & Frise, P., (2005). Linearizing the equations of motion for multibody systems using an orthogonal complement method, *J. Vib. Control*, Vol. 11, (51-66), 1077-5463

Müller, A., (2004). Elimination of redundant cut joint constraints for multibody system models, *ASME J. Mech. Design*, Vol. 126, No. 3, (488-494), 1050-0472

Negrut, D. & Ortiz, JL, (2006). A practical approach for the lnearization of the constrained multibody dynamics equations, *J. Comput. Nonlin. Dyn.*, Vol. 1, No. 3, (230-239), 1555-1415

Pott, A., Kecskeméthy, A., Hiller, M., (2007). A simplified force-based method for the linearization and sensitivity analysis of complex manipulation systems, *Mech. Mach. Theory*, Vol. 42, No. 11, (1445-1461), 0094-114X

Richard, MJ, McPhee, JJ & Anderson, RJ, (2007). Computerized generation of motion equations using variational graph-theoretic methods, *Appl. Math. Comput.*, Vol. 192, No. 1, (135-156), 0096-3003

Roy, D. & Kumar, R., (2005). A multi-step transversal linearization (MTL) method in non-linear structural dynamics, *J. Sound Vib.*, Vol. 287, No. 1-2, (203-226), 0022-460X

Rui, XT, Wang, GP, Lu, YQ, et al., (2008). Transfer matrix method for linear multibody system, *Multibody Syst. Dyn.*, Vol. 19, No. 3, (179-207), 1384-5640

Schiehlen, W., Guse, N. & Seifried, R., (2006). Multibody dynamics in computational mechanics and engineering applications, *Comput. Method Appl. Mech. Eng.*, Vol. 195, No. 41-43, (5509-5522), 0045-7825

Sliva, G., Brezillon, A., Cadou, JM, et al., (2010). A study of the eigenvalue sensitivity by homotopy and perturbation methods, *J. Computat. Appl. Math.*, Vol. 234, No. 7, (2297-2302), 0377-0427

Sohl, GA & Bobrow, JE, (2001). A Recursive Multibody Dynamics and Sensitivity Algorithm for Branched Kinematic Chains, ASME J. Dyn. Syst. Meas. Control, Vol. 123, (391-399), 0022-0434

Valasek, M., Sika, Z. & Vaculin, O., (2007). Multibody formalism for real-time application using natural coordinates and modified state space, *Multibody Syst. Dyn.*, Vol. 17, No. 2, (209-227), 1384-5640

Van Keulen, F., Haftk, RT & Kim, NH, (2005). Review of options for structural design sensitivity analysis. part 1: linear systems, *Comput. Methods Appl. Mech. Eng.*, Vol. 194, (3213-3243) , 0045-7825

Wasfy, TM & Noor, AK, (2003). Computational strategies for flexible multibody systems, *Appl. Mech. Rev.*, Vol. 56, No. 6, (553-613), 0003-6900

Wittbrodt, E., Adamiec-Wójcik, I. & Wojciech, S., (2006). *Dynamics of flexible multibody systems: rigid finite element method*, Springer-Verlag, 9783540323518, Berlin

Wittenburg, J., (2008). *Dynamics of multibody systems*, Springer-Verlag, 9780521850117, Berlin

Xu, ZH, Zhong, HX, Zhu, XW, et al., (2009). An efficient algebraic method for computing eigensolution sensitivity of asymmetric damped systems, *J. Sound Vib.*, Vol. 327, (584–592), 0022-460X

21

Non-Linear Periodic and Quasi-Periodic Vibrations in Mechanical Systems - On the use of the Harmonic Balance Methods

Emmanuelle Sarrouy and Jean-Jacques Sinou
Laboratoire de Tribologie et Dynamique des Systèmes UMR-CNRS 5513
Ecole Centrale de Lyon, 36 avenue Guy de Collongue 69134 Ecully Cedex
France

1. Introduction

Due to the fact that non-linear dynamical structures are encountered in many areas of science and engineering, strong developments in the treatment of non-linear differential equations have been proposed and various computational techniques are commonly applied in a wide range of mechanical engineering problems.

The most common techniques for predicting the non-linear behaviour of systems are based on numerical integration over time. However, the use of these methods for non-linear systems with many degrees of freedom can be rather expensive and requires considerable resources both in terms of computation time and data storage. Due to the complexity of non-linear systems and to save time, approximate methods for the study of non-linear oscillating systems described by ordinary non-linear differential equations are usually required. In this category, the most popular methods for approximating the stationary non-linear responses of systems are the harmonic balance methods. The principal idea for these non-linear methods is to replace the non-linear responses and the non-linear forces in the dynamical systems by constructing linear functions such as Fourier series. The main objective of these non-linear methods is to extract and characterize the non-linear behaviours of mechanical systems by using non-linear approximations.

In this chapter, the general formulation and extensions of the harmonic balance method will be presented. The chapter is divided into four parts. Firstly we propose to present the general formulation and the basic concept of the harmonic balance method to find periodic oscillations of non-linear systems. Secondly a generalization of the method is exposed to treat quasi-periodic solutions. Thirdly, a condensation procedure that keeps only the non-linear degrees of freedom of the mechanical system is described. This technique may be of great interest to reduce the original non-linear system and to calculate the dynamical behaviour of non-linear systems with many degrees of freedom. The last part presents the classical continuation procedures that let us follow the evolution of a solution as a system parameter varies. For these two steps procedures, several prediction methods (secant, tangent and

Lagrange polynomial methods) and correction methods (arc length, pseudo arc length and Moore-Penrose methods) are detailed.

2. General theory of the harmonic balance method

The most general formulation for a non-linear dynamical system is

$$\underline{\underline{M}}\,\ddot{\underline{q}} + \underline{\underline{C}}\,\dot{\underline{q}} + \underline{\underline{K}}\,\underline{q} + \hat{\underline{f}}(t, \underline{q}, \dot{\underline{q}}) = \underline{f}_e(t) \quad (1)$$

where $\underline{\underline{M}}$, $\underline{\underline{C}}$ and $\underline{\underline{K}}$ are respectively the mass, damping (including gyroscopic effects if any) and stiffness matrices, $\hat{\underline{f}}(t, \underline{q}, \dot{\underline{q}})$ stands for the non-linear effects in the system and $\underline{f}_e(t)$ the external forces. $\underline{q}$ is the displacement vector with size n. Looking for periodic solutions $\underline{q}(t)$ with a determined period T, it is legitimate to look for the signal as a Fourier series which is truncated for the sake of the numerical application. Thus we assume that the non-linear dynamical response of the system may be approximated by finite Fourier series with $\omega = \frac{2\pi}{T}$ the fundamental pulsation:

$$\underline{q}(t) = \frac{\underline{a}_0}{\sqrt{2}} + \sum_{k=1}^{m} \left(\underline{a}_k \cos(k\omega t) + \underline{b}_k \sin(k\omega t)\right) \quad (2)$$

where m is the order of the Fourier series. $\underline{a}_0$, $\underline{a}_k$ and $\underline{b}_k$ define the unknown coefficients of the finite Fourier series. It should be noted that these coefficients define $\dot{\underline{q}}$ and $\ddot{\underline{q}}$ too.

The number of harmonic coefficients is selected on the basis of the number of significant harmonics expected in the non-linear dynamical response. Generally speaking, harmonic components become less significant when m increases. This formulation includes only harmonic and super-harmonic responses of the system. Some terms can be added to take sub-harmonics (with pulsation $\frac{k'}{l}\omega$) into account. So as to keep simple equations these terms will not be included in the following sections.

In order to determine the value of the $n \times (2m + 1)$ unknowns, the decomposition (2) is reinjected in (1); the time variable is then removed by projecting the resulting system onto the basis $(1/\sqrt{2}, \cos(k\omega t), \sin(k\omega t))_{(k=1,\ldots,m)}$ using the scalar product:

$$< f, g >_T = \frac{2}{T} \int_0^T f(t) g(t) dt \quad (3)$$

This leads to a set of $n \times (2m + 1)$ non-linear (non-differential) equations that can be solved using a dedicated algorithm such as Broyden method (Broyden, 1965):

$$\underline{H}(\tilde{\underline{x}}) = \underline{\underline{H}}_l\, \tilde{\underline{x}} + \hat{\underline{H}}(\tilde{\underline{x}}) - \underline{H_e} = \underline{0} \quad (4)$$

where $\tilde{\underline{x}}$ regroups the unknowns $\underline{a}_0$, $\underline{a}_k$ and $\underline{b}_k$,

$$\tilde{\underline{x}} = \left\{ \begin{array}{cccccc} \underline{a}_0^T & \underline{a}_1^T & \underline{b}_1^T & \ldots & \underline{a}_m^T & \underline{b}_m^T \end{array} \right\}^T, \quad (5)$$

$\underline{\underline{H}}_l$ contains the contribution of the linear part of (1), $\hat{\underline{H}}(\tilde{\underline{x}})$ is the projection of the non-linear part and $\underline{H_e}$ the one of the external forces. For further use, the following quantities are defined: first, the blocks of the $\underline{\underline{H}}_1$ (block diagonal) matrix

$$\underline{\underline{H}}_1 = \begin{bmatrix} \underline{\underline{\Lambda}}_0 & \underline{\underline{0}} & \dots & & \\ \underline{\underline{0}} & \underline{\underline{\Lambda}}_1 & \underline{\underline{0}} & \dots & \\ \dots & \underline{\underline{0}} & \underline{\underline{\Lambda}}_k & \underline{\underline{0}} & \\ & \dots & \underline{\underline{0}} & \underline{\underline{\Lambda}}_m & \end{bmatrix}, \tag{6a}$$

$$\underline{\underline{\Lambda}}_0 = \underline{\underline{K}} \in \mathcal{M}_n(\mathbb{R}) \tag{6b}$$

$$\forall\, k \in \{1, \dots, m\},\ \underline{\underline{\Lambda}}_k = \begin{bmatrix} \underline{\underline{K}} - (k\omega)^2 \underline{\underline{M}} & (k\omega)\underline{\underline{C}} \\ -(k\omega)\underline{\underline{C}} & \underline{\underline{K}} - (k\omega)^2 \underline{\underline{M}} \end{bmatrix} \in \mathcal{M}_{2n}(\mathbb{R}) \tag{6c}$$

Then, the approximation of the non-linear contribution using its projections $\underline{c}_0$, $\underline{c}_k$ and $\underline{d}_k$ onto $1/\sqrt{2}$, $\cos(k\omega t)$ and $\sin(k\omega t)$ respectively is written as follow:

$$\underline{\hat{f}}(t, \underline{q}, \underline{\dot{q}}) \approx \frac{\underline{c}_0}{\sqrt{2}} + \sum_{k=1}^{m} \left(\underline{c}_k \cos(k\omega t) + \underline{d}_k \sin(k\omega t) \right) \tag{7}$$

Using this notations, $\underline{\hat{H}}(\underline{\tilde{x}})$ is the vector

$$\underline{\hat{H}}(\underline{\tilde{x}}) = \left\{ \begin{array}{cccccc} \underline{c}_0^T & \underline{c}_1^T & \underline{d}_1^T & \dots & \underline{c}_m^T & \underline{d}_m^T \end{array} \right\}^T \tag{8}$$

Cameron and Griffin (Cameron & Griffin, 1989) suggested to compute these quantities using an alternate frequency/time domain (AFT) method. First, an Inverse Fast Fourier Transform (IFFT) is used to recompose $\underline{q}(t_j)$ and $\underline{\dot{q}}(t_j)$ from $\underline{a}_0$, $\underline{a}_k$, $\underline{b}_k$ coefficients for some $t_j \in [0, T]$. Then, for each time step t_j the $\underline{\hat{f}}(t_j, \underline{q}(t_j), \underline{\dot{q}}(t_j))$ vectors are computed and $\underline{c}_0$, $\underline{c}_k$ and $\underline{d}_k$ projections are finally obtained using a Fast Fourier Transform (FFT) to switch back into the frequency space.
Usually, the external forces are T-periodic and there is no numerical computation required to obtain the $\underline{H_e}$ vector.

3. Extension of the Harmonic Balance Method for multiple excitations

Now, the general case in which the structural system is excited by several incommensurable frequencies $\omega_1, \omega_2, \dots, \omega_p$ is discussed. The previous non-linear dynamical equation (1) is considered with multiple excitations contained in the external excitation forces $\underline{f}_e(t)$. So, non-linear responses are no longer periodic when oscillatory systems are subjected to p incommensurable frequencies. The non-linear oscillations contain the frequency components of any linear combination of the incommensurable frequency components

$$k_1\omega_1 + k_2\omega_2 + \dots + k_j\omega_j + \dots + k_p\omega_p$$
$$\text{with} \quad k_j = -m, -m+1, \dots, -1, 0, 1, \dots, m-1, m \tag{9}$$

where m defines the order for each fundamental frequency and p the number of incommensurable frequencies.

Thus the approximation of the dynamic non-linear response of equation (1) can be expressed with a generalized Fourier series in the following form

$$\underline{q}(t) = \sum_{k_1=-m}^{m} \sum_{k_2=-m}^{m} \cdots \sum_{k_p=-m}^{m} \left(\underline{a}_{k_1,k_2,\ldots,k_p} \cos\left(\sum_{j=1}^{p} k_j \omega_j t \right) \underline{b}_{k_1,k_2,\ldots,k_p} \sin\left(\sum_{j=1}^{p} k_j \omega_j t \right) \right) \quad (10)$$

where $\underline{a}_{k_1,k_2,\ldots,k_p}$ and $\underline{b}_{k_1,k_2,\ldots,k_p}$ define the unknown Fourier coefficients of any linear combinations of the incommensurable frequency components $\omega_1, \omega_2, \ldots, \omega_p$. For the reader comprehension, it may be noted that a definition for retaining m harmonics in a multiple Fourier series can be given by (Kim & Choi, 1997)

$$\sum_{j=1}^{p} |k_j| \leq m \quad (11)$$

Considering that all harmonics at negative combination frequencies can be replaced by harmonic terms at positive combination frequencies due to the following trigonometric relation

$$\cos\left(\sum_{j=1}^{p} k_j \omega_j t \right) = \cos\left(\sum_{j=1}^{p} -k_j \omega_j t \right) \quad (12)$$

$$\sin\left(\sum_{j=1}^{p} k_j \omega_j t \right) = -\sin\left(\sum_{j=1}^{p} -k_j \omega_j t \right) \quad (13)$$

it may be concluded that only terms at positive combination frequencies (i.e. $\sum_{j=1}^{p} k_j \omega_j t \geq 0$) can be retained in the non-linear response and non-linear force expressions.
So, the previous expression (10) can be rewritten in a condensed form

$$\underline{q}(t) = \frac{\underline{a}_0}{\sqrt{2}} + \sum_{\underline{k} \in \mathbb{Z}^p} \underline{a}_{\underline{k}} \cos\left(\underline{k}.\underline{\omega}\right) t + \sum_{\underline{k} \in \mathbb{Z}^p} \underline{b}_{\underline{k}} \sin\left(\underline{k}.\underline{\omega}\right) t \quad (14)$$

where the (.) denotes the dot product, $\underline{k}$ is the harmonic number vector of each frequency direction and $\underline{\omega}$ is the vector of the p incommensurable frequencies considered in the solution. The contributions $\underline{a}_{\underline{k}}$ and $\underline{b}_{\underline{k}}$ contain the new Fourier decomposition of cosine and sine terms corresponding to the positive frequency combinations.
For convenience, it is wise to deal with a multiple time parameter. By introducing a non dimensional multiple time parameter $\underline{\tau} = \underline{\omega} t$ that refers to hyper-time concept proposed by (Kim & Choi, 1997), the approximated non-linear expression (14) is composed from elements of cosine and sine terms such as

$$\underline{q}(\underline{\tau}) = \frac{\underline{a}_0}{\sqrt{2}} + \sum_{\underline{k} \in \mathbb{Z}^p} \underline{a}_{\underline{k}} \cos\left(\underline{k}.\underline{\tau}\right) + \sum_{\underline{k} \in \mathbb{Z}^p} \underline{b}_{\underline{k}} \sin\left(\underline{k}.\underline{\tau}\right) \quad (15)$$

Injecting this in Eq. (1), one gets

$$\underline{\underline{K}}\frac{\underline{a}_0}{\sqrt{2}} + \sum_{\underline{k}\in\mathbb{Z}^p}\left(\left(\underline{\underline{K}} - (\underline{k}.\underline{\omega})^2\,\underline{\underline{M}}\right)\underline{a}_{\underline{k}} + \left((\underline{k}.\underline{\omega})\,\underline{\underline{C}}\right)\underline{b}_{\underline{k}}\right)\cos(\underline{k}.\underline{\tau})$$
$$+ \sum_{\underline{k}\in\mathbb{Z}^p}\left(\left(\underline{\underline{K}} - (\underline{k}.\underline{\omega})^2\,\underline{\underline{M}}\right)\underline{b}_{\underline{k}} - \left((\underline{k}.\underline{\omega})\,\underline{\underline{C}}\right)\underline{a}_{\underline{k}}\right)\sin(\underline{k}.\underline{\tau})$$
$$+\hat{\underline{f}}(\tilde{\underline{x}}) = \underline{f}_e(t) \quad (16)$$

where the non-linear forces vector $\hat{\underline{f}}(t,\underline{q},\dot{\underline{q}})$ is approximated by the generalized Fourier series in a condensed form

$$\hat{\underline{f}}(t,\underline{q},\dot{\underline{q}}) = \frac{\underline{c}_0}{\sqrt{2}} + \sum_{\underline{k}\in\mathbb{Z}^p}\underline{c}_{\underline{k}}\cos(\underline{k}.\underline{\omega})\,t + \sum_{\underline{k}\in\mathbb{Z}^p}\underline{d}_{\underline{k}}\sin(\underline{k}.\underline{\omega})\,t \quad (17)$$

Thus, the non-linear equations (16) can be rewritten in the form of an algebraic equation system similar to (4) for unknown vector of harmonic coefficients with only terms at positive frequency combinations. In this case $\tilde{\underline{x}}$ denotes the unknown vector of harmonic coefficients $\underline{a}_0$, $\underline{a}_{\underline{k}}$ and $\underline{b}_{\underline{k}}$

$$\tilde{\underline{x}} = \left\{\begin{array}{cccccccc} \frac{\underline{a}_0^T}{\sqrt{2}} & \underline{a}_{\underline{1}}^T & \underline{b}_{\underline{1}}^T & \underline{a}_{\underline{2}}^T & \underline{b}_{\underline{2}}^T & \ldots & \underline{a}_{\underline{N}}^T & \underline{b}_{\underline{N}}^T \end{array}\right\}^T \quad (18)$$

where N represents the total number of frequency components including all harmonic terms up to m of each frequency direction and all the coupling frequencies chosen by using (11). $\underline{H_e}$ and $\hat{\underline{H}}(\tilde{\underline{x}})$ contain the projection of the external forces $\underline{f}_e(t)$ and the non-linear part $\hat{\underline{f}}(t,\underline{q},\dot{\underline{q}})$, respectively. $\hat{\underline{H}}(\tilde{\underline{x}})$ is given by

$$\hat{\underline{H}}(\tilde{\underline{x}}) = \left\{\begin{array}{cccccccc} \frac{\underline{c}_0^T}{\sqrt{2}} & \underline{c}_{\underline{1}}^T & \underline{d}_{\underline{1}}^T & \underline{c}_{\underline{2}}^T & \underline{d}_{\underline{2}}^T & \ldots & \underline{c}_{\underline{N}}^T & \underline{d}_{\underline{N}}^T \end{array}\right\}^T \quad (19)$$

The non-linear treatment of Fourier coefficients is performed by extending the generalization of the AFT to a p-dimensional frequency domain with a p-dimensional FFT. $\underline{\underline{H}}_l$ contains the contribution of the linear part of (1) and refers to the block diagonal matrix:

$$\underline{\underline{H}}_l = \begin{bmatrix} \underline{\underline{K}} & \underline{\underline{0}} & \ldots & & \\ \underline{\underline{0}} & \underline{\underline{\Lambda}}_1 & \underline{\underline{0}} & \ldots & \\ \ldots & \underline{\underline{0}} & \underline{\underline{\Lambda}}_k & \underline{\underline{0}} & \\ & \ldots & \underline{\underline{0}} & \underline{\underline{\Lambda}}_N & \end{bmatrix} \quad (20)$$

with

$$\underline{\underline{\Lambda}}_k = \begin{bmatrix} \underline{\underline{K}} - (\underline{k}.\underline{\omega})^2\,\underline{\underline{M}} & (\underline{k}.\underline{\omega})\,\underline{\underline{C}} \\ -(\underline{k}.\underline{\omega})\,\underline{\underline{C}} & \underline{\underline{K}} - (\underline{k}.\underline{\omega})^2\,\underline{\underline{M}} \end{bmatrix} \text{ for } k \in [1, N] \quad (21)$$

4. Condensation procedure

If the considered non-linear system has n degrees of freedom but only q of them are used in the formula of the non-linear forces $\hat{f}(t,\underline{q},\underline{\dot{q}})$, then it is possible to work with a system similar to (4) but with size $q(2m+1)$ instead of $n(2m+1)$. For systems with localized non-linearities, this kind of condensation is very interesting (Hahn & Chen, 1994; Sinou, 2008). To achieve this, one has to partition the variables into the p linear ones, denoted $\underline{q}^{\mathrm{l}}$, and the q non-linear ones, denoted $\underline{q}^{\mathrm{nl}}$. This implies later a partition of $\underline{\tilde{x}}$ into $\underline{\tilde{x}}^{\mathrm{l}}$ and $\underline{\tilde{x}}^{\mathrm{nl}}$, reflecting the harmonic components of linear degrees of freedom and non-linear ones respectively. A relation can then be established that let us express $\underline{\tilde{x}}^{\mathrm{l}}$ as function of $\underline{\tilde{x}}^{\mathrm{nl}}$. First, this relationship is exposed and used to get the reduced non-linear system to solve. In a second part, the link between $\underline{q}$ partition and $\underline{\tilde{x}}$ partition is detailed in order to get the expressions of the partitioned HBM elements. The procedure is exposed in the case of a simple Harmonic Balance Method but can easily be extended to the case of quasi-periodic solutions.

4.1 Working with a smaller system

Once the partition is achieved, (4) can be rewritten

$$\begin{bmatrix} \underline{\underline{H}}_{\mathrm{l}}^{\mathrm{ll}} & \underline{\underline{H}}_{\mathrm{l}}^{\mathrm{lnl}} \\ \underline{\underline{H}}_{\mathrm{l}}^{\mathrm{nll}} & \underline{\underline{H}}_{\mathrm{l}}^{\mathrm{nlnl}} \end{bmatrix} \begin{Bmatrix} \underline{\tilde{x}}^{\mathrm{l}} \\ \underline{\tilde{x}}^{\mathrm{nl}} \end{Bmatrix} + \begin{Bmatrix} \underline{\hat{H}}^{\mathrm{l}}(\underline{\tilde{x}}^{\mathrm{nl}}) \\ \underline{\hat{H}}^{\mathrm{nl}}(\underline{\tilde{x}}^{\mathrm{nl}}) \end{Bmatrix} - \begin{Bmatrix} \underline{H}_{\mathrm{e}}^{\mathrm{l}} \\ \underline{H}_{\mathrm{e}}^{\mathrm{nl}} \end{Bmatrix} = \underline{0} \tag{22}$$

The first set of lines provides a relationship between $\underline{\tilde{x}}^{\mathrm{l}}$ and $\underline{\tilde{x}}^{\mathrm{nl}}$:

$$\underline{\tilde{x}}^{\mathrm{l}} = \underline{\underline{H}}_{\mathrm{l}}^{\mathrm{ll}^{-1}} \left[\underline{H}_{\mathrm{e}}^{\mathrm{l}} - \underline{\underline{H}}_{\mathrm{l}}^{\mathrm{lnl}} \underline{\tilde{x}}^{\mathrm{nl}} - \underline{\hat{H}}^{\mathrm{l}}(\underline{\tilde{x}}^{\mathrm{nl}}) \right] \tag{23}$$

This expression is used to replace $\underline{\tilde{x}}^{\mathrm{l}}$ in the second set of equations, leading to a non-linear system with size $q(2m+1)$ depending on $\underline{\tilde{x}}^{\mathrm{nl}}$ only:

$$\left[\underline{\underline{H}}_{\mathrm{l}}^{\mathrm{nlnl}} - \underline{\underline{H}}_{\mathrm{l}}^{\mathrm{nll}} \underline{\underline{H}}_{\mathrm{l}}^{\mathrm{ll}^{-1}} \underline{\underline{H}}_{\mathrm{l}}^{\mathrm{lnl}} \right] \underline{\tilde{x}}^{\mathrm{nl}} + \left(\underline{\hat{H}}^{\mathrm{nl}}(\underline{\tilde{x}}^{\mathrm{nl}}) - \underline{\underline{H}}_{\mathrm{l}}^{\mathrm{nll}} \underline{\underline{H}}_{\mathrm{l}}^{\mathrm{ll}^{-1}} \underline{\hat{H}}^{\mathrm{l}}(\underline{\tilde{x}}^{\mathrm{nl}}) \right) - \left(\underline{H}_{\mathrm{e}}^{\mathrm{nl}} - \underline{\underline{H}}_{\mathrm{l}}^{\mathrm{nll}} \underline{\underline{H}}_{\mathrm{l}}^{\mathrm{ll}^{-1}} \underline{H}_{\mathrm{e}}^{\mathrm{l}} \right) = \underline{0} \tag{24}$$

For any solution $\underline{\tilde{x}}^{\mathrm{nl}^*}$ of (24), $\underline{\tilde{x}}^{\mathrm{l}^*}$ is obtained thanks to equation (23).

4.2 Getting the expressions of the partitioned elements

Let us denote $\underline{\underline{R}}$ the matrix that reorders the degrees of freedom from the partitioned ones to the initial ones:

$$\underline{q} = \underline{\underline{R}} \begin{Bmatrix} \underline{q}^{\mathrm{l}} \\ \underline{q}^{\mathrm{nl}} \end{Bmatrix} \tag{25}$$

Injecting this in (1) and pre-multiplying the equation by $\underline{\underline{R}}^T$ to reorder the equations, one can write

$$\begin{bmatrix} \underline{\underline{M}}^{\mathrm{ll}} & \underline{\underline{M}}^{\mathrm{lnl}} \\ \underline{\underline{M}}^{\mathrm{nll}} & \underline{\underline{M}}^{\mathrm{nlnl}} \end{bmatrix} \overset{\cdot\cdot}{\overbrace{\begin{Bmatrix} \underline{q}^{\mathrm{l}} \\ \underline{q}^{\mathrm{nl}} \end{Bmatrix}}} + \begin{bmatrix} \underline{\underline{C}}^{\mathrm{ll}} & \underline{\underline{C}}^{\mathrm{lnl}} \\ \underline{\underline{C}}^{\mathrm{nll}} & \underline{\underline{C}}^{\mathrm{nlnl}} \end{bmatrix} \overset{\cdot}{\overbrace{\begin{Bmatrix} \underline{q}^{\mathrm{l}} \\ \underline{q}^{\mathrm{nl}} \end{Bmatrix}}} + \begin{bmatrix} \underline{\underline{K}}^{\mathrm{ll}} & \underline{\underline{K}}^{\mathrm{lnl}} \\ \underline{\underline{K}}^{\mathrm{nll}} & \underline{\underline{K}}^{\mathrm{nlnl}} \end{bmatrix} \begin{Bmatrix} \underline{q}^{\mathrm{l}} \\ \underline{q}^{\mathrm{nl}} \end{Bmatrix} + \begin{Bmatrix} \underline{f}^{\mathrm{l}}(t, \underline{q}^{\mathrm{nl}}, \dot{\underline{q}}^{\mathrm{nl}}) \\ \underline{f}^{\mathrm{nl}}(t, \underline{q}^{\mathrm{nl}}, \dot{\underline{q}}^{\mathrm{nl}}) \end{Bmatrix} = \begin{Bmatrix} \underline{f}_e^{\mathrm{l}}(t) \\ \underline{f}_e^{\mathrm{nl}}(t) \end{Bmatrix} \quad (26)$$

Projecting this set of equations onto the basis $(1/\sqrt{2}, \cos(k\omega t), \sin(k\omega t))_{(k=1,\ldots,m)}$, one obtains a set of equations relative to $\underline{\tilde{x}}'$ vector of unknowns while equation (4) is relative to $\underline{\tilde{x}}$ and partitioned equation (22) uses $\underline{\tilde{x}}''$:

$$\underline{\tilde{x}} = \begin{Bmatrix} \underline{a}_0 \\ \underline{a}_1^{\mathrm{l}} \\ \underline{b}_1 \\ \vdots \end{Bmatrix}, \ \underline{\tilde{x}}' = \begin{Bmatrix} \underline{a}_0^{\mathrm{l}} \\ \underline{a}_0^{\mathrm{nl}} \\ \underline{a}_1^{\mathrm{l}} \\ \underline{a}_1^{\mathrm{nl}} \\ \underline{b}_1^{\mathrm{l}} \\ \underline{b}_1^{\mathrm{nl}} \\ \vdots \end{Bmatrix} \text{ and } \underline{\tilde{x}}'' = \begin{Bmatrix} \underline{\tilde{x}}^{\mathrm{l}} \\ \underline{\tilde{x}}^{\mathrm{nl}} \end{Bmatrix} = \begin{Bmatrix} \begin{Bmatrix} \underline{a}_0^{\mathrm{l}} \\ \underline{a}_1^{\mathrm{l}} \\ \underline{b}_1^{\mathrm{l}} \\ \vdots \end{Bmatrix} \\ \begin{Bmatrix} \underline{a}_0^{\mathrm{nl}} \\ \underline{a}_1^{\mathrm{nl}} \\ \underline{b}_1^{\mathrm{nl}} \\ \vdots \end{Bmatrix} \end{Bmatrix} \quad (27)$$

This vector is related to $\underline{\tilde{x}}$ by

$$\underline{\tilde{x}} = (\underline{\underline{I}}_{2m+1} \otimes \underline{\underline{R}})\underline{\tilde{x}}' \quad (28)$$

where $\otimes$ denotes the Kronecker product.
The last step consists in linking $\underline{\tilde{x}}'$ to the vector $\underline{\tilde{x}}''$ where all the linear unknowns harmonics are at the top and all the non-linear unknowns are at the bottom:

$$\underline{\tilde{x}}' = \left[\underline{\underline{I}}_{2m+1} \otimes \begin{bmatrix} \underline{\underline{I}}_p \\ \underline{\underline{0}}_{q,p} \end{bmatrix}, \underline{\underline{I}}_{2m+1} \otimes \begin{bmatrix} \underline{\underline{0}}_{p,q} \\ \underline{\underline{I}}_q \end{bmatrix} \right] \underline{\tilde{x}}'' \quad (29)$$

Finally, partitioned and initial harmonics vectors are linked by a matrix $\underline{\underline{T}}$ with the following expression:

$$\underline{\tilde{x}} = \underbrace{\left[\underline{\underline{I}}_{2m+1} \otimes \left(\underline{\underline{R}} \begin{bmatrix} \underline{\underline{I}}_p \\ \underline{\underline{0}}_{q,p} \end{bmatrix} \right), \underline{\underline{I}}_{2m+1} \otimes \left(\underline{\underline{R}} \begin{bmatrix} \underline{\underline{0}}_{p,q} \\ \underline{\underline{I}}_q \end{bmatrix} \right) \right]}_{\underline{\underline{T}}} \underline{\tilde{x}}'' \quad (30)$$

The partitioned elements of equation (22) are obtained from the initial elements of equation (4) using

$$\begin{bmatrix} \underline{\underline{H}}_{\mathrm{l}}^{\mathrm{ll}} & \underline{\underline{H}}_{\mathrm{l}}^{\mathrm{lnl}} \\ \underline{\underline{H}}_{\mathrm{l}}^{\mathrm{nll}} & \underline{\underline{H}}_{\mathrm{l}}^{\mathrm{nlnl}} \end{bmatrix} = \underline{\underline{T}}^T \underline{\underline{H}}_{\mathrm{l}} \underline{\underline{T}} \tag{31a}$$

$$\left\{ \begin{array}{c} \underline{\hat{H}}^{\mathrm{l}}(\underline{\tilde{x}}^{\mathrm{nl}}) \\ \underline{\hat{H}}^{\mathrm{nl}}(\underline{\tilde{x}}^{\mathrm{nl}}) \end{array} \right\} = \underline{\underline{T}}^T \underline{\hat{H}} \tag{31b}$$

$$\left\{ \begin{array}{c} \underline{H}_{\mathrm{e}}^{\mathrm{l}} \\ \underline{H}_{\mathrm{e}}^{\mathrm{nl}} \end{array} \right\} = \underline{\underline{T}}^T \underline{H_{\mathrm{e}}} \tag{31c}$$

5. Path following: continuation

It may be useful to track the evolution of the system behaviour as one of its parameter, μ, varies. For instance in the field of rotating machinery, the behaviour of systems is often calculated for different operational speeds of interest while all the other parameters are kept constant. Special algorithms should then be implemented for two main reasons: first, such methods let us take advantage of the fact that if two values of the parameter are close, solutions of the non-linear system have good chances to be close from one another too. Second, following the path in the $(\underline{\tilde{x}}, \mu)$ space helps to find coexisting solutions for the same μ parameter value. This case is illustrated on Fig. 1 which depicts the evolution of maximum cycle amplitude versus a continuation parameter. This curve can be obtained by studying the classical Duffing oscillator which has a cubic stiffness, μ being the excitation frequency. The resonnance peak is bent on the right: this oscillator belongs to the hardening systems. Plus, there are two points B and E - turning points - at which the path folds, creating a range where multiple solutions can coexist. Without a proper continuation scheme, one would obtain at best parts A to B and C to D by looking for solutions with a positive increment in μ or parts D to E and F to A with a negative one. The B-E part of the curve would be missed in every case. Continuation algorithms are based on two main steps applied recursively for each point: first a prediction is done based on the point(s) previously obtained, then a correction step provides the new point. Different methods exist for these two steps and are exposed in the next subsections; the third subsection is dedicated to the step length adaptation techniques which basically control the distance between two consecutive points. The last part summarizes the steps and gives a global overview of the entire procedure.

The following notations will be used throughout the current section:

- $\underline{y}^{(i)} = (\underline{\tilde{x}}^{(i)}, \mu^{(i)})$: i-th converged point;
- $\underline{y}^{(i+1,0)} = (\underline{\tilde{x}}^{(i+1,0)}, \mu^{(i+1,0)})$: prediction for $(i+1)$-th point;
- $\underline{y}^{(i+1,j)} = (\underline{\tilde{x}}^{(i+1,j)}, \mu^{(i+1,j)})$: $(i+1)$-th point after j correction steps.

Moreover, to emphasize the dependency of the equation set (4) with the parameter μ, $\underline{H}(\underline{\tilde{x}})$ will be noted $\underline{H}(\underline{\tilde{x}}, \mu)$.

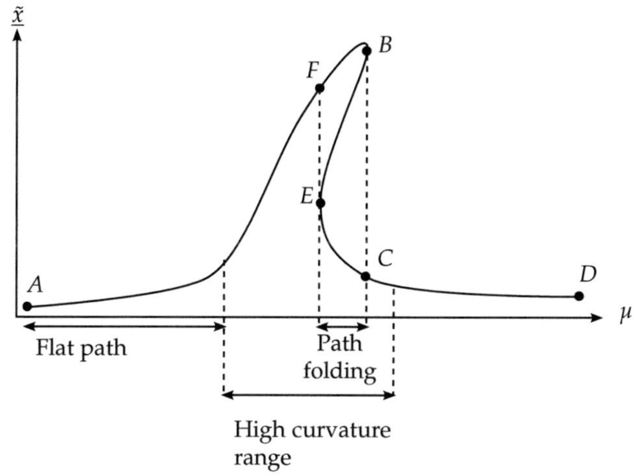

Fig. 1. Continuation applied to a typical hardening oscillator

5.1 Prediction methods

To predict a point, one needs a direction and a distance. The usual way to mesure the distance between two points is to use the following approximation of the curvilinear abscissa increment:

$$\Delta s^{(i+1)} = \sqrt{(\underline{\tilde{x}}^{(i+1)} - \underline{\tilde{x}}^{(i)})^T(\underline{\tilde{x}}^{(i+1)} - \underline{\tilde{x}}^{(i)}) + (\mu^{(i+1)} - \mu^{(i)})^2} \tag{32}$$

The way to optimize this length is discussed in subsection 5.3.
Different methods provide a direction. Among all of them one finds the secant method, the tangent method and the use of Lagrange polynomials. For most methods, one needs to have a few points already converged. To obtain these points, different values $\mu^{(i)}$, $1 \leq i \leq i_{ini}$ are chosen and usual algorithms are applied to find the corresponding solutions $\underline{\tilde{x}}^{(i)}$. The three methods exposed are illustrated on Fig. 2.

5.1.1 Secant method

The secant method (Fig. 2 (a)) uses the two previous points $(\underline{\tilde{x}}^{(i-1)}, \mu^{(i-1)})$ and $(\underline{\tilde{x}}^{(i)}, \mu^{(i)})$ to predict $(\underline{\tilde{x}}^{(i+1,0)}, \mu^{(i+1,0)})$:

$$(\underline{\tilde{x}}^{(i+1,0)}, \mu^{(i+1,0)}) = (\underline{\tilde{x}}^{(i)}, \mu^{(i)}) + \Delta s^{i+1}\left((\underline{\tilde{x}}^{(i)}, \mu^{(i)}) - (\underline{\tilde{x}}^{(i-1)}, \mu^{(i-1)})\right) \tag{33}$$

It is a very cheap predictor but it does not suit paths with small curvature radii.

5.1.2 Tangent method

This predictor (Fig. 2 (b)) uses only one previous point $(\underline{\tilde{x}}^{(i)}, \mu^{(i)})$ to predict $(\underline{\tilde{x}}^{(i+1,0)}, \mu^{(i+1,0)})$ but requires the evaluation of derivatives which can have a prohibitive cost. To evaluate the tangent vector $\overrightarrow{t} = (\overrightarrow{t_{\tilde{x}}}, t_\mu)$ to the curve at point $(\underline{\tilde{x}}^{(i)}, \mu^{(i)})$ the following steps can be used.
First, get the $\overrightarrow{z}$ vector:

$$D_{\tilde{x}}\underline{H}(\underline{\tilde{x}}^{(i)}, \mu^{(i)})\overrightarrow{t_{\tilde{x}}} + D_\mu \underline{H}(\underline{\tilde{x}}^{(i)}, \mu^{(i)})t_\mu = \underline{0}$$

$$\Leftrightarrow \overrightarrow{t_{\tilde{x}}} = \underbrace{-D_{\tilde{x}}\underline{H}(\underline{\tilde{x}}^{(i)},\mu^{(i)})^{-1} D_{\mu}\underline{H}(\underline{\tilde{x}}^{(i)},\mu^{(i)})}_{\overrightarrow{z}}\, t_{\mu} \tag{34}$$

where $D_{\tilde{x}}\underline{H}(\underline{\tilde{x}}^{(i)},\mu^{(i)})$ and $D_{\mu}\underline{H}(\underline{\tilde{x}}^{(i)},\mu^{(i)})$ denotes the derivatives of $\underline{H}$ with respect to $\underline{\tilde{x}}$ and μ variables respectively at point $(\underline{\tilde{x}}^{(i)},\mu^{(i)})$.
Then normalize the tangent vector:

$$\overrightarrow{t_{\tilde{x}}}^{T}\overrightarrow{t_{\tilde{x}}} + t_{\mu}^{2} = 1$$

$$\Leftrightarrow t_{\mu} = \pm 1/\sqrt{1+\overrightarrow{z}^{T}\overrightarrow{z}} \tag{35}$$

The sign depends on the direction chosen to depict the curve; the positive direction is usually used. The next point can finally be predicted using

$$(\underline{\tilde{x}}^{(i+1,0)},\mu^{(i+1,0)}) = (\underline{\tilde{x}}^{(i)},\mu^{(i)}) + \Delta s^{i+1}(\overrightarrow{t_{\tilde{x}}},t_{\mu}) \tag{36}$$

5.1.3 Lagrange polynomials

This last predictor uses Lagrange polynomials of degree d to extrapolate the curve defined by the $d+1$ previous points (Fig. 2 (c)). The variable used for this polynomials is the curvilinear abscissa s. Let us redefine it locally by taking its origin at the last converged point:

$$\begin{cases} s^{(i)} = 0, \\ \forall k \in \{1,\ldots,d\}, \quad s^{(i-k)} = s^{(i-k+1)} - \Delta s^{(i-k+1)} \end{cases} \tag{37}$$

where $s^{(i)}$ denotes curvilinear abscissa of point $(\underline{\tilde{x}}^{(i)},\mu^{(i)})$). The Lagrange polynomials define then the unique polynomial $\mathbf{P}_d$ with degree d such that $\mathbf{P}_d(s^{(i-k)}) = \underline{y}^{(i-k)}$, $0 \le k \le d$. A classical expression of these polynomials is

$$\mathbf{P}_d(s) = \sum_{k_0=0}^{d}\left(\prod_{\substack{k=0\\k\neq k_0}}^{d}\frac{s-s^{(i-k)}}{s^{(i-k_0)}-s^{(i-k)}}\right)\underline{y}^{(i-k_0)} \tag{38}$$

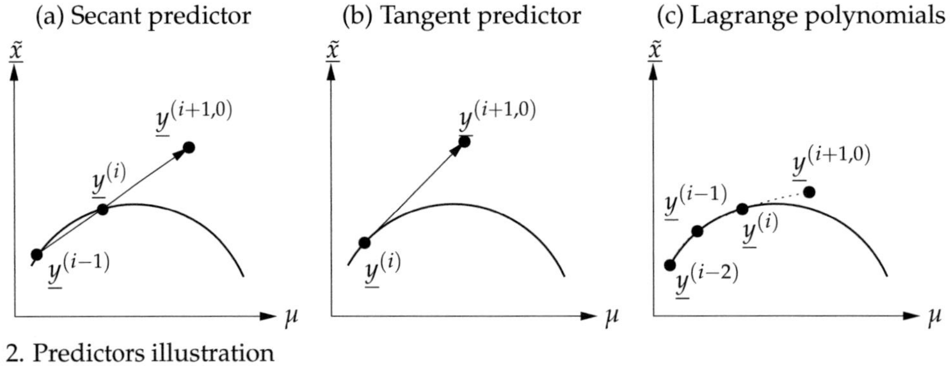

Fig. 2. Predictors illustration

The predicted point whose curvilinear abscissa is $0 + \Delta s^{(i+1)}$ is finally evaluated using

$$(\underline{\tilde{x}}^{(i+1,0)}, \mu^{(i+1,0)}) = \mathbf{P}_d(\Delta s^{(i+1)}) \tag{39}$$

Orders d greater than 2 or 3 are usually avoided because they require more operations and draw oscillating paths.

5.2 Correction methods

The aim of the correction step is to move from the predicted point that usually does not satisfy $\underline{H}(\underline{\tilde{x}}^{(i+1,0)}, \mu^{(i+1,0)}) = \underline{0}$ towards $\underline{y}^{(i+1)} = (\underline{\tilde{x}}^{(i+1)}, \mu^{(i+1)})$ that does. This is done recursively by writing

$$(\underline{\tilde{x}}^{(i+1,j+1)}, \mu^{(i+1,j+1)}) = (\underline{\tilde{x}}^{(i+1,j)} + \underline{\Delta\tilde{x}}, \mu^{(i+1,j)} + \Delta\mu) \tag{40a}$$

$$\begin{aligned}\underline{H}(\underline{\tilde{x}}^{(i+1,j+1)}, \mu^{(i+1,j+1)}) \approx\ & \underline{H}(\underline{\tilde{x}}^{(i+1,j)}, \mu^{(i+1,j)}) \\ & + D_{\tilde{x}}\underline{H}(\underline{\tilde{x}}^{(i+1,j)}, \mu^{(i+1,j)})\underline{\Delta\tilde{x}} \\ & + D_{\mu}\underline{H}(\underline{\tilde{x}}^{(i+1,j)}, \mu^{(i+1,j)})\Delta\mu\end{aligned} \tag{40b}$$

The second equation linearizes the problem around $\underline{y}^{(i+1,j)}$ to get a linear approximation that algorithms can solve to get corrections $\underline{\Delta\tilde{x}}$ and $\Delta\mu$:

$$D_{\tilde{x}}\underline{H}(\underline{\tilde{x}}^{(i+1,j)}, \mu^{(i+1,j)})\underline{\Delta\tilde{x}} + D_{\mu}\underline{H}(\underline{\tilde{x}}^{(i+1,j)}, \mu^{(i+1,j)})\Delta\mu = -\underline{H}(\underline{\tilde{x}}^{(i+1,j)}, \mu^{(i+1,j)}) \tag{41}$$

This provides $n(2m+1)$ scalar equations but there are $n(2m+1)+1$ unknowns to determine (the additional unknown being $\Delta\mu$). To get a square system, one has to add an equation. The name of the corrector depends on this equation.
Moreover, one usually limits the number of correction steps allowed to move from the predicted point towards a converged one; if this number $j_{\max}$ is reached, the correction process is aborted and a new prediction, closer to the previous one is made.

5.2.1 Constant continuation parameter

The easiest equation to add is

$$\forall j \geq 1,\ \mu^{(i+1,j)} = \mu^{(i+1,0)} \tag{42}$$

That is, the continuation parameter is kept constant and equal to the predicted value. This very simple additional constraint does not suits folding paths: using such a corrector, the path portion between the two turning points on Fig. 1 would be missed.

5.2.2 Arc length constraint

This method adds a distance condition between the corrected point and the previous converged point:

$$\forall j \geq 1,\ ||\underline{\tilde{x}}^{(i+1,j+1)} - \underline{\tilde{x}}^{(i)}||^2 + |\mu^{(i+1,j+1)} - \mu^{(i)}|^2 = (\Delta s^{(i+1)})^2 \tag{43}$$

As illustrated on Fig. 3 (a), it forces the successive point $\underline{y}^{(i+1,j)}$ to lie on the (hyper)sphere with center $\underline{y}^{(i)}$ and radius $\Delta s^{(i+1)}$.

This constraint being quadratic, it is not verified exactly and it is in fact the tangent system that is used to complete (41):

$$2\underline{\Delta\tilde{x}}^T(\underline{\tilde{x}}^{(i+1,j)} - \underline{\tilde{x}}^{(i)}) + 2\Delta\mu(\mu^{(i+1,j)} - \mu^{(i)}) = (\Delta s^{(i+1)})^2 - \left(||\underline{\tilde{x}}^{(i+1,j)} - \underline{\tilde{x}}^{(i)}||^2 + |\mu^{(i+1,j)} - \mu^{(i)}|^2\right) \quad (44)$$

5.2.3 Pseudo arc length constraint

This method adds an orthogonality condition between the prediction vector $\underline{y}^{(i+1,0)} - \underline{y}^{(i)}$ and the corrected points as depicted on Fig. 3 (b):

$$\forall j \geq 1,\ (\underline{\tilde{x}}^{(i+1,0)} - \underline{\tilde{x}}^{(i+1,j+1)})^T(\underline{\tilde{x}}^{(i+1,0)} - \underline{\tilde{x}}^{(i)}) + (\mu^{(i+1,0)} - \mu^{(i+1,j+1)})(\mu^{(i+1,0)} - \mu^{(i)}) = 0 \quad (45)$$

In terms of $\underline{\Delta\tilde{x}}$ and $\Delta\mu$ it gives the exact linear condition:

$$\forall j \geq 1,\ \underline{\Delta\tilde{x}}^T(\underline{\tilde{x}}^{(i+1,0)} - \underline{\tilde{x}}^{(i)}) + \Delta\mu(\mu^{(i+1,0)} - \mu^{(i)}) = 0 \quad (46)$$

5.2.4 Moore-Penrose pseudo-inverse

A last way to add a constraint is to use the Moore-Penrose pseudo-inverse matrix. This matrix provides a way to solve underconstrained systems $\underline{\underline{A}}x = \underline{b}$ where $\underline{\underline{A}}$ has less rows than columns. The pseudo-inverse matrix of this system is

$$\underline{\underline{A}}^+ = \underline{\underline{A}}^T(\underline{\underline{A}}\underline{\underline{A}}^T)^{-1}. \quad (47)$$

In fact, no explicit extra condition is added but it adds implicitely an orthogonality condition with the kernel of matrix $\underline{\underline{A}}$; this is illustrated in Fig. 3 (c) where $\overrightarrow{k_j}$ are representative of the kernel for the j-th correction step.
In the case of the studied system (41),

$$\underline{\underline{A}} = \left[\ D_{\tilde{x}}\underline{H}(\underline{\tilde{x}}^{(i+1,j)}, \mu^{(i+1,j)}), D_{\mu}\underline{H}(\underline{\tilde{x}}^{(i+1,j)}, \mu^{(i+1,j)})\ \right]. \quad (48)$$

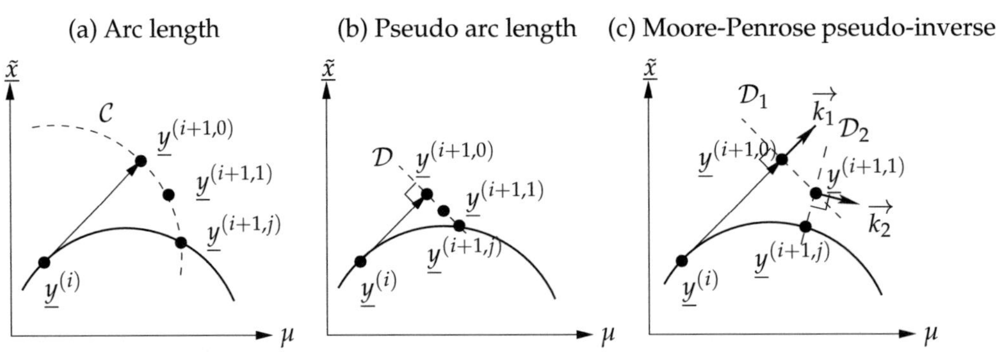

Fig. 3. Correctors illustration

5.3 Step length adaptation

As mentionned in subsection 5.1, a prediction distance has to be decided. Using a constant distance does not suit paths with regions of different curvature. Within a region of low curvature ("Flat path" on Fig. 1), the distance can be great because $\underline{\tilde{x}}$ does not vary a lot with μ; on the contrary, it is more efficient to make small steps in regions with high curvature ("High curvature range" on Fig. 1) in order to avoid a lot of correction steps.
The step length $\Delta s^{(i)}$ is adapted in two cases: if the correction procedure did not converge towards a point on the path or everytime a point is obtained. Different methods can be used to determine the step length correction (Allgower & Georg, 2003). One of the cheapest but still efficient is to monitor the number of correction steps j used to get $\underline{y}^{(i+1)}$:

- if $j = j_{\max}$, a new prediction is made with a smaller step length;
- if $j \leq j_1$, the correction process converged very quickly and $\Delta s^{(i+1)}$ can be increased;
- if $j_2 \leq j < j_{\max}$, the correction process converged slowly and a smaller step length is used for the next prediction;
- if $j_1 < j \leq j_2$, step length is left unchanged for the next prediction .

The ratios used to increase or decrease $\Delta s^{(i+1)}$ are arbitrary, as j_1, j_2 and $j_{\max}$ values. A typical set of values would be $j_1 = 2$, $j_2 = 10$, $j_{\max} = 15$ and divide $\Delta s^{(i+1)}$ by 2 to decrease it or multiply it by 2 to increase it.

5.4 Sum up

Fig. 4 summarizes the different steps of a continuation algorithm. Besides the choice of the prediction and correction methods, a few parameters have to be chosen: initial step length $\Delta s^{(1)}$, maximum number of correction iterations $j_{\max}$, as well as parameters j_1 and j_2 mentionned in section 5.3 that drive the recast of the step length. Note that the proposed convergence criterion can be slightly different. Anyhow a convergence criterion requires the choice of an ϵ value.
Finally, one needs to add a criterion for stopping the algorithm: usually a range of interest for μ is known and the algorithm is stopped as soon as a point outside this range is obtained. This is then checked each time a new point is found.

6. Conclusion

This chapter aims to provide a comprehensive overview of the basic theory of the harmonic balance methods and continuation for non-linear periodic and quasi-periodic vibrations in mechanical systems. In the past decades, these approaches have been at the focus of attention of many researchers in order to obtain very efficient tools of great importance for mechanical engineering communities. This is why, in recent years, various developments of the harmonic balance techniques have been extensively published not only to estimate the non-linear vibration of mechanical systems but also to better assess and understand some specific non-linear behaviors in mechanical systems. Moreover, the need to be able to propose more practical and commonly implemented techniques in the early stage in complex mechanical structures has led to the increase of the harmonic balance methods and new developments. Giving an exhaustive list of illustrative examples and applications showing efficiency and robustness of the harmonic balance methodology is not possible. However,

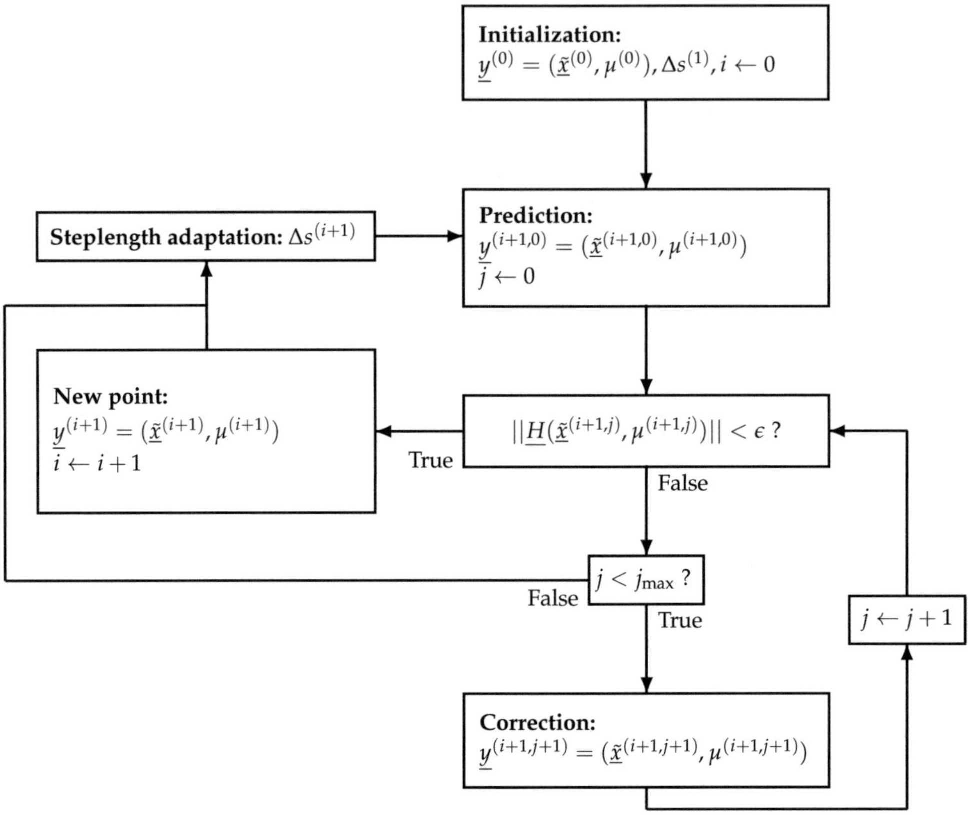

Fig. 4. General algorithm of a continuation procedure

for the interested reader, some non-exhaustive studies that have been previously published by the authors, can be found in practical cases of mechanical applications, aeronautics and car manufacturers communities, rotating machinery or structural health monitoring such as: the non-linear periodic vibration of a flexible rotor supported by ball bearings (Sinou, 2009; Villa et al., 2008), multi-dimensional harmonic balance applied to rotor dynamics (Guskov et al., 2008), the steady-state responses of autonomous mechanical systems with frictional interfaces for single or multiple input frequencies linked to unstable modes (Coudeyras, Nacivet & Sinou, 2009; Coudeyras, Sinou & Nacivet, 2009), damage detection in mechanical systems from changes in the measurement of non-linear vibrations (Sinou, 2007; 2008; Sinou & Lees, 2005; 2007), periodic non-linear response of blisks with friction ring dampers (Laxalde et al., 2007), periodic non-linear vibration for bolted structures (Jaumouillé et al., 2010), use of the Harmonic Balance Method to realize a global analysis of the dynamical behaviour of a simplified rotor supported by a squeeze-film damper (Sarrouy & Thouverez, 2010).

7. Nomenclature

$\underline{q}$	displacement vector
$\dot{\underline{q}}$	velocity vector
$\ddot{\underline{q}}$	acceleration vector
$\underline{\underline{M}}$	mass matrix
$\underline{\underline{K}}$	stiffness matrix
$\underline{\underline{C}}$	damping matrix (including gyroscopic effects if any)
$\underline{f}_e(t)$	vector of external forces
$\hat{\underline{f}}(t, \underline{q}, \dot{\underline{q}})$	vector of non-linear forces
m	order of the Fourier series
$\underline{a}_k$	Fourier coefficients of the cosine function for the kth order
$\underline{b}_k$	Fourier coefficients of the sine function for the kth order
$\underline{q}^{\mathrm{l}}$	p linear degrees of freedom of the system
$\underline{q}^{\mathrm{nl}}$	q non-linear degrees of freedom of the system
$\underline{\underline{A}} \otimes \underline{\underline{B}}$	Kronecker product of matrices $\underline{\underline{A}}$ and $\underline{\underline{B}}$
$D_{\tilde{x}}\underline{H}(\underline{\tilde{x}}, \mu)$	Derivatives of $\underline{H}$ function with respect to $\underline{\tilde{x}}$ at point $(\underline{\tilde{x}}, \mu)$: $\left[D_{\tilde{x}}\underline{H}(\underline{\tilde{x}}, \mu)\right]_{ij} = \frac{\partial \underline{H}_i}{\partial \underline{\tilde{x}}_j}(\underline{\tilde{x}}, \mu)$
$D_{\mu}\underline{H}(\underline{\tilde{x}}, \mu)$	Derivatives of $\underline{H}$ function with respect to μ at point $(\underline{\tilde{x}}, \mu)$: $\left\{D_{\mu}\underline{H}(\underline{\tilde{x}}, \mu)\right\}_{i} = \frac{\partial \underline{H}_i}{\partial \mu}(\underline{\tilde{x}}, \mu)$
x^T	vector or matrix transposition

8. References

Allgower, E. L. & Georg, K. (2003). *Introduction to Numerical Continuation Methods*, Springer-Verlag.

Broyden, C. G. (1965). A class of methods for solving nonlinear simultaneaous equations, *Mathematics of Computation* 19: 2024–2040.

Cameron, T. M. & Griffin, J. H. (1989). An alternating frequency/time domain method for calculating the steady-state response of nonlinear dynamic systems, *Journal of Applied Mechanics* 56(1): 149–154.

Coudeyras, N., Nacivet, S. & Sinou, J.-J. (2009). Periodic and quasi-periodic solutions for multi-instabilities involved in brake squeal, *Journal of Sound and Vibration* 328(4-5): 520–540.

Coudeyras, N., Sinou, J.-J. & Nacivet, S. (2009). A new treatment for predicting the self-excited vibrations of nonlinear systems with frictional interfaces: the constrained harmonic balance method, with application to disc brake squeal, *Journal of Sound and Vibration* 319: 1175–1199.

Guskov, M., Sinou, J.-J. & Thouverez, F. (2008). Multi-dimensional harmonic balance applied to rotor dynamics, *Mechanics Research Communications* 35: 537–545.

Hahn, E. J. & Chen, P. Y. P. (1994). Harmonic balance analysis of general squeeze film damped multidegree-of-freedom rotor bearing systems, *Journal of Tribology* 116(3): 499–507.

Jaumouillé, V., Sinou, J.-J. & Petitjean, B. (2010). An adaptive harmonic balance method for predicting the nonlinear dynamic responses of mechanical systems – application to bolted structures, *Journal of Sound and Vibration* 329(19): 4048–4067.

Kim, Y.-B. & Choi, S.-K. (1997). A multiple harmonic balance method for the internal resonant vibration of a non-linear jeffcot rotor, *Journal of Sound and Vibration* 208(3): 745–761.

Laxalde, D., Thouverez, F., Sinou, J.-J. & Lombard, J.-P. (2007). Qualitative analysis of forced response of blisks with frinction ring dampers, *European Journal of Mechanics A/Solids* 26(4): 676–687.

Sarrouy, E. & Thouverez, F. (2010). Global search of non-linear systems periodic solutions: A rotordynamics application, *Mechanical Systems and Signal Processing* 24(6): 1799–1813.

Sinou, J.-J. (2007). Effects of a crack on the stability of a non-linear rotor, *International Journal of Nonlinear Mechanics* 42(7): 959–972.

Sinou, J.-J. (2008). Detection of cracks in rotor based on the 2x and 3x super-harmonic frequency components and the crack-unbalance interactions, *Communications in Nonlinear Science and Numerical Simulation* 13(9): 2024–2040.

Sinou, J.-J. (2009). Non-linear dynamics and contacts of an unbalanced flexible rotor supported on ball bearings, *Mechanism and Machine Theory* 44(9): 1713–1732.

Sinou, J.-J. & Lees, A. (2005). Influence of cracks in rotating shafts, *Journal of Sound and Vibration* 285(4-5): 1015–1037.

Sinou, J.-J. & Lees, A. (2007). A non-linear study of a cracked rotor, *Journal of European Mechanics - A/Solids* 26(1): 152–170.

Villa, C., Sinou, J.-J. & Thouverez, F. (2008). Stability and vibration analysis of a complex flexible rotor bearing system, *Communications in Nonlinear Science and Numerical Simulation* 13(4): 804–821.

22

Support Vector Machine Classification of Vocal Fold Vibrations Based on Phonovibrogram Features

Michael Döllinger[1], Jörg Lohscheller[2], Jan Svec[3], Andrew McWhorter[4] and Melda Kunduk[5]
[1]University Hosptital Erlangen,
[2]University of Applied Sciences Trier,
[3]Palacky University Olomouc,
[4]Our Lady of the Lake Voice Center,
[5]Louisiana State University,
[1,2]Germany
[3]Czech Republic
[4,5]USA

1. Introduction

Voice is invaluable for our livelihood, as it takes place in humans everyday lives, like talking, laughing, crying, singing, screaming, shouting etc. Over the past 200 000 years, humans use the lung, larynx, tongue, and lips, to produce and modify the highly intricate arrays of voice (Titze, 2006) for realizing verbal communication and emotional expression. Among the participating tissues, the vocal folds within the human larynx have evolved to be a key organ in the creation of human voice. Their vibrations serve as origin of the primary voice signal. The process of voice production is called phonation (Titze, 2006), and is the preliminary stage for speech.

In our knowledge-based societies, communication skills have become more and more important. Communication disorders became a socio economic factor: A study in the year 2000 estimated losses within the Gross National Product of the USA being up to $186 billion annually (Ruben, 2000), on the basis that approx. 10% of the entire population suffers from communication disturbances. To increase the quality of life of the people concerned on one hand and to keep the economic costs under control on the other, appropriate technologies have to be developed to disclose all factors conducive to communication disorders. Also, analysis methods have to be applied to objectively quantify grades of disease, document therapy, and to guide surgical interventions. A high number of communication disorders are due to a disturbance in voice, i.e. disturbed vocal fold vibrations.

Examination of vocal fold vibrations (*100 Hz – 300 Hz*) and the acoustic signal are the basic components of clinical voice assessment. It is widely held that vocal fold vibration irregularities lead to an impairment of the voice signal. Irregularities being present in vocal fold vibrations during sound production can be determined by direct (i.e. endoscopic

laryngeal imaging) or indirect (i.e. acoustic and aerodynamic) assessment techniques. However, detailed quantitative knowledge about interrelations between acoustic signal and vibrations of the voice generator is still in its infancies.
Currently, videostroboscopy is a commonly used clinical laryngeal imaging tool to investigate the vocal fold vibratory dynamics. However, videostroboscopy is just suitable for periodic vocal fold vibrations since the image sensor captures only one frame per oscillation cycle and thus does not fulfil the Nyquist sampling theorem (Kendall et al., 2005; Svec et al., 2008). Hence, videostroboscopy has severe limitations when it comes to investigating pathological voices which frequently exhibit non-periodic vibrations. State-of-the-art technology in investigating of vocal fold vibrations is high-speed digital imaging (HSI). Current systems are equipped with a 2D image sensor delivering images at frame rates up to *2,000-8,000 fps,* which can capture the vibration patterns of vocal folds at their usual frequencies of up to *300 Hz* along the entire visible glottal length (Schade & Mueller, 2005; Hertegard, 2005; Bonilha & Deliyski, 2008; Deliyski, et al., 2008). Thus, HSI allows visualizing regular and irregular vibration patterns which are found in normal and pathological voices (Kendall et al., 2005; Doellinger, 2009), Fig. 1.

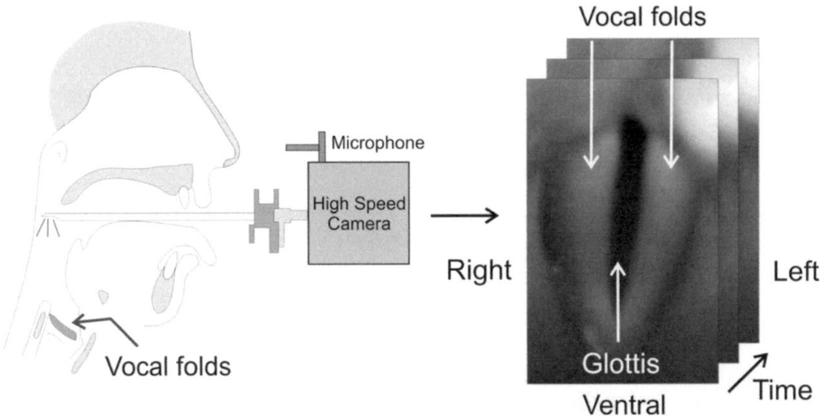

Fig. 1. Schematic representation of performing endoscopic high-speed recordings. Left, the recording situation including camera and endoscope are shown. On the left, the recorded area (vocal folds and opening and closing glottis) can be seen.

Even though high-speed videos deliver a novel insight into laryngeal vibrations, the investigation of vocal fold vibrations demands a sophisticated quantitative analysis of the video data (Doellinger, 2009). To reach this objective, different approaches have been developed to analyze vocal fold vibrations (Doellinger, 2009). Commonly, from the endoscopic HSI data the time varying opening between the vocal folds (i.e. glottis) is analyzed or trajectories are extracted at specific positions of the vocal folds (Braunschweig et al., 2008). To quantify the obtained motion data, several measures have been introduced describing the symmetry and regularity of vocal fold vibrations (Qiu et al., 2003, Yan et al., 2005). Instabilities of fundamental frequencies, amplitude and phase asymmetries as well as regularity parameters were detected in pathological voices (Bonilha & Deliyski, 2008). Other approaches automatically adapt biomechanical models to vocal fold vibrations extracted from HSI videos applying parameter optimization strategies (Doellinger et al., 2002; Doellinger et al. 2003; Tokuda et al., 2007, Yang et al., 2010). These obtained parameters represent the degree of laryngeal asymmetry and vibration stability (Schwarz et al., 2008; Wurzbacher et al., 2006; Wurzbacher et al., 2008). However, up to the present there is still no established feature extraction strategy describing the entire vibration patterns of vocal fold dynamics adequately.

Recently, the novel Phonovibrogram (PVG) approach was suggested to quantify the entire visible vocal fold vibrations (Lohscheller et al., 2007; Lohscheller et al., 2008a) expanding formerly introduced spatio-temporal plots (Westphal & Childers, 1983; Neubauer et al., 2001). In the PVG approach, contours of the oscillating vocal folds are segmented from video data and are transformed into a single colour coded PVG image. Depending on the underlying vocal fold vibrations, characteristic geometric patterns occur within a PVG which can be used for further clinical interpretation (Lohscheller & Eysholdt, 2008). PVG images can be regarded as fingerprints of vocal fold vibrations, enabling intuitional assessment of vocal fold vibrations (Eysholdt & Lohscheller, 2008). PVG analysis demonstrates that the complex two-dimensional vibratory patterns of vocal folds can robustly be described (Eysholdt & Lohscheller, 2008). It further establishes an objective basis for novel automatic analysis and classification approaches (Doellinger et al., 2009; Lohscheller et al. 2008b, Kunduk et al. 2010).

Within this work we propose a novel approach to achieve a fully automatic analysis of PVG images for detecting even slight alterations within underlying vocal fold vibrations: After segmenting the vocal fold vibrations from HSI and computing the appropriate PVG image matrix a set of novel PVG features will be introduced which describe the main characteristics of vocal fold dynamics. For investigating the sensitivity of the proposed PVG analysis approach the following physiological conditions were considered:

Vocal fold vibrations show individual patterns for each subject and can thus be highly variable between different patients. However, during voice production for a single subject the vocal fold vibrations show at specific voice intensity and fundamental frequency a reproducible dynamical behaviour. Within a subject, alterations of the fundamental frequency and/or intensity result into slight changes within vocal fold vibrations (Rovirosa et al., 2008). To obtain clinically relevant information about the physiology of a subject's voice the changes of vocal fold vibrations need to be traced. Accordingly, a computerized analysis procedure has to be sensitive enough to capture the individual changes within a subject. Hence, the validation of sensitivity of a computerized analysis approach needs to be performed within one single subject as changes of vocal fold vibrations between different subjects are not comparable.

According to the fulfilments above the sensitivity of PVG analysis was investigated by applying the PVG approach extensively to data sets obtained from a single healthy female subject. For data acquisition the subject was instructed to phonate at nine specified combinations of fundamental frequencies (low, normal, and high) and voice intensities (soft, normal, and loud). For each of these nine phonatory tasks twelve different high-speed sequences were obtained. Totally, 108 HSI sequences from this single subject could be processed. To obtain reliable results it is further of great importance to examine a healthy subject with no signs of voice disorders. Only for a single healthy subject it can be assumed that during the repeated examinations of a phonatory task the vocal fold vibrations are reproducible and do not change. Hence, the presence of pathologically caused and thus arbitrarily induced alterations of vocal fold vibrations can be excluded between recordings.

For further validation, simultaneously to the video data the emitted acoustic signal was recorded. From the acoustic data clinically used acoustic quality measures like Jitter, Shimmer, HNR, and SNR (Murphy, 1999; Zhang & Jiang, 2008) were computed allowing indirect conclusions about the vibrational behavior of vocal folds.
The results of this work will show that using PVG features in combination with a Support Vector Machine (SVM) even minor changes of vocal fold vibrations - caused by frequency and intensity alterations - can be highly robustly detected. Comparing the classification results gained by PVG features with results obtained from conventionally applied glottal as well as acoustic features will show the superiority of the novel PVG analysis approach.

2. Methods

2.1 Data collection

The KAY Elemetrics, High–Speed Digital Video System, Model 97, was used for data collection. Recordings were performed at a *2,000 fps* rate by using a specially designed, multi-port, super sensitive camera for eight seconds of recording. Gray scaled images were captured at 384Mb/sec into high-speed video RAM with a spatial resolution of *128 x 256* pixels. Images were obtained with a rigid *70°* endoscope (KAY Elemetrics, 9106) with a 300-watt-coldlight source (Olympus CLV-U20). The rigid laryngoscope was coupled to the high-speed digital camera head and endoscopy was performed as in conventional videostroboscopy. A microphone was placed *15 cm* from the lips to obtain the acoustic signal. This signal was fed through the KAY Elemetrics System for simultaneous recording of the endoscopic and acoustic signals (*50 KHz*). KAY Elemetrics, Rhino-Laryngeal Stroboscope (RLS 9100 B) and its microphone was used to determine F_o and the volume of the voice signal. The visual display on the system directed the subject for the maintenance and consistency of the desired F_o and volume for each phonatory task.

2.2 Subject and phonatory tasks

One female subject's voice was recorded with HSI for this study. The subject was non smoker and had no known history of neurological disease, laryngeal surgery, prior/or existing laryngeal disorders, voice problems at the time of data collection nor observed neither reported speech/language impairment. The HSI and acoustic recordings were simultaneously acquired while the subject was producing the vowel /i/ at the following fundamental frequency (F_0) / intensity (I) combinations:

- low F_0 (F1) at soft (I1), normal (I2), and high (I3) intensity,
- normal F_0 (F2) at soft (I1), normal (I2), and high (I3) intensity,
- high F_0 (F3) at soft (I1), normal (I2), and high (I3) intensity,

resulting in 9 different phonatory tasks. For all F_0/I combinations four phonation trails were performed. Within each recorded trail three different intervals of phonation were present. Each interval contained a voice onset followed by sustained phonation of at least one second being divided by short periods of silence. Hence, for each F_0/I combination 12 phonation sequences were available. For later analysis purposes the following class system is introduced (F→ Frequency, I→Intensity):
3 Frequency classes

$$CF1:=\{F1I1, F1I2, F1I3\}; CF2:=\{F2I1, F2I2, F2I3\}; CF3:=\{F3I1, F3I2, F3I3\}. \quad (1)$$

3 Intensity classes

$$CI1:=\{F1I1, F2I1, F3I1\};\ CI2:=\{F1I2, F2I2, F3I2\};\ CI3:=\{F1I3, F2I3, F3I3\}. \quad (2)$$

9 Combined Frequency/Intensity classes

$$\begin{aligned} &CS1 := \{F1I1\};\ CS2 := \{F1I2\};\ CS3 := \{F1I3\}; \\ &CS4 := \{F2I1\};\ CS5 := \{F2I2\};\ CS6 := \{F2I3\}; \\ &CS7 := \{F3I1\};\ CS8 := \{F3I2\};\ CS9 := \{F3I3\}. \end{aligned} \quad (3)$$

2.3 Selection of sequences

Within the acoustic signals the intervals of sustained phonation were identified by visual inspection. Within each interval a time section of 1 second was selected. The identical section was analyzed in high speed video data. The sequence length of one second time (> 150 glottal cycles) was in accordance with previous studies who suggested approx. 130 - 190 cycles (Karnell, 1991). Thus, altogether 108 pairs of high-speed and acoustic data sets were available (Tab. 1), reflecting isochronal information about vibratory characteristics of the voice generator (high-speed data) and the acoustic outcome (voice signal). Only in four cases the video data could not be further processed due to low image quality. To ensure, that possible occurring differences between recordings were only induced by the different phonation task, the recordings were performed within a day. As far as we know these data represent the most exhaustive examination of a single subject's vocal fold dynamics using HSI.

Intensity/F0	Low(F1)	Normal(F2)	High(F3)	**CI1-CI3**
Soft(I1)	4(12)	4(12)	4(12)	**12(36)**
Normal(I2)	4(9)	4(11)	4(12)	**12(32)**
Loud(I3)	4(12)	4(12)	4(12)	**12(36)**
CF1-CF3	**12(33)**	**12(35)**	**12(36)**	**36(104)**

Table 1. Applied Data. Overview of the performed 36 recordings which equals 108 sequences. From these sequences 104 could be analysed for acoustic and dynamical data.

2.4 PVG parameters describing vocal fold dynamics

2.4.1 Image processing

The vibrating edges of both vocal folds were extracted alongside their entire glottal length to analyze the laryngeal vibrations during phonation (Lohscheller et al., 2007). Information at each specific position of vocal folds is required to obtain detailed information about the vibration characteristics at dorsal, medial and ventral parts of vocal folds. For this purpose an extensively evaluated image segmentation procedure was applied (Lohscheller et al., 2007). The procedure delivers the left/right vocal fold edge contours $c_{L/R}(t)$, the glottal area $a(t)$, the location of anterior/posterior glottal ending $A(t)$ and $P(t)$ as well as the glottal main axis $l(t)$. A typical result of a segmented high-speed image is shown in Fig. 2.
Since the segmentation accuracy highly affects the following analysis, the quality of the results was visually monitored. For this purpose, within a movie viewer the segmented vocal fold contours were displayed. Further, for identifying potential faulty segmented

images (outliers) the glottal area *a(t)* was displayed within a diagram, see Fig. 2. Thus, in case of imprecise results, a re-segmentation of the high speed videos could be performed.

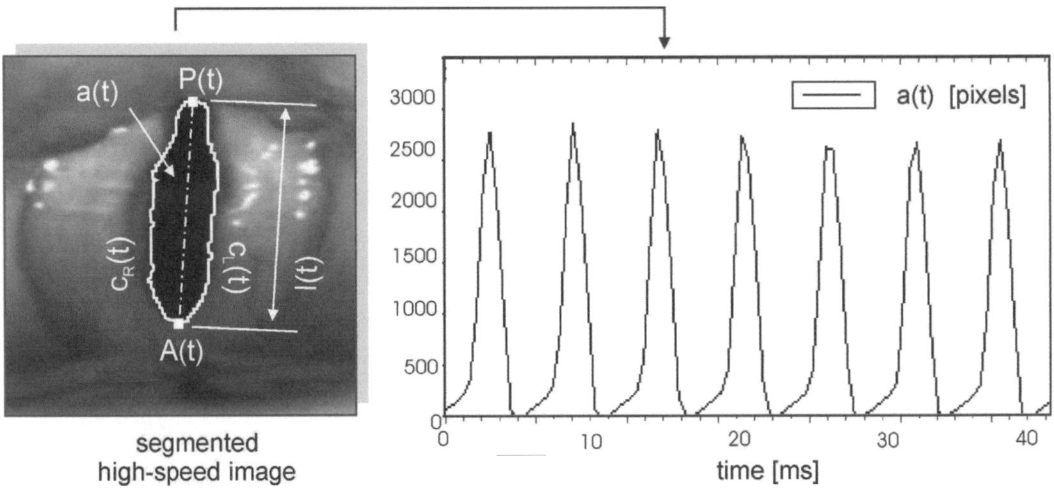

Fig. 2. Glottal area function. Left: Segmented image of a high-speed video. The extracted vocal fold edges are superimposed and are used to verify visually the accuracy of the segmentation results. Right: The glottal area waveform *a(t)* is monitored to detect faulty segmented images within a segmented video sequence.

In this study, the image processing procedure was applied only when the glottal length was fully visible during one second. From all 108 data sets 104 sequences each containing 2,000 consecutive images were successfully processed resulting in 208,000 segmented images. In all cases satisfactory segmentation accuracy were obtained, which are comparable to the example shown in Fig. 3.

2.4.2 Generation of phonovibrograms

For visualizing the entire vibration characteristics of both vocal folds the Phonovibrogram (PVG) was applied which was described in detail before (Lohscheller et al., 2008a). The principles of PVG computation are shortly summarized in Fig. 3. For each image of a high-speed video, the segmented glottal axis is longitudinally split and the left vocal fold contour is turned *180°* around the posterior end. Following, the distances $d^{L,R}(y,t)$ between the glottal axis and the vocal fold contours are computed; $y \in [1,\ldots,Y]$ with $Y=256$ denotes the spatial sampling of glottal axis. The distance values are stored as column entries of a vector and become color coded. The distance magnitudes are represented by the pixel intensities and two different colors. If vocal fold edges cross the glottal axis during an oscillation cycle the pixel is encoded by the color blue, otherwise the color red was used to indicate the distance from the glottal axis. A grayscale representation (black: vocal fold edges are at the glottal midline, white vocal fold edges have a distance to the glottal midline) of the originally colored PVG is given in Fig. 3. The entire vibration characteristics of both vocal folds are captured within one single PVG image by iterating the described procedure for an entire sequence and consecutively arranging the obtained vectors to a two-dimensional matrix. The left vocal fold is represented in the upper and the right vocal fold in the lower horizontal plane of the PVG, respectively. The PVG enables at the same time an assessment of the individual vibration characteristics for each vocal fold and gives evidence about left/right and posterior/anterior vibration asymmetries as well as predications about the temporal stability of the vibration pattern.

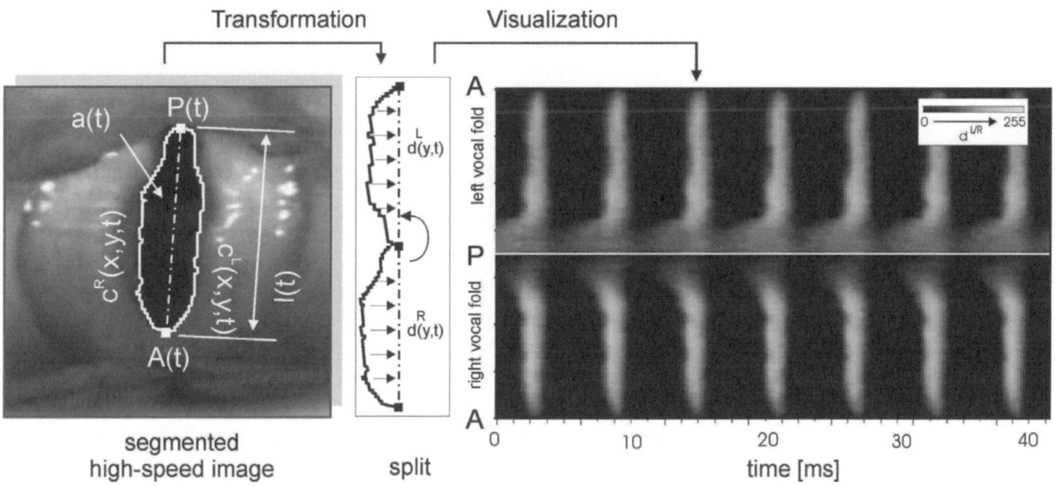

Fig. 3. PVG generation. 1) Segmentation of HS video. 2) Transformation of extracted vocal fold contours and computation of the distance values $d^{L,R}(y,t)$ which represent the distances from the vocal fold edges to the glottal midline. 3) Color coding of distance values for an entire high-speed video result into a PVG image comprising the entire vibration dynamics of both vocal folds in a single image (PVG is shown as grayscale image).

2.4.3 Analysis of vocal fold vibrations

PVG pre-processing: Phonovibrograms obtained from high speed sequences contain multiple reoccurring geometric patterns representing consecutive oscillation cycles of vocal folds. In order to describe the vibratory characteristics of vocal folds objectively, the 104 PVGs were pre-processed as follows: Firstly, for the left and right vocal fold unilateral PVGs are computed, denoted as $uPVG^{L/R}$ which are in the following regarded as two-dimensional functions $v^L(k,y)$ and $v^R(k,y)$ with $k \in \{1,\ldots,K\}$ and $K=2{,}000$ representing the number of frames within a sequence. From the unilateral PVGs the Glottovibrogram (*GVG*) is derived $v^G(k,y)= v^L(k,y) + v^R(k,y)$ which represents the glottal width (distances between the vocal folds) at each vocal fold position y over time, Fig. 4. In a subsequent step, the *uPVGs* and the *GVG* are automatically subdivided into a set of single *PVG/GVG* cycles, Fig. 4 right. A frequency analysis and peak picking strategy in the image domain is performed for the cycle identification (Lohscheller et al., 2008a).

Finally, the obtained single cycle PVGs are normalized to a constant width and height which are denoted $sPVG^L_i$, $sPVG^R_i$, $sGVG_i$, with $i \in \{1,\ldots,I^{L,R,G}\}$ and $I^{L,R,G}$ representing the number of cycles within the corresponding Phonovibrogram. Hence, vocal fold vibrations can be described by a set of the three functions

$$d_i^L(t,y) := sPVG_i^L\,,\; d_i^R(t,y) := sPVG_i^R\,,\; g_i(t,y) := sGVG_i \tag{4}$$

with $t \in \{1,\ldots,T\}$ where $T=256$ represents the normalized cycle length. In the following, the index $\alpha:=\{L,R\}$ is introduced to distinguish the functions $d^{\alpha}_i(t,y)$ representing the left and right vocal fold. Both, the unilateral as well as the normalized PVGs form the basis for the following analysis to obtain detailed information about vocal fold dynamics.

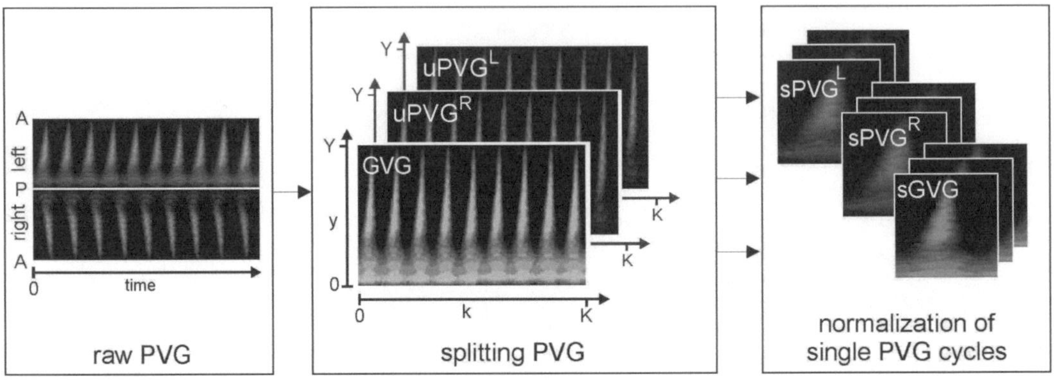

Fig. 4. Pre-Processing. From a raw PVG (left) so-called unilateral PVGs are computed (middle) which are further subdivided into a set of normalized single cycle *PVGs* (right).

Extraction of symmetry features: In order to describe the overall behavior of vocal fold dynamics the PVGs are analyzed as follows. At each glottal position y the 1D-power spectrum

$$\mathbf{P}^{\alpha}(f,y) := | FFT\{v^{\alpha}(k,y)\} | \quad \forall y \tag{5}$$

is calculated by Fast Fourier Transform algorithm (*FFT*). Due to settings, corresponding frequency resolution of the spectral components were 1 *Hz*. Fundamental frequencies $\mathbf{f}_0^{\alpha}$ are estimated by identifying the maxima within the discrete power spectra

$$\mathbf{f}_0^{\alpha} := \arg\max_{f} P^{\alpha}(f,y) \quad \forall y. \tag{6}$$

By defining the feature vector

$$\boldsymbol{\theta} := \theta(y) := \frac{\mathbf{f}_0^{L}}{\mathbf{f}_0^{R}} \quad \forall y \tag{7}$$

frequency differences between the left and right vocal fold as well as differences alongside the glottal axis are captured. If lateral (i.e. left/right) fundamental frequencies are identical the feature vector

$$\boldsymbol{\upsilon} := \upsilon(y) := \varphi\{\mathbf{P}^{L}(\mathbf{f}_0^{L},y)\} - \varphi\{\mathbf{P}^{R}(\mathbf{f}_0^{R},y)\} \quad \forall y \tag{8}$$

describes the phase delays between the left and right vocal fold.

The left/right vibration asymmetry is further described by introducing the mean relative amplitude ratios $\bar{a}(y)$ which are computed as follows. Within the $sPVG^{L,R}$ the points in time

$$\mathbf{T}_{y,i}^{\alpha\,\max} := \arg\max_{t} d_i^{\alpha}(t,y) \quad \forall \alpha,y,i \tag{9}$$

along the vocal fold length are identified when the maximum vocal fold deflections occur. By identifying the time points of minimal vocal fold deflection

$$\mathbf{T}_{y,i}^{\alpha\,\min} := \arg\min_{t} d_i^{\alpha}(t,y) \quad \forall \alpha, y, i \tag{10}$$

the relative peak-to-peak amplitudes

$$\mathbf{A}_{y,i}^{\alpha} := d_i^{\alpha}(\mathbf{T}_{y,i}^{\alpha\,\max}, y) - d_i^{\alpha}(\mathbf{T}_{y,i}^{\alpha\,\min}, y) \quad \forall \alpha, y, i \tag{11}$$

can be defined which are independent from the absolute position of the glottal axis. The mean relative amplitude ratios

$$\bar{\mathbf{a}} := \bar{a}(y) = \overline{\left(\frac{\mathbf{A}_{y,i}^{L}}{\mathbf{A}_{y,i}^{R}}\right)} \quad \forall y \tag{12}$$

and corresponding standard deviations $\boldsymbol{\sigma}_{\mathbf{a}}:=\sigma_a(y)$ serve as features to describe left/right asymmetries as well as the stability of vibrations at each position of the vocal folds. The obtained parameters are merged to the symmetry feature vector s (Eqs. (7),(8),(12)):

$$\mathbf{s} := [\boldsymbol{\theta}, \boldsymbol{\upsilon}, \bar{\mathbf{a}}, \boldsymbol{\sigma}_{\mathbf{a}}]. \tag{13}$$

Extraction of glottal features g: In order to capture characteristics of the glottal dynamics within the oscillation cycles, the following parameters are extracted from the normalized GVG matrices $g_i(t,y)$. Firstly, the maximum glottal area of each oscillation cycle i is determined as

$$\boldsymbol{\rho}_i = \max_{t} \sum_{y=1}^{Y} g_i(t,y) \quad \forall t, i. \tag{14}$$

The feature

$$\sigma_{\rho} = \sqrt{Var(\boldsymbol{\rho}_i)} \tag{15}$$

describes the stability of the glottal vibratory cycles over time. Subsequently, the open quotients $\boldsymbol{OQ}_{y,i}$ are defined for each glottal position i as duration of open phase divided by duration of complete glottal cycle and are computed as

$$\mathbf{OQ}_{y,i} = \left(\sum_{t} \hat{g}_i(t,y)\right) / T \quad \forall y, i; \tag{16}$$

with

$$\hat{g}_i = \begin{cases} 1 & g_i(t,y) > 0 \quad \forall t. \\ 0 & otherwise. \end{cases} \tag{17}$$

The mean values

$$\overline{\mathbf{oq}} = \frac{1}{I}\sum_{i}^{I}\mathbf{OQ}_{y,i} \quad \forall y \tag{18}$$

and standard deviations

$$\boldsymbol{\sigma}_{oq} = \sqrt{Var(\mathbf{OQ}_{y,i})} \quad \forall y \tag{19}$$

are used as features describing the stability of the glottal opening behavior at each position alongside the glottal axis (*Var* symbolizes the variance). Analogously, the mean speed quotients $\overline{\mathbf{sq}}$ and the corresponding standard deviations $\boldsymbol{\sigma}_{sq}$ are computed describing the mean glottal vibratory shape and its stability over time (Jiang et al., 1998).
Finally, the glottal closure insufficiencies

$$\mathbf{gci}_i = \frac{\min\limits_t \sum_{y}^{Y} \hat{h}_i(t,y)}{Y} \quad \forall t,i. \tag{20}$$

are derived using

$$\hat{h}_i = \begin{cases} 1 & g_i(t,y) > 0 \quad \forall y. \\ 0 & otherwise. \end{cases} \tag{21}$$

which are identifiable for each oscillation cycle i. The supplemental features $\overline{gci}$ and σ_{gci} describe the mean glottal closure insufficiency and its stability for the entire high-speed sequence. The glottal parameters are merged to the glottal feature vector (Eqs. (15),(18),(19)):

$$\mathbf{g} := [\sigma_\rho, \overline{\mathbf{oq}}, \boldsymbol{\sigma}_{oq}, \overline{\mathbf{sq}}, \boldsymbol{\sigma}_{sq}, \overline{gci}, \sigma_{gci}]. \tag{22}$$

Extraction of geometric PVG feature ω: Besides the conventional symmetry and glottal parameters we propose a novel way for describing vocal fold vibrations by quantifying the geometric structure within $sPVG^\alpha$ images. The main vibration characteristics of a vocal fold can be described by extracting representative contour lines from the $sPVG^\alpha$ images. This is done by determining the oscillatory states n during the opening ($t < \mathbf{T}_{y,i}^{\alpha\max}$) and closing ($t > \mathbf{T}_{y,i}^{\alpha\max}$) phases where vocal folds reach a certain percentage of relative deflection

$$\mathbf{A}_{y,i}^{\alpha n} := \frac{n}{100}\mathbf{A}_{y,i}^{\alpha}, \quad n \in [0,100]. \tag{23}$$

Hence, the set of vectors

$$\mathbf{O}_{y,i}^{\alpha n} := \arg_x(d_i^\alpha(x,y) = \mathbf{A}_{y,i}^{\alpha n}), \text{ with } t < \mathbf{t}_i^{\alpha\max} \quad \forall \alpha, y, i. \tag{24}$$

$$\mathbf{C}_{y,i}^{\alpha n} := \arg_x(d_i^\alpha(x,y) = \mathbf{A}_{y,i}^{\alpha n}), \text{ with } t > \mathbf{t}_i^{\alpha\max} \quad \forall \alpha, y, i. \tag{25}$$

describe temporal and spatial propagation of each vocal fold at different oscillation states during glottal opening $\mathbf{O}_{y,i}^{\alpha n}$ and closing $\mathbf{C}_{y,i}^{\alpha n}$. In order to get a comprehensive

understanding of the entire vibration cycle, multiple contour lines are extracted at different oscillation states. Fig. 5 shows exemplarily extracted contour lines at *n=(30,60,90)* for the left and right vocal fold during a single oscillation cycle.
The functional characteristics

$$\mathbf{PO}_{y,i}^{\alpha\, n} := d_i^{\alpha}(t,y)\Big|_{\mathbf{O}_i^{\alpha\, n}} \quad \mathbf{PC}_{y,i}^{\alpha\, n} := d_i^{\alpha}(t,y)\Big|_{\mathbf{C}_i^{\alpha\, n}} \quad \forall \alpha, y, i \tag{26}$$

of $sPVG^{\alpha}$ at positions $\mathbf{O}_{y,i}^{\alpha\, n}$ and $\mathbf{C}_{y,i}^{\alpha\, n}$ of the contour lines give precise information on actual deflection of the vocal folds. As features which describe the average vibratory pattern of vocal folds, the means for the contour lines *n=(30,60,90)*, the deflection characteristics and their time indices

$$\overline{\mathbf{O}_{y,i}^{\alpha\, n}}, \overline{\mathbf{PO}_{y,i}^{\alpha\, n}}, \overline{\mathbf{C}_{y,i}^{\alpha\, n}}, \overline{\mathbf{PC}_{y,i}^{\alpha\, n}}, \tag{27}$$

are computed for all cycles *i*. The vibration stability is captured by the corresponding standard deviations

$$\boldsymbol{\sigma}(\mathbf{O}_{y,i}^{\alpha\, n}), \boldsymbol{\sigma}(\mathbf{PO}_{y,i}^{\alpha\, n}), \boldsymbol{\sigma}(\mathbf{C}_{y,i}^{\alpha\, n}), \boldsymbol{\sigma}(\mathbf{PC}_{y,i}^{\alpha\, n}). \tag{28}$$

The Euclidian-Norm $\| \ \|_2$ between the mean positions of the contour lines

$$N_{O,C}^{n} = \left\| \overline{\mathbf{O}_{y,i}^{L n}} - \overline{\mathbf{O}_{y,i}^{R n}} \right\|_2 \quad \forall n \tag{29}$$

describes deviations between the mean left and right vocal fold vibration patterns. Finally, all parameters (Eqs. (27),(28),(29)) are merged to the PVG feature vector

$$\boldsymbol{\omega} := [\overline{\mathbf{O}_{y,i}^{\alpha\, n}}, \overline{\mathbf{PO}_{y,i}^{\alpha\, n}}, \overline{\mathbf{C}_{y,i}^{\alpha\, n}}, \overline{\mathbf{PC}_{y,i}^{\alpha\, n}}, \boldsymbol{\sigma}(\mathbf{O}_{y,i}^{\alpha\, n}), \boldsymbol{\sigma}(\mathbf{PO}_{y,i}^{\alpha\, n}), \boldsymbol{\sigma}(\mathbf{C}_{y,i}^{\alpha\, n}), \boldsymbol{\sigma}(\mathbf{PC}_{y,i}^{\alpha\, n}), N_{O,C}^{n}]. \tag{30}$$

The entire vocal fold dynamics extracted from one high speed sequence can be described by merging the introduced features for left-right symmetry, glottal and PVG characteristics (Eqs. (13),(22),(30)) to the feature vector

$$\boldsymbol{\beta} := [\mathbf{s}, \mathbf{g}, \boldsymbol{\omega}]. \tag{31}$$

The feature vector $\boldsymbol{\beta}$ represents vocal fold dynamics at each position y along the glottal axis with $y \in \{1,\ldots,Y\}$. In order to reduce the dimensionality of the parameter space for further analysis, the feature vector is reduced to $y \in \{1,\ldots,12\}$ by computing average values. Hence, for an effective vocal fold length of *1 cm* the feature vector represents the average oscillation dynamics within *0.9 mm* sections of the vocal length which constitutes sufficient accuracy.
Acoustic voice quality measures: For the nine frequency/intensity phonatory tasks also the acoustic voice signals were analyzed. The selected acoustic sequences correspond to the time intervals of the analyzed video data. From the selected intervals 10 voice quality measures were derived using Dr.Speech-Tiger-Electronics/Voice-Assessment-3.2 software (www.drspeech.com). The computed parameters describe temporal voice properties as cycle duration stability (Jitter, STD F_0, STD Period, F_0 tremor), amplitude stability (Shimmer, STD

Ampl., Amp. Tremor), harmonic to noise ratio (HNR), signal to noise ratio (SNR), and normalized noise energy (NNE). The nine different frequency/intensity classes are given by the measured sound pressure level (*SPL[dB]*) and mean fundamental frequency (Mean $F_0[Hz]$), Tab. 2.

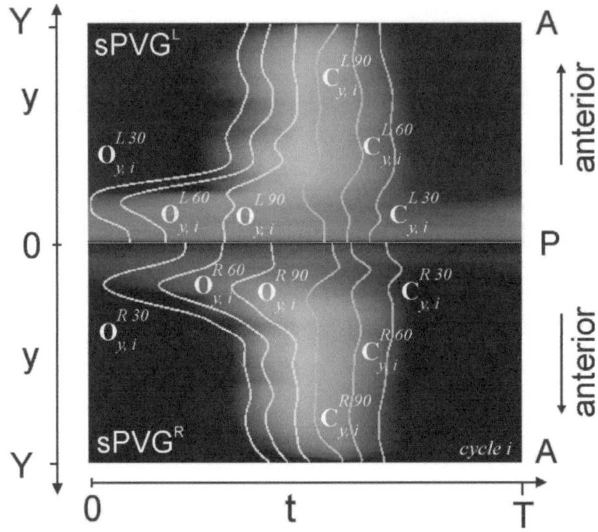

Fig. 5. The contour lines **O** (opening phase) and **C** (closing phase) describe the main characteristics of $sPVG^{\alpha}$ geometry. The contours represent the spatio-temporal positions of vocal fold edges at the oscillation states *n=(30,60,90)* for the left and right vocal fold. The *n* value corresponds to the percentage of open and closed positions.

	CS1	CS2	CS3	CS4	CS5	CS6	CS7	CS8	CS9
No.Sequ.	12	9	12	12	11	12	12	12	12
SPL(dB)	*59,0 ±0,8*	*63,3 ±0,5*	*72,5 ±1,7*	*58 ±0*	*63 ±0*	*75 ±0*	*58,3 ±0,5*	*64,3 ±1,4*	*71 ±0,9*
Mean F0 (Hz)	*153 ±3*	*160 ±4*	*201 ±2*	*182 ±4*	*193 ±4*	*231 ±8*	*318 ±5*	*328 ±8*	*328 ±5*

Table 2. Mean values and standard deviations for the different fundamental frequencies [mean F_0] and voice intensities [sound pressure level (*SPL[dB]*)] representing the nine different phonatory tasks CS1-CS9.

Classification of different phonation conditions: Due to the high number of PVG parameters conventional statistics and correlation analysis is not appropriate to identify potential parameter changes between the different phonation conditions. Thus, to explore the influence of intensity and frequency alterations within the parameter sets a nonlinear classification approach was applied (Hild et al., 2006; Selvan & Ramakrishnan, 2007; Lin, 2008).

The following hypothesis was investigated: if a classifier is capable of distinguishing between different phonatory classes it can be concluded that intensity and frequency variations are actually present within the observed vocal fold dynamics represented by the introduced feature sets.

For classification of the PVG features, a nonlinear support vector machine (SVM) was used (Duchesne et al., 2008; Kumar & Zhang, 2006). For the SVM, a Gaussian radial basis function kernel (RBF) was chosen (Vapnik, 1995). Appropriate SVM parameters were determined by an evolutionary strategy optimization procedure (Beyer & Schwefel, 2002). The parameter space of SVM, cost parameter and the width of the RBF kernel was automatically searched in order to obtain best classification results (Hsu et al., 2003). The models' classification accuracy was evaluated via 10-fold cross-validation with stratification (Kohavi, 1995).
In order to compare PVG result with conventionally used measures the classifier was also applied to traditional glottal and symmetry parameters as well as to the ten acoustic voice quality measures.

3. Results

3.1 Validation of data acquisition

For a reliable interpretation of the later classification results it is essential to verify that the data acquisition representing the nine different phonatory tasks effectively succeeded. Tab. 2 shows the means and standard deviations for the different sound pressure levels (*SPL*) and fundamental frequencies (mean F_0) for all nine phonatory tasks. Already the very small standard deviations of the *SPL* and mean F_0 within the classes CS1-CS9 prove the high consistency of the data acquisition which included the repeated recording of the different phonatory tasks. Applying statistical analysis (Kolmogorov-Smirnov-Tests following *t*-Tests or Mann-Whitney-U-Tests) it could be shown that for frequency classes LOW (CF1), NORMAL (CF2), and HIGH (CF3) (Eq. (1)) the fundamental frequencies were significantly ($p<0.05$) different. Also for intensity classes SOFT (CI1), NORMAL (CI2), and LOUD (CI3) (see Eq. (2)) the intensity values were computed significantly ($p<0.05$) different.

3.2 SVM classification of vocal fold vibrations

Exemplarily, Tab. 3 shows SVM classification results obtained for frequency classes CF1-CF3. The *Class Precision* reflects the percentage of the correct allocation: 30 out of 104 sequences were predicted as low (CF1). From these 30, three sequences were wrongly assigned to the class low (being actually in class CF2) resulting in 90% *Class Precision*. In contrast, the *Class Recall* reflects the percentage of how many members of the class were allocated towards the class. Here, 35 out of 38 normal sequences were correctly assigned to class CF2 whereas three sequences were predicted to class CF1. This results in a *Class Recall* accuracy of 92.1%. The *Overall Accuracy* for all classes is 94.18% ±6.53% which represents the mean performance of the classifier which is in the following used for interpretation purpose.

	True Low	True Normal	True High	***Class Precision***
Low (CF1)	27	3	0	90.0%
Normal (CF2)	3	35	0	92.1
High (CF3)	0	0	36	100.0%
Class Recall	90.0%	92.1%	100.0%	

Table 3. Classification result of the SMV of the intensity class problem CF1-CF3 using the entire feature vector from eq. (31). The overall classification accuracy amounts approx. 94%.

Using the parameters captured within the feature vector $\boldsymbol{\beta}:=[\mathbf{s},\mathbf{g},\boldsymbol{\omega}]$ (Eq. (31)) the SVM reached a classification accuracy of 95.1%±6.7% for the frequency class problem (CF1-3),

97.3%±4.2% for the intensity class problem (CI1-3), and 94.2%±9.1% for the nine class problem (CS1-CS9). This very high classification accuracy was obtained just by parameters describing vocal fold dynamics extracted from the high speed videos.

In order to investigate which parameters can be made responsible for the high performance of the classifier, the SVM was individually applied to components **[s]**, **[g]** and **[ω]** as well as to the combinations **[s,g]**, **[g,ω]**, **[s,ω]**. The results are summarized in Fig. 6. The conventional symmetry **[s]** and glottal parameters **[g]** achieved classification accuracy of only 15.5%±4.9% and 40.5%±10.5% for the nine class problem. Likewise, the classification accuracies for the frequency and intensity class problems were significantly reduced. Contrarily, very high classification accuracy was obtained using the new introduced PVG features **[ω]**. Applying exclusively the PVG features **[ω]** a classification accuracy of 85.5%±7.7% for the nine class problem, 96.2%±4.7% for the frequency class problem, and 91.6%±7.6% for the intensity class problem was obtained.

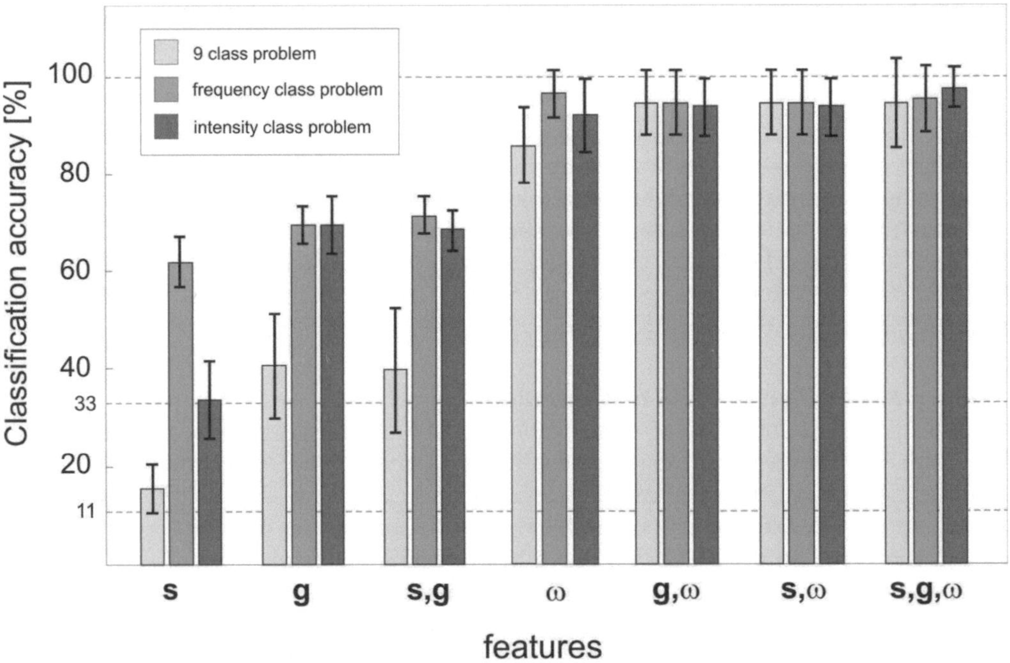

Fig. 6. Mean classification accuracies and standard deviations achieved by applying conventional symmetry **[s]**, glottal **[g]** and PVG **[ω]** parameters using a support vector machine (SVM) classification approach with stratified 10-fold cross-validation. The highest classification accuracy is obtained by the new introduced PVG features **[ω]**.

As the PVG feature vector contains information derived from different oscillation states ($\mathbf{O}^{\alpha n}_{y,i}, \mathbf{C}^{\alpha n}_{y,i}$) it was further investigated which oscillation state delivers the most valuable information needed for classifying vocal fold vibrations. For this purpose, the SVM was applied to different oscillation parts *n={30,60,90}* of the feature vector **[ω]**. Fig. 7 summarizes the achieved classification accuracies obtained by *n={[30,60],[60,90],[30,60,90]}*. Using the single oscillation states *n={[30],[60],[90]}*, already a mean classification accuracy of 58.2%±9.9% could be obtained for the nine class problem which exceeds considerably the classification rates obtained by the conventional symmetry **[s]** and glottal **[g]** parameters as shown in Fig. 6. The classification accuracies by applying combined oscillation states *n={[30,60],[60,90] ,[30,60,90]}* are significantly improved.

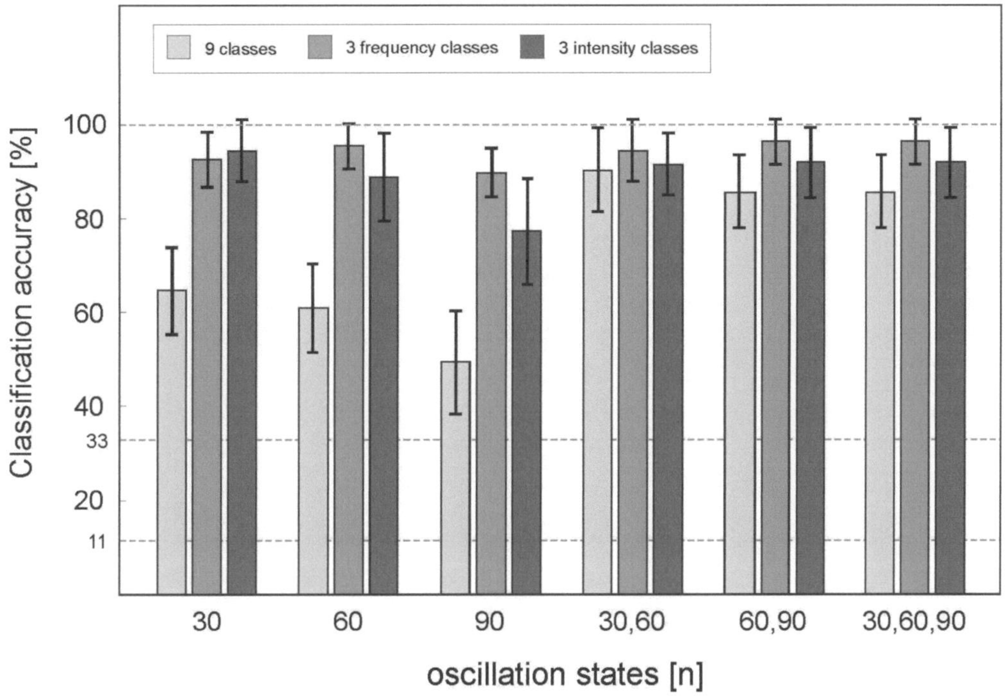

Fig. 7. Mean SVM classification accuracies and standard deviations achieved by applying part of the PVG features vector [ω] representing different oscillation states *n={30,60,90}*. Highest classification accuracy is obtained by a combination of the different oscillation states.

In a final step it was investigated which PVG components contribute most to the classification accuracy. For this purpose the feature vector [ω] (eq. (30)) was divided into parameter groups representing the average vibration type $[\omega_1] := (\overline{\mathbf{O}_{y,i}^{\alpha n}}\ \overline{\mathbf{C}_{y,i}^{\alpha n}})$, the average deflection characteristics $[\omega_2] := (\overline{\mathbf{PO}_{y,i}^{\alpha n}}\ \overline{\mathbf{PC}_{y,i}^{\alpha n}})$, the average lateral vibration symmetry $[\omega_3]{:=}(N_{O,C}^{n})$, and the average temporal stability of vocal fold vibrations $[\omega_4]{:=}(\sigma(\mathbf{O}_{y,i}^{\alpha n}), \sigma(\mathbf{PO}_{y,i}^{\alpha n}), \sigma(\mathbf{C}_{y,i}^{\alpha n}), \sigma(\mathbf{PC}_{y,i}^{\alpha n}))$. Figure 8 shows the classification accuracies obtained by the different parts of the feature vector [ω].

The isolated consideration of the average vibration type [ω₁] results into the highest classification accuracy of 52.8%±6.8% for the nine class problem and a mean accuracy of 85.1%±10.58% for the frequency and intensity class problems. By comparing the results in Fig. 6 and Fig. 8, it can be seen, that information about the mean vibration type (Fig. 8) already gives better classification results than information about the conventional parameters as speed quotient, open quotient, glottal closure insufficiency (Fig. 6). Information about vocal fold deflection amplitudes $[\omega_2]$, left/right discrepancies $[\omega_3]$ and vibration instabilities [ω_4] do not reach the same level of classification accuracy. However, combining all PVG features increases considerably the classification accuracy of up to 96.2%±4.7% for the frequency class problem.

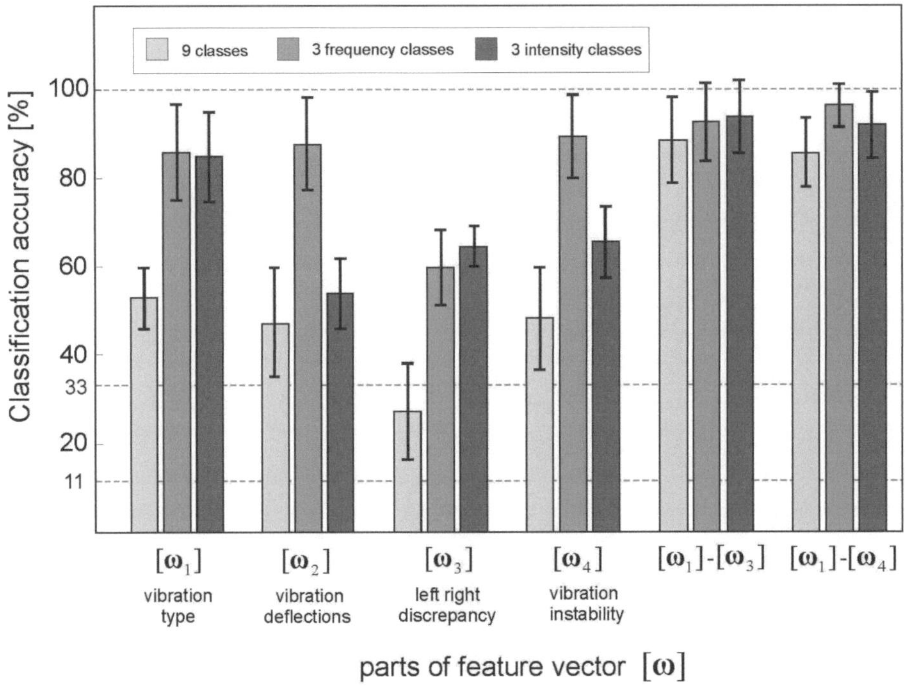

Fig. 8. Results show the comparison between the different features within the PVG parameters. The PVG parameters are split into groups representing the spatio-temporal vibration type, information about vibration amplitudes and symmetry as well as vibration instabilities. Fusing all information ω_1-ω_4 to a common feature vector results highest classification performance (i.e. frequency classes). The performance of the different classification results shows that the more precisely the vocal fold dynamics is described using a combination of several PVG features the better the dynamical changes of vocal fold dynamics can be captured.

3.3 SVM classification of the acoustic signal

To give an overview of the acoustic measures, Tab. 4 shows the means and standard deviations for all 10 computed acoustic voice quality parameters used for classification.

Table 5 summarizes the classification results for acoustic parameters. The best classification performance (93.45%) was achieved for the frequency class problem (CF1-CF3). The accuracy for the three class intensity problem (85.64%) was just slightly higher than accuracy for the combined nine class problem (83.73%). In contrast to the classification results obtained using the PVG parameters the acoustic parameters reached lower classification accuracies. Nevertheless, for the nine class problem still a classification accuracy of more than 80% could be achieved. It proves that even for a single subject frequency and intensity changes of the voice signal influence voice quality outcome measures.

	CS1	CS2	CS3	CS4	CS5	CS6	CS7	CS8	CS9
Jitter (%)	0,30 ±0,06	0,13 ±0,02	0,12 ±0,03	0,21 ±0,04	0,21 ±0,08	0,10 ±0,03	0,21 ±0,05	0,11 ±0,02	0,12 ±0,05
Shimmer (%)	2,17 ±0,44	1,07 ±0,17	0,98 ±0,35	1,69 ±0,36	1,48 ±0,25	0,88 ±0,45	1,68 ±0,42	0,85 ±0,22	0,84 ±0,11
HNR (%)	23,4 ±1,6	30,2 ±1,0	33,2 ±1,7	27,6 ±1,9	28,8 ±1,3	32,4 ±3,0	28,6 ±2,2	34,6 ±1,7	28,4 ±1,6
SNR (%)	23,4 ±1,6	30,2 ±1,0	33,2 ±1,7	27,6 ±1,9	28,8 ±1,3	32,4 ±3,0	28,7 ±2,2	34,6 ±1,7	28,4 ±1,6
NNE (%)	-3,0 ±1,5	-13,6 ±1,9	-17,1 ±2,7	-8,6 ±4,0	-11,1 ±2,9	-21,2 ±1,1	-9,3 ±2,6	-13,1 ±2,5	-21,5 ±2,3
STD F0 (Hz)	1,4 ±0,5	1,0 ±0,3	1,4 ±0,5	1,4 ±0,4	1,5 ±0,7	1,5 ±0,4	2,8 ±1,5	2,3 ±0,6	1,6 ±0,3
STD Period (ms)	0,06 ±0,02	0,04 ±0,01	0,04 ±0,01	0,04 ±0,01	0,04 ±0,02	0,03 ±0,01	0,03 ±0,01	0,02 ±0,01	0,02 ±0,01
Mean Amp (%)	86 ±4,8	92 ±2,1	91 ±3,1	86 ±4,1	90 ±3,1	90 ±3,1	85 ±5,5	88 ±4,3	93 ±2,3
STD Amp. (%)	5,9 ±1,7	3,5 ±1,3	4,4 ±1,5	6,1 ±1,5	5,4 ±2,1	4,8 ±1,8	6,2 ±2,1	5,1 ±1,5	2,7 ±0,9
F0 Tremor (Hz)	4,0 ±2,6	2,6 ±1,2	2,8 ±1,3	3,3 ±1,3	2,7 ±1,3	2,1 ±0,8	2,8 ±1,5	2,5 ±1,8	1,8 ±0,7
Amp. Tremor (Hz)	2,5 ±1,3	2,1 ±1,2	2,4 ±1,5	2,6 ±1,0	3,0 ±1,3	2,2 ±1,1	2,6 ±1,4	2,4 ±1,2	4,9 ±3,8

Table 4. Mean values and standard deviations of the 10 acoustic measured parameters (Dr.Speech 3.2) grouped for the nine paradigms. The vertical grey shadings correspond to the frequency classes.

SVM accuracy for acoustic parameters			
	Intensity	Frequency	Frequency/Intensity
Accuracy (%)	85.64	93.45	83.73
STD (%)	6.14	8.25	8.60

Table 5. Overall accuracy of the acoustic SVM classification results.

4. Discussion

The endoscopic imaging of vocal fold vibrations is an essential part of clinical examination of voice disorders. Digital high-speed videolaryngoscopy is the state-of-the-art technology for investigation of asymmetric and irregular vocal fold vibrations (Doellinger, 2009). Similar to stroboscopy, high-speed videos are frequently evaluated by visual inspection relying on the experience of the investigator. There is still no objective or standardized procedure for describing the entire vibration patterns of vocal folds. Besides the description of vocal fold vibrations, the acoustic analysis of the voice signal gives valuable information for describing the severity of voice disorders. However, in most of the applied methods the acoustic properties and the laryngeal vibrations are separately examined. Thus, there is still little knowledge about the direct relation between the acoustic voice signal and the vibration pattern of vocal folds.

In this work, we presented a novel approach, called Phonovibrography, allowing an objective analysis of the visible vocal fold dynamics. Here, quantitative features are derived from PVG images which describe precisely the entire characteristics of vocal fold dynamics. For validation purpose Phonovibrography was applied to 108 high-speed sequences recorded from a single healthy female subject with normal voice. The female subject was instructed to produce 9 different phonatory tasks, i.e. phonation at different frequency and intensity combinations. A sequence length of one second time (> 150 glottal cycles) was chosen. The simultaneously recorded acoustic signals were analyzed using established voice quality measures (www.drspeech.com). Thus, besides evaluating the PVG analysis approach the effect of different phonation conditions on both the laryngeal vibrations and the acoustic voice signal could be studied.

Choosing just a single subject for validating the accuracy of the proposed PVG approach is mandatory as only within a healthy subject the phonatory tasks related changes of vocal fold vibration patterns can be interpreted in a correct way. For a single subject the extensive data acquisition comprising the recording of 108 repeated phonatory tasks is very time-consuming and potentially incriminating for the subject. Thus, collecting such a full data set from several subjects is difficult to achieve. As far as we know this examination presents the worldwide most detailed analysis of vocal fold vibrations within a single subject. Besides evaluating the performance of novel analysis approaches, the data set can further be used to investigate very precisely the fundamental principles of voice production in normal voice.

In the present study we applied methods from the field of machine learning towards recognition of different phonatory tasks within vocal fold dynamics as well as within the simultaneously recorded acoustic signals. Even though endoscopic and voice data represent different physical properties describing voice production (tissue vibrations vs. acoustic sound pressure), both modalities could be used to individually classify the nine different phonatory tasks within normal voice of one female.

4.1 Classification of vocal fold vibrations

The results given in Fig. 6 clearly show that a very high SVM classification accuracy (up to 96%) could be obtained using the new introduced PVG features. Even the classification of the nine class problem showed a very high performance of 85.5% which is in the same range as the results obtained using the acoustic measures, Tab. 5. It can be concluded from the results that the investigated frequency and intensity variations can be quantitatively traced back to alterations of the laryngeal dynamics. Furthermore, changes of vocal folds dynamics induce alterations of the acoustic signal as shown in Tabs. 4 and 5. To our knowledge, this is the first time that vocal fold vibrations could be quantitatively described so precisely during different phonation tasks and that the different phonatory task could automatically be classified at the vocal fold level.

The results obtained by the PVG parameters were further compared to symmetry/glottal parameters (Eqs. (13) and (22): [s], [g]) which are frequently used to describe vocal fold vibrations. Fig. 6 shows, that using the conventionally used glottal and symmetry parameters the performance of the classification is highly reduced. Using the feature vector [s] only a classification accuracy of approx. 15% for the nine class problem could be obtained. The glottal features [g] show a better performance with approx. 40% but are still far worse than the classification accuracy (94%) obtained using PVG parameters $\boldsymbol{\omega}$. The low classification results obtained by the glottal parameters show, that the reduction of the complex 2D vocal fold vibration pattern to a few parameters based on 1D glottal area

waveform signal is not sufficient for analyzing the laryngeal vibrations completely. Likewise, putting the focus only onto specific features as vocal fold symmetry (amplitude, phase, frequency) - which is frequently evaluated within the subjective assessment of stroboscopic or high speed movies - is not sufficient to fully describe vocal fold vibrations.
Having a closer look at PVG features at different oscillation states $n=\{30,60,90\}$, similar results were found for $n=30$ and $n=60$ state (Fig. 7). While the three class problems could still be classified with a high accuracy, for the nice class problem a classification accuracy of only approx. 60% was obtained. For $n=90$ the classification results show a similar behavior with a slightly reduced performance. However, when fusing all information obtained from the three oscillation states, the highest classification results were obtained. The increase of the performance documents that a precise analysis of vocal fold dynamics demands to describe the entire vibration pattern very comprehensively as it is done by PVG parameters which describe the temporal and spatial propagation of vocal fold vibrations.
Splitting up PVG parameters in different features ([ω_1]: vibration type, [ω_2]: deflection information, ω_3: symmetry, and ω_4: instabilities) further proves the benefit of including all extracted parameters together. Considering the parameter features separately (Fig. 8) the classification accuracy is reduced. Nevertheless, despite the feature reduction the classification accuracy using PVG parameters ω_1 - which comprises only information about the mean spatio-temporal vibration propagation of vocal folds - still shows a better performance than glottal [g] and symmetry [s] parameters together. Combining all features together results into highest classification accuracy of up to 96%. This again suggests the necessity of considering a combination of all features types as deflections, discrepancy, and instability.

4.2 Comparison of acoustics and vocal fold vibration classification

The highly consistent results obtained from acoustic and motion data show that within a subject vocal fold vibrations as well as the acoustic voice signal obtained from different trials can only be compared if they are recorded at similar intensity levels and similar fundamental frequencies. Recordings at significantly different intensity levels or frequencies will definitely cause different perturbations measures (e.g. Jitter, Shimmer, HNR, SNR, NNE) as well as changes within the laryngeal vibrations (Rovirosa et al., 2008). The results suggest that in clinical practice the repeated examination of a subject's voice needs to be performed at a comparable phonatory condition. Otherwise, the clinical value of measurements as objective and representative voice quality measures is highly limited.
In this work it could be shown that PVG analysis is a sufficiently sensitive approach to successfully identify even subtle changes in vocal fold vibratory characteristics induced by different phonatory tasks. As the sensitivity of the PVG approach could successfully be demonstrated, it can be used in ongoing studies to investigate vocal fold vibrations in presence of voice disorders. For studying pathologically induced alterations of vocal fold dynamics within a subject it must be considered that the examinations should be done under similar phonation conditions to exclude examination dependent influences.

5. Conclusion

Digital high-speed videolaryngoscopy is the state-of-the-art technology for investigating normal and pathological vocal fold vibrations. However, without adequate image analysis there is hardly an additional benefit comparing to the currently used stroboscopy technique in sense of evidenced based medicine. The Phonovibrogram (PVG) has the potential to overcome the subjective or semi-automatic assessment of high-speed videos (Kunduk et al., 2010). Within this study it was proven that PVG image analysis has the necessary sensitivity to capture even minor alterations within vocal fold vibrations induced just by frequency and intensity variations. It was further shown that alterations of vocal fold vibrations are also detectable within acoustic perturbation measures. The high accordance between the results further proves that changes within the acoustic signal can directly be traced back to alterations of vocal fold vibrations. In respect to future clinical application, PVG analysis may be a useful tool to standardize the description of healthy and abnormal vocal fold vibrations. Objective Phonovibrography can directly be applied after examination and the obtained PVG images can easily be documented and stored on a hard-disc-drive using a lossless image data format which is essential for evidenced based medicine. An objective endoscopic image analysis tool, such as PVG, describing the vocal fold dynamics, could not only enhance voice assessment techniques but also help to objectively determine the outcome following an intervention in voice disorders (Voigt et al., 2010).

6. Acknowledgements

This work was supported by Louisiana State University, Faculty Research Grant 2008/2009, Deutsche Forschungsgemeinschaft (DFG) grant no. FOR894/1 and no. LO1413/2-1.

7. References

Beyer H.G. Schwefel H.P. (2002). Evolution strategies - a comprehensive introduction. *Natural Computing*, vol. 1, pp. 3-52.

Braunschweig T., Flaschke J., Schelhorn-Neise P., Doellinger M. (2008). High-speed video analysis of the phonation onset, with an application to the diagnosis of functional dysphonia. *Med Phys Eng*, vol. 30,no. 1, pp. 59-66.

Doellinger M., Hoppe U., Hettlich F., Lohscheller J., Schuberth S., Eysholdt U. (2002). Vibration parameter extraction from endoscopic image series of the vocal folds. *IEEE T Biomed Eng*, vol. 49, no. 8, pp. 773–781.

Doellinger M., Braunschweig T, Lohscheller J, Eysholdt U., Hoppe U. (2003). Normal voice production: computation of driving parameters from endoscopic digital high speed images. *Methods Inf Med*, vol. 42, no.3, pp. 271-276.

Doellinger M. (2009). The next Step in voice assessment: High-Speed digital endoscopy and objective evaluation. *Current Bioinformatics*, vol. 60, no. 2, pp. 101-111.

Doellinger M., Lohscheller J., McWhorter A., Kunduk M. (2009). Variability of Normal Vocal Fold Dynamics for Different Vocal Loading in One Healthy Subject Investigated by Phonovibrograms. *J Voice*, vol. 23, no. 2, pp. 175-181.

Deliyski D.D., Petrushev P.P., Bonilha H.S., Gerlach T., Martin-Harris B., Hillman R.E. (2008). Clinical Implementation of Laryngeal High-Speed Videoendoscopy: Challenges and Evolution. *Folia Phoniatr Logop*, vol. 60, no. 1, pp. 33-44.

Duchesne S., Caroli A., Geroldi C., Barillot C., Frisoni G.B., Collins D.L. (2008). MRI-based automated computer classification of probabile AD versus normal controls. *IEEE Trans Med Imaging*, vol. 27, no. 4, pp. 509-520.

Eysholdt U. & Lohscheller J. (2008). Phonovibrogram: vocal fold dynamics integrated within a single image. *HNO*, vol. 56, no. 12, pp. 1207-1212.

Hild K.E., Erdogmus D., Torkkola K., Principe J.C. (2006). Feature extraction using information-theoretic learning. *IEEE Trans Pattern Anal Mach Intell*, vol. 28, no. 9, pp. 1385-1392.

Hsu C.W., Chang C.C., Lin C.J. (2003). A practical guide to support vector classification. Technical report, Department of Computer Science and Information Engineering, National Taiwan University.

Jiang J.J., Tangc S., Dalal M., Wu C.H, Hanson D.G. (1998). Integrated analyzer and classifier of glottographic signals. *IEEE Trans Rehabil Eng*, vol. 6, no. 2, pp. 227-234.

Karnell M.P. (1991). Laryngeal perturbation analysis: minimum length of analysis window. *J Speech Hear Res*, vol. 34, no. 4, pp. 544-548.

Kohavi R. (1995). A Study of Cross-Validation and Bootstrap for Accuracy Estimation and Model Selection. *IJCAI*, pp. 1137-1145.

Kumar A. & Zhang D. (2006). Personal recognition using hand shape and texture. *IEEE Trans Image Process*, vol. 15, no. 8, pp. 2454-2461.

Kunduk M., Doellinger M., McWhorter A., Lohscheller J. (2010). Assessment of the variability of vocal fold dynamics with and between recordings with high-speed imaging and by Phonovibrogram. *Laryngoscope*, vol. 120, no. 5, 981-987.

Lin H. (2008). Identification of spinal deformity classification with total curvature analysis based on coded structured light. *IEEE Trans Biomed Eng*, vol. 55, no. 1, pp. 376-382.

Lohscheller J., Toy H., Rosanowski F., Eysholdt U., Doellinger M. (2007). Clinically evaluated procedure for the reconstruction of vocal fold vibrations from endoscopic digital high-speed videos. *Med Image Anal*, vol. 11, no. 4, pp. 400-413.

Lohscheller J., Eysholdt U., Toy H., Doellinger M. (2008a). Phonovibrography: mapping high-speed movies of vocal fold vibrations into 2-d diagrams for visualizing and analyzing the underlying laryngeal dynamics. *IEEE Trans Med Imaging*, vol. 27, no. 3, pp. 300-309.

Lohscheller J., Doellinger M., McWhorter A., Kunduk M. (2008b). Quantitative analysis of vocal loading effects on vocal fold dynamics using Phonovibrograms. *Ann Otol Rhinol Laryngol*, vol. 117, no. 7, pp. 484-493.

Lohscheller J. & Eysholdt U. (2008). Phonovibrogram Visualization of Entire Vocal Fold Dynamics. *Laryngoscope*, vol. 118, no. 4, pp. 753-758.

Murphy P.J. (1999). Perturbation-free measurement of the harmonics-to-noise ratio in voice signals using pitch synchronous harmonic analysis. *J Acoust Soc Am*, vol. 105, no. 5, pp. 2866-2881.

Neubauer J., Mergell P., Eysholdt U., Herzel H. (2001). Spatio-temporal analysis of irregular vocal fold oscillations: biphonation due to desynchronization of spatial modes. *J. Acoust. Soc. Am.*, vol. 110, no. 6, pp. 3179-3192.

Qiu Q., Schutte H.K., Gu L., Yu Q. (2003). An automatic method to quantify the vibration properties of human vocal folds via videokymography. *Folia Phoniatr Logop*, vol. 55, no. 3, pp. 128–136.

Rovirosa A., Ascaso C., Abellana R., Martínez-Celdrán E., Ortega A., Velasco M., Bonet M., Herrero T., Arenas M., Biete A. (2008). Acoustic voice analysis in different phonetic contexts after larynx radiotherapy for T1 vocal cord carcinoma. *Clin Transl Oncol*, vol. 10, no. 3, pp. 168-174.

Ruben R.J. (2000). Redefining the survival of the fittest: Communication disorders in the 21st century. *Laryngoscope*, vol. 110, no. 6, pp. 241-245.

Schwarz R., Doellinger M., Wurzbacher T., Eysholdt U., Lohscheller J. (2008). Spatio-temporal quantification of vocal fold vibrations using high-speed videoendoscopy and a biomechanical model. *J Acoust Soc Am*, vol. 123, no. 5, pp. 2717-2732.

Selvan S. & Ramakrishnan S. (2007). SVD-based modeling for image texture classification using wavelet transformation. *IEEE Trans Image Process*, vol. 16, no. 11, pp. 2688-2696.

Titze, I.R. (2006). *The Myoelastic Aerodynamic Theory of Phonation*. National Center for Voice and Speech, Iowa City, IA 52242, USA, ISBN 978-0-87414-156-6

Tokuda I., Horácek J., Svec J.G., Herzel H. (2007). Comparison of biomechanical modeling of register transitions and voice instabilities with excised larynx experiments. *J Acoust Soc Am*, vol. 122, no. 1, pp. 519-531.

Vapnik V.N. (1995). *The nature of statistical learning theory*. Springer-Verlag New York, Inc., ISBN-10: 0387987800, New York, NY, USA.

Voigt D., Doellinger M., Braunschweig T., Yang A., Eysholdt U., Lohscheller J. (2010). Classification of functional voice disorders based on Phonovibrograms. *Artif Intell Med*, vol. 49, no. 1, 51-59.

Westphal L & Childers D. (1983). Representation of glottal shape data for signal processing. *IEEE Trans Acoust Speech*, vol. 31, pp. 766–769.

Wurzbacher T, Schwarz R., Doellinger M, Hoppe U., Eysholdt U., Lohscheller J. (2006). Model-based classification of non-stationary vocal fold vibrations. *J Acoust Soc Am*, vol. 120, no. 2, pp. 1012–1027.

Wurzbacher T., Doellinger M., Schwarz R., Hoppe U., Eysholdt U., Lohscheller J. (2008). Spatiotemporal classification of vocal fold dynamics by a multi mass model comprising time-dependent parameters. *J Acoust Soc Am*, vol. 123, no. 4, pp. 2324-2334.

Yan Y., Ahmad K., Kunduk M., Bless D. Analysis of vocal-fold vibrations from high-speed laryngeal images using a hilbert transform-based methodology. *J Voice*, vol. 19, no. 2, pp. 161–175.A.

Yang A., Lohscheller J., Berry D.A., Becker S., Eysholdt U., Voigt D., Döllinger M. (2010). Biomechanical Modeling of Human Vocal Fold Dyanmics by a 3D-Multi-Mass-Model. *J Acoust Soc Am*, vol.127, no. 2, pp. 1014-1031.

Zhang Y &. Jiang J.J. (2008). Acoustic analyses of sustained and running voices from patients with laryngeal pathologies. *J Voice*, vol. 22, no. 1, pp. 1-9.